Y0-CEN-761

ANNUAL REVIEW OF NANO RESEARCH

VOLUME 1

ANNUAL REVIEW OF NANO RESEARCH

Series Editors: Guozhong Cao (*University of Washington, USA*)
C Jeffrey Brinker (*University of New Mexico & Sandia National Laboratories, USA*)

Vol. 1: ISBN-13 978-981-270-564-8
ISBN-10 981-270-564-3
ISBN-13 978-981-270-600-3 (pbk)
ISBN-10 981-270-600-3 (pbk)

ANNUAL REVIEW OF NANO RESEARCH

VOLUME 1

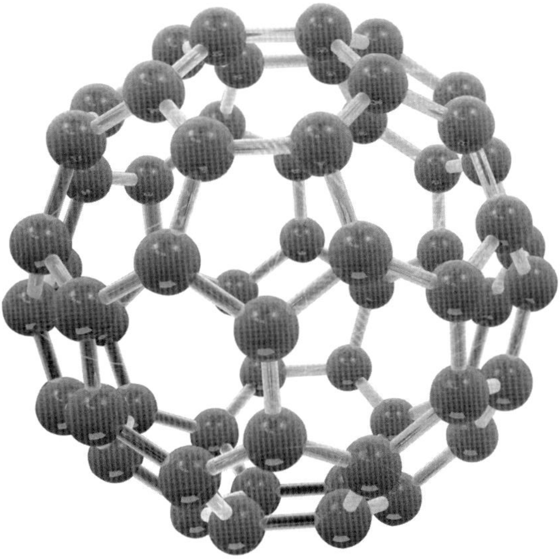

EDITORS

GUOZHONG CAO
University of Washington, USA

C. JEFFREY BRINKER
University of New Mexico and Sandia National Laboratories, USA

World Scientific

NEW JERSEY • LONDON • SINGAPORE • BEIJING • SHANGHAI • HONG KONG • TAIPEI • CHENNAI

Published by

World Scientific Publishing Co. Pte. Ltd.
5 Toh Tuck Link, Singapore 596224
USA office: 27 Warren Street, Suite 401-402, Hackensack, NJ 07601
UK office: 57 Shelton Street, Covent Garden, London WC2H 9HE

British Library Cataloguing-in-Publication Data
A catalogue record for this book is available from the British Library.

ANNUAL REVIEW OF NANO RESEARCH, Vol. 1

Copyright © 2006 by World Scientific Publishing Co. Pte. Ltd.

All rights reserved. This book, or parts thereof, may not be reproduced in any form or by any means, electronic or mechanical, including photocopying, recording or any information storage and retrieval system now known or to be invented, without written permission from the Publisher.

For photocopying of material in this volume, please pay a copying fee through the Copyright Clearance Center, Inc., 222 Rosewood Drive, Danvers, MA 01923, USA. In this case permission to photocopy is not required from the publisher.

ISBN-13 978-981-270-564-8
ISBN-10 981-270-564-3
ISBN-13 978-981-270-600-3 (pbk)
ISBN-10 981-270-600-3 (pbk)

Printed by Mainland Press Pte Ltd

TABLE OF CONTENTS

PREFACE

Nanotechnology is so pervasive; it has an impact on all research fields. Scientists, engineers, pharmacists, and physicians from all different disciplines, and all branches of physical and life sciences and engineering, are either actively contributing to the rapid progress of the nanotechnology field, or will soon be affected by evolving nanotechnology. There are many archive journals and monographs dedicated to publishing the most recent research progress in nanotechnology with varied focus. Many well-established journals in more traditional disciplines have an increasing coverage of nanotechnology research. Not only are the specialty journals making increasing efforts to keep up with the rapid advancement of nanotechnology, but the popular science journals and books are also keeping up the pace to inform the general population and policymakers. More and more symposia, conferences, and workshops focusing on nanotechnology have been organized. Similarly, more and more short courses and regular courses on nanotechnology are offered to educate both graduate and undergraduate students.

In spite of the broad impact and evolving development of nanotechnology, it is most unlikely its own industry will form. Instead, it penetrates into every sector of industry and technology, and all aspects of mankind's activity. Consequently, there is a constant flux of emerging new research directions in nanotechnology. It becomes obvious that nanotechnologists face great difficulty in keeping themselves embraced with the most recent development in the nanotechnology field. Yet many developments and understanding achieved in different focal areas can be directly applicable to other sub-fields of nanotechnology. Review articles by authority experts in their own sub-fields play a significant role in summarizing the recent development in selected topic areas, and thus serve two vital purposes: (1) to present a comprehensive and coherent description of the state-of-the-art understanding by distilling the most valuable experimental results and theories from otherwise segmented and scattered literature, and (2) to offer critical opinions regarding the challenges, promises, and possible future developing directions. Review articles are published in most archive journals, and are thus associated with the focal areas that journals are devoted to. This annual review collects excellent review articles authored by the authority experts and their colleagues in various sub-fields of nanotechnology and thus offer easy access of critical opinions of the most recent developments in nanotechnology.

This volume of *Annual Review of Nano Research* includes 13 review articles and offers a concise review detailing recent advancements in a few selected topics in nanotechnology. Topics in this volume consist of inverse opals and macroporous materials, photonic crystals, and molecular electronics. Other areas involving functional nanostructures, oxide nanofibers, carbon nanotubes, and nanocomposites are also covered in great detail. In addition, many different areas related to organic materials, nano-patterned materials, and biological materials are also described extensively.

Dr. Tammy P. Chou has put in many hours and has placed great effort in editing and formatting the review articles published in this volume. Mr. Yeow-Hwa Quek from World Scientific Publishing coordinated a lot of effort to make the publication of this volume possible.

Guozhong Cao
Seattle, WA

C. Jeffrey Brinker
Albuquerque, NM

CONTRIBUTING AUTHORS

Almairac, R.
 * Université Montpellier II, France

Alvarez, L.
 * Université Montpellier II, France

Bantignies, J-L.
 * Université Montpellier II, France

Blanco, Álvaro
 * Instituto de Ciencia de Materiales de Madrid (CSIC), Spain
 * Unidad Asociada CSIC-U, Spain

Bockstaller, Michael R.
 * Carnegie Mellon University, USA

Bombalski, Lindsay
 * Carnegie Mellon University, USA

Brinker, C. Jeffrey
 * Sandia National Laboratories, USA
 * University of New Mexico, USA

Brucale, Marco
 * University of Bologna Via Irnerio, Italy

Chang, Walter H.
 * Chung Yuan Christian University, Taiwan

Duclaux, L.
 * Polytech'Savoie, Université de Savoie, France

Fan, Hongyou
 * Sandia National Laboratories, USA
 * University of New Mexico, USA

Fritzsche, Wolfgang
*Institute for Physical High Technology, Germany

Fujita, Norifumi
* Kyushu University, Japan

Hou, Shimin
* Peking University, China

Jiang, Lei
* Chinese Academy of Sciences, China
* National Center for NanoScience and Technology, China

Li, Dan
* University of Washington, USA
* Philip Morris, USA

Li, Jimmy K.
* Chung Yuan Christian University, Taiwan

Li, Rui
* Peking University, China

Lin, Cheng-An J.
* Ludwig Maximilians University, Germany
* Chung Yuan Christian University, Taiwan

Listak, Jessica
* Carnegie Mellon University, USA

Liu, Huan
* National Center for NanoScience and Technology, China

López, Cefe
* Instituto de Ciencia de Materiales de Madrid (CSIC), Spain
* Unidad Asociada CSIC-U, Spain

Lytle, Justin C.
* University of Minnesota, USA

Manna, Liberato
  * Universita delgi studi Lecce, Italy

Marquez, Manual
  * Center for Theoretical and Computational Nanosciences, USA
  * Arizona State University, USA
  * Philip Morris, USA

McCann, Jesse T.
  * University of Washington, USA

Möller, Robert
  * Friedrich-Schiller-University, Germany

Mukhopadhyay, Pritam
  * Kyushu University, Japan

Parak, Wolfgang J.
  * Ludwig Maximilians University, Germany

Qian, Zekan
  * Peking University, China

Samorì, Bruno
  * University of Bologna Via Irnerio, Italy

Sauvajol, J-L.
  * Université Montpellier II, France

Schubert, Ulrich S.
  * Eindhoven University of Technology, Netherlands
  * Ludwig Maximilians University, Germany

Shinkai, Seiji
  * Kyushu University, Japan

Sperling, Ralph A.
  * Ludwig Maximilians University, Germany

Stein, Andreas
* University of Minnesota, USA

Vinelli, Alessandra
* University of Bologna Via Irnerio, Italy

Wang, Shutao
* Chinese Academy of Sciences, China

Wouters, Daan
* Eindhoven University of Technology, Netherlands

Xia, Younan
* University of Washington, USA

Zuccheri, Giampaolo
* University of Bologna Via Irnerio, Italy

CHAPTER 1

RECENT PROGRESS IN SYNTHESES AND APPLICATIONS OF INVERSE OPALS AND RELATED MACROPOROUS MATERIALS PREPARED BY COLLOIDAL CRYSTAL TEMPLATING

Justin C. Lytle and Andreas Stein*

Department of Chemistry, University of Minnesota,
207 Pleasant Street SE, Minneapolis, MN 55455
**E-mail: stein@chem.umn.edu*

This comprehensive review chapter provides an overview of colloidal crystal templating methods to prepare three-dimensionally ordered macroporous (3DOM) materials or inverse opals, highlighting progress in the last few years. Assembly of colloidal crystal templates from monodisperse spheres and syntheses of 3DOM structures with a wide range of compositions are described. The current state of knowledge of the mechanical, optical, magnetic, electronic and reactive properties of inverse opal materials is reviewed. Advances towards applications of 3DOM materials are discussed, including photonic crystals, responsive pigments, sensors, bioactive materials, catalysts, electrodes and battery materials.

1. General Introduction

A valuable synthetic paradigm originated in the late 1990's when researchers recognized that colloidal crystals could serve as templates for a new class of three-dimensionally ordered macroporous (3DOM) materials [1-3]. The concept of preparing 3DOM materials is simple: close-packed colloidal crystal templates (CCT) are infiltrated with solution or vapor-phase chemical precursors, followed by the removal of templates by thermal processing, solvent extraction, or chemical etching

(Figure 1). This templating process is adaptable to produce powders, thin films, monoliths, and rods, depending on material composition and the specific preparation method. The resulting macroporous products are comprised of solid walls tens of nm in thickness that surround close-packed spherical voids hundreds of nm in diameter (Figure 2). This geometry has been described as an inverse opal, since it is the inverse replica of the close-packed spherical colloids that constitute natural opal gemstones. Researchers worldwide have responded to this development with stunning breakthroughs that combine this ordered nanoarchitecture with diverse chemical compositions. A variety of potential applications have been investigated for materials built upon inverse opal structures, ranging from applications that benefit from the periodic structures (e.g., photonic crystals and optical sensors) to those that benefit from easy access to the relatively large surface areas provided by the interconnected macropores (e.g., bioactive glasses and fuel cells).

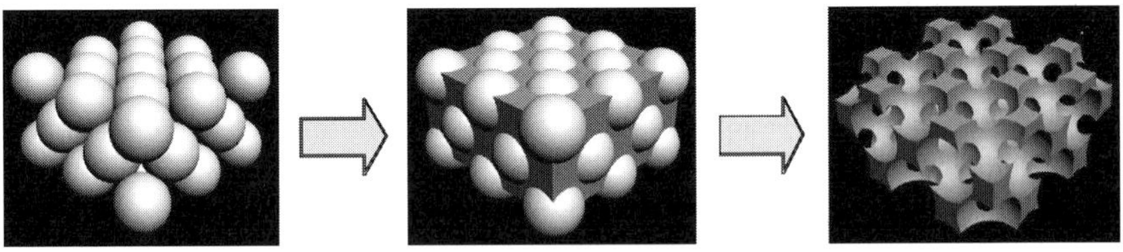

Figure 1. General steps in the preparation of 3DOM solids by colloidal crystal templating.

Figure 2. Electron microscopy images of various 3DOM structures. From left to right: TEM image of 3DOM silica with amorphous walls; TEM image of 3DOM iron oxide showing large grains; SEM image of 3DOM tungsten with thin, rod-like walls. The SEM image of a 3DOM oxide on the right shows the three-dimensional nature of the skeleton.

Scientific contributions to the fabrication and characterization of 3DOM materials have previously been chronicled in several review articles, including a special issue of Advanced Materials that was devoted to photonic crystal structures in 2001 [4-16]. The reviews have emphasized new synthetic approaches, as well as a wide range of original applications for inverse opal materials. Colloidal crystal research has also been reviewed several times since 1998 [17-20], and most recently twice in 2005 [21,22]. Progress in this field has been accelerating during the last few years and a better understanding of the structures, together with increased sophistication in the synthetic control over nanomaterials, have resulted in the development of more complex inverse opal materials and a number of new applications. Hence it is appropriate to reexamine the literature related to colloidal crystal templating and 3DOM or inverse opal materials. The purpose of this review chapter is to document the most recent accomplishments in these fields. The chapter is structured into two major segments, one emphasizing the synthetic aspects of colloidal crystal templating, and the other reviewing some of the physical properties of inverse opals in the context of recently reported applications.

2. Synthesis of Colloidal Crystals

The preparation of a template, a colloidal crystal with sufficient interstitial volume for subsequent infiltration of a precursor, is the first step in the fabrication of 3DOM materials. Most research has involved monodisperse spheres as the building blocks for colloidal crystals, appropriate for producing "true" inverse opals. Polystyrene (PS) [23-28], poly(methyl methacrylate) (PMMA) [29-31], and silica spheres with diameters that range from tens of nm to hundreds of μm have been the most commonly used colloids in the preparation of CCT and inverse opal materials. Polymer latex spheres are typically produced via one-pot syntheses through emulsion free-radical polymerizations with or without added surfactant. Silica spheres are made through adaptations of the Stöber-Fink-Bohn method [32]. Polymer latex and silica spheres are simple to prepare, amenable to seed-growth routes that are necessary to prepare larger spheres, have exceptional monodispersity, and are easy to remove. Polymer and silica sphere CCT are often used in complementary roles during the fabrication of 3DOM solids. Polymer spheres can be removed by thermal processing (combustion, thermal depolymerization) and by solvent extraction, whereas silica spheres offer structural integrity even at relatively high temperatures but are removed by chemical etching.

Other compositions of spherical colloids can also be fabricated by the decomposition of molecular precursors [33], the mechanical disruption of large particles, chemical methods in conjunction with emulsion templates [34], and nebulizing spray depositions. A wide range of amorphous and crystalline metal oxide colloids have been prepared by these methods including alumina [35-40], ceria [41], copper oxides [42,43], manganese oxide [44], titania [39,41,45-48], zirconia [39,41], and ternary metal oxides such as titanium silicates [39], titanium zirconates, and tungstosilicic acid [49]. Several researchers have demonstrated the ability to produce hollow and dense metallic spheres of copper [43,50], gold [50], nickel [34,50], palladium [50], silver [51,52], and low-melting point metals including lead, indium, tin, cadmium, and related alloys [33]. Composite spherical particles with designed internal mesoporosity [39], rare-earth doping for enhanced photoluminescence [53], embedded magnetic nanoparticles [54], and hollow zeolitic walls [55] have also been reported. Within the past decade, numerous core-shell spherical particles have been assembled. Core-shell products are spherical particles in which single or alternating nanoscale layers of alternative compositions have been deposited on the surface of a seed particle. This technique is capable of combining materials with synergistic properties in addition to tailoring the electronic, chemical, and physical nature of particle surfaces. Such materials are prepared via layer-by-layer (LBL) deposition using polyelectrolyte coatings [56], sputtering [57], nanoparticle deposition [58], and seed-growth techniques. Biocompatible solids [59], silica [60,61], gold [56,57], various metals and metal sulfides [62], several polymeric compositions [60,63,64], semiconducting alloys with high refractive indices [65], and titania [66] can be coated either symmetrically or asymmetrically on sphere surfaces.

Monodisperse spherical colloids can be assembled into various packing geometries, including fcc [67], hcp, random hcp, bcc [68-71], and diamond [72]. Of these geometries, fcc is the most energetically-favorable arrangement, having a slight entropic advantage over hcp [73,74]. Several original approaches to colloidal crystallization have recently been developed. The buoyancy of polystyrene (PS) colloids in ethylene glycol (ρ=1.113 g mL^{-1}) has been used as the driving force to float and self-assemble monodisperse spheres into a hexagonal crystal lattice that is later harvested upon solvent evaporation [75]. Binary ionic colloidal crystals, such as CsCl types, have been prepared from oppositely charged polymer spheres with appropriate radius ratios in near-density-matching solvents that also matched the refractive indices of the spheres [76]. Colloidal crystallization can also be driven by hydrophobic interactions at the oil-water interface within water droplets, resulting in mm-sized CCT whose spherical or elliptical shapes reflect the contour of water droplets. These shapes are tuned by applying electric fields and by adding surfactants [77]. New approaches to assembling CCT films have been developed. Monodisperse PS spheres can be transported by capillary action up capillary tubes and dispersed via capillary forces between thinly separated glass substrates [78]. Flow-controlled vertical deposition (FCVD) has been used to deposit PS CCT films by gradually sweeping a colloidal suspension's meniscus across a vertically aligned substrate by draining the suspension with a metering pump [79]. This technique results in ordered domains of several μm, and can linearly control CCT film thickness through surface tension forces inherent to the suspension solvent. A recent study has demonstrated that the choice of colloidal crystal composition is essential to controlling the mechanical strength of the resulting inverse opal material. 3DOM TiO_2 films prepared from PS CCT materials are more defective and brittle than their analogous 3DOM TiO_2 films templated with silica CCT [80]. Annealed silica sphere templates increase the mechanical robustness of the densifying film, whereas polymer templates decompose at temperatures that are significantly lower than the solidification of the pore-wall skeleton. Surface interactions between CCT and sol–gel precursors can be adjusted to promote either surface- or volume-templating, as demonstrated in the titanium precursor infiltration of either acidic or neutralized sulfonated-PS core-shell materials [81].

Colloidal crystals have been prepared from non-spherical particles. This deviation from high symmetry particles generates new periodic dielectric geometries with potentially tunable photonic bandgaps. Titanium dioxide colloids with oblate morphologies are fabricated by infiltrating compressed/distorted 3DOM hydrogels with tetrabutyl titanate precursor mixtures [82]. Colloidal crystals of polyhedral PS colloids are generated by compressing self-assembled thin films of PS CCT between glass substrates at a temperature slightly below the glass transition temperature (T_g=110 °C) of the polymer spheres [83]. The polyhedral particles maintain face-centered cubic packing, but significantly shift the photonic bandgap of the resulting opal film by *ca.* 90 nm relative to their spherical CCT preform.

2.1. Synthesis of 3DOM Structures

After preparation of the colloidal crystal template as a thin film or shaped particle, the next processing steps involve infiltration of the CCT with a fluid precursor, solidification of the precursor and template removal. Specific details of these synthetic steps depend critically on the desired sample composition, and even for a given composition they vary with the desired phase or application. Here we will highlight progress in the preparation of 3DOM simple and complex oxides, non-oxides, metals, semiconductors, polymers, hydrogels, hybrid materials and hierarchical structures.

2.1.1. Synthesis of Simple Oxides

Simple metal oxide inverse opals have been fabricated since the first reports of colloidal crystal templating [1-3]. A rich assortment of synthetic techniques has since been developed to fabricate numerous compositions of binary metal oxide inverse opals, each with a unique morphology that is formed as a function of the specific chemical route employed [11]. The arrangement of spherical voids in the inverse opal structure has been characterized by transmission electron microscopy

[84,85]. Bright-field imaging and Fourier imaging analysis of 3DOM metal oxides and silica re-confirmed the face-centered cubic packing nature of macropores in colloidal-crystal-templated solids. Furthermore, stacking faults in the inverted opal are observed in real-space, which is relevant in order to fully characterize the optical properties of these materials. Recent literature about 3DOM TiO_2 is one example of the interplay between the synthetic approach and the product morphology. In particular, variations in experimental conditions can produce either surface-templated or volume-templated materials. Surface templating occurs with strong wetting of spheres by the precursor and yields shell-like structures in which the macroporous products follow mainly the curvature of the original spheres. Small holes are often observed at the contact points between multiple spheres. With weaker attractive forces between spheres and precursors, volume-templated structures tend to be produced, where the skeletal walls become more rod-like (see also Figure 4). 3DOM TiO_2 powders were initially prepared by infiltrating polymer latex CCT with neat or dilute titanium alkoxides, which results in volume-templated, ordered inverse opal regions that are *ca.* several micrometers in size [2,67]. In other studies using a similar chemical approach with titanium alkoxides diluted in ethanol, either surface-templated [86] or volume-templated structures are observed [87]. Although the chemical nature of the PS sphere surfaces was not reported, different electrostatic interactions between the CCT and polar precursor solutions are likely responsible for the different morphologies. The solid-volume filling fraction of 3DOM TiO_2 powders is increased by using multiple infiltrations of dilute alkoxide [88], although this can lead to surface crust formation and blocked macroporous domains. Non-alkoxide chemical routes have successfully produced titania inverse opals. Smooth macropore walls are observed upon the electrochemical deposition of $TiCl_3$ [89]. Liquid-phase precipitation of amorphous titania from $(NH_4)_2TiF_6/H_3BO_3$ generates inverse opals that are covered with a dense precipitated crust [89,90].

Inverse opal TiO_2 materials prepared in all of the above ways typically have ordered regions of <10 μm that are separated by large cracks and planar defects. Cracks form as the solid skeleton shrinks by *ca.* 30 vol. % during inorganic densification/crystallization, and as the infiltrated-latex composite outgases volatile species during decomposition/combustion. Cracks can also propagate in solvent-extracted 3DOM TiO_2 materials, due to the refluxing and agitation of the solvent [91]. Inverse opals fabricated by the co-sedimentation of PS CCT and titania nanoparticles are less defective, because the pre-formed nanoparticles shrink less than 7 vol. % [92,93]. Co-sedimentation routes are useful for the preparation of optically functionalized inverse opals with luminescent Er-doped titania nanoparticles [94]. Additional crack suppression is achieved with annealed silica CCT, which are more thermally stable than latex and avoid outgassing [95,96]. The finest control over the inverse opal structure is perhaps achieved with ALD techniques. For example, 3DOM TiO_2 and TiO_2/ZnS:Mn multilayer composite films fabricated by ALD are extremely high quality photonic crystal materials with surface-templated pore walls, defect-free domains covering several μm, macropore wall surface roughness values <1 nm, and solid-volume filling fractions approaching 88 vol.% of the interstitial void volume [97-100]. The ALD approach deposits uniformly thin coatings on CCT, which enables the tuning of pore wall thicknesses in order to reproducibly create triangular defects in unfilled tetrahedral spaces between close-packed spherical templates.

Similar comparisons of synthesis and structure can be made with ZnO inverse opals. 3DOM ZnO was first prepared in 2000 via the precipitation of zinc oxalate from soluble zinc acetate that is wetted within PS CCT [101]. Zinc oxide inverse opals have since been fabricated by four separate electrodeposition approaches, based on the reduction of zinc salt species within CCT films that are supported on conductive substrates [102-105]. The morphological texture of electrodeposited 3DOM ZnO films is smoothed by controlling the rate of deposition via the applied potential. Chemical vapor deposition (CVD) techniques have additionally been used to prepare ZnO and ZnS photonic crystals with inverse opal geometries [106]. Vapor deposition approaches permit structural refinement on an atomic length scale, and facile control of photonic crystal film thickness. The richness of chemistry, morphology, and properties in inverse opal powders and films is again illustrated by a variety of other transition metal oxides (ZrO_2 [107], WO_3 [108], Cr_2O_3 [101], Mn_2O_3, Fe_2O_3, Co_3O_4, NiO), alkaline earth metals

(MgO [101], $CaCO_3$), rare earth oxides (CeO_2 [109,110], $La_2O_3/La_2O(CO_3)_2$) [109], Eu_2O_3 [111], Nd_2O_3, Sm_2O_3), SiO_2 [1,112-116], Al_2O_3 [117,118], and SnO_2 [119-121].

2.1.2. Synthesis of Ternary Oxides and Higher Compositions

Advanced synthetic control has been demonstrated by the development of important ternary and mixed oxide inverse opal materials. $BaTiO_3$ [122,123], $PbTiO_3$ [124,125], $LiFePO_4$ [126], $LiCoO_2$ [127], $LiNiO_2$ [128,129], $LiNbO_3$ [130], Mn^{2+}-doped Zn_2SiO_4 [131], and $Li_4Ti_5O_{12}$ [132], inverse opal ternary phases have been prepared as powders and films by sol-gel infiltrations of CCT with various metal alkoxides and salts. These reports demonstrate the delicate balance between forming the desired crystalline polymorphs, often at temperatures in excess of 500 °C, and maintaining ordered macroporosity by resisting morphological changes that are caused by crystallite growth and sintering. Recent 3DOM $BaTiO_3$ syntheses illustrate how the fabrication of complex oxide inverse opals is promoted by judicious choices of chemical precursors, thermal processing temperatures, heating rates, and processing atmospheres. Seemingly minor variations in barium precursors and heating rates determine whether the $BaTiO_3$ material has a relatively smooth and ordered $BaTiO_3$ macropore wall structure [122], or is densified by sintering [123]. Surface-templated ternary oxide inverse opal structures are promoted via the layer-by-layer (LBL) deposition of polyelectrolytes on CCT surfaces, as was demonstrated with $LiNbO_3$ [130]. Phase purity is achieved in mixed metal oxide inverse opals by altering the chemical environment via metal-ligand chelation and the pre-hydrolysis of alkoxides, as shown in the case of dip-coated Si:Ti and Si:Al binary oxide inverse opal films [133]. Processing temperatures must be appropriate in order to limit crystallite growth and to control phase transformations, which have been shown to alter the luminescence of Mn^{2+}-doped Zn_2SiO_4 inverse opals [131].

Complex oxides with hierarchical porosities have been prepared by infiltrating latex CCT with metal salts and alkoxides to form $CsAlTiO_4$ wall skeletons with tridymite-like microporous frameworks [134] and mesoporous $Mg_2Al(OH)_6(CO_3)_{0.5}$ layered double hydroxides [135]. In the latter case, double hydroxide inverse opals are directly formed by removing latex CCT via solvent extraction, but can also be indirectly formed by exposing calcined products to atmospheric humidity in order to regenerate the hydrotalcite-like structure. Hierarchical inverse opals are likewise assembled by dispersing solvent-exfoliated magnesium organosilicates within CCT [136]. Researchers have actively pursued inverse opal perovskites in order to combine photonic and ferroelectric properties for non-linear optics applications. Typical perovskite precursors are metal acetates, alkoxides, and nitrates that are dissolved in polar solvents, such as 2-methoxyethanol [137], glycol monomethyl ether [125,138], and acetic acid [139-141], which are capable of metal-ligand complexation. $La_{0.7}Ca_{0.3}MnO_3$ [137], $(Pb_{1-x}La_x)(Zr_{1-y}Ti_y)O_3$ [125,138-141], $Sr_xBa_{1-x}Nb_2O_6$, Pb-doped $Ba_{0.7}Sr_{0.3}TiO_3$ [142] ferroelectrics and $Li_{0.35}La_{0.55}TiO_3$ [143] ion-conducting inverse opals have been prepared in this fashion, with ordered macroporous morphologies that are controlled by annealing CCT and by adjusting the filling fraction of CCT via multiple precursor infiltrations. A similar complexation strategy has been reported for perovskite $La_{1-x}Sr_xFeO_3$ inverse opals that are synthesized by transforming metal nitrates into metal glyoxylates via a reaction with a mixture of ethylene glycol and methanol [144].

2.1.3. Synthesis of Non-Oxides

Inverse opal carbon materials have been fabricated by CVD, melt-infiltration, and the uptake of resorcinol-formaldehyde (RF) polymers. Graphitic carbon films and monoliths are prepared by CVD of propylene within CCT [145] and by CVD of benzene on the pore-wall surfaces of inverse opal silica templates [146]. Carbons that are vapor deposited onto 3DOM silica templates show good transfer of morphological detail from their inverse opal molds. Diamond inverse opals have been achieved by reacting pre-infiltrated diamond seed nanocrystallites with $H_2(g)$ and $CH_4(g)$ [145]. High-surface-area graphitic carbons are additionally generated by melt-incorporating silica sphere colloids within mesophase pitch, which is graphitizable at 2500 °C in Ar(*g*) as demonstrated by TEM and Raman spectroscopy [147,148]. Melt infiltration of bis-*ortho*-

diynyl arene (BODA) monomers in silica sphere CCT, followed by in-situ polymerization and pyrolysis, has similarly been used to prepare glassy carbon inverse opals [149]. This method resists shrinkage due to the high loading of the BODA monomer to form polyarenes.

Glassy carbon inverse opal carbons are typically fabricated by infiltrating and pyrolyzing resorcinol-formaldehyde (RF) precursors within CCT. Latex CCT are thermally degraded and are removed by volatilization during high temperature pyrolysis; silica sphere CCT are subsequently etched by mineralizing acids or bases after pyrolysis. Other monomers, such as divinylbenzene, have alternatively served as the carbon source in ordered porous carbons [150]. The surface chemistry of CCT strongly directs the formation of RF carbon inverse opals. Aluminum-incorporated silica spheres have strong acid sites that direct surface-templating of RF polymers, whereas volume-templating is promoted with pure silica CCT even in the presence of dilute sulfuric acid [151]. Ordered macroporous RF carbons have been prepared from silica sphere CCT as small as 11 nm [152], and have approached the lower limit of the macroporous size regime, exhibiting nitrogen sorption-desorption hysteresis that is usually associated with mesopores [153]. Textural mesoporosity can be tailored into the pore wall structure of inverse opal RF carbons via supercritical carbon dioxide extraction after polymerization but prior to pyrolysis [154]. Volume-templated RF carbon pore wall surfaces have been functionalized with Pt-Ru catalytic nanoparticles after silica sphere CCT are etched by hydrofluoric acid [155-157], and with SnO_2 nanoparticles by wetting 3DOM carbon monoliths with $SnSO_4$(*aq*) [158].

Further compositional complexity has been achieved with metal and silicon carbide, nitride, and chalcogenide inverse opals. In general, these more exotic compositions are more susceptible to oxidation and decomposition reactions that occur during thermal processing, necessitating the chemical etching of silica spheres and the solvent extraction of latex CCT. ALD approaches have successfully prepared surface-templated WN [159] and Ta_2N_5 [160] inverse opal thin films by volatilizing and thermally degrading organometallic precursors within silica CCT. Voids in CCT can become obstructed and limit further ALD precursor infiltration, as was observed in WN inverse opals reported with large uninfiltrated pockets *ca.* tens of μm in size. Optical quality ZnS [161-163] inverse opals are instead achieved by optimizing the vapor deposition rate in order to maximize solid-volume filling fractions. Similar CVD methods fabricate 3DOM SnS_2 films with high filling fractions over 5-μm ordered areas [164,165]. Metal sulfide inverse opals are also fabricated by reacting solution-infiltrated metal salts with sulfur-bearing species, such as thioacetamide, within CCT, creating nanocrystallite-textured pore wall surfaces that are substantially smoothed by using poly(styrene-acrylic) CCT [161,166,167]. This type of chemical bath deposition has been used to form CdS and Sb_2S_3 [168]. Finer structural control is afforded by solvothermally growing 4-nm ZnS crystallites within CCT. Other inverse opal fabrication routes include the synthesis of GaN via gas-phase nitridation of Ga_2O_3 [169], SiC by the carburization of organosilane precursors [170,171], melt-infiltration of selenium into silica sphere CCT [172], and the sol-gel syntheses of magnetic $M_4Co_4[Fe(CN)_6]_4$ (M = alkali metal) inverse opals from aqueous solutions of $CoCl_2$ and $K_3[Fe(CN)_6]$ [173].

Recently, colloidal templating has been shown to provide a means of patterning the surface of single crystals on all faces. As an example, surface porosity was introduced into calcite single crystals by templating with polymer latex particles which had been functionalized with surface carboxylate groups [174]. The authors provided good evidence that the $CaCO_3$-latex composite particles were single crystals with rhombohedral morphology (Figure 3). Most of the latex particles were found near the surface of the crystals and could be removed either by solvent extraction or by calcination to induce the surface porosity. $CaCO_3$ crystals preferentially nucleated on the carboxylate-functionalized latex particles and grew into $CaCO_3$-coated latex particles. Subsequent growth is thought to occur by the continuous addition of mostly Ca^{2+} and CO_3^{2-} ions. At a later stage, when the Ca^{2+} concentration was significantly reduced, free latex particles attached themselves to the surface and were embedded in the surface layer. The structures shown in Figure 3 were only obtained with carboxylate-functionalized latex particles, and not with $-SO_3^-$ or $-PO_3^{2-}$ surface groups.

2.1.4. Synthesis of Metals

Several new metallic inverse opal compositions have been prepared by innovative synthetic approaches since 3DOM metals were first reviewed in 2000 [175]. Transition metal (Ni [176], Co [177], Fe) and metal alloy (Ni_xCo_{1-x} [177,178], Mn_3Co_7 [178]) inverse opal powders were originally prepared via the precipitation of metal oxalates from soluble salts within

CCT interstices. Ordered macroporous metals were generated by calcining metal oxalate-latex composites in air and then reducing the oxidized powders in flowing $H_2(g)$, or by directly reducing the infiltrated composites via thermal processing in either $H_2(g)$ or $N_2(g)$. Macropore walls were significantly textured because of substantial densification after direct reduction in hydrogen, although the carbon content of hydrogen-processed products was <2 wt% in contrast to the carbon-rich products generated in $N_2(g)$.

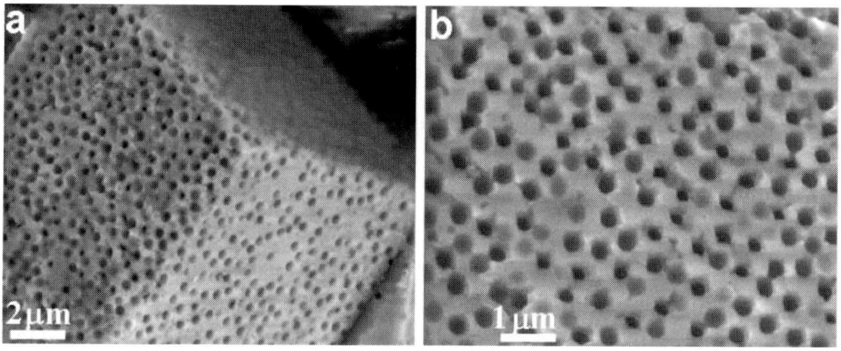

Figure 3. SEM images of surface patterned $CaCO_3$ particles obtained by templating and THF extraction of 380 nm latex particles. Reprinted with permission from [Ref. 174], C. Lu *et al.* Chem. Mater. 17, 5218, (2005), Copyright @ American Chemical Society.

Electrochemical depositions of Ni [179,180], Co, Fe, and a Ni-Fe alloy offer an alternative synthetic approach that avoids densification of the pore wall skeleton, as observed in the reduction of metal oxalates. Metal sulfates/sulfamates were electrochemically reduced within PS CCT that were assembled on planar current collecting electrode surfaces; template spheres were removed via solvent extraction. Metal films prepared in this way exhibit higher degrees of ordered macroporosity than salt-precipitated inverse opals, but their pore windows often appear blocked by excessive metal deposition. Macropore interconnectivity is improved by using pulsed potential deposition schemes that generate volume-templated Zn inverse opals because of effective rest periods after nucleation duty cycles [181]. In contrast, surface-templated Zn inverse opals with comparatively larger crystallites are induced via voltammetric electrodeposition conditions. Ordered macroporous Au films are fabricated by reducing 0.1 M $HAuCl_4$ within silica sphere and latex CCT films grown on indium tin oxide (ITO) current collectors [182]. Larger ordered areas and enhanced inverse replication were achieved with silica sphere CCT, because latex-templated products result in low-aspect ratio Au flakes that contain approximately one layer of macropores.

Likewise, electroless deposition has been successfully utilized to prepare metallic inverse opal materials. In the first report of this type, Au nanoparticles were chemically bonded to thiol-modified silica sphere CCT surfaces. Organic species were removed via calcination, exposing pure Au surfaces that initiate the electroless reduction of Ni, Cu, Ag, Au, and Pt salt baths, respectively [183]. Such films can be highly ordered and up to several μm thick, although product morphologies are significantly roughened without careful control of the salt composition, concentration, and deposition time [184,185]. Close-packed, but hollow, opal morphologies are obtained by surface-templating PS CCT with tin salts, which serve as nucleation points for the electroless deposition of thin metallic Ag coatings [51]. Hollow Ag spheres with smooth surface morphologies are subsequently produced by solvent extracting the PS CCT.

Vapor deposition techniques are another tool that enables the fabrication of metallic inverse opals. Tungsten inverse opals prepared via CVD of tungsten hexacarbonyl are surface-templated, exhibiting distorted and roughened macropore windows within ordered regions of <10 μm [186]. The solid-volume filling fraction of the 3DOM W film is adjustable by this method, but high loadings increasingly limit the photonic properties of the material. The morphology of 3DOM W powders and films depends on the synthesis method (CVD vs. wet solution chemistry) and the choice of tungsten precursor [187]. Nearly spherical walls with relatively small windows are generated by CVD of tungsten hexacarbonyl within silica colloidal crystals, whereas very open structures with low solid volume fractions and large interconnecting windows are fabricated from acetylated peroxotungstic acid by wet infiltration of PMMA spheres (Figure 4).

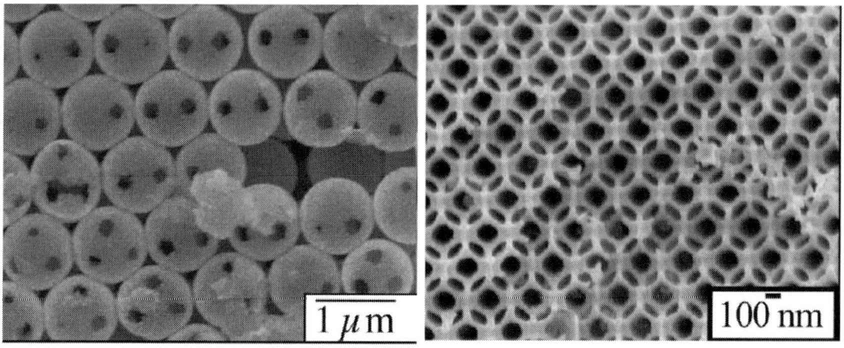

Figure 4. SEM images illustrating the differences between structures produced by surface templating (left) and volume templating (right). The tungsten inverse opal structure on the left was prepared by CVD of tungsten hexacarbonyl within a silica CCT, the open 3DOM tungsten structure on the right by wet methods using a PMMA CCT. Reprinted with permission from [Ref. 187], N. R. Denny *et al.* Proc. of SPIE 6005, 600505-1, (2005), Copyright @ Society of Photo-Optical Instrumentation Engineers.

Noble metal inverse opals of Au and Pt have also been created by infiltrating CCT with sol-gel and pre-formed nanoparticles, respectively, to create high surface area catalysts and high-dielectric photonic materials. The densification of metallic gold and platinum from gold chloride and platinic acid precursors, respectively, creates Au [188,189] and Pt [190] inverse opals with low filling fractions that can be increased by wetting the remaining interstitial void volume with polypropylene and by multiple precursor infiltrations. The shrinkage of the pore wall framework is minimized by co-sedimenting Au nanoparticles (15–25 nm) with latex spheres via sweeping meniscus deposition or static deposition within a confining cell [191-193]. The resulting 3DOM Au films owe their opalescence to defect-free regions that are *ca.* several μm in size, and offer surface plasmons for enhanced Raman spectroscopy of various molecules [194]. Surface-enhanced Raman spectroscopy (SERS) measurements that use 3DOM Au films are prime examples of the importance of structure-property relationships between ordered macroporous structures, surface texture on the metallic pore walls, and the intensity of the detected Raman signal [195,196].

2.1.5. Synthesis of Semiconductors

Semiconducting inverse opals couple the electronic bandgap and dielectric constants of semiconductor materials with photonic crystal structures in order to create hybrid compositions for opto-electronic applications. Silicon inverse opals have been particularly interesting to the optics community, and were recently reviewed in 2004 [197]. Vapor deposition of silanes within silica sphere CCT yields surface-templated Si pore wall surfaces that are extremely smooth, creating a complete three-dimensional photonic bandgap [198,199]. The filling fraction of these materials is increased by elevating the deposition temperature, and polycrystalline Si is formed by annealing at 800 °C under vacuum [200,201]. Recently, Si inverse opals were fabricated by wetting silica CCT with molten Si that was generated by localized laser-imprinting [202,203]. After removing the silica sphere CCT via chemical etching with HF(*aq*), this method resulted in volume-templated Si inverse opals with smooth macropore surfaces. Metal organic CVD (MOCVD) has been used to prepare 3DOM GaP and InP films, because of the interest in their high refractive indices for photonic waveguide applications [204]. Germanium is also important to optical research because it has such a high dielectric constant (ε=16) [8]. 3DOM germanium is generated by wetting silica CCT with neat tetramethoxygermane or trichlorogermanium propanoic acid [205], followed by hydrolysis, reduction in flowing $H_2(g)$, and template removal. This technique results in a highly textured pore skeleton, because of the grainy texture of the intermediate GeO_2 and the inorganic densification that occurs as a result of chemical reduction. Surface-templated Ge inverse opals have alternatively been prepared by CVD of digermane (Ge_2H_6) under $H_2(g)$ [206].

2.1.6. Synthesis of Polymers

A wide range of hydrocarbon polymers have been fabricated with an inverse opal morphology, including polyurethane, PS, and PMMA inverse opals that are polymerized from neat monomers within silica CCT, creating thin films with well-interconnected macropores [48]. More complex compositions, such as bisphenol A-epichlorohydrine copolymer epoxy resins [207], polyacrylamide-based polymers [208], mesoporous poly(vinyl alcohol) hosts for organizing fluorescent proteins [209], polyurethane [210], and solvent-cast polyvinylidene fluoride-trifluoroethylene ferroelectric copolymers [211] are synthesized within either silica or latex CCT, yielding products that are typically volume-templated with smooth macropore surface morphologies, but sometimes with occluded pore windows that limit the three-dimensionally interconnected porosity of the material. Inverse opal polymers with directed tacticity have been developed. Syndiotactic PS is synthesized by wetting silica CCT with *p*-methylstyrene in the presence of a β-diketonate titanium catalyst [212-214]; atactic PS is prepared directly from styrene and an azo-initiator within silica CCT, demonstrating that polymer tacticity is not directed within the confined CCT voids [215]. Disordered macroporous polystyrene is assembled from crosslinked-microgels that are prepared by *R*eversible *A*ddition *F*ragmentation Chain *T*ransfer (RAFT) polymerization [216]. Perhaps one of the most original fabrication approaches in the assembly of polymer inverse opals is a method which avoids the conventional infiltration of precursor species into a CCT. Ordered arrays of monodisperse poly(styrene-*co*-divinylbenzene) (PSDVB) spheres are immersed in cyclohexane; the polystyrene component of the copolymer is selectively dissolved to form poly(divinylbenzene) inverse opals [217].

Inverse opals have additionally been formed from electronically conducting and semiconducting polymers. Polyaniline inverse opal materials are deposited by electropolymerization [218] and chemical infiltration [219] routes that each result in unique porous morphologies. Electropolymerized porous architectures are tuned by incorporating dopants of anionic polyelectrolytes and acidic aniline derivatives, which can vary the pore wall thickness and pore window size and promote polar interactions with biological molecules. In comparison, self-supporting inverse opals with smooth pore walls >100 nm thick are synthesized by oxidizing centrifuge-infiltrated aniline hydrochloride in the presence of potassium persulfate. Inverse opal copolymers, such as π-conjugated poly[2-(9',9'-dihexylfluorenyl)-1,4-phenylenevinylene] [220], are melt-infiltrated and silica CCT are subsequently etched to yield volume-templated photonic crystal films that have a thin crust deposited on their exterior surface. Conductive inverse opal polythiophenes are fabricated by photocuring either terthiophene [221] or poly(3-alkylthiophene) [222] within silica CCT; surface-templated 3DOM polypyrrole and polythiophene films with highly ordered macropores are electropolymerized within PS CCT that are assembled on ITO conductive sheets [223]. The refractive index of polythiophene inverse opals is shifted as a function of applied voltage, resulting in tunable Bragg reflections. The enzyme creatinine deiminase has been incorporated into electrodeposited polypyrrole inverse opals for catalytic detection of the molecule creatinine [224]. In addition, an electropolymerized polythiophene copolymer inverse opal covalently binds the biological molecule biotin, shifting the photonic stop band slightly in the visible range [225].

Organometallic polymers enrich the diversity of inverse opal compositions and properties by offering unique redox activities and electronic conductivities. High-filling-fraction poly(ferrocenylsilane) inverse opals are synthesized by the thermal ring opening of silaferrocenophane monomers within silica CCT [226,227]. The fine structure of the inverse opal and its corresponding photonic properties are subsequently tuned by sequentially depositing water-soluble polyelectrolytes onto the 3DOM surfaces. The properties of the structure are further altered by pyrolyzing the poly(ferrocenylsilane) skeleton in order to yield macroporous iron oxide ceramics. Similarly, the pyrolysis of organosilanes and molybdenum chloride deposits a molybdenum carbide/silicide composite within silica sphere CCT, which are then removed by chemical etching to yield a multifunctional semiconducting and catalytic host inverse opal material [228].

2.1.7. Synthesis of Hydrogels

The dynamic tuning of 3DOM pore sizes in response to external stimuli, such as pH or temperature, has been studied for potential chromatographic and drug release applications. There is precedent in the literature for using inverse opal solids for timed drug release in response to body pH, temperature, or chemical conditions [229,230]. Furthermore, chemical and biological sensing capabilities have been demonstrated by hydrogel-infiltrated CCT films that are co-assembled from cross-linked polar monomers and latex colloids, followed by subsequent photopolymerization [231-234]. Partially reversible swelling and deswelling has been shown for inverse opal polyacrylamide hydrogels, which have been examined as humidity sensors [235] and biocompatible supports for cell culturing and tissue regeneration [236]. Polyacrylamide inverse opals are polymerized within PMMA CCT by thermal, photo, or redox initiations of acrylamide in the presence of free-radical initiators and cross-linking agents. Photopolymerization has been successfully employed in the syntheses of other inverse opal hydrogel compositions, including poly(boronic acid)-based materials [237], poly(ethylene glycol) methacrylates with compression moduli similar to body tissues for cell culturing [238], and hydroxymethacrylate-acrylic acid co-polymers [239]. In the latter case, hierarchical mesopores are formed within the slightly distorted, yet robust, ordered macropore walls, and the sensitivity of the hydrogel to pH is optimized by adjusting the ratio of methacrylate:acrylic acid in the mixed monomer precursor. Temperature-sensing poly(N-isopropylacrylamide) inverse opal hydrogels are fabricated by infiltrating silica CCT with a thermally initiated precursor mixture composed of monomer, crosslinker, and a free-radical initiator [240-242]. The monomer:crosslinker ratio determines the degree of hydrogel swelling, and influences the Bragg reflection conditions such that 3DOM hydrogels can exhibit nearly the full visible spectrum of colors simply by varying the temperature between 15–30 °C. Inverse opal hydrogels have additionally found usefulness as scaffolds that template the deposition of titania replicas on polar polymer surfaces [243].

2.1.8. Synthesis of Hybrid Compositions and Composites

Ordered macropore wall surfaces have been decorated with various molecular and nanoparticle functionalities in order to diversify the interactions of inverse opals with their surroundings. At molecular length scales, alkylthiol groups are chemically incorporated into 3DOM SiO_2 [244], TiO_2 [91], and ZrO_2 pore walls by directly mixing either (mercaptoalkyl)alkoxides or mercaptoalkyl sulfonic acids with structure-forming precursor solutions during colloidal-crystal templating. At sizes of *ca.* 1–2 nm, pre-synthesized tungstosilicate polyoxometalate clusters are included in the organosiloxane precursors that are used to prepare silica inverse opals [245-247]. The clusters are homogeneously dispersed throughout the ordered macroporous architecture after latex sphere solvent extraction, but bulk tungsten oxide domains are formed upon calcination. Inverse opal carbon monoliths have been further functionalized with TiO_2 nanoparticles via the stepwise layer-by-layer (LBL) deposition and hydrothermal treatment of a ligand-stabilized titanium salt onto polyelectrolyte-coated pore walls [248]. Likewise, surface-templated 3DOM $HgTe/TiO_2$ composites with dense surface crusts are assembled by the sequential deposition of HgTe nanocrystals and a titanium alkoxide onto polyelectrolyte-coated PS CCT films [65]. Nanoparticles can themselves be functionalized, as was demonstrated in the encapsulation of 15 nm gold colloids with a nanoscale silica coating, and have been co-sedimented with monodisperse spherical colloids to form highly-textured 3DOM $Au@SiO_2$ composites [249]. Other inverse opal composites have been prepared by infiltrating 3DOM solids with sol-gel precursors, as in the decoration of 3DOM SiO_2 with Ag nanoparticles by an in-situ Tollens' reaction [250] the salt-imbibition and thermal decomposition of $SnSO_4$ to form SnO_2-coated 3DOM carbon monoliths, the wetting of 3DOM $La_{0.7}Ca_{0.3}MnO_3$ with various ternary and higher metal titanate precursor fluids [137], and the infiltration of silsesquioxanes in order to generate highly ordered organosilica glass inverse opals [251]. Additional approaches to composite inverse opals include CVD and electrodeposition, which prepare organosilica glasses with disordered mesopores and ordered macropores by the vapor deposition of organosilane precursors within PS CCT films [252] and surface-templated polypyrrole-CdTe inverse opals, respectively [253].

2.1.9. Nanocasting with 3DOM Templates

Inverse opals, having been molded from opal templates, can themselves be used as molds for secondary replica structures with altered composition, if desired. When applied to mesoporous materials, this replication technique is now well known as "nanocasting" [254-257]. In relation to macroporous solids, the approach has been called "micromolding in inverted polymer opals" (MIPO) [258], "micromolding in inverse silica opals" (MISO) [259], or a "lost-wax" approach [260]. One example of MIPO involved the preparation of 3DOM polystyrene around a silica opal, elimination of the silica by fluoride-based etching and then infiltration of the macropores with a silicatropic liquid crystal precursor for mesoporous silica in the void spaces. Calcination of this composite material removed the polystyrene, leaving a replica opal structure composed of mesoporous silica spheres [258]. This approach is quite versatile. 3DOM polymer (PS or PMMA) molds have been applied to form colloidal crystals of filled or hollow titania spheres, hollow zirconia-alumina colloids, polypyrrole colloidal crystals, and other opals composed of alumina, cadmium sulfide, silver chloride and diverse metals [260,261]. The colloidal crystals can be broken up into smaller particles and discrete spheres by ultrasonication in appropriate solvents. If the 3DOM polymer template is uniformly deformed during molding, ellipsoidal particles can be formed [260]. 3DOM silica is a suitable mold for MISO of silicon spheres by CVD of Si_2H_6 [259], and 3DOM carbon has been used as a hard template to prepare hollow SiC and filled SiCN sphere assemblies, using low molecular weight polymethylsilane and polysilazane as precursors [262]. The silica template was removed by controlled HF etching, and the carbon template by heating in air.

While the above studies generally targeted filled or hollow spheres/sphere assemblies, the intermediate composite structure may also be of interest. For example, during MISO of silicon in a silica mold, a composite Si/SiO_2 phase with an approximate inverse opal structure is obtained [259]. An even more intimate nanocomposite of mixed graphitic carbon/amorphous carbon phases was recently obtained from 3DOM carbon monoliths containing macropores surrounded by mesoporous, glassy carbon walls (3DOM/m C) [263]. 3DOM/m C had, in turn, been synthesized by nanocasting from monolithic silica with a hierarchical pore structure. Subsequent introduction of more graphitic, nitrogen-doped carbon into the carbon mesopores by chemical vapor deposition (CVD) produced a monolithic nanocomposite material (3DOM/m C/C). With appropriate control during CVD, mainly the mesopores were filled, while macropores could remain relatively open. As a result, the product was not an opal replica, but remained an inverted opal. The mechanical strength and electronic conductivity of the nanocomposite were significantly higher than before addition of the second carbon phase. The nanocomposite suppressed formation of a solid-electrolyte interface layer during lithiation and had higher lithiation capacity than 3DOM RF-derived carbon monoliths at high discharge rates.

2.1.10. Hierarchical Structuring

Colloidal crystal templating provides a useful avenue for preparing porous materials with two or more levels of hierarchy in the pore structure [13,264]. Hierarchical structuring has been applied mostly to silicate systems and more recently to carbon materials employing nanocasting methods via hierarchical silica templates [155,156,255,263,265]. In the reported syntheses, latex spheres (PS or PMMA) have provided the macropores whose diameters have ranged from 90–800 nm. Secondary and possibly tertiary porosity has been introduced within the silica skeleton by applying other templating techniques that result in mesopores (2–50 nm) (Figure 5), micropores (0.2–2 nm), or a subset of the latter, supermicropores (1–2 nm) [266]. Yet another length scale can be incorporated into a porous material by applying lithographic processing or by creating monolithic structures.

The motivation for synthesizing materials with hierarchical porosity is that such fractal structures can rapidly distribute guests to a high surface area while minimizing pressure build-up. The specific architecture of porous solids, including pore size ranges and distributions, as well as pore interconnectivity, can significantly influence diffusion and reactivity of guests [267]. Relatively large molecules pass readily through a connected macropore system, whereas smaller molecules can be differentiated through size- and/or shape-selective uptake in the smaller pores. Although, the smaller pores provide the largest fraction of the surface area, they also pose the greatest diffusional resistance. Diffusion limitations are reduced by making micropores more accessible and by keeping diffusion paths in micropores relatively short within the walls of, for example, an inverse opal structure. Hierarchical structuring by colloidal crystal templating therefore yields opportunities for improved heterogeneous catalysts and separation materials with narrow pore size distributions on several length scales. Several approaches to assemble relatively well-ordered macroporous structures with multiple porosity are outlined below.

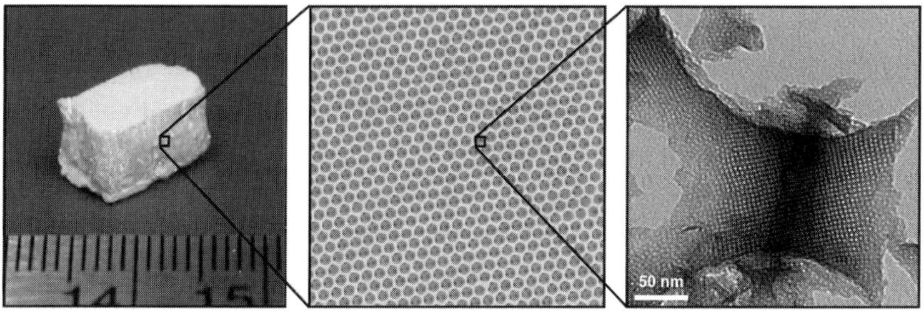

Figure 5. Monolithic silica with hierarchical porosity. The photograph on the left shows the monolith with millimeter edge lengths. The macropore arrays are apparent in the SEM image in the middle. The TEM image on the right shows cubic mesopores in the silica walls that surround macropores.

2.1.10.1. Infiltration of CCT with Sol-Gel Precursors

According to an early report of combined colloidal crystal/surfactant templating, the addition of a cationic surfactant to the silicate sol-gel precursor produced macroporous silica with uniform mesopores within the walls [268]. Addition of the surfactant template increased the surface area of the silica inverse opal from 200 m^2 g^{-1} (without template) to 1300 m^2 g^{-1} (with template) in the calcined products. At the same time, the

BJH pore volume doubled to 0.80 mL g^{-1} and the pore size distribution tightened significantly with a median mesopore diameter of 2.3 nm. Block-copolymers, widely used in the synthesis of mesoporous solids, can also be employed in dual templating processes. Latex sphere opals have been infiltrated with silica precursor gels mixed with an amphiphilic triblock copolymer, such as $EO_{20}PO_{70}EO_{20}$ or $EO_{106}PO_{70}EO_{20}$. In aqueous solutions of these triblock copolymers, hydrophobic poly(propylene oxide) (PPO) chains form the core of micelles and hydrophilic poly(ethylene oxide) (PEO) chains stay in the water region around the core. In hierarchical silica prepared with $EO_{20}PO_{70}EO_{20}$, macropores were surrounded by relatively thin walls (up to ca. 80 nm) with uniformly sized mesopores ca. 7.7 nm in diameter [269]. Mesopores obtained with $EO_{106}PO_{70}EO_{20}$, were slightly smaller (4–5 nm) [270,271]. The mesopores provided sets of connecting paths between adjacent macropores, and larger windows created from the close-packing of spheres generated additional interconnections. Ordering at the mesopore length scale was not perfect because of spatial confinement between spheres. Further investigations of one-pot syntheses that combined latex spheres with ionic block-copolymers and alcohols as cosurfactant, revealed that porosity at a third length scale was present: disordered micropores were templated by individual polymer chains (the EO tails) from the triblock copolymer [272,273]. Depending on synthesis conditions, macropores were either interconnected or not interconnected. Amorphous regions were noted around windows. These regions corresponded to regions of the interstitial space in the colloidal crystal template which was too restricted for micelle templating. Using similar methods, silica films with macropores and mesopores have been prepared by spin coating the silica precursor/triblock copolymer mixture on an PS opal thin film [274].

The simultaneous application of PS opals and amphiphilic ionic liquids (1-hexadecyl-3-methylimidazolium-chloride) as dual templates produces a bimodal porous silica with macropores and a lamellar nanostructure within the walls that forms super-micropores (ca. 1.3 nm thick slit pores) [275]. By combining PS or PMMA colloidal crystals, mixtures of block copolymers with strong hydrophilic-hydrophobic contrast, and an ionic liquid, porous silica with trimodal hierarchical porosity was synthesized [276]. The latex spheres created macropores (90–360 nm diameters), block copolymer mixtures produced large mesopores (12–15 nm) and the ionic liquid smaller mesopores (2.5 nm). Patterned, hierarchically-structured silica films have been prepared by combining PDMS micromolding, polymer sphere templating and cooperative assembly of amphiphilic triblock copolymers with sol-gel precursors [277].

With tetrapropylammonium hydroxide as a structural agent, a dual templating method was employed to synthesize macroporous zeolites [268]. The periodic order of the colloidal crystal template was not fully maintained during the synthesis; however, mercury porosimetry measurements demonstrated that the macropores were interconnected. The walls comprised of silicalite were very thin (average: 113 nm), resulting in short diffusion paths for potential guest species. As a result, one might expect improved reaction efficiencies and reduced blocking of channels.

2.1.10.2. Infiltration of Colloidal Crystals with Nanocrystals

In a few reports, microporous nanoparticles were preformed and then infiltrated into colloidal crystal templates. For example, calcined nanocrystalline silicalite (30–80 nm grains) was suspended in ethanol and infiltrated into an ordered array of 300 nm PS spheres [278]. The silicalite crystallites interacted via surface silanol groups, possibly aided by the ethanol. Calcination resulted in further condensation of the macroporous zeolite skeleton, which exhibited a final macropore size of 160 nm. A similar approach was used for ZSM-5 and silicalite nanocrystals, although in this case, an aqueous suspension of the zeolite was employed with colloidal crystals consisting of relatively large spheres (2600 ± 100 nm) [279]. In this system, little shrinkage was observed, and void diameters were similar to the sphere sizes. In addition to zeolite micropores and templated macropores, this sample contained intercrystalline voids between agglomerated zeolite grains.

2.1.10.3. Sedimentation and Aggregation

As an alternative to infiltrating preformed opals, sphere sedimentation and aggregation processes can be combined in one step. When a mixture of TEOS and PS spheres dispersed in ethanol was briefly stirred and then permitted to settle, a layer of the precursor formed on the beads before the beads agglomerated as they settled [280]. A relatively high TEOS/PS ratio was required in order to achieve sufficient coverage of the spheres to maintain a well-formed structure. Addition of cetyltrimethylammonium surfactant to the synthesis gel with a composition identical to that used for conventional MCM-48 samples [281] resulted in macrostructured silica with a cubic arrangement of mesopores in the walls [282]. In this synthesis, silica exhibited an egg-shell morphology with relatively small windows between macropores. A much more open skeleton structure was obtained by a variation of this method leading to macrostructured MCM-41 [283]. In that report, dried PS spheres were stirred in a cetyltrimethylammonium surfactant solution followed by addition of sodium silicate solution. The products obtained after hydrothermal reaction and calcination possessed an fcc ordered array of voids, implying that spheres sedimented in a close-packed arrangement during hydrothermal processing.

2.1.10.4. Assembly of Core-Shell Structures

Additional control of the wall thickness is possible by growing porous nanoparticles on the surface of latex spheres via layer-by-layer deposition, followed by assembly of the composite spheres into colloidal crystals. TPA-silicalite-1 nanoparticles were assembled on the surface of monodisperse PS spheres with polycationic poly(diallyldimethylammonium) chloride interlayers [284,285]. The thickness of the resulting core-shell structure was controlled by depositing alternating layers of negatively charged zeolite nanoparticles and positively charged polyelectrolytes. Zeolite nanoparticles are fused during calcination via the condensation of surface silanols. Uncalcined core-shell particles could be used as building blocks for colloidal crystallization via centrifugation and sintering. Hierarchical porosity was provided by the microporosity of silicalite and by the random mesoporosity between the zeolite grains. In these structures, large windows between macropores are generally absent.

The layer-by-layer growth approach was combined with reactive shape transformations to form three-dimensionally ordered macroporous zeolite monoliths [286]. Mesoporous silica spheres were coated with polyelectrolyte films to allow attachment of nanocrystalline silicalite-1 seeds. For the preparation of hollow zeolite capsules, the seeded spheres were exposed to vapor phase transport treatment with amines, which formed a zeolite crust around a hollow core. In this case, the mesoporous silica spheres acted as a nutrient pool for zeolite shell growth. Assembly of seed-coated silica spheres into 3D arrays followed by hydrothermal transformation in a silicalite-producing gel resulted in macroporous zeolite monoliths with dimensions extending over centimeters. The monoliths, whose shapes depended on the mold used, were described as crack-free and mechanically stable. Although the macropores produced in this synthesis were closed, the method presents a significant step forward in the shaping of materials with hierarchical pore structures.

2.1.10.5. Sequential Multidimensional Structures

In all of the above examples, smaller pores occupy the space throughout the macropore structure. In contrast to this parallel pore architecture, a sequential 2D channel/3D void structure has recently been described [287]. 2D/3D composite porous alumina was prepared by combining colloidal crystal templating with anodization processes. The interstitial space within a PS colloidal crystal film on an Al substrate was filled with amorphous alumina by anodization. After removal of the PS template by toluene extraction, a second anodization process produced a 2D channel array underneath the 6–10 layer-thick inverse opal alumina structure. The pore periodicity and film thickness were controlled through the diameter of template particles and anodization conditions (voltage, time).

2.1.11. Two-Dimensional Pore Arrays

A technique related to colloidal crystal templating, but aimed at 2D surface patterning, is nanosphere lithography or "natural lithography" [288-290]. A monolayer or double-layer colloidal crystal acts as an etching mask or a deposition mask to outline a 2D ordered array of columnar structures inside the interstitial spaces between spheres. Numerous structural motifs are accessible by this technique [291]. The spheres can act as microlenses for nearfield photolithography to generate microstructures in a photoresist [292]. While nanosphere lithography is typically carried out using only one or two layers of close-packed spheres, 3D colloidal crystal arrays can also be applied as templates to create 2D surface gratings with submicron periodicity (Figure 6) [293].

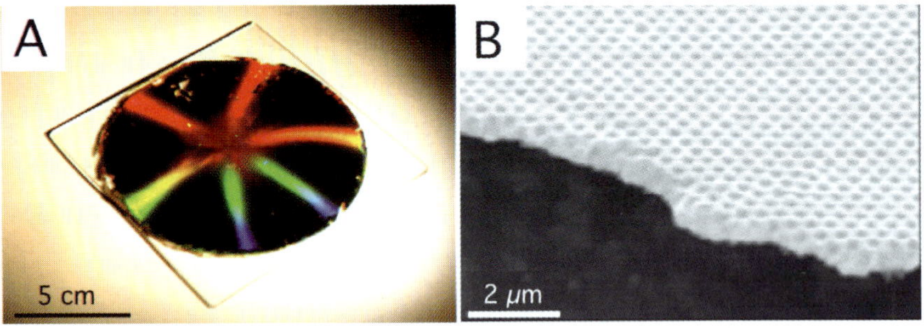

Figure 6. Photograph A) and SEM image B) of a wafer-scale surface grating of platinum deposited on a spin-coated silica surface and transferred to a glass substrate. Reprinted with permission from [Ref. 293], P. Jiang, Angew. Chem. Int. Ed. 43, 5625, (2004), Copyright @ Wiley-VCH.

Such gratings can be either free-standing or attached to a substrate. The synthesis involves growth of a colloidal crystal film on a planar substrate by convective self-assembly. Onto this template, a thin layer (up to ca. 1 μm) of material is deposited by physical vapor deposition methods. The grating with periodic wells on a denser layer of material is obtained after dissolution of the colloidal crystal template or by delamination of the film using sticky tape. The depth of the wells is approximately one-third of the radius of the templating spheres. Surface gratings of this type are of interest for a range of optical applications (*e.g.*, selective optical absorbers, broadband waveguide polarizers) and as picoliter containers for microanalysis. Similar 2D arrays of pores in a ZnO matrix were fabricated by potentiostatic electrochemical deposition using a 2D colloidal monolayer template on a conducting substrate [294]. The pore morphology depended on the deposition potential and included hemispherical or well-like structures.

3. Properties and Applications of 3DOM Materials

The unique 3DOM architecture benefits a wide range of potential applications, as has been clearly demonstrated in the literature during the past few years. Some of these applications rely on the periodic wall structure with unit cell dimensions that overlap with the wavelength of light. The resulting multiple light scattering is utilized in photonic crystals, pigments and sensors. Other uses of inverse opals benefit from the nanosize features of the skeleton, which provide confinement effects, short diffusion paths, and large surfaces available for chemical interactions. An advantage over discrete nanoparticles is that the grains in the skeleton are interconnected and are therefore not readily released into the surrounding environment. This may have ramifications for health and safety issues. Another advantage over packed nanoparticles is that the skeletal surface in the open 3DOM structure remains fully accessible to reagents without significant pressure build up, a benefit to applications involving reagent transport through the porous structure. This section will review recent progress towards applications of inverse opaline materials.

3.1. Mechanical Characterization

The mechanical properties of inverse opals have not yet been studied widely. Given the open nature of these structures, which are typically built from connected nanocrystals or from thin amorphous walls, many 3DOM materials are relatively fragile and can be crushed under small loads. This is certainly true for oxides and some metallic 3DOM structures. However, monolithic 3DOM carbon and inverse opal silicon or polymer films are strong enough for handling. For example, when used as electrodes, 3DOM carbon monoliths withstand the pressure exerted by binder clips [295]. The first study to quantify the mechanical strength of inverted opals employed depth-sensing indentation (DSI) with 3DOM silica samples having two different pore sizes [296]. In these experiments, contact loads are measured as a diamond indentation tip is lowered into the sample and then raised again. The resulting load–displacement traces can be used to quantify the mechanical integrity of the porous structures. The characteristics of the load-displacement responses changed from a behavior associated with elastic deformation by cell-wall bending at very small loads to periodic discontinuities, associated with cell wall fracture and pore collapse, and then to smooth and increased stiffness, as a result of densification due to accumulation of material under the indentation tip. Modulus and hardness values obtained from standard load-displacement traces could not be directly applied to these materials because of their nonelastic and nonvolume-conserving responses. However, when load-displacement responses were translated into pressure-volume space, the average pressure during indentation became a measure of the crushing pressure of the cell walls. Recently, DSI has also been applied to monolithic 3DOM carbon samples with various skeletal structures: a purely macroporous skeleton (3DOM C), a macroporous skeleton with mesoporous walls (3DOM/m C), and a 3DOM/m C/C nanocomposite sample in which mesopores had been filled with a more graphitic carbon phase by chemical vapor deposition [263]. The monolith strength followed the order 3DOM C > 3DOM/m C/C > 3DOM/m C for all indentation volumes. It was confirmed that filling mesopores with a graphitic phase strengthened the composite material (3DOM/m C/C) in comparison to its precursor (3DOM/m C).

3.2. Optical Applications

3.2.1. Photonic Crystals

Inverse opal structures are being considered as relatively inexpensive photonic crystal materials with potential value for a range of optical and optoelectronic applications. Photonic crystals are periodic structures in which the refractive index is spatially modulated [297-299]. They comprise a class of dielectric, metallic or metallodielectric materials in which the basic electromagnetic interactions can be altered in a controllable way over certain wavelengths and length scales that typically fall in a range close to the period of modulation. These periodic structures give rise to strong coherent multiple scattering of electromagnetic waves within the material, which produces a band structure for photons. Two major effects contribute to the band structure: (1) Bragg scattering resonance associated with the periodicity of the dielectric microstructure on the scale of the macropores; and (2) scattering resonance of the dielectric structure in a single unit cell of the periodic system, on the scale of the components forming the photonic crystal skeleton [244,300]. The scattering effects can lead to stop bands (spectral frequency regions in which the photonic density of states vanishes, that is, light cannot propagate). If scattering is strong enough, stop gaps in all directions overlap to form a full photonic band gap (PBG), in which propagation of electromagnetic waves is forbidden in all directions. As a result of these effects, it is possible to localize light within the photonic crystal structure, leading to phenomena such as enhancement or inhibition of spontaneous emission of light and control of spontaneous emission rates. At high frequencies, and for a limited range of input angles and photonic crystal compositions, light beams entering a photonic crystal can undergo large angular dispersion (the so-called superprism effect), even if a complete PBG is not present [301]. These properties can be useful in a variety of potential applications, including low threshold microlasers, ultrafast optical switches, microtransistors, anti-reflective coatings, sensors and filters of electromagnetic radiation, and optochemical sensing.

Theoretical calculations have shown that inverted opals with high refractive index walls can exhibit a full photonic band gap in the visible range of light if the refractive index contrast between the wall skeleton and the macropore content is large enough and the voids are centered on a perfect fcc lattice [300,302,303]. The filling fraction of the material surrounding spherical voids influences the size of the bandgap and the minimum refractive index contrast required to sustain a full PBG. For example, the minimum refractive index contrast requirement for a full inverse opal is ~3.0, but it is only ~2.85 if the walls occupy only 97.2% of the maximum possible space [304]. Simulations have indicated that direct gap structures are much more dependent on the refractive index contrast than are indirect gap structures. The direct gap region is dominated by the change in the filling fraction and the refractive index. The indirect gap region is dominated by the change in the air network topology [304]. The photonic band structure depends critically on the morphology of the macropore walls. For example, based on calculations for silicon, inverse opals with an egg-shell like structure can possess a complete PBG between the 8th and 9th bands, whereas a skeletal structure with cylinders connected by tetrahedral and octahedral centers can have a second PBG between the 5th and 6th bands [87,305,306]. These observations point at the importance of controlling shape and thickness of the skeletal walls.

The first inverse opal photonic crystals were prepared using titania as the wall component [3]. 3DOM titania photonic crystals have been synthesized both with egg-shell like structures [88] and with cylindrical skeletal structures [67,87,305,306]. Macroporous titania is one of the few photonic crystals that does not absorb in the visible region of the spectrum, yet has a refractive index (RI) contrast near the threshold value required to form a complete photonic bandgap (PBG). With an RI contrast of ca. 2.5, 3DOM titania in the anatase phase (the one more readily synthesized with a well-ordered macropore structure) does not have a complete PBG, but it has stop bands in the visible that overlap for a wide range of angles covering over 55% of all directions [3,88]. These stopbands can also be designed to fall into the UV region by employing 3DOM TiO_2 with sufficiently small void diameters (e.g., λ = 380 nm with a 240 nm lattice constant for fcc symmetry) [307].

Silicon is of great interest for photonic crystal devices, because silicon is already widely used as the major material in advanced devices, and processing techniques for this element have been highly developed. Because silicon absorbs strongly in the visible, it is not suitable for that frequency range. Experimental evidence for a complete 3D photonic bandgap in the near infrared range (1.46 μm) has been offered with silicon inverse opals [198,199]. An optical study of the band structure of 3DOM Si in multiple directions revealed a certain frequency range that produced a reflectance peak regardless of orientation and could be assigned to a PBG [308]. Such a PBG could be achieved with various photonic crystal topologies, including those obtained from direct Si-CVD infiltration of silica opals [198] and those derived from preformed inverse silica opals which were used as micromolds for CVD deposition of silicon [259]. Normally, amorphous silicon walls are obtained by these methods. However, tempering can transform silicon into a polycrystalline state. Time-resolved, two-color pump-probe experiments on freestanding 3DOM Si showed that the conversion to the polycrystalline form modified the non-linear optical response of the photonic crystals from an absorptive response to a mainly dispersive response [309]. This non-linear response is of interest in the context of ultrafast optical switching. In addition to inverted opal Si thin films, photonic band gap fibers have recently been prepared by coating a cylindrical silica fiber (125 μm diameter) with a silica colloidal photonic crystal, which was then infiltrated with silicon by CVD [310]. The colloidal crystal was assembled around the fiber core by evaporation-induced self-assembly, covering several centimeters of the fiber. After removal of both the silica template and the original silica fiber by etching, an air-core waveguide structure was obtained. Although this structure did not result in a complete photonic band gap, the study demonstrated that curved surfaces, such as optical fibers, can be enveloped by 3D photonic crystals.

3.2.2. Modification of Spontaneous Emission

When a light source is embedded in an inverse opal photonic crystal, the spontaneous emission of light originating from the interior of the photonic crystal can be strongly modified [311]. Changes may involve the stop-band position, its intensity and its width. Scattering of emitted light that propagates through the photonic crystal can change in an angle-dependent manner and parameters, such as lifetime or total emitted power, in an angle-independent manner [312]. These properties allow one to tune the emission spectrum, directionality of emission and the rate of emission. According to calculations from a finite-difference time-domain method with perfectly matched layer boundary conditions, emission rates depend strongly on the position of an emitting dipole and its polarization within the unit cell [313]. Strong enhancement effects are predicted at interfaces between the high-index and low-index material. The inhibition of spontaneous emission within the bandgap was calculated to be on the order of two magnitudes, even for relatively small crystallites of periodic structures.

Modified spontaneous emission has been experimentally observed with dye molecules, rare earth ions and semiconductor nanoparticles confined in inverse opals and also in the case of intrinsic emission by the 3DOM skeleton. It has been pointed out that in such investigations, it is important to choose nonphotonic reference host materials that provide the same chemical environment as the system under study, so as to ascertain unmodified transition dipole moments, nonradiative decay rates and inhomogeneous broadening [314].

A broadband blue emission intrinsic to the skeleton was observed under UV excitation of 3DOM SiO_2 derived from sol-gel processes and polystyrene templating [315]. The fluorescence was clearly influenced by the stop-band. If the void space was filled with a solvent that shifted the stopband away from the emission, no inhibition in photoluminescence was observed. Inhibition was also eliminated by destroying the 3DOM structure through mechanical grinding. Similarly, the photoluminescence spectrum of a π-conjugated polymer inverse opal was modified by the optical stopband of the photonic crystal at a specific viewing angle [220]. More recently these effects have been studied in materials that efficiently photoluminesce, such as 3DOM ZnO [316] and TiO_2/ZnS:Mn multilayer inverse opals, both prepared by atomic layer deposition [99]. In samples where stopbands overlapped the emission spectrum in the near-UV to visible region, emission was modified and suppressed due to modified densities of photonic states. In the multilayer inverse opals, optical properties could be tuned by backfilling the structures with TiO_2, which caused a shift in the stop band [99]. Changes in optical properties do not always arise from photonic effects but can also be attributed to other properties related to the sample architecture. For example, 3DOM ZnO prepared by electrodeposition exhibited a more significant red-shift and broadening of the UV photoluminescence peak with increasing excitation power compared to compact ZnO nanocrystal films [317]. This behavior was attributed to the large fraction of air in the ordered ZnO nanostructures, where convective cooling is more inefficient and thermal conductivity is reduced.

Spontaneous emission spectra of laser dyes have been studied in 3DOM TiO_2 [314,318], 3DOM alumina [312] and 3DOM SnS_2 photonic crystals [165]. Compared to free-space dyes, within the photonic crystals, internal Bragg diffraction of the luminescence produced stop bands that suppressed the spectra in angle-dependent frequency ranges. When the lattice parameter of a photonic crystal was increased, the inhibition shifted to larger wavelengths. In some cases, a strong modulation of the local photonic density of states was noted, due to spatially nonuniform distribution of dye molecules in the unit cell [314]. In emission experiments of R6G dye molecules embedded in 3DOM alumina, enhanced emission was observed at the blue side of the stop bands, attributed to the escape of diffuse light from the photonic crystal [312].

Semiconductor nanoparticles are suitable for tuning emission within a photonic bandgap, because their emission spectra are strongly size-dependent. In a system of 3DOM TiO_2 containing 3 nm CdTe nanoparticles, the intensity of the CdTe photoluminescence was suppressed within the directional gap of the photonic crystal, following the PBG anisotropy [319]. This spectral modification resulted from both the photonic bandgap of the inverted opal and from energy exchange between CdTe nanoparticles and the TiO_2 framework.

3.2.3. Tunable Photonic Crystals

For many device applications, it is desirable to tune the band structure of photonic crystals in real time. Tuning is possible by dynamic variation of the refractive index, periodicity or space filling factor of a photonic crystal under the control of electric field, temperature, strain, or magnetic field effects.

Because the voids constitute the largest volume fraction in inverse opals, modulations in the refractive index (RI) of the void content have a significant influence on the overall refractive index. Such RI changes can be effected readily with infiltrated liquid crystals, which exhibit large optical anisotropy. In normal opals with smaller void fractions, these changes are less pronounced. Several theoretical [320] and experimental studies have highlighted the efficacy of liquid crystals in tuning photonic properties via applied fields, temperature variations or photochemical reorientation. Electric field tuning of the photonic stop band position was demonstrated in polymeric as well as ceramic inverse opal structures.

For instance, in polymer inverse opals sandwiched between two conducting glass substrates and infiltrated with nematic liquid crystals, an electric field reoriented the liquid crystal molecules, changing the effective refractive index of the void space [321-323]. A discontinuous transition of the stop band position was observed upon a change in the applied voltage. The response was faster for liquid crystals in the connected cavity network in a 3DOM structure, compared to isolated, micron-sized liquid crystal droplets trapped within a polymer matrix. At high voltages, a memory effect was detected, which was thought to originate from anchoring effects of the confined liquid crystal molecules. The hysteresis effect may be linked to surface treatment of the photonic crystal, as demonstrated for thin films of 3DOM TiO_2 infiltrated with nematic liquid crystals [100]. In these systems, the optical spectra were modified by variations in the electric field amplitude and in the film structure, with shifts in Bragg peak position up to 20 nm (for a field of 50 kV cm^{-1}). Infiltration of the inverse opal with nematic liquid crystal was facilitated by hydrophobic surface treatment. This surface treatment produced hysteresis for the Bragg peak position under the influence of an electric field, but hysteresis was not seen in the untreated sample. As in the case of the polymer photonic crystal mentioned above, this effect was explained as surface pinning of liquid crystal molecules onto the titania surface in the hydrophobic material. The ability to electrically switch 3D Bragg diffraction may be useful in applications such as non-mechanical beam steering or optical beam splitting.

Another approach of tuning the optical band structure of photonic crystals takes advantage of a phase transition in photoresponsive liquid crystals [324-326]. The pores of inverse opal silica films were infiltrated with liquid crystals containing azobenzene derivatives, which could undergo a phase transition induced by photoisomerization with UV light. This process was reversible with visible light. A change from an isotropic to a nematic liquid crystal phase modified the stop band position and intensity in both reflection and transmission spectra. This behavior varied with temperature and light intensity. Based on these effects, liquid-crystal-infiltrated inverse opals could store and display images created by irradiation with UV light through a photomask. The display color could be selected by varying the lattice distance of the inverse opal structure (Figure 7).

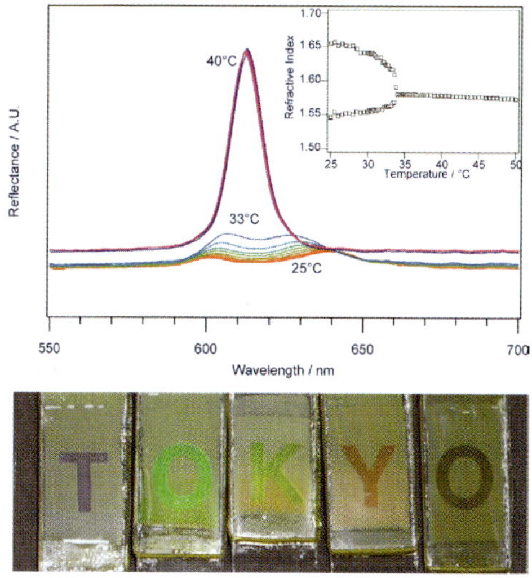

Figure 7. Top: Reflection spectra of nematic-liquid crystal-infiltrated inverse opal film as a function of temperature. The inset shows changes in effective refractive index with temperature [325]. Bottom: Photographs of stored patterns in inverse opal silica films that had been infiltrated with photoresponsive liquid crystals and irradiated with UV light through a patterned photomask. The different samples contain macropores with increasing pore diameters, ranging from 200–300 nm. Reprinted with permission from [Ref. 326], S. Kubo *et al.*, Chem. Mater. 17, 2298, (2005), Copyright @ American Chemical Society.

By combining photoinduced transitions with electric field control, it was possible to switch reversibly among three different states: the initial state in which light was scattered because of random orientation of the confined liquid crystals; the state induced by application of the electric field in which light of a specific wavelength was reflected; and the state induced by irradiation with UV light in which light of another specific wavelength was reflected. The response time of this system was on the order of tens of seconds and depended on temperature and irradiation intensity.

Tuning of optical properties without liquid crystal guest molecules was demonstrated for ferroelectric, electrochromic and magnetic inverse opals. In ferroelectric 3DOM (Pb,La)(Zr,Ti)O_3 (PLZT) prepared by the sol-gel method, optical stopbands shifted continuously and reversibly with application of an electric field, although the shift of the mid-gap position covered only a small wavelength range (4 nm) [140]. The shift resulted from a small variation in refractive index of PLZT inverse opals via an electro-optic effect. In electrochromic WO_3 inverse opals, Li intercalation into the crystals caused the first reflection peak to shift gradually to shorter wavelengths with its intensity decreasing, while the second reflection peak shifted toward longer wavelength for a 900 V range of applied voltage [108]. With magnetite inverse opals suspended in an aqueous medium, a magnetic field was used to rotate the inverse opals with respect to a light detector [327]. As different crystallographic planes were exposed to the detector, the measured stop-band positions changed.

3.2.4. Metallic and Metallodielectric Photonic Crystals via Colloidal Crystal Templating

Resistively heated photonic crystals can produce blackbody radiation which appears to be modified by the photonic crystal structure. Such effects have been demonstrated experimentally for metallic photonic crystals with a woodpile structure [328] and by theoretical calculations for finite silicon inverted opal PCs [329]. For the latter structures, it was suggested that angle-sensitive enhancement of thermal emission/absorption of radiation might be achieved. These observations suggest further applications for photonic crystals, such as efficient lighting technologies, tunable IR emitters and thermophotovoltaic devices with increased efficiency, as well as frequency-sensitive and angle-sensitive PC absorbers/emitters. Thermal emission can be modified by log structure tungsten photonic crystals, which particularly attenuate the long-wavelength components of emission [328]. With the goal of developing more economical methods of fabricating metallic photonic crystals for thermal emission, tungsten inverse opals have been fabricated by chemical vapor deposition of $W(CO)_6$ within opaline silica films [186]. Recently, 3DOM tungsten monoliths have been prepared by colloidal crystal templating, using wet chemical methods. These monoliths were found suitable to study thermal emission effects [187]. Metallodielectric inverse opals have been prepared by co-crystallizing silica-coated gold-nanoparticles and polymer spheres. Their optical properties could be engineered by varying the nanoparticle morphology (core size, shell thickness) and nanoparticle volume-filling ratio of the composite material [330].

3.2.5. Defects and Deformations in Photonic Crystals

Full photonic bandgaps are easily destroyed by crystal imperfections, even with a moderate amount of disorder in the positions and sizes of the macropores [331-334]. Such defects arise from both the CCT and from the infiltration and processing steps. CCT are usually polycrystalline, and exhibit ordered domains of various sizes that are separated by combinations of point, line, and planar defects. It is possible to reduce the number of defects found in CCT by utilizing careful assembly methodologies [10]. Gravity sedimentation of colloidal suspensions can produce well-ordered templates, since this process slowly constructs the template over several weeks. Significantly shorter assembly times can be accomplished via centrifugation [2], electrophoretic deposition [335], filtration [1], freezing [336,337], physical confinement [18,338,339], spin coating [290,340], and convective self-assembly methods [25,26,28,121,341]. However, faster assembly of colloids generally translates into increasingly defective CCT materials. Most commonly, templates for inverse opals as photonic crystals are prepared in the form of thin films by vertical deposition [342] or variations of this method. Once the colloidal crystal has formed, further defect formation can be minimized by strengthening connections between adjacent spheres via thermal annealing, and by reducing shrinkage during any subsequent processing and drying steps.

Several theoretical investigations have addressed effects of stacking faults on optical properties of inverse opals with fcc, hcp and doubly hcp stacking [343,344]. Stacking faults destroy periodicity in one direction but preserve periodicity in the plane normal to this direction. Regardless of the stacking sequence, the position and width of the lowest stop gap along the direction normal to a hcp plane was found to be invariant and robust to stacking disorder [344]. However, at higher frequencies, the propagation properties were strongly affected by stacking disorder, especially near edges of the absorption band gap. Disorder leads to interesting light localization. Frequency gaps, as seen in optical transmission experiments generally appear wider in the presence of stacking faults.

One-, two- and three-dimensional defects have been engineered into 3DOM solids via CCT in order to promote the propagation and collection of light within photonic crystals. Interstitial point defects have been introduced into silicon inverse opals films by adding trace amounts of silica spheres of another size to the initial opal template before infiltration [199]. Whereas this method leads to random positioning of defects, more controlled defect placement is possible by writing defect structures into photosensitive polymer photonic crystals using confocal imaging [345]. Two-dimensional planar defects have been formed by sedimenting a monolayer of silica spheres within two halves of a PS latex CCT and subsequently infiltrating the template with a silicon alkoxide precursor [346]. PS spheres are removed during calcination, producing a silica inverse opal film with a planar silica defect. Optical transmission measurements reveal a defect state that promotes transmission of infrared light, and the position of the defect band is adjustable by varying the size/thickness of the silica sphere monolayer within the CCT. Recent theoretical estimates have hypothesized the existence of triply degenerate optical states that localize light within inverted opals that contain substitutional point defects [347]. Interstitial defects were also considered, but did not yield major changes in the optical band structure of photonic crystals. Electron-beam lithography and subsequent solvent dissolution has been used to selectively pattern micrometer-sized SiO_2 photonic crystal domains within CVD-infiltrated CCT films [348]. A combination of photolithography and self-assembly has formed linear channels as extrinsic defects within inverse opal films to serve as waveguides for the propagation of light [349,350]. In all these cases, the spectral positions of localized planar defects are tuned by varying the wall thickness of vapor-deposited Si inverse opals [351], but great care must be made during fabrication not to introduce >2 % structural deviation from the lattice constant of the inverse opal geometry or risk complete closure of the photonic bandgap, even at high refractive-indice contrasts [352].

Shearing and stretching provide possible pathways to modifying the photonic band structure of inverted opals [353]. Numerical calculations showed that shearing lowers the gap width and slightly changes the midgap frequency. The midgap frequencies of *partial* PBGs depend almost linearly on the stretch amplitude, allowing one to tune the structure to a desired operational frequency by deformation of the photonic crystal. *Full* PBGs are not very sensitive to small stretching or shearing deformations up to 10%. However, the midgap frequencies of full PBGs can be tuned if the volume of the unit cell is not preserved under the stretch. Deformation of elastomeric inverse opals functionalized with luminescing PbS nanoparticles has been applied to tune stopbands in real time [354]. In these systems, PbS nanoparticles emitted in the near-IR spectral range. When the stop band of the photonic crystal was designed to overlap with the PbS photoluminescence peak, the spontaneous emission curve was strongly modified. This effect could be adjusted by compressing the photonic crystal, which enhances or suppresses the emission with respect to a reference sample where the stopband did not overlap with the photoluminescence peak. Suggested applications for tunable elastomeric photonic crystals include fingerprinting materials, piezo-electric driven color displays, accelerometer sensors, sensors to monitor explosion shockwaves, and strain sensors for buildings.

3.2.6. 3DOM Pigments

The structurally derived color of 3DOM materials mimics the beautiful patterns found, for example, in butterfly wings [355]. Inorganic pigments based on 3DOM zirconia or titania are not only stable to light and many chemicals, so that their colors do not fade; their colors can also be tuned across the whole spectrum by modification of pore sizes, wall thicknesses, compositions and filling of voids with fluids of variable refractive index (Figure 8) [244,356]. In order to obtain more brilliant colors, it is important to control the fine structure of the wall skeleton. The most intense colors were obtained in samples with smooth surfaces composed of very small nanoparticles. Surface roughening produced more pastel-colors or even white samples, in spite of the periodic macropore structure which normally brings forth Bragg scattering [244]. The loss in coloration has been attributed to random scattering by larger grains. In samples with very smooth, amorphous walls (e.g., 3DOM silica), light can be transmitted through the sample or reflected by it, the two processes resulting in complementary colors. While 3DOM oxides typically have the form of discrete particles or supported thin coatings, polymeric photonic crystals are suitable for producing self-supporting films with structural color [354]. Colloidal crystals embedded in an elastomer have been employed to produce photonic paper/ink systems [357], and similar approaches are suitable for elastomeric films with an inverse opal structure.

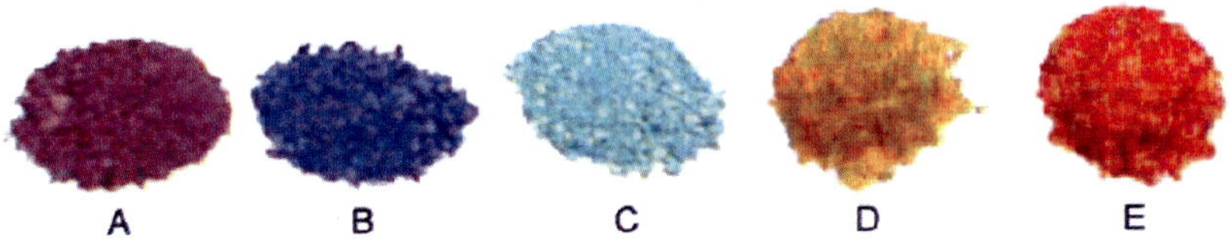

Figure 8. Photograph of 3DOM ZrO_2 powders with optical reflections spanning the colors of the visible spectrum with pore size equal to A) 200 nm, B) 250 nm, C) 285 nm, D) 250 nm (pores filled with methanol), and E) 285 nm (pores filled with methanol). Reprinted with permission from [Ref. 244], R. C. Schroden *et al.* Chem. Mater. 14, 3305, (2002), Copyright @ American Chemical Society.

3.2.7. Dye-Sensitized Titania Photonic Crystals

Titania photonic crystals have been considered as components in dye-sensitized photoelectrochemical cells designed to increase the efficiency of the solar cells [358-361]. In these cells, light absorption by a Ru-based dye produces an electron-hole pair (an exciton). When the exciton reaches the titania surface, it injects an electron into the conduction band of this wide-bandgap semiconductor, generating a photocurrent. A redox couple present in the cell regenerates the dye. Some problems in conventional dye-sensitized solar cells include charge recombination before the exciton diffuses to the surface and poor absorption by efficient dye-sensitizers in the red region. The 3DOM architecture can benefit the efficiency of these solar cells in several ways that involve both structural and optical properties of the photonic crystals. The interconnected voids permit penetration of the dye and, in heterojunction solar cells, penetration of a solid-state hole-conductor throughout the porous material [360]. A large interface is available between the dye and the semiconductor. Confinement of the dye minimizes the separation of excitons from the surface to diminish recombination processes. Because of their unique optical properties, macroporous titania films can significantly enhance backscattering of light, which ultimately leads to increased photocurrent efficiencies. In dye-sensitized photoelectrodes which incorporate a 3DOM titania layer coupled with a conventional titania nanoparticle film, the short-circuit photocurrent efficiency was increased by 26% compared to a non-structured dye-sensitized photoelectrode prepared only from titania nanocrystals [358]. The enhancement was originally attributed to the dielectric mirror effects of the photonic crystals which could cause a change in dye absorbance depending on the position of the stop band. A more detailed mechanistic study revealed that the bilayer architecture (porous titania/nanocrystalline titania) was mainly responsible for enhanced light conversion efficiency in the red spectral range, where the dye-sensitizer was a poor absorber [359]. Several factors contributed to the enhanced backscattering, including localization of heavy photons near the edges of photonic gap, Bragg diffraction in the periodic inverse opal lattice and multiple scattering events at disordered regions, such as structural defects. Of these factors, the multiple scattering was the major influence. As a result, enhancement in incident photon-to-current conversion efficiency was also achieved with bilayers involving disordered porous films combined with nanocrystal films. However, in a study of nanostructured heterojunction solar cells based on single layers of inverse opal TiO_2 films, photovoltaic responses were significantly higher in ordered areas than in non-ordered areas of the titania photonic crystal [361]. It should be noted that the absolute conversion efficiencies in these early systems adopting inverted opal architectures were still low, and the systems require further optimization to fully exploit the advantages of the photonic crystals.

3.2.8. Surface-Enhanced Raman Spectroscopy

When molecules are adsorbed to gold or silver particles, their Raman scattering can be strongly enhanced through the influence of highly localized electric fields near the precious metal surface. This so-called SERS (surface-enhanced Raman spectroscopy) effect can be modified by structuring gold with an inverse opal architecture [192,193]. The SERS activity for thin, ordered macroporous gold films (ca. two void layers thick) was examined, using *trans*-1,2-bis(4-pyridyl)ethylene as a model compound. The enhancement for characteristic Raman peaks was reported to be ca. 10^4 for ordered macroporous gold relative to glass. The Raman signal was twice as intense for monolayers of regularly structured porous films as for gold particles randomly adsorbed on glass. With multilayers, the signal was further enhanced. The performance was comparable to that of lithographically produced silver gratings, but fabrication is much cheaper by the colloidal crystal templating process. The SERS application can tolerate domain boundaries and crystal defects that are detrimental to many other photonic crystal applications. The colloidal-crystal-templated gold films have been integrated into an on-line flow chamber and used to detect sodium cyanide in water under limited conditions [194]. A systematic characterization of this system demonstrated that textural porosity between gold particles in the skeleton was critical for obtaining strong Raman signal enhancement [195].

3.3. Sensors

Sensors based on an inverse opal structure rely either on optical changes, such as changes in refractive index, pore geometry, or surface plasmons, or on electronic changes facilitated by the large reactive and accessible surface in inverted opals, the nanoparticle nature of the skeleton, and sometimes by the presence of well dispersed electrocatalysts. In certain cases, both optical and electronic interactions can be used for signal responses. It has been demonstrated that inverse opals can be used to sense changes in mechanical pressure, gas pressure, solvent composition, refractive index, pH, ionic strength, humidity, as well as uptake of molecules, including biomolecules.

3.3.1. Response Based on Changes in Refractive Index

As noted earlier, the periodic 3DOM structure causes light to be diffracted and can impart materials that are normally white or colorless with sometimes intense coloration. Optical stopbands can be observed even if a photonic crystal lacks full photonic band gaps. Ignoring dynamical diffraction effects, an approximate expression for the spectral positions of stop bands for 3DOM structures combines Bragg's law of diffraction with Snell's law of refraction to provide an estimate of the peak position for the stop band:

$$\lambda = \frac{2d_{hkl}}{m}\sqrt{n_{avg}^2 - \sin^2\theta} \qquad (1)$$

where λ is the wavelength of the stop band peak, d_{hkl} is the spacing between (hkl) diffraction planes, m is the order of the Bragg diffraction, θ is the angle measured from the normal to the planes, and n_{avg} is the average refractive index of the macroporous material [244]. The contributions to the average refractive index arise from the skeletal walls, from the voids and from any analyte present in the system:

$$n_{avg} = \phi_{walls}n_{walls} + \phi_{voids}n_{voids} + \phi_{analyte}n_{analyte} \qquad (2)$$

where ϕ_i is the volume fraction of each component, and $\phi_{walls} + \phi_{voids} + \phi_{analyte} = 1$. Because the largest volume fraction is that of the voids, changes in the void content can result in significant changes in stop band position and in color. For example, the minimum in the stop band position of 3DOM ZrO_2 with 285 nm pores could be shifted by 420 nm per unit increase in the refractive index of the void content, when macropores were filled with various solvents [244]. The response of 3DOM C (233 nm pores) to solvent filling with H_2O/acetonitrile mixtures was even steeper: nearly 1200 nm per unit increase in refractive index [149]. The stop bands are less responsive to changes in refractive index of the walls, simply because the corresponding volume fraction is so low. Nonetheless, measurable shifts have been reported when analyte molecules changed the effective refractive index near a pore surface. For example, when 3DOM polystyrene films were used to immobilize proteins on the pore surface, shifts of 2–4 nm in the diffraction peak position in the visible region were observed as the refractive index of the solution near the pore surface was changed during binding of biomolecule analytes [362]. In biotinylated 3DOM copolymer films, affinity-binding of avidin (a molecule with a diameter of ca. 8 nm, i.e., 2.5% of the lattice constant of the film) changed the pore wall thickness and shifted the stopband by up to 15 nm [225]. The detection limit for this system was estimated to be ca. 0.4 nm for the thickness of an organic coating.

To prepare a label-free biosensor system, the refractive index response was combined with changes in the localized surface plasmon resonance (LSPR) of gold nanoparticles deposited on 3DOM titania films [363]. The LSPR depends on the void content, including solvents and surface-adsorbed compounds. Further surface modification of the gold-modified 3DOM films with an activated ester and human immunoglobulin-G (hIgG) permitted selective capture of anti-hIgG from solution. As this occurred, shifts in the LSPR could be employed for real-time detection of the immunoreaction. Some size-selectivity was noted and the authors suggested that these films might be used for simultaneous separation and sensing.

3.3.2. Response Based on Changes in Pore Spacing or Pore Geometry

Inverse opals prepared from hydrogels can undergo changes in pore spacing with humidity or pH. Hydrogels can also be functionalized to respond to specific chemical analytes, such as glucose, by changing the geometric dimensions of the inverse opal structure [364]. These dimensional changes produce shifts in optical diffraction bands (see equation 1) and they can be readily detected by monitoring color changes. As an example of a mechanically robust pH-sensor, inverse opal hydrogels based on 2-hydroxyethyl methacrylate (HEMA)-acrylic acid copolymers have tunable pH sensitivity of potentially 0.01 pH units

[239]. The magnitude of the stop-band shifts could be tailored by varying the acrylic acid concentration. In spite of the open nature of these materials, the swelling or contraction kinetics followed a delayed diffusion-limited process with relatively long equilibration times of ca. 20 min. The optical response time of a polyacrylamide inverse opal hydrogel to changes in humidity was significantly faster, on the order of tens of seconds [235]. The response was slower for a drying cycle than for a water absorption cycle as a result of the hydrophilicity of the polyacrylamide hydrogel. It was noted that optical responses in this study included not only peak positions, but also peak widths and intensities. Functionalization of inverse opal HEMA hydrogels with 3-acrylamidophenylboronic acid (APBA) provided reversible binding sites for glucose at physiological glucose concentrations and ionic strength, producing shifts in the stopband position [237]. As in the case of proton detection, the response for glucose was slow and controlled by diffusion-limited kinetics. Changes in pore geometry can result if pressure or tension are applied to compressible inverse opal structures. For example, uniaxial elongation of a polymer inverse opal causes the distance between the (111) planes parallel to the film surface to decrease, resulting in a shift of the stop band to shorter wavelength [365]. This is coupled with a shift to longer wavelength of the stop band associated with periodicity in the stretching direction. The ability of elastomeric photonic crystal films to undergo readily reversible, local changes in stop-band position, reflection intensity and peak width upon compression and decompression has been applied to biometric recognition of fingerprints [354].

3.3.3. Response Based on Changes in Surface Electronic States

Responses based on changing surface electronic states may be used to monitor the concentrations of adsorbed gases for gas-sensing applications, as has been demonstrated with colloidal-crystal-templated SnO_2, a metal oxide commonly used for gas detection. The conductivity of SnO_2 increases when a gas reacts with surface oxygen species, forming ionized oxygen vacancies in the subsurface region, and releasing electrons trapped by the oxygen adsorbate [119,366]. The sensitivity benefits from the small (4–10 nm) tin oxide grains present in 3DOM SnO_2 [121]. As a result of the interconnected pore structure that provides ready access to these grains, a sensor for CO and toluene fabricated from 3DOM SnO_2 dispersed on an interdigitated electrode array showed rapid and reliable response at 400 °C [119]. When the inverted SnO_2 opal structures were grown directly within interdigitated gold microelectrodes, good reproducibility of the response from one sensor to another was observed [120]. The response for CO was further enhanced by dispersing Ru or Pt centers on the 3DOM SnO_2 structure [366]. A notable difference that was cited in comparison with nanosized powders was the stronger tendency of 3DOM SnO_2 to repair damage from oxygen defects at the surface by migration of subsurface oxygen atoms from their regular sites [366]. Rebuilding of the surface may be driven by the very high surface:volume ratio of nanoparticles that form the skeleton of the 3DOM structure. For CO sensing, the electrical response of the inverted opal structure was not quantitatively different from that of less structured nanosized materials [366]. However, responses to NO(*g*) or NO_2(*g*) were measured to be significantly stronger than for conventional SnO_2 thick film sensors and also in comparison with mesoporous SnO_2 materials [367]. The sensitivity of 3DOM SnO_2 to NO_2(*g*) was further tuned by varying the pore size in the range from 15–800 nm, with the highest response being observed for 250 nm pores.

3.3.4. *Electrochemical Response*

A potentiometric inverse opals biosensor was fabricated by electrochemically depositing creatinine deiminase (CD)-doped poly(pyrrole) around a PS colloidal crystal template [224]. The enzyme catalyzes the conversion of creatinine into 1-methylhydantoin and ammonia. The electric potential at the porous electrode surface is adjusted by modulating the pH around the conducting polymer by exposing the products to ammonia. The porosity and permeability of the inverse opal structures benefited the response by providing a larger surface area in contact with the reagents. In the open structure, the initial potential was restored within seconds of analyte injection compared to minutes for non-macroporous films.

An amperometric, nonenzymatic glucose sensor system was based on 3DOM Pt films [368]. 3DOM Pt displayed improved electrocatalytic activity toward glucose oxidation compared to directly deposited Pt, as a result of the large effective surface area and interconnectivity of pores in the macroporous metal. The electrocatalytic activity of the 3DOM Pt depended on the macropore size, the largest activity being observed for the smallest pore size in the range from 200–500 nm. Because the oxidation mechanism varies for different analytes, 3DOM Pt was more sensitive to glucose than to common interfering species, such as ascorbic and uric acids.

3.4. Magnetic Properties

Unusual magnetic properties of 3DOM metals and oxides were first reported by Krivorotov et al., who studied the exchange bias in a 3DOM Co sample, whose wall skeleton was composed of 3–4 nm Co nanoparticles with a CoO surface skin [369]. As a consequence of the synthesis, many of these nanoparticles were embedded in a graphite-like carbon matrix. Hysteresis loops in magnetic moment vs. magnetic field measurements of the field-cooled samples of 3DOM Co were shifted along the field axis, the shift being consistent with the exchange bias phenomenon (Figure 9).

The exchange bias was attributed to antiferromagnetic CoO, which formed when the samples were exposed to the ambient atmosphere. The temperature dependence of the coercivity was nonmonotonic, which was explained in terms of a spin-glass-like state of Co nanoparticles that forms below 45 K. Subsequent magnetic measurements on 3DOM Co, Fe, Ni, Ni/Fe alloys and periodic nickel sphere arrays templated from inverse opal PMMA revealed large coercivity enhancements of the macroporous metals in comparison to nonporous thin films [179,261,370-372]. For macroporous films of constant thickness, the coercive field was inversely proportional to the diameter of voids [370]. This increase in coercivity has been related to particle size and shape effects, because the 3DOM skeleton pins individual nanoparticle domains together [179]. The remanence of 3DOM Ni was considerably lower than for a nonporous Ni film with a magnetic field parallel to the film.

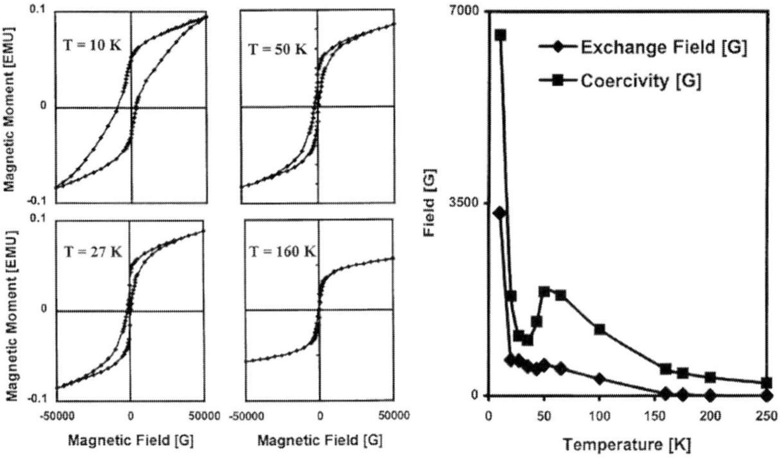

Figure 9. Hysteresis loops (left) of 3DOM Co cooled from 300K to the indicated measurement temperature in an external field of 1 T. Temperature dependence (right) of exchange bias and coercivity of 3DOM Co. Reprinted with permission from [Ref. 369], I. N. Krivorotov *et al.*, J. Magnetism Magnetic Mater. 226, 1800, (2001), Copyright @ Elsevier.

The magnetic properties of a few ceramic macroporous systems have also been investigated. In 3DOM perovskite manganites of composition $La_{0.7}Ca_{0.3-x}Sr_xMnO_3$, the ferromagnetic transition temperature (T_c) gradually rose to 350 K as the Sr content was increased [373]. For x = 0.1, a metal-insulator transition was observed near T_c and a large, temperature-independent negative magnetoresistance below T_c. The large magnetoresistance was ascribed to spin polarized tunneling between weakly bound grains within the 3DOM skeleton. Inverse opal ceramics containing spherical γ-Fe_2O_3 particles were ferromagnetic [226]. These were synthesized pyrolyzing 3DOM polyferrocenylsilanes, which had been prepared by the thermal ring-opening polymerization of spirocyclic [1] silaferrocenophanes confined within the interstitial void spaces of silica opals.

3.5. Catalysis

Surprisingly few investigations of 3DOM materials as catalysts or catalyst support have been reported. This is probably related to the comparatively larger cost of creating macropores by templating methods than by compacting of powders and perhaps to the concern that the extremely open inverse opal structures can be fragile. Nonetheless, 3DOM structures possess some unique advantages over pressed powders, including pressed powders of nanoparticles. The 3DOM structure provides the benefits of nanoparticles (short diffusion lengths, high specific surface areas, size-dependent properties) without the potential health risks related to free nanoparticles, because the nanoparticles are interconnected in 3DOM pore walls. The open, continuous macropore structure (1) provides good access for relatively big molecules to a large surface area, (2) reduces pressure-build up and (3) permits short contact times that eliminate side reactions at rapid flow rates. For example, short contact times help prevent over-oxidation in oxidation reactions. Though perhaps not immediately obvious, the structural periodicity of 3DOM materials can also be beneficial in the context of catalysis, because it ensures that voids are interconnected and all void space is accessible. Another benefit of colloidal crystal templating for catalysts is that the method allows hierarchical structuring, as noted earlier (section 2.2.10). Inverse opals with hierarchical pore structure combine features of easy access through macropores with potential size selectivity in mesopores or micropores. In spite of the large void fraction in 3DOM solids, some compositions are strong enough to be handled, in particular 3DOM carbon [263]. Even in the case of oxides, the macropore structure remains when 3DOM particles are crushed into smaller fragments [315]. After multiple crushing and grinding cycles, intact skeletal fragments can still be observed via TEM.

Studies of potential 3DOM catalysts have involved oxides, including functionalized silica, zirconia and perovskites, as well as conducting inverse opals (polyaniline, carbon). Functionalization of 3DOM silica is possible by many of the methods that have been developed for other silica morphologies and can introduce catalytic activity. For example, 3DOM silica functionalized with highly dispersed polyoxometalate

(POM) clusters (lacunary γ-decatungstosilicate clusters) exhibited catalytic activity for epoxidation of cyclooctene with an anhydrous H_2O_2/*t*-BuOH solution at room temperature [245]. The epoxidation of cyclohexene to cyclohexene oxide in the presence of isobutyraldehyde was catalyzed by 3DOM silica modified with transition-metal-substituted polyoxometalates of the type $[Co^{II}(H_2O)PW_{11}O_{39}]^{5-}$ and $[SiW_9O_{37}\{Co^{II}(H_2O)\}_3]^{10-}$ [374]. When clusters were attached datively to the amine-modified surface, they were not lost during washing or reaction processes. However, the catalytic activities of the functionalized 3DOM silica samples were comparable to those of identically modified mesoporous silica and amorphous silica. 3DOM silica functionalized with monovacant Keggin-type polyoxometalates $(X^{n+}W_{11}O_{39})^{(12-n)-}$ (X^{n+} = P^{5+}, Si^{4+}, Ge^{4+}, B^{3+}) exhibited photocatalytic activity to degrade aqueous malic acid under irradiation in the near UV region, also with negligible leakage of POM clusters from the support [246,247].

3DOM sulfated zirconia with surface areas up to 120 m^2 g^{-1} was evaluated and optimized for *n*-butane isomerization [107]. The catalytic activity was correlated to the structure and composition of this porous catalyst. Template removal and calcination conditions had a strong effect on physicochemical conditions of the catalyst. The calcination temperature affected the sulfate content, surface area, size and phase of crystallites forming the wall skeleton, catalytic activity, and even the appearance of 3DOM ZrO_2. The catalytic activity exhibited a bell-shaped behavior as a function of sulfate density and calcination temperature. While a clear advantage of the inverse opal geometry was not demonstrated for *n*-butane isomerization, one might expect benefits for reactions involving larger reactant or product molecules.

3DOM perovskite-type materials ($LaFeO_3$) have been evaluated for combustion of nanosize carbon, a model of particulate matter exhausted from diesel engines [144]. The 3DOM architecture influences the temperature at which half of a colloidal carbon sample decomposed when it was in contact with the perovskite. This temperature is related to the combustion activity of carbon particles on a support. The temperature was lower for 3DOM $LaFeO_3$ than for nonporous $LaFeO_3$, due to enhanced contact between the catalyst and nanosize carbon particles on the porous support.

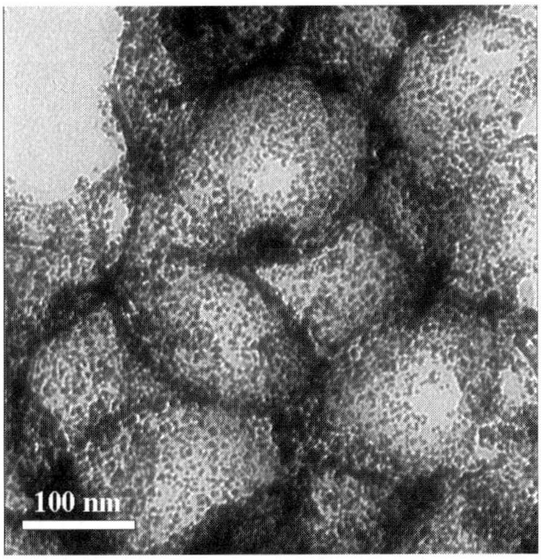

Figure 10. TEM image of Pt-Ru alloy clusters supported on 3DOM carbon. The small dots correspond to the alloy particles, the continuous lines to the carbon skeleton. Reprinted with permission from [Ref. 156], G. S. Chai *et al.*, J. Phys. Chem. B 108, 7074, (2004), Copyright @ American Chemical Society.

Conductive 3DOM structures are of interest, for example, as electrocatalysts for biosensing applications. Doped 3DOM polyaniline films were employed in the electrocatalytic oxidation of reduced beta-nicotinamide adenine dinucleotide (NADH) [218]. Their efficiencies were more than an order of magnitude higher than for unpatterned films, as a consequence of the higher surface area in the inverted opal geometry. Other applications of conductive inverse opals relate to energy storage and production systems. 3DOM carbon, prepared by carbonizing templated phenol-formaldehyde polymers, was used as a catalyst support for a Pt(50)-Ru(50) alloy catalyst to study the effect of this structure on anodic performance of the catalyst in a direct methanol fuel cell [155,156]. The high degree of catalyst dispersion on the porous carbon support (see Figure 10) and optimal conditions for diffusing reagents through the 3D interconnected macropore network, benefited the catalytic activity for methanol oxidation in this fuel cell. 3DOM carbon supports with smaller pore sizes yielded higher surface areas and better catalytic activity for methanol oxidation. Carbon with ca. 25 nm pores showed the highest performance among the materials tested, with power densities of ca. 58 and ca. 167 mW cm^{-2} at 30 and 70 °C, respectively. According to the investigators, these values correspond to significant increases compared to those of a commercially available Pt-Ru alloy catalyst [155,156].

3.6. Electrode and Battery Applications

The inverse opal geometry offers important morphological advantages over conventional electrode designs, which commonly consist of microcrystalline materials formed in bulk film geometries with thicknesses of tens to hundreds of μm. Commercial electrode dimensions create extremely long ion diffusion pathlengths, hindering the electrode's ability to rapidly charge and discharge, since solid-state ionic diffusion is a major rate-limiting factor among battery materials. In addition, conventional electrodes have relatively small interfacial areas in contact with electrolytes, and therefore conceal the majority of their active material within their bulk. Ionic accessibility to these bulk active materials is blocked by the formation of bands of reacted electrode material at electrode surfaces that serve as diffusion bottlenecks. In contrast, mass- and volume-normalized energy and power densities are substantially improved by the line-of-sight ionic diffusion pathways through three-dimensionally interconnected macropores and by the relatively large electrode/electrolyte interfacial areas that inverse opals provide.

Based on these advantages over conventional electrode designs, the inverse opal structure and other nanoarchitectured electrode geometries have been interesting subjects of recent Li-ion electrode and battery studies [375]. Three simple oxide inverse opal electrode compositions have recently been reported. Relatively small concentration polarization gradients are observed upon cycling 3DOM vanadia ambigel thin films as Li-ion cathodes, because of the material's inherently open macroporous network [376]. Alternatively, crystalline Shcherbinaite V_2O_5 powders have been fabricated by the colloidal-crystal-templating of proton-exchanged sodium metavanadate, although successive vanadic acid infiltrations resulted in surface crust formation but not increasing wall thickness [128]. Anatase TiO_2 inverse opals tend to polarize during cycling because of the relatively few electronic pathways available in the low-density macroporous structure and because anatase TiO_2 is an insulator [89,377-380]. Polarization is diminished by templating the inverse opal with smaller colloids in order to densify the material. Thin films of 3DOM SnO_2 undergo massive morphological changes during lithiation, in which the SnO_2 unit cell irreversibly swells during the formation of a series of Li-Sn alloys [121]. This is in contrast to the observed shape retention and extremely fast charge and discharge performance of SnO_2 nanofibril electrodes fabricated by infiltrating one-dimensional anodic alumina membrane templates [381].

Higher oxide inverse opals with spinel, olivine, and cubic structures have also been studied, because these phases are common cathode materials in Li-ion batteries. The degree of structural order and the crystallite size strongly affect the electrochemical performances of 3DOM $LiNiO_2$ [128] and 3DOM $LiCoO_2$ [127] powders. Thermal processing conditions have been optimized for 3DOM $LiNiO_2$ in order to avoid diffusion limitations in poorly crystalline $LiNiO_2$ phases that have mixed Li^+/Ni^{2+} cations in designated Li-ion sites. The cycling performance of 3DOM $LiCoO_2$ is improved by modifying mixed metal acetate precursors with poly(ethylene glycol), which increases the size of the nanocrystallites and makes the pore walls more robust to volumetric expansions during cycling. $LiCoO_2$, $LiNiO_2$, and $LiMn_2O_4$ inverse opals are additionally prepared by using metal citrate precursor solutions [129]. Fabricating $LiFePO_4$ powders with an ordered macroporous architecture has proven more challenging than other compositions, but has resulted in relatively high surface area cathode materials that deliver *ca.* 100 mAh g^{-1} at a discharge current density of 5.9 C [126]. Li-ion anode materials, such as 3DOM $Li_4Ti_5O_{12}$, have also been reported with relatively high rate performance because of the open macroporous structure and the large interface between the electrode and electrolyte phases [132]. 3DOM $Li_4Ti_5O_{12}$ exhibits little capacity fading during cycling because it is a zero-strain insertion material that has small volumetric changes during reversible lithiation. Colloidal-crystal-templated $Li_{0.35}La_{0.55}TiO_3$ perovskite inverse opals have shown promising solid-state Li-ion conductivities as large as $2x10^{-4}$ S cm^{-1}, although it is not clear whether the inverse opal geometry remained after milling and pressing the pellets used for conductivity measurements [143]. Metallic inverse opal electrodes have been developed for biological electrochemistry, as was demonstrated by 3DOM Au films that reversibly bind hemoglobin with up to 88 % coverage of the macropore wall surface and exhibit one-electron transfer with the heme center [382].

Carbon inverse opals are substantially more conductive than 3DOM inorganic oxide electrodes, and unlike most other 3DOM compositions can be prepared as films and monoliths that are useful for self-supported electrochemical and battery applications. Glassy carbon monoliths prepared from resorcinol-formaldehyde (RF) precursors have Li-ion insertion/de-insertion rate advantages over chemically similar, but non-templated, carbon monoliths because of relatively short ion-diffusion pathlengths through 3DOM macropore walls, high surface areas, and reasonable electronic conductivity values [158]. The rate performance of inverse opal carbons is further improved by depositing SnO_2 on macropore wall surfaces and by increasing the graphitic character and electronic conductivity of the pore walls via CVD of benzene within silica sphere CCT [146]. The latter 3DOM graphitic carbons deliver almost twice the capacity (260 mAh g^{-1}) of RF-derived 3DOM carbon monoliths at specific currents (1000 mA g^{-1}) that are more than six times greater than reported for 3DOM RF carbons [146]. Inverse opal carbon materials have recently served as the porous foundation for a three-dimensionally interpenetrated, all-solid-state, Li-ion electrochemical cell, in which the macropore walls of the porous carbon anode are sequentially deposited with an ultrathin polyphenol-based electrolyte/separator and a vanadia cathode gel (Figure 11) [295]. Three-dimensional microbatteries incorporate more energy- and power-density per geometrical footprint than existing battery designs that are constructed around two-dimensional thin-film technologies, which will enable power-on-chip applications such as MEMS devices for distributed autonomous sensing needs [375]. Double-layer capacitor and fuel cell applications also benefit from the large surface areas of inverse opal carbons. Three-dimensionally ordered mesoporous RF-carbons generate large double-layer capacitances up to 190 F g^{-1} and 20 $\mu F\ cm^{-2}$, at discharge rates >100 mV s^{-1} [383,384].

3.7. Sorption and Wetting Behavior

Studies of 3D ordered macroporous materials as sorbents have involved ionic, liquid and gaseous sorbates. In the case of gas adsorption, 3DOM solids that do not have secondary porosity in the skeleton, typically show type II adsorption behavior, which is associated with monolayer–multilayer adsorption on a relatively open surface. Type II isotherms show no plateau at high relative pressures, unlike Type IV isotherms that are obtained with mesoporous materials. The IUPAC classification has set the division between macropores and mesopores at 50 nm [385]. However, recent work by Jaroniec and coworkers showed that ordered macroporous carbon with spherical pores 62 nm in diameter exhibits a Type IV isotherm for nitrogen adsorption [386]. This observation suggests that the definition of mesopores and macropores may need to be revisited as new materials are synthesized with pore sizes near the borderline of the two pore regimes.

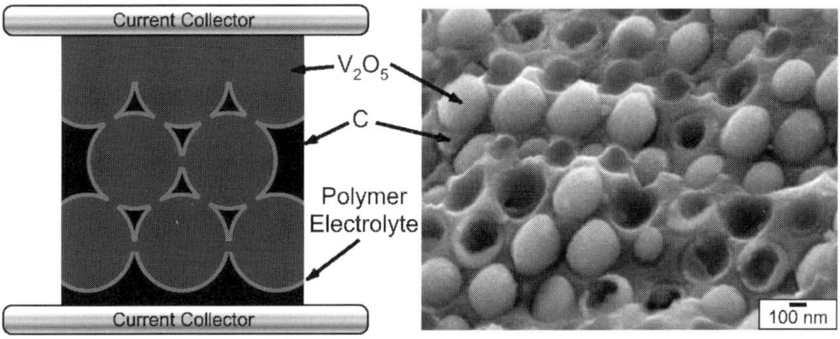

Figure 11. Schematic diagram and SEM image of a 3D nanostructured interpenetrating electrochemical cell. Macropores in 3DOM carbon were coated with a poly(phenylene oxide) layer and infiltrated with vanadia. After lithiation, this cell was charged to 3.2 V and was electrochemically cycled. Reprinted with permission from [Ref. 295], N. S. Ergang *et al.*, Adv. Mater. 18, 1750, (2006), Copyright @ Wiley-VCH.

Wettability is important during the synthesis of inverse opals, when reagents must infiltrate a colloidal crystal. It is also a critical parameter in applications of these porous materials that require pore filling by a liquid. Unmodified 3DOM silica is relatively hydrophobic. This is a consequence of the silica surface, as well as one of the nanostructure. This so-called lotus effect also enhances the hydrophobicity of butterfly wings. The degree of hydrophobicity/hydrophilicity can be varied through surface functionalization. A "superhydrophobic" material was obtained, for example, by modifying 3DOM silica with a fluoroalkylsilane (contact angle: 155 ° compared to 100° on a flat glass substrate modified with the same fluoroalkylsilane) [355]. Recently it has been demonstrated that the wettability of silica inverse opals can be switched with light if the pore surface is coated with a monolayer of photoresponsive azobenzene [387]. An azobenzene-functionalized polyelectrolyte film was self-assembled on the surface of the macropores by electrostatic layer-by-layer self-assembly. Azobenzene derivatives undergo reversible transitions from *trans* to *cis* form upon irradiation with UV/visible irradiation. This transition, in turn, induces changes in wettability, because the two states differ in dipole moment, surface free energy and water contact angle. It was found that on the modified 3DOM silica film, the contact angle decreased with increasing void diameter. Changes in contact angle were enlarged by the curved surfaces in comparison with planar substrates. Furthermore, the contact angle could be switched reversibly over several cycles (see Figure 12). This ability to control wettability has implications for many applications that rely on stimuli-responsive surfaces, such as microfluidic devices.

Metal adsorption on the surface of macroporous solids can be applied to impart functionality to the material (catalytic, magnetic or optically absorbing sites), but it is also relevant for removal of toxic metals from waste water. Adsorption sites can be incorporated by surface functionalization, a process which is very well developed for silicates [374,388]. Some recent work has also described functionalization of 3DOM carbon materials [389], as well as 3DOM titania and zirconia with thiol groups [91]. 3DOM carbon was modified by a multi-step oxidation, reduction, bromination and thiolation procedure. The absorption capacity for Pb(II) was improved nearly threefold compared to the unmodified carbon, but remained a modest 0.2 mmol g^{-1}. Higher metal absorption capacities of 0.33-1.41 mmol g^{-1} for Hg(II) and 0.27-1.24 mmol g^{-1} for Pb(II) were measured for 3DOM titania and zirconia functionalized with thiol groups via siloxane or sulfonate linkages [91].

These hybrid materials remained effective for metal adsorption after regeneration by an acid wash, with metal loading capacities of the recycled materials being on average two-thirds those of the original capacities.

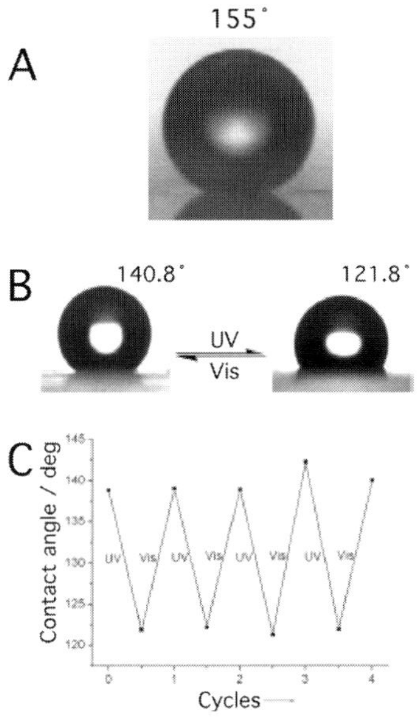

Figure 12. Water droplet (A) on fluoroalkylsilane-treated 3DOM SiO_2 [355]. Shapes (B) of water drops on 3DOM silica with void diameters of 263 nm, fabricated with a photoresponsive monolayer, on UV and visible irradiation. Reversible wettability transitions (C) are observed after UV or visible irradiation as shown in the graph of contact angle vs. irradiation cycle. Reprinted with permission from [Ref. 387], H. Ge *et al.*, Chem. Phys. Chem. 7, 575, (2006), Copyright @ Wiley-VCH.

3DOM SiO_2 films have been considered as components for inorganic/organic hybrid membranes to improve the mechanical strength of a polymer membrane. One investigation targeted polymer electrolyte membranes as proton conductors for fuel cells [390]. The composite membrane was prepared by multiple infiltration of 3DOM SiO_2 macropores with solutions of sulfonated polysulfones (SPSU, a chemically stable, inexpensive proton conductor) until a final composition of ca. 70 mass% SPSU : 30 mass% SiO_2 was attained. At this stage a continuous phase of the polymer permeated through a continuous phase of the silica skeleton. The high-frequency conductivity of the composite membrane was reduced by ca. 40% compared to the pure polymer. However, it was anticipated that the lifetime and performance of the composite membrane might be improved due to enhanced mechanical strength and rigidity.

3.8. Bioactive Materials

Syntheses and reactivities of bioactive glasses derived from inverse opal structures (3DOM-BG) have been investigated in detail and their biocompatibility verified [229,391-394]. Bioactive glasses form a biologically compatible hydroxycarbonate apatite (HCA) layer at the interface with bone *in vivo* or *in vitro* when the glasses are soaked in simulated body fluid (SBF) [395]. The accessible surface areas and hence the ability of bioglasses to be resorbed are significantly enhanced by incorporating interconnected macropores that are large enough to readily accommodate biofluids. 3DOM-BG have been templated from calcium, silicate and, in certain samples, phosphate precursors, using PMMA colloidal crystals to produce the continuous void system. Typical molar compositions were 20–21% CaO: 75–80% SiO_2: 0–4% P_2O_5. Pore sizes in these materials ranged from 300 to 800 nm and were uniform for a given material. When exposed to SBF at body temperature, the 3DOM BG was converted to calcium-deficient, bone-like apatite. The macroporous materials exhibited faster growth of apatite than a nontemplated control sample, resulting from the more accessible surface. Interesting morphological changes were noted during the conversion reactions, including the formation of amorphous calcium phosphate nodules on the originally smooth 3DOM-BG surface, an intermediate stage of spherical clusters of HCA crystals and the final conversion to larger HCA aggregates that no longer resembled the original inverse opal morphology (Figure 13).

Subsequent heating of a phosphate-containing bioactive glass converted the bone-like material to highly crystalline hydroxyapatite with a minor tricalcium phosphate phase. It was shown that thermal treatment permitted control over degradation properties [391]. Cell culture studies of 3DOM-BG with osteoblastic cells indicated that leachates from 3DOM-BG particles are nontoxic, the 3DOM-BG particles themselves are not cytotoxic, and they are compatible with these cells in vitro [393,394]. The cells could attach, spread and proliferate on and around 3DOM-BG particles. These studies point at the viability of using 3DOM-BG as a novel bone filler.

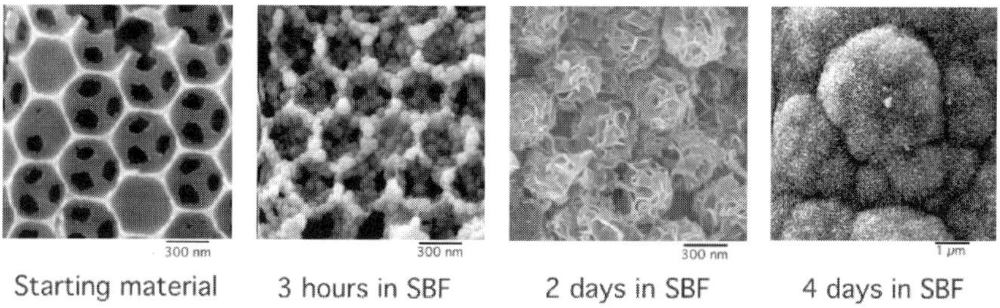

Figure 13. SEM images showing the conversion of 3DOM-BG to hydroxy carbonate apatite. 3DOM-BG with 20 mol% CaO–80 mol% SiO_2 was calcined at 600°C and soaked in SBF for the indicated times. Reprinted with permission from [Ref. 229], H. Yan *et al.*, Chem. Mater. 13, 4314, (2001), Copyright @ American Chemical Society.

Macroporous supports also provide suitable scaffolds for cell growth and tissue engineering. Submicrometer sized pores in biocompatible hosts are sufficient to incorporate small biological molecules. For example, a poly(vinyl alcohol) film containing 2D hexagonal arrays of 500 nm pores was shown to localize 3×4 nm^2 cylindrical proteins in the regularly spaced voids [209]. However, relatively large macropore sizes from 10–500 μm in diameter are necessary to accommodate cells. The formation of colloidal crystal templates for 3DOM materials normally relies on Brownian motion of the colloidal particles, as they assemble into periodic arrays. With sphere sizes near 100 μm, Brownian mobility is too low and sedimentation rates are too high to form well-crystallized structures. This limitation was recently overcome by use of a mixed-solvent evaporation techniques, in which 10–240 μm PS spheres were floated on the surface of ethylene glycol (a liquid with higher density than that of the polymer spheres) and self-assembled into close-packed structures through an interplay of buoyancy, gravity and immersion lateral capillary forces [75]. This colloidal crystal was used as a template for a large-pore 3D silicate scaffold for cells, a scaffold that was biocompatible with human hepatocellular carcinoma and human bone marrow cell cultures. An alternative approach to colloidal crystals composed of large polymer spheres has been to employ ethanol and gentle agitation during colloidal crystallization of 100 μm PMMA spheres [236]. The resulting colloidal crystals were used to template 3DOM polyacrylamide hydrogels with ca. 100 μm interconnected voids, which acted as 3D cell scaffolds for tissue growth and regeneration. The inverse opal architecture led to much greater swelling ratios than in bulk hydrogels and to swelling kinetics an order of magnitude faster than in the bulk. Cell culture experiments with two human cell lines demonstrated that the 3DOM hydrogel scaffold facilitates infiltration of cells (though only to a depth of *ca.* 3 pore layers), and that cells remain viable within the structure, thereby enabling tissue regeneration. The transparent nature of the hydrogel allowed for continuous high-resolution optical monitoring of cell proliferation and cell-cell interactions within the scaffold.

3.9. Pseudomorphic Transformation of 3DOM Materials

Pseudomorphic transformations are chemical reactions that change the composition of a solid preform without altering the preform's shape and morphological features [396]. While this class of reaction has historically been applied to macroscale objects, recent studies have indicated that shape-retaining chemical transformation reactions are applicable to biogenic materials with micrometer-sized features [397,398] and nanostructured materials [399,400]. The inverse opal geometry has a wide range of feature sizes that make 3DOM solids relevant as test materials to determine the structural and chemical limitations of pseudomorphic transformation reactions on a sub-micrometer length scale. 3DOM SiO_2 powders are pseudomorphically converted to 3DOM TiO_2 via a two-step gas-phase reaction that generates 3DOM $TiOF_2$ as an intermediate composition [401]. Uniform cubic crystallites with edgelengths of *ca.* 100 nm nucleate on the macroporous framework during the initial reaction of $TiF_4(g)$ and 3DOM SiO_2; the oxyfluoride species is then hydrolyzed in a flow of moist air and retains the morphology of the cube surfaces on a sub-100 nm length scale (Figure 14). Cubic crystallite nucleation is promoted at pore rims and on other sharp features, but dense surface crust regions on 3DOM SiO_2 and within colloidal silica spheres are incompletely converted. Other inverse opal compositions have been pseudomorphically transformed via variations of the gas-solid reaction method. Tungsten carbide inverse opal powders are synthesized by reacting 3DOM WO_3 with flowing $H_2(g)$ and $CH_4(g)$ at elevated temperature [Lytle, #962]. ZnO inverse opal films are directly converted into 3DOM ZnS by heating ZnO preforms within sealed ampules containing sulfur [106]. The chemical reaction is driven by the build up of sulfur vapor pressure (500 torr) within the ampules at 450 °C, and the degree of sulfidation is apparent from the spectral shift of the film's Bragg reflection and the difference in XRD patterns before and after the reaction. 3DOM Ta_2O_5 films are converted in flowing ammonia gas at 800 °C to form 3DOM Ta_3N_5 with a similar surface-templated morphology as its ALD-derived Ta_2O_5 preform [160].

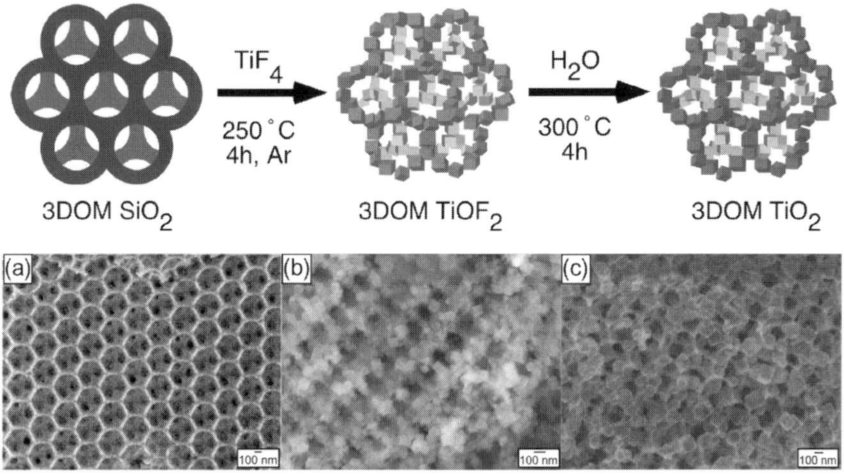

Figure 14. Schematic diagrams and SEM images of the pseudomorphic transformation of (a) 3DOM SiO_2 via (b) 3DOM $TiOF_2$ to (c) 3DOM TiO_2. Reproduced with permission from [Ref. 401], J. C. Lytle *et al.*, Chem. Mater., 16, 3829, (2004), Copyright @ American Chemical Society.

4. Conclusion

During its first decade, research involving colloidal crystal templating has advanced at a rapid pace. Improved syntheses have not only added to the variety and complexity of inverse opal compositions, but have also permitted greater control over structural morphology from the nanometer length scale of the wall skeletons to the micrometer-to-centimeter dimensions of the overall macroporous solids. New methods to functionalize the internal surfaces of 3DOM materials have resulted in the development of unique composite structures. Many of the proposed applications have reached the proof-of-concept stage, and for some of these, the transition from "potential" to "actual" application may not be far away. Yet opportunities for further progress remain. The types of monolithic 3DOM shapes are still limited, and lessons may be learned from the field of ceramics. Minimization of unwanted defects and controlled introduction of desired defects continue to be important goals for photonic crystals. Preparations of non-fcc periodic structures are still in their infancy. Active shape and size control over pore openings will be beneficial to various applications. One of the most important steps towards the realization of devices will be process integration into existing technologies. Given the current momentum in the field, we can expect new breakthroughs in the upcoming years. The future of 3DOM materials and colloidal crystal templating looks bright.

Acknowledgements

The authors are grateful for support by the Office of Naval Research (grant number N00014-01-1-0810, subcontracted from NWU), and the Petroleum Research Foundation administered by the American Chemical Society (ACS-PRF grant number 42751-AC10), which allowed us to prepare this review.

References

1. O. D. Velev, T. A. Jede, R. F. Lobo and A. M. Lenhoff, Nature 389, 447 (1997).
2. B. T. Holland, C. F. Blanford and A. Stein, Science 281, 538 (1998).

3. J. E. G. Wijnhoven and W. L. Vos, Science 281, 802 (1998).
4. O. D. Velev and A. M. Lenhoff, Curr. Opin. Colloid Interface Sci. 5, 56 (2000).
5. O. D. Velev and E. W. Kaler, Adv. Mater. 12, 531 (2000).
6. Special Issue on Photonic Crystals, Adv. Mater. 13, (2001).
7. C. F. Blanford, H. Yan, R. C. Schroden, M. Al-Daous and A. Stein, Adv. Mater. 13, 401 (2001).
8. H. Miguez, A. Blanco, F. Garcia-Santamaria, M. Ibisate, C. Lopez, F. Meseguer and F. Lopez-Tejeira, NATO ASI Ser., Ser. C. 563, 219 (2001).
9. D. J. Norris and Y. A. Vlasov, Adv. Mater. 13, 371 (2001).
10. A. Stein, Micropor. Mesopor. Mat. 44-45, 227 and references therein (2001).
11. A Stein and R. C. Schroden, Curr. Opin. Solid St. Mater. Sci. 5, 553 and references therein (2001).
12. F. Meseguer, A. Blanco, H. Miguez, F. Garcia-Santamaria, M. Ibisate and C. Lopez, Colloids Surf., A. 202, 281 (2002).
13. A. Stein and R. C. Schroden, in Nanostructured Catalysts edited S. L. Scott, C. M. Crudden and C. W. Jones, Kluwer-Plenum, Dordrecth, Netherlands (2003), Vol. 3, p. 257.
14. Y. Xia, Y. Lu, K. Kamata, B. Gates and Y. Yin, in The Chemistry of Nanostructured Materials edited P. Yang, World Scientific Publishing Co., Singapore (2003), p. 69.
15. R. C. Schroden and A. Stein, in Colloids and Colloid Assemblies: Synthesis, Modification, Organization and Utilization of Colloid Particles edited F. Caruso, Wiley-VCH, Weinheim, Germany (2004), p. 465.
16. Z. C. Zhou and X. S. Zhao, in Series on Chemical Engineering edited (2004), Vol. 4, p. 206.
17. A. D. Dinsmore, J. C. Crocker and A. G. Yodh, Curr. Opin. Colloid Interface Sci. 3, 5 (1998).
18. Y. Xia, B. Gates, Y. Yin and Y. Lu, Adv. Mater. 12, 693 (2000).
19. J. Texter, Comptes Rendus Chimie 6, 1425 (2003).
20. D. Wang and H. Moehwald, J. Mater. Chem. 14, 459 (2004).
21. N. V. Dziomkina and G. J. Vancso, Soft Mater. 1, 265 (2005).
22. J. McLellan, Y. Lu, X. Jiang and Y. Xia, in Nanoscale Assembly edited W. T. S. Huck, Springer, New York, NY (2005), p. 187.
23. J. W. Goodwin, J. Hearn, C. C. Ho and R. H. Ottewill, Brit. Polym. J. 5, 347 (1973).
24. J. W. Goodwin, R. H. Ottewill, R. Pelton, G. Vianello and D. E. Yates, Brit. Polym. J. 10, 173 (1978).
25. N. D. Denkov, O. D. Velev, P. A. Kralchevsky, I. B. Ivanov, H. Yoshimura and K. Nagayama, Langmuir 8, 3183 (1992).
26. C. D. Dushkin, H. Yoshimura and K. Nagayama, Chem. Phys. Lett. 204, 455 (1993).

27. G. T. D. Shouldice, G. A. Vandezande and A. Rudin, Eur. Polym. J. 30, 179 (1994).
28. A. S. Dimitrov and K. Nagayama, Langmuir 12, 1303 (1996).
29. D. Zou, S. Ma, R. Guan, M. Park, L. Sun, J. J. Aklonis and R. Salovey, J. Polym. Sci. Pol. Chem. 30, 137 (1992).
30. S. Shen, E. D. Sudol and M. S. El-Aasser, J. Polym. Sci. Pol. Chem. 31, 1393 (1993).
31. T. Tanrisever, O. Okay and I. C. Sonmezoglu, J. Appl. Polym. Sci. 61, 485 (1996).
32. W. Stober, A. Fink and E. Bohn, J. Colloid Interface Sci. 26, 62 (1968).
33. Y. Wang and Y. Xia, Nano Lett. 4, 2047 (2004).
34. J. Bao, Y. Liang, Z. Xu and L. Si, Adv. Mater. 15, 1832 (2003).
35. R. Brace and E. Matijevic, J. Inorg. Nucl. Chem. 35, 3691 (1973).
36. D. L. Catone and E. Matijevic, J. Colloid Interf. Sci. 48, 291 (1974).
37. W. B. Scott and E. Matijevic, J. Colloid Interf. Sci. 66, 447 (1978).
38. T. Ogihara, H. Nakajima, T. Yanagawa, N. Ogata, K. Yoshida and N. Matushita, J. Am. Ceram. Soc. 74, 2263 (1991).
39. A. Dong, N. Ren, Y. Tang, Y. Wang, Y. Zhang, W. Hua and Z. Gao, J. Am. Chem. Soc. 125, 4976 (2003).
40. D. G. Shchukin and R. A. Caruso, Chem. Mater. 16, 2287 (2004).
41. D. Grosso, G. J. d. A. A. S. Illia, E. L. Crepaldi, B. Charleux and C. Sanchez, Adv. Funct. Mater. 13, 37 (2003).
42. S.-H. Lee, Y.-S. Her and E. Matijevic, J. Colloid Interf. Sci. 186, 193 (1997).
43. Z. C. Orel, E. Matijevic and D. V. Goia, J. Mater. Res. 18, 1017 (2003).
44. M. Ocana, Colloid Polym. Sci. 278, 443 (2000).
45. E. A. Barringer and H. K. Bowen, J. Am. Ceram. Soc. 65, C199 (1982).
46. E. A. Barringer and H. K. Bowen, Langmuir 1, 414 (1985).
47. M. T. Harris and C. H. Byers, J. Non-Cryst. Solids 103, 49 (1988).
48. X. Jiang, T. Herricks and Y. Xia, Adv. Mater. 15, 1205 (2003).
49. A. Koliadima, L. A. Perez-Maqueda and E. Matijevic, Langmuir 13, 3733 (1997).
50. D. V. Goia and E. Matijevic, New J. Chem. 1203 (1998).
51. Z. Chen, P. Zhan, Z. Wang, J. Zhang, W. Zhang, N. Ming, C. T. Chan and P. Sheng, Adv. Mater. 16, 417 (2004).
52. I. M. Yakutik and G. P. Shevchenko, Surf. Sci. 566-568, 414 (2004).
53. M. J. A. de Dood, L. H. Sloof, A. Polman, A. Moroz and A. van Blaaderen, Appl. Phys. Lett. 79, 3585 (2001).
54. P. Tartaj, T. G. Carreno, M. L. Ferrer and C. J. Serna, Angew. Chem. Int. Ed. 43, 6304 (2004).
55. A. Dong, N. Ren, W. Yang, Y. Wang, Y. Zhang, D. Wang, J. Hu, Z. Gao and Y. Tang, Adv. Funct. Mater. 13, 943 (2003).
56. P. T. Miclea, A. S. Susha, Z. Liang, F. Caruso, C. M. S. Torres and S. G. Romanov, Appl. Phys. Lett. 84, 3960 (2004).

57. Y. Lu, H. Xiong, X. Jiang, Y. Xia, M. Prentiss and G. M. Whitesides, J. Am. Chem. Soc. 125, 12724 (2003).
58. L. Nagle, D. Ryan, S. Cobbe and D. Fitzmaurice, Nano Lett. 3, 51 (2003).
59. A. J. Khopade and F. Caruso, Chem. Mater. 16, 2107 (2004).
60. K. Kamata, Y. Lu and Y. Xia, J. Am. Chem. Soc. 125, 2384 (2003).
61. Y. Lu, J. McLellan and Y. Xia, Langmuir 20, 3464 (2004).
62. P. Schuetz and F. Caruso, Chem. Mater. 16, (2004).
63. O. Kalinina and E. Kumacheva, Macromolecules 32, 4122 (1999).
64. M.-K. Park, K. Onishi, J. Locklin, F. Caruso and R. C. Advincula, Langmuir 19, 8550 (2003).
65. D. Wang, A. L. Rogach and F. Caruso, Chem. Mater. 15, 2724 (2003).
66. A. Imhof, Langmuir 17, 3579 (2001).
67. B. T. Holland, C. F. Blanford, T. Do and A. Stein, Chem. Mater. 11, 795 (1999).
68. T. Okubo, J. Chem. Soc. Faraday Trans. 86, 2871 (1990).
69. A. E. Larsen and D. G. Grier, Nature 385, (1997).
70. R. H. Baughman, S. O. Dantas, S. Stafstrom, A. A. Zakhidov, T. B. Mitchell and D. H. E. Dublin, Science 288, 2018 (2000).
71. J. Z. Wu, D. Bratko, H. W. Blanch and J. M. Prausnitz, J. Chem. Phys. 113, 3360 (2000).
72. F. Garcia-Santamaria, H. T. Miyazaki, A. Urquia, M. Ibisate, M. Belmonte, N. Shinya, F. Meseguer and C. Lopez, Adv. Mater. 14, 1144 (2002).
73. L. V. Woodcock, Nature 385, 141 (1997).
74. L. V. Woodcock, Nature 388, 235 (1997).
75. Y. Liu, S. Wang, J. W. Lee and N. A. Kotov, Chem. Mater. 17, 4918 (2005).
76. M. E. Leunissen, C. G. Christova, A.-P. Hynninen, P. Royall, A. I. Campbell, A. Imhof, M. Dijkstra, R. van Roij and A. van Blaaderen, Nature 437, 235 (2005).
77. G.-R. Yi, J. H. Moon and S.-M. Yang, Chem. Mater. 13, (2001).
78. H.-L. Li, W. Dong, H.-J. Bongard and F. Marlow, J. Phys. Chem. B 109, 9939 (2005).
79. Z. C. Zhou and X. S. Zhao, Langmuir 21, 4717 (2005).
80. S.-L. Kuai, V.-V. Truong, A. Hache and X.-F. Hu, J. Appl. Phys. 96, 5982 (2004).
81. J. Rong, J. Ma and Z. Yang, Macromol. Rapid Commun. 25, 1786 (2004).
82. L. Ji, J. Rong and Z. Yang, Chem. Commun. 1080 (2003).
83. Z. Q. Sun, X. Chen, J. H. Zhang, Z. M. Chen, K. Zhang, X. Yan, Y. F. Wang, W. Z. Yu and B. Yang, Langmuir 21, 8987 (2005).
84. C. F. Blanford, C. B. Carter and A. Stein, J. Microscopy 216, 263 (2004).
85. C. F. Blanford, C. B. Carter and A. Stein, J. Phys.: Conference Series 26, 264 (2005).
86. N. P. Johnson, D. W. McComb, A. Richel, B. M. Treble and R. M. De La Rue, Synthetic Met. 116, 469 (2001).
87. W. Dong, H. J. Bongard and F. Marlow, Chem. Mater. 15, 568 (2003).

88. J. E. G. Wijnhoven, L. Bechger and W. L. Vos, Chem. Mater. 13, 4486 (2001).
89. M. Zukalova, M. Kalbac and L. Kavan, Mater. Res. Symp. Proc. 820, 363 (2004).
90. Y. Aoi, S. Kobayashi, E. Kamijo and S. Deki, J. Mater. Sci. 40, 5561 (2005).
91. R. C. Schroden, M. Al-Daous, S. Sokolov, B. J. Melde, J. C. Lytle, A. Stein, M. C. Carbajo, J. T. Fernandez and E. E. Rodriguez, J. Mater. Chem. 12, 3261 (2002).
92. Q. Meng, C. H. Fu, Y. Einaga, Z. Z. Gu, A. Fujishima and O. Sato, Chem. Mater. 14, 83 (2002).
93. G. Subramania, K. Constant, A. Biswas, M. M. Sigalas and K.-M. Ho, J. Am. Ceram. Soc. 85, 1383 (2002).
94. S. Jeon and P. V. Braun, Chem. Mater. 15, 1256 (2003).
95. S.-L. Kuai, S. Badilescu, G. Bader, R. Bruning, X. Hu and V.-V. Truong, Adv. Mater. 15, 73 (2003).
96. S.-L. Kuai, G. Bader, A. Hache, V.-V. Truong and X.-F. Hu, Thin Solid Films 483, 136 (2005).
97. J. S. King, D. Gaillot, T. Yamashita, C. Neff, E. Graugnard and C. J. Summers, Mater. Res. Symp. Proc. 846, 315 (2005).
98. J. S. King, E. Graugnard and C. J. Summers, Adv. Mater. 17, 1010 (2005).
99. J. S. King, E. Graugnard and C. J. Summers, Appl. Phys. Lett. 88, 081109/1 (2006).
100. E. Graugnard, J. S. King, S. Jain, C. J. Summers, Y. Zhang-Williams and I. C. Khoo, Phys. Rev. B 72, 233105/1 (2005).
101. H. Yan, C. F. Blanford, B. T. Holland, W. H. Smyrl and A. Stein, Chem. Mater. 12, 1134 (2000).
102. T. Sumida, Y. Wada, T. Kitamura and S. Yanagida, Chem. Lett. 38 (2001).
103. M. Fu, J. Zhou, R. Zong and L. Li, Key. Eng. Mater. 280-283, 569 (2005).
104. H. Yan, Y. Yang, Z. Fu, B. Yang, Z. Wang, L. Xia, S. Yu, S. Fu and F. Li, Chem. Lett. 34, 976 (2005).
105. K. H. Yeo, L. K. Teh and C. C. Wong, J. Cryst. Growth 287, 180 (2006).
106. B. H. Juarez, P. D. Garcia, D. Golmayo, A. Blanco and C. Lopez, Adv. Mater. 17, 2761 (2005).
107. M. A. Al-Daous and A. Stein, Chem. Mater. 15, 2638 (2003).
108. S.-L. Kuai, G. Bader and P. V. Ashrit, Appl. Phys. Lett. 86, 221110 (2005).
109. Q. Z. Wu, Y. Shen, J. F. Liao and Y. G. Li, Mater. Lett. 2688 (2004).
110. X.-Y. Zhang, T. W. Wang, W. Q. Jiang, D. Wu, L. Liu and A. H. Duan, Chin. Chem. Lett. 16, 1109 (2005).
111. Y. Zhang, Z. Lei, J. Li and S. Lu, New J. Chem. 25, 1118 (2001).
112. O. D. Velev, T. A. Jede, R. F. Lobo and A. M. Lenhoff, Chem. Mater. 10, 3597 (1998).
113. Y.-H. Ye, S. Badilescu and V.-V. Truong, Appl. Phys. Lett. 81, 616 (2002).
114. A. Cabanas, E. Enciso, M. C. Carbajo, M. J. Torralvo, C. Pando and J. A. R. Renuncio, Chem. Mater. 17, 6137 (2005).

115. A. Cabanas, E. Enciso, M. C. Carbajo, M. J. Torralvo, C. Pando and J. A. R. Renuncio, Chem. Commun. 2618 (2005).
116. Y. W. Chung, I. C. Leu, J. H. Lee and M. H. Hon, J. Cryst. Growth 275, e2389 (2005).
117. S. Sokolov, D. Bell and A. Stein, J. Am. Ceram. Soc. 86, 1481 (2003).
118. L.-X. Liu, P. Dong, D.-J. Wang, Q. Zhou and B.-Y. Cheng, Chinese Phys. Lett. 22, 741 (2005).
119. R. W. J. Scott, S. M. Yang, G. Chabanis, N. Coombs, D. E. Williams and G. A. Ozin, Adv. Mater. 13, 1468 (2001).
120. R. W. J. Scott, S. M. Yang, D. E. Williams and G. A. Ozin, Chem. Comm. 6, 688 (2003).
121. J. C. Lytle, H. Yan, N. S. Ergang, W. H. Smyrl and A. Stein, J. Mater. Chem. 14, 1616 (2004).
122. I. Soten, H. Miguez, S. M. Yang, S. Petrov, N. Coombs, N. Tetreault, N. Matsuura, H. E. Ruda and G. A. Ozin, Adv. Funct. Mater. 12, 71 (2002).
123. A. Xiang, J. P. Gao, H. K. Chen, J. G. Yu and R. X. Liu, Chin. Chem. Lett. 15, 228 (2004).
124. G. Gundiah and C. N. R. Rao, Solid State Sci. 2, 877 (2000).
125. S. Huang, A.-D. Li, Y. Gao, Y.-F. Tang and N.-B. Ming (2005). Preparation and Characterization of Ferroelectric Inverse Opal Via a Sol-gel Process, Proc. of SPIE p. 366.
126. J. Lu, Z. Tang, Z. Zhang and W. Shen, Mater. Res. Bull. 40, 2039 (2005).
127. N. S. Ergang, J. C. Lytle, H. Yan and A. Stein, J. Electrochem. Soc. 152, A1989 (2005).
128. H. Yan, S. Sokolov, J. C. Lytle, A. Stein, F. Zhang and W. Smyrl, J. Electrochem. Soc. 150, A1101 (2003).
129. K. S. Seo, H.-K. Jung and Y. Son, Han'guk Chaelyo Hakhoechi 14, 587 (2004).
130. D. Wang and F. Caruso, Adv. Mater. 15, 205 (2003).
131. B. Li, J. Zhou, R. Zong and L. Li, Key. Eng. Mater. 280-283, 541 (2005).
132. E. M. Sorensen, S. J. Barry, H.-K. Jung, J. R. Rondinelli, J. T. Vaughey and K. R. Poeppelmeier, Chem. Mater. 18, 482 (2006).
133. S. M. Hant, G. S. Attard, R. Riddle and K. M. Ryan, Chem. Mater. 17, 1434 (2005).
134. S. Madhavi, C. Ferraris and T. White, J. Solid State Chem. 179, 866 (2006).
135. E. Geraud, V. Prevot, J. Ghanbaja and F. Leroux, Chem. Mater. 18, 238 (2006).
136. B. Lebeau, J. Brendle, C. Marichal, A. J. Patil, E. Muthusamy and S. Mann, J. Nanosci. Nanotech. 6, 352 (2006).
137. Y. N. Kim, E. O. Chi, J. C. Kim, E. K. Lee and N. H. Hur, Solid State Commun. 128, 339 (2003).
138. S. Huang, A.-D. Li, Y. Gao, D. Wu and Y. Zhang, Int. J. Mod. Phys. B 19, 2769 (2005).

139. B. Li, J. Zhou, L. Hao, W. Hu, R. Zong, M. Cai, M. Fu, Z. Gui, L. Li and Q. Li, Appl. Phys. Lett. 82, 3617 (2003).
140. B. Li, J. Zhou, L. Li, X. Wang, X. H. Liu and Z. Jian, Appl. Phys. Lett. 83, 4704 (2003).
141. B. Li, J. Zhou, Q. Li, L. Li and Z. Gui, J. Am. Ceram. Soc. 86, 867 (2003).
142. K. Matsumoto, O. Kutowy and C. E. Capes, Powder Technol. 31, 197 (1982).
143. K. Dokko, N. Akutagawa, Y. Isshiki, K. Hoshina and K. Kanamura, Solid State Ionics 176, 2345 (2005).
144. M. Sadakane, T. Asanuma, J. Kubo and W. Ueda, Chem. Mater. 17, 3546 (2005).
145. A. A. Zakhidov, R. H. Baughman, Z. Iqbal, C. Cui, I. Khayrullin, S. O. Dantas, J. Marti and V. G. Ralchenko, Science 282, 897 (1998).
146. F. Su, X. S. Zhao, Y. Wang, J. Zeng, Z. Zhou and J. Y. Lee, J. Phys. Chem. B 109, 20200 (2005).
147. Z. Li, M. Jaroniec, Y. J. Lee and L. R. Radovic, Chem. Commun. 1346 (2002).
148. S. B. Yoon, G. S. Chai, S. K. Kang, J. S. Yu, K. P. Gierszal and M. Jaroniec, J. Am. Chem. Soc. 127, 4188 (2005).
149. M. W. Perpall, K. P. U. Perera, J. DiMaio, J. Ballato, S. H. Foulger and D. W. Smith, Jr., Langmuir 19, 7153 (2003).
150. J.-S. Yu, S. J. Lee and S. B. Yoon, Mol. Cryst. Liq. Cryst. Sci. Tech. A 371, 107 (2001).
151. J.-S. Yu, S. Kang, S. B. Yoon and G. Chai, J. Am. Chem. Soc. 124, 9382 (2002).
152. H. Take, T. Matsumota, S. Hiwatashi, T. Nakayama, K. Niihara and K. Yoshino, Jap. J. Appl. Phys. 43, 4453 (2004).
153. K. Kanamura, M. Fujisaki, Y.-H. Rho and T. Umegaki, Key. Eng. Mater. 216, 123 (2002).
154. T. F. Baumann and J. H. Satcher, Jr., J. Non-Cryst. Sol. 350, 120 (2004).
155. G. Chai, S. B. Yoon, S. Kang, J. H. Choi, Y. E. Sung, Y. S. Ahn, H. S. Kim and J. S. Yu, Electrochim. Acta 50, 823 (2004).
156. G. S. Chai, S. B. Yoon, J.-S. Yu, J.-H. Choi and Y.-E. Sung, J. Phys. Chem. B 108, 7074, (2004).
157. J.-S. Yu, Rev. Adv. Mater. Sci. 10, 347 (2005).
158. K. T. Lee, J. C. Lytle, N. S. Ergang, S. M. Oh and A. Stein, Adv. Funct. Mater. 15, 547 (2005).
159. A. Rugge, J. S. Becker, R. G. Gordon and S. H. Tolbert, Nano Lett. 3, 1293 (2003).
160. A. Rugge, J.-S. Park, R. G. Gordon and S. H. Tolbert, J. Phys. Chem. B 109, 3764 (2005).
161. L. Y. Hao, M. You, X. Mo, W. Q. Jiang, Y. Zhou, Y. R. Zhu, Y. Hu, X. M. Liu and Z. Y. Chen, Chin. Chem. Lett. 13, 181 (2002).
162. J. S. King, C. W. Neff, C. J. Summers, W. Park, S. Blomquist, E. Forsythe and D. Morton, Appl. Phys. Lett. 83, 2566 (2003).

163. J. M. Cao, X. Chang, M. B. Zheng, H. B. Huang, Y. L. Cao and X. F. Ke, Stud. Surf. Sci. Catal. 156, 273 (2005).
164. S. G. Romanov, T. Maka, C. M. Sotomayor Torres, M. Muller and R. Zentel, Synthetic Met. 116, 475 (2001).
165. S. G. Romanov, T. Maka, C. M. Sotomayor Torres, M. Muller and R. Zentel, Appl. Phys. Lett. 79, 731 (2001).
166. A. Blanco, H. Miguez, F. Meseguer, C. Lopez, F. Lopez-Tejeira and J. Sanchez-Dehesa, Appl. Phys. Lett. 78, 3181 (2001).
167. H. Lingyun, M. You, X. Mo, W. Jian, Y. Zhu, Y. Zhou, Y. Hu, X. Liu and Z. Chen, Mater. Res. Bull. 38, 723 (2003).
168. B. H. Juarez, S. Rubio, J. Sanchez-Dehesa and C. Lopez, Adv. Mater. 14, 1486 (2002).
169. G. Gajiev, V. G. Golubev, D. A. Kurdyukov, A. B. Pevtsov, A. V. Selkin and V. V. Travnikov, Phys. Status Solidi B 231, R7 (2002).
170. H. Wang, X.-D. Li, J.-S. Yu and D.-P. Kim, J. Mater. Chem. 14, 1383 (2004).
171. I.-K. Sung, S.-B. Yoon, J.-S. Yu and D.-P. Kim, Chem. Commun. 1480 (2002).
172. P. V. Braun, R. W. Zehner, C. A. White, M. K. Weldon, C. Kloc and S. S. Patel, Adv. Mater. 13, 721 (2001).
173. S. Vaucher, E. Dujardin, B. Lebeau, S. R. Hall and S. Mann, Chem. Mater. 13, 4408 (2001).
174. C. Lu, L. Qi, H. Cong, X. Wang, J. Yang, D. Zhang, J. Ma and W. Cao, Chem. Mater. 17, 5218 (2005).
175. K. M. Kulinowski, P. Jiang, H. Vaswani and V. L. Colvin, Adv. Mater. 13, (2000).
176. H. Yan, C. F. Blanford, B. T. Holland, M. Parent, W. H. Smyrl and A. Stein, Adv. Mater. 11, 1003 (1999).
177. H. Yan, C. F. Blanford, J. C. Lytle, B. Carter, W. H. Smyrl and A. Stein, Chem. Mater. 13, 4314 (2001).
178. H. Yan, C. F. Blanford, W. H. Smyrl and A. Stein, Chem. Commun. 1477 (2000).
179. T. S. Eagleton and P. C. Searson, Chem. Mater. 16, 5027 (2004).
180. P. N. Bartlett, M. A. Ghanem, I. S. El Hallag, P. de Groot and A. Zhukov, J. Mater. Chem. 13, 2596 (2004).
181. B. H. Juarez, C. Lopez and C. Alonso, J. Phys. Chem. B 108, 16708 (2004).
182. J. E. G. Wijnhoven, S. J. M. Zevenhuizen, M. A. Hendriks, D. Vanmaekelbergh, J. J. Kelly and W. L. Vos, Adv. Mater. 12, 888 (2000).
183. P. Jiang, J. Cizeron, J. F. Bertone and V. L. Colvin, J. Am. Chem. Soc. 121, 7957 (1999).
184. H. Cong and W. Cao, J. Colloid Interf. Sci. 278, 423 (2004).
185. H. Cong and H. Cao, Adv. Funct. Mater. 15, (2005).
186. G. Von Freymann, S. John, M. Schulz-Dobrick, E. Vekris, N. Tetreault, S. Wong, V. Kitaev and G. A. Ozin, Appl. Phys. Lett. 84, 224 (2004).

187. N. R. Denny, S. Han, R. T. Turgeon, J. C. Lytle, D. J. Norris and A. Stein (2005). Synthetic Approaches Toward Tungsten Photonic Crystals For Thermal Emission, Proc. of SPIE p. 600505/1.
188. P. B. Landon, B. G. Kim, R. H. Baughman, R. Glosser, A. A. Zakhidov, V. Kamaev and V. Vardeny, NATO Science Series, II: Mathematics, Physics and Chemistry 100, 483 (2003).
189. B. C. Jarvis, C. L. Gilleland, T. Renfro, J. Guitierrez, K. Parikh, R. Glosser and P. B. Landon (2005). Inverse Opal Photonic Crystals With Photonic Band Gaps in the Visible and Near-Infrared, Proc. of SPIE p. 592414/1.
190. G. L. Egan, J.-S. Yu, C. H. Kim, S. J. Lee, R. E. Schaak and T. E. Mallouk, Adv. Mater. 12, 1040 (2000).
191. O. D. Velev, P. M. Tessier, A. M. Lenhoff and E. W. Kaler, Nature 401, 548 (1999).
192. P. M. Tessier, O. D. Velev, A. T. Kalambur, J. F. Rabolt, A. M. Lenhoff and E. W. Kaler, J. Am. Chem. Soc. 122, 9554 (2000).
193. P. M. Tessier, O. D. Velev, A. T. Kalambur, A. M. Lenhoff, J. F. Rabolt and E. W. Kaler, Adv. Mater. 13, 396 (2001).
194. P. M. Tessier, K. K. O. Christesen, E. M. Clemente, A. M. Lenhoff, E. W. Kaler and O. D. Velev, Appl. Spectrosc. 56, 1524 (2002).
195. D. M. Kuncicky, S. D. Christesen and O. D. Velev, Appl. Spectrosc. 59, 401 (2005).
196. D. M. Kuncicky, B. B. Prevo and O. D. Velev, J. Mater. Chem. 16, 1207 (2006).
197. N. Tetreault, H. Miguez and G. A. Ozin, Adv. Mater. 16, 1471 (2004).
198. A. Blanco, E. Chomski, S. Grabtchak, M. Ibisate, S. John, S. W. Leonard, C. Lopez, F. Meseguer, H. Miguez, J. P. Mondia, G. A. Ozin, O. Toader and H. M. van Driel, Nature 405, 437 (2000).
199. Y. A. Vlasov, X.-Z. Bo, J. C. Sturm and D. J. Norris, Nature 414, 289 (2001).
200. V. G. Golubev, V. A. Kosobukin, D. A. Kurdyukov, A. V. Medvedev and A. B. Pevtsov, Semiconductors 35, 680 (2001).
201. V. G. Golubev, J. L. Hutchison, V. A. Kosobukin, D. A. Kurdyukov, A. V. Medvedev, A. B. Pevtsov, J. Sloan and L. M. Sorokin, J. Non-Cryst. Sol. 299-302, 1062 (2002).
202. K. K. Mendu, J. Shi, Y. F. Lu, L. P. Li, N. Batta, D. W. Doerr and D. R. Alexander, Nanotechnology 16, (2005).
203. H. Wang, Y. F. Lu and Z. Y. Yang (2006). Fabrication, Characterizatio and Simulation of 3-D Inverse-Opal Photonic Crystals Using Laser-Assisted Imprinting, Proc. of SPIE p. 61070E/1.
204. H. M. Yates, D. E. Whitehead, M. G. Nolan, M. E. Pemble, E. Palacios-Lidon, S. Rubio, F. J. Meseguer and C. Lopez, Proc.-Electrochem. Soc. 2003-8, (2003).
205. L. Zhang, L. Xu, Y. Long and K. Xie, Guisuanyan Xuebao 34, 358 (2006).

206. H. Miguez, E. Chomski, F. Garcia-Santamaria, M. Ibisate, S. John, C. Lopez, F. Meseguer, J. P. Mondia, G. A. Ozin, O. Toader and H. M. Van Driel, Adv. Mater. 13, 1634 (2001).
207. H. Miguez, F. Meseguer, C. Lopez, F. Lopez-Tejeira and J. Sanchez-Dehesa, Adv. Mater. 13, 393 (2001).
208. H. Bu, J. Rong and Z. Yang, Macromol. Rapid Comm. 23, 460 (2005).
209. D. Batra, S. Vogt, P. D. Laible and M. A. Firestone, Langmuir 21, 10301 (2005).
210. Y. Xu, G. J. Schneider, E. D. Wetzel and D. W. Prather, J. Microlith. Microfab. Microsys. 3, 168 (2004).
211. T. B. Xu, Z. Y. Cheng, Q. M. Zhang, R. H. Baughman, C. Cui, A. A. Zakhidov and J. Su, J. Appl. Phys. 88, 405 (2000).
212. W. Yan, H. Li and X. Shen, Eur. Polym. J. 41, 992 (2005).
213. W. Yan, H. Li and X. Shen, Macromol. Rapid Comm. 26, 564 (2005).
214. X. Zhang, W. Yan, H. Li and X. Shen, Polymer 46, 11958 (2005).
215. J. Rong and Z. Yang, Macromolecular Materials and Engineering 287, 11 (2002).
216. H. T. Lord, J. F. Quinn, S. D. Angus, M. R. Whittaker, M. H. Stenzel and T. P. Davis, J. Mater. Chem. 13, 2819 (2003).
217. D. K. Yi and D.-Y. Kim, Nano Lett. 3, 207 (2003).
218. S. Tian, J. Wang, U. Jonas and W. Knoll, Chem. Mater. 17, (2005).
219. D. Wang and F. Caruso, Adv. Mater. 13, 350 (2001).
220. H. Kayashima, K. Sumioka, S. H. Lee, B.-B. Jang, K. Fujita and T. Tsutsui, Jap. J. Appl. Phys. 42, 5731 (2003).
221. D. Allard, S. Allard, M. Brehmer, L. Conrad, R. Zentel, C. Stromberg and J. W. Schultze, Electrochim. Acta 48, 3137 (2003).
222. R. Ootake, N. Takamoto, A. Fujii, M. Ozaki and K. Yoshino, Synthetic Met. 137, 1417 (2003).
223. T. Cassagneau and F. Caruso, Adv. Mater. 14, 34 (2002).
224. T. Cassagneau and F. Caruso, Adv. Mater. 14, 1837 (2002).
225. T. Cassagneau and F. Caruso, Adv. Mater. 14, 1629 (2002).
226. J. Galloro, M. Ginzburg, H. Miguez, S. M. Yang, N. Coombs, A. Safa-Sefat, J. E. Greedan, I. Manners and G. A. Ozin, Adv. Funct. Mater. 12, 382 (2002).
227. A. Arsenault, J. Galloro, M. Ginsburg, H. Miguez, S. M. Yang, G. A. Ozin and I. Manners, Polym. Prepr. (Am. Chem. Soc., Div. Polym. Chem.) 43, 33 (2002).
228. H. Wang, X.-D. Li and D.-P. Kim, Appl. Organomet. Chem. 19, 742 (2005).
229. H. Yan, K. Zhang, C. F. Blanford, L. F. Francis and A. Stein, Chem. Mater. 13, 1374 (2001).
230. B. J. Melde and A. Stein, Chem. Mater. 14, 3326 (2002).
231. J. H. Holtz and S. A. Asher, Nature 389, 829 (1997).
232. M. Kamenjicki, I. Lednev and S. A. Asher, Adv. Funct. Mater. 15, 1401 (2005).
233. J. P. Walker and S. A. Asher, Anal. Chem. 77, 1596 (2005).
234. M. Ben-Moshe, V. Alexeev and S. Asher, Anal. Chem. 78, 5149 (2006).
235. R. A. Barry and P. Wiltzius, Langmuir 22, 1369 (2006).

236. Y. Zhang, S. Wang, M. Eghtedari, M. Motamedi and N. A. Kotov, Adv. Funct. Mater. 15, 725 (2005).
237. Y.-J. Lee, S. A. Pruzinsky and P. V. Braun, Langmuir 20, 3096 (2004).
238. A. N. Stachowiak, A. Bershteyn, E. Tzatzalos and D. J. Irvine, Adv. Mater. 17, 399 (2005).
239. Y.-J. Lee and P. V. Braun, Adv. Mater. 15, 563 (2003).
240. X.-Z. Zhang and R.-X. Zhuo, Eur. Polym. J. 36, 643 (2000).
241. Y. Takeoka and M. Watanabe, Langmuir 18, 5977 (2002).
242. Y. Takeoka and M. Watanabe, Langmuir 19, 9104 (2003).
243. Z.-Z. Yang, J.-H. Rong and D. Li, Chin. J. Polym. Sci. 21, 175 (2003).
244. R. C. Schroden, M. Al-Daous, C. F. Blanford and A. Stein, Chem. Mater. 14, 3305 (2002).
245. R. C. Schroden, C. F. Blanford, B. J. Melde, B. J. S. Johnson and A. Stein, Chem. Mater. 13, 1074 (2001).
246. Y. Guo, Y. Yang, C. Hu and C. Guo, J. Mater. Chem. 12, 3046 (2002).
247. Y.-H. Guo and C.-W. Hu, J. Cluster Sci. 14, 505 (2004).
248. Z. Wang, N. S. Ergang and M. A. Al-Daous, Chem. Mater. 17, 6805 (2005).
249. D. Wang, V. Salgueirino-Maceira, L. M. Liz-Marzan and F. Caruso, Adv. Mater. 14, 908 (2002).
250. Q.-Z. Wu, Q. Yin, J.-F. Liao, J.-H. Deng and Y.-G. Li, Chin. J. Chem. 23, 689 (2005).
251. Z. Zhou, X. Bao and X. S. Zhao, Chem. Commun. 1376 (2004).
252. Q. Wu, A. D. Ross and K. K. Gleason, Plasma Process. Polym. 2, 401 (2005).
253. A. Yu, F. Meiser, T. Cassagneau and F. Caruso, Nano Lett. 4, 177 (2004).
254. B. Lebeau, J. Parmentier, M. Soulard, C. Fowlder, R. Zana, C. Vix-Guterl and J. Patarin, C. R. Chemie 8, 597 (2005).
255. A. H. Lu, J. H. Smatt and M. Linden, Adv. Funct. Mater. 15, 865 (2005).
256. A. H. Lu and F. Schueth, C. R. Chemie 8, 609 (2005).
257. H. Yang and D. Zhao, J. Mater. Chem. 15, 1217 (2005).
258. S. M. Yang, N. Coombs and G. A. Ozin, Adv. Mater. 12, 1940 (2000).
259. H. Miguez, N. Tetreault, S. Yang, V. Kitaev and G. A. Ozin, Adv. Mater. 15, 597 (2003).
260. P. Jiang, J. F. Bertone and V. L. Colvin, Science 291, 453 (2001).
261. L. Xu, L. D. Tung, L. Spinu, A. A. Zakhidov, R. H. Baughman and J. B. Wiley, Adv. Mater. 15, (2003).
262. H. Wang, J.-S. Yu, X.-D. Li and D.-P. Kim, Chem. Commun. 2352 (2004).
263. Z. Wang, F. Li, N. S. Ergang and A. Stein, submitted (2006).
264. Z. Y. Yuan and B. L. Su, J. Mater. Chem. 16, 663 (2006).
265. A. Taguchi, J. H. Smatt and M. Linden, Adv. Mater. 15, 1209 (2003).
266. K. S. W. Sing, D. H. Everett, R. A. W. Haul, L. Moscou, R. A. Pierotti, J. Rouquerol and T. Siemieniewska, Pure Appl. Chem. 57, 603 (1985).
267. U. A. El-Nafaty and R. Mann, Chem. Eng. Sci. 54, 3475 (1999).

268. B. T. Holland, L. Abrams and A. Stein, J. Am. Chem. Soc. 121, 4308 (1999).
269. Q. Luo, L. Li, B. Yang and D. Zhao, Chem. Lett. 378 (2000).
270. Y. S. Yin and Z. L. Wang, Microsc. Microanal. 5 (Suppl. 2: Proceedings), 818 (1999).
271. J. Y. Yin, C. P. Mehnert and M. S. Wong, Angew. Chem. Int. Ed. 38, 56 (1999).
272. T. Sen, G. J. T. Tiddy, J. L. Casci and M. W. Anderson, Angew. Chem. Int. Ed. 42, 4649 (2003).
273. T. Sen, G. J. T. Tiddy, J. L. Casci and M. W. Anderson, Chem. Mater. 16, 2044 (2004).
274. L. A. Villaescusa, A. Mihi, I. Rodriguez, A. E. Garcia-Bennett and H. Miguez, J. Phys. Chem. B 109, 19643 (2005).
275. Y. Zhou and M. Antonietti, Chem. Commun. 2564 (2003).
276. O. Sel, D. Kuang, M. Thommes and B. Smarsly, Langmuir 22, 2311 (2006).
277. P. Yang, T. Deng, D. Zhao, P. Feng, D. Pine, B. F. Chmelka, G. M. Whitesides and G. D. Stucky, Science 282, 2244 (1998).
278. L. Huang, Z. Wang, J. Sun, L. Miao, Q. Li, Y. Yan and D. Zhao, J. Am. Chem. Soc. 122, 3530 (2000).
279. Y. J. Wang, Y. Tang, Z. Ni, W. M. Hua, W. L. Yang, X. D. Wang, W. C. Tao and Z. Gao, Chem. Lett. 510 (2000).
280. S. Vaudreuil, M. Bousmina, S. Kaliaguine and L. Bonneviot, Micropor. Mesopor. Mat. 44-45, 249 (2001).
281. J. C. Vartuli, K. D. Schmitt, C. T. Kresge, W. J. Roth, M. E. Leonowicz, S. B. McCullen, S. D. Hellring, J. S. Beck, J. L. Schlenker, D. H. Olson and E. W. Sheppard, Chem. Mater. 6, 2317 (1994).
282. C. Danumah, S. Vaudreuil, L. Bonneviot, M. Bousmina, S. Giasson and S. Kaliaguine, Micropor. Mesopor. Mat. 44-45, 241 (2001).
283. C. G. Oh, Y. Baek and S. K. Ihm, Adv. Mater. 17, 270 (2005).
284. F. Caruso, R. A. Caruso and H. Mohwald, Chem. Mater. 11, 3309 (1999).
285. F. Caruso, R. A. Caruso and H. Mohwald, Science 282, 1111 (1998).
286. A. Dong, Y. Wang, Y. Tang, Y. Zhang, N. Ren and Z. Gao, Adv. Mater. 14, 1506 (2002).
287. H. Asoh and S. Ono, Appl. Phys. Lett. 87, 103102/1 (2005).
288. U. C. Fischer and H. P. Zingsheim, J. Vac. Sci. Technol. 19, 881 (1981).
289. H. W. Deckman and J. H. Dunsmuir, Appl. Phys. Lett. 41, 377 (1982).
290. J. C. Hulteen and R. P. Van Duyne, J. vac. Sci. Technol. A 13, 1553 (1995).
291. C. L. Haynes and R. P. V. Duyne, J. Phys. Chem. B 105, 5599 (2001).
292. M. H. Wu, C. Park and G. M. Whitesides, J. Colloid Interf. Sci. 265, 304 (2003).
293. P. Jiang, Angew. Chem. Int. Ed. 43, 5625 (2004).
294. X. Cao, Y. Xie and L. Li, J. Solid State Chem. 177, 202 (2004).
295. N. S. Ergang, J. C. Lytle, K. T. Lee, S. M. Oh, W. H. Smyrl and A. Stein, Adv. Mater. 18, 1750 (2006).
296. Y. Toivola, A. Stein and R. F. Cook, J. Mater. Res. 19, 260 (2004).

297. E. Yablonovitch, Phys. Rev. Lett. 58, 2059 (1987).
298. S. John, Phys. Rev. Lett. 58, 2486 (1987).
299. E. Yablonovitch, Sci. Am. 285, (2001).
300. O. Toader and S. John, Phys. Rev. E 70, 046605/1 (2004).
301. T. Prasad, V. L. Colvin and D. M. Mittleman, Phys. Rev. B 67, 165103/1 (2003).
302. K. Busch and S. John, Phys. Rev. E 58, 3896 (1998).
303. D. P. Aryal, J. C. Hermann and O. Hess (2005). Photonic Material Properties of Core-Shell Inverted Opals, Proc. of SPIE p. 59500Y/1.
304. D. Gaillot, T. Yamashita and C. J. Summers, Phys. Rev. B 72, 205109/1 (2005).
305. W. Dong, H. Bongard, B. Tesche and F. Marlow, Adv. Mater. 14, 1457 (2002).
306. W. Dong and F. Marlow, Physica E 17, 431 (2003).
307. P. Ni, B. Cheng and D. Zhang, Appl. Phys. Lett. 80, 1879 (2002).
308. E. Palacios-Lidon, A. Blanco, M. Ibisate, F. Meseguer, C. Lopez and J. Sanchez-Dehesa, Appl. Phys. Lett. 81, 4925 (2002).
309. C. Becker, S. Linden, G. von Freymann, M. Wegener, N. Tetreault, E. Vekris, V. Kitaev and G. A. Ozin, Appl. Phys. Lett. 87, 091111/1 (2005).
310. J. Li, P. R. Herman, C. E. Valdivia, V. Kitaev and G. A. Ozin, Optics Express 13, 6454 (2005).
311. W. L. Vos and A. Polman, MRS Bulletin 26, 642 (2001).
312. L. Bechger, P. Lodahl and Vos, J. Phys. Chem. B 109, 9980 (2005).
313. C. Hermann and O. Hess, J. Opt. Soc. Am. B 19, 3013 (2002).
314. A. F. Koenderink, L. Bechger, A. Lagendijk and W. L. Vos, Physica Status Solidi A 197, (2003).
315. R. C. Schroden, M. Al-Daous and A. Stein, Chem. Mater. 13, 2945 (2001).
316. M. Scharrer, X. Wu, A. Yamilov, H. Cao and R. P. H. Chang, Appl. Phys. Lett. 86, 51113 (2005).
317. Y. Yang, H. Yan, Z. Fu, B. Yang, L. Xia, Y. Xu, J. Zuo and F. Li, J. Phys. Chem. B 110, 846 (2006).
318. H. P. Schriemer, H. M. van Driel, A. F. Koenderink and W. L. Vos, Phys. Rev. A 63, 011801/1 (2001).
319. V. G. Solovyev, S. G. Romanov, C. M. Sotomayor Torres, M. Muller, R. Zentel, N. Gaponik, A. Eychmuller and A. L. Rogach, J. Appl. Phys. 94, 1205 (2003).
320. H. Takeda and K. Yoshino, J. Appl. Phys. 92, 5659 (2002).
321. M. Ozaki, Y. Shimoda, M. Kasano and K. Yoshino, Adv. Mater. 14, 514 (2002).
322. P. Mach, P. Wiltzius, M. Megens, D. A. Weitz, K.-H. Lin, T. C. Lubensky and A. G. Yodh, Europhys. Lett. 58, 679 (2002).
323. P. Mach, P. Wiltzius, M. Megens, D. A. Weitz, K.-H. Lin, T. C. Lubensky and A. G. Yodh, Phys. Rev. E 65, 031720/1 (2002).
324. S. Kubo, Z.-Z. Gu, K. Takahashi, Y. Ohko, O. Sato and A. Fujishima, J. Am. Ceram. Soc. 124, 10950 (2002).
325. S. Kubo, Z.-Z. Gu, K. Takahashi, A. Fujishima, H. Segawa and O. Sato, J. Am. Chem. Soc. 126, 8314 (2004).

326. S. Kubo, Z.-Z. Gu, K. Takahashi, A. Fujishima, H. Segawa and O. Sato, Chem. Mater. 17, 2298 (2005).
327. B. Gates, Y. Yin and Y. Xia, Adv. Mater. 13, 1605 (2001).
328. J. G. Fleming, S. Y. Lin, I. El-Kady, R. Biswas and K. M. Ho, Nature 417, 52 (2002).
329. M. Florescu, H. Lee, A. J. Stimpson and J. Dowling, Phys. Rev. A. 72, 033821/1 (2005).
330. D. Wang, J. Li, C. T. Chan, V. Sagueirino-Maceira, L. M. Liz-Marzan, S. Romanov and F. Caruso, Small 1, 122 (2005).
331. Z. Y. Li and Z. Q. Zhang, Phys. Rev. B 62, 1516 (2000).
332. R. Biswas, M. M. Sigalas, G. Subramania, C. M. Soukoulis and K. M. Ho, Phys. Rev. B 61, 4549 (2000).
333. Y. A. Vlasov, M. Deutsch and D. J. Norris, Appl. Phys. Lett. 76, 1627 (2000).
334. A. D'Andrea, L. Pilozzi, D. Schiumarini and N. Tomassini, Proc. SPIE 5036, 424 (2003).
335. J. Hamagami, K. Hasegawa and K. Kanamura, Key. Eng. Mater. 248, 195 (2003).
336. C. A. Murray and D. G. Grier, Am. Sci. 83, 238 (1995).
337. S. K. Im and O. O. Park, Appl. Phys. Lett. 80, 4133 (2002).
338. A. van Blaaderen, R. Ruel and P. Wiltzius, Nature 385, 321 (1997).
339. Y. Xia, Y. Yin, Y. Lu and J. McLellan, Adv. Funct. Mater. 13, 907 (2003).
340. H. W. Deckman, J. H. Dunsmuir, S. Garoff, J. A. McHenry and D. G. Peiffer, J. Vac. Sci. Technol. B6, 333 (1988).
341. D. J. Norris, E. G. Arlinghaus, L. Meng, R. Heiny and L. E. Scriven, Adv. Mater. 16, 1393 (2004).
342. P. Jiang, J. F. Bertone, K. S. Hwang and V. L. Colvin, Chem. Mater. 11, 2132 (1999).
343. V. Yannopapas, N. Stefanou and A. Modinos, Phys. Rev. Lett. 86, 4811 (2001).
344. Z. L. Wang, C. T. Chang, W. Y. Zhang, Z. Chen, N. B. Ming and P. Sheng, Phys. Rev. E 67, 016612/1 (2003).
345. W. Lee, S. A. Prazinsky and P. V. Braun, Adv. Mater. 271-274, (2002).
346. L. Wang, Q. Yan and X. S. Zhao, Langmuir 22, 3481 (2006).
347. D. L. C. Chan, E. Lidorikis and J. D. Joannopoulos, Phys. Rev. E 71, 056602/1 (2005).
348. B. H. Juarez, D. Golmayo, P. A. Postigo and C. Lopez, Adv. Mater. 16, 1732 (2004).
349. Y.-H. Ye, T. S. Mayer, I.-C. Khoo, I. B. Divliansky, N. Abrams and T. E. Mallouk, J. Mater. Chem. 12, 3637 (2002).
350. Q. Yan, Z. Zhou and X. S. Zhao, Chem. Mater. 17, 3069 (2005).
351. E. Palacios-Lidon, J. F. Galisteo-Lopez, B. H. Juarez and C. Lopez, Adv. Mater. 16, 341 (2004).
352. Z.-Y. Li and Z.-Q. Zhang, Adv. Mater. 13, 433 (2001).

353. V. Babin, P. Garstecki and R. Holyst, Appl. Phys. Lett. 82, 1553 (2003).
354. A. C. Arsenault, T. J. Clark, G. von Freymann, L. Cademartiri, R. Sapienza, J. Bertolotti, E. Vekris, S. Wong, V. Kitaev, I. Manners, R. Z. Wang, S. John, D. Wiersma and G. A. Ozin, Nature Mater. 5, 179 (2006).
355. Z.-Z. Gu, H. Uetsuka, K. Takahashi, R. Nakajima, H. Onishi, A. Fujishima and O. Sato, Angew. Chem. Int. Ed. 42, 894 (2003).
356. C. F. Blanford, R. C. Schroden, M. Al-Daous and A. Stein, Adv. Mater. 13, 26 (2001).
357. H. Fudouzi and Y. Xia, Langmuir 19, 9653 (2003).
358. S. Nishimura, N. Abrams, B. A. Lewis, L. I. Halaoui, T. E. Mallouk, K. D. Benkstein, J. van de Lagemaat and A. J. Frank, J. Am. Chem. Soc. 125, 6306 (2003).
359. L. I. Halaoui, N. M. Abrams and T. E. Mallouk, J. Phys. Chem. B 109, 6334 (2005).
360. P. R. Somani, C. Dionigi, M. Murgia, D. Palles, P. Nozar and G. Ruani, Sol. Energy Mater. Sol. Cells 87, 513 (2005).
361. C. L. Huisman, J. Schoonman and A. Goosens, Solar Energy Materials & Solar Cells 85, (2005).
362. W. Qian, Z. Gu, A. Fujishima and O. Sato, Langmuir 18, 4526 (2002).
363. H. H. Chen, H. Suzuki, O. Sato and Z. Z. Gu, Appl. Phys. A 81, (2005).
364. S. A. Asher, J. Holtz, J. Weissman and G. S. Pan, MRS Bull. 23, 44 (1998).
365. K. Sumioka, H. Kayashima and T. Tsutsui, Adv. Mater. 14, 1284 (2002).
366. M. Acciarri, R. Barberini, C. Canevali, M. Mattoni, C. M. Mari, F. Morazzoni, L. Nodari, S. Polizzi, R. Ruffo, U. Russo, M. Sala and R. Scotti, Chem. Mater. 17, 6167 (2005).
367. M. Egashira, Y. Shimizu and T. Hyodo, Mater. Res. Symp. Proc. 828, 3 (2005).
368. Y.-Y. Song, D. Zhang, W. Gao and X.-H. Xia, Chem. Eur. J. 11, 2177 (2005).
369. I. N. Krivorotov, H. Yan, D. E. Dahlberg and A. Stein, J. Magnetism Magnetic Mater. 226-230, (2001).
370. P. N. Bartlett, M. A. Ghanem, I. S. El Hallag, P. De Groot and A. Zhukov, J. Mater. Chem. 13, 2596 (2003).
371. A. A. Zhukov, A. V. Goncharov, P. A. J. de Groot, P. N. Bartlett, M. A. Ghanem, H. Kuepfer, R. J. Pugh and G. J. Tomka, IEE Proc.-Sci. Meas. Technol. 150, 257 (2003).
372. A. A. Zhukov, A. V. Goncharov, P. A. J. de Groot, P. N. Bartlett and M. A. Ghanem, J. Appl. Phys. 93, (2003).
373. Y. N. Kim, S. J. Kim, E. K. Lee, E. O. Chi, N. H. Hur and C. S. Hong, J. Mater. Chem. 14, 1774 (2004).
374. B. J. S. Johnson and A. Stein, Inorg. Chem. 40, 801 (2001).
375. J. W. Long, B. Dunn, D. R. Rolison and H. S. White, Chem. Rev. 104, 4463 (2004).
376. J. Sakamoto and B. Dunn, J. Mater. Chem. 12, 2859 (2002).

377. L. Kavan, M. Zukalova, M. Kalbac and M. Graetzel, J. Electrochem. Soc. 151, A1301 (2004).
378. H. Yamada, T. Yamato, I. Moriguchi and T. Kudo, Solid State Ionics 175, 195 (2004).
379. H. Yamada, T. Yamato, I. Moriguchi and T. Kudo, Chem. Lett. 33, 1548 (2004).
380. Z. Bing, Y. Yuan, Y. Wang and Z.-W. Fu, Electrochem. Sol. St. Lett. 9, A101 (2006).
381. N. Li, C. R. Martin and B. Scrosati, J. Electrochem. Soc. 148, A164 (2001).
382. C. Wang, C. Yang, Y. Song, W. Gao and X. Xia, Adv. Funct. Mater. 15, 1267 (2005).
383. I. Moriguchi, F. Nakawara, H. Furukawa, H. Yamada and T. Kudo, Electrochem. Solid State Lett. 7, A221 (2004).
384. I. Moriguchi, F. Nakawara, H. Yamada and T. Kudo, Stud. Surf. Sci. Catal. 156, 589 (2005).
385. F. Rouquerol, J. Rouquerol and K. Sing, Adsorption By Powders and Porous Solids: Principles, Methodology, Applications, Academic Press,New York (1999).
386. S. Kang, J.-S. Yu, M. Kruk and M. Jaroniec, Chem. Commun. 1670 (2002).
387. H. Ge, G. Wang, Y. He, X. Wang, Y. Song, L. Jiang and D. Zhu, Chem. Phys. Chem. 7, 575 (2006).
388. A. Stein, B. J. Melde and R. C. Schroden, Adv. Mater. 12, 1403 (2000).
389. D. Yu, Z. Wang, N. S. Ergang and A. Stein, Stud. Surf. Sci. Catal. in press, (2006).
390. S.-L. Chen, K.-Q. Xu and P. Dong, Chem. Mater. 17, 5880 (2005).
391. K. Zhang, H. Yan, D. C. Bell, A. Stein and L. F. Francis, J. Biomed. Mater. Res., Part A 66A, 860 (2003).
392. K. Zhang, N. R. Washburn, J. M. Antonucci and C. G. Simon, Jr., Key. Eng. Mater. 284-285, 655 (2005).
393. K. Zhang, N. R. Washburn and C. G. Simon, Jr., Biomaterials 26, 4532 (2005).
394. K. Zhang, L. F. Francis, H. Yan and A. Stein, J. Am. Ceram. Soc. 88, 587 (2005).
395. L. L. Hench, J. Am. Ceram. Soc. 81, 1705 (1998).
396. C. Frondel, Bull. Am. Mus. Nat. Hist. 67, 389 (1935).
397. R. I. Harker, Science 173, 235 (1971).
398. K. H. Sandhage, M. B. Dickerson, P. M. Huseman, M. A. Caranna, J. D. Clifton, T. A. Bull, T. J. Heibel, W. R. Overton and M. E. A. Schoenwaelder, Adv. Mater. 14, 429 (2002).
399. T. Martin, A. Galarneau, F. Di Renzo, F. Fajula and D. Plee, Angew. Chem. Int. Ed. 41, 2590 (2002).
400. Y. Sun and Y. Xia, J. Am. Chem. Soc. 126, 3892 (2004).
401. J. C. Lytle, H. Yan, R. T. Turgeon and A. Stein, Chem. Mater. 16, 3829 (2004).
402. J. C. Lytle, R. T. Turgeon, N. R. Denny and A. Stein, unpublished Results.

CHAPTER 2

PHOTONIC CRYSTALS: FUNDAMENTALS AND APPLICATIONS

Álvaro Blanco and Cefe López*

Instituto de Ciencia de Materiales de Madrid (CSIC) and Unidad Asociada CSIC-U, Vigo C/ Sor Juana Inés de la Cruz 3, 28049 Madrid Spain
**E-mail: cefe@icmm.csic.es*

Photonics is the technology of photons (as electronics is the technology of electrons) and promises to be the new century's driving force in the advancement of, mainly but not only, communications and computing, the information technology. This technology was initiated with the advent of lasers and optical fibers which, for various reasons, embody the best choice of source and channel of the information carrier: the photon. Many more components are needed not only in the transport section of the technology but also, and principally, in the logic section: signal processing. An answer is needed to many of these demands by the potentiality of the new photonics era: the photonic band gap (PBG) materials, otherwise known as photonic crystals (PC). In the present article a general perspective is presented on the state of the art in photonic crystals technology providing a broad audience-oriented description of fundamentals and properties and centring in self-assembly.

1. Introduction

Most of the notions relating to PCs, as they are nowadays understood, are rooted in basic solid state physics concepts. It is, however, worth pointing that they were originally borrowed by solid state physics from the theory of electromagnetism and are, thus, returning to their source. There is still a long way to go in the materials aspects of the problem which imposes a clear need in the focus of this compilation. Further, we must also admit that, among the approaches to the problem, the ones involving more materials issues and unknowns are the three-dimensional photonic crystals and the production methods other than those adopted from microelectronics. Colloidal systems lend themselves to be used as natural starting points for the purpose of creating and using photonic crystals and, as such, occupy an important place in current research in this field. In their exploitation and characterization a broad range of techniques and different sources of knowledge contribute and the interdisciplinary nature of the subject further enriches it. The review is thus oriented to an audience of physicists, chemists, engineers etc. and tries to bring together concepts from various fields for the benefit of the widest possible audience. For instance, due to their inherent porosity, these materials, and the techniques involved, attract interest from other areas such as catalysis, templating other materials etc. A whole wealth of materials properties is involved that spring from their architectural scale. From the molecular level, and up to the macroscopic structure, micro- and mesoscopic regimes involve new properties, processes, and phenomena like synthesis in new environments, mechanical, optical, electronic, magnetic etc.

In his seminal paper of 1987 Yablonovitch [1] sensed that "further materials development" might be required before the benefits of spontaneous emission inhibition were fully felt. Likewise, John [2] already suggested some materials that based upon their refractive index (RI) could provide a test-bed for the study of strong localization of light. These two very different approaches to a common subject revealed a consciousness of the materials aspects in basic science. A huge effort has since been devoted to developing materials and structures that could support and expand their initial hypotheses. Besides, because RI contrast is a key parameter in this class of structures a materials dilemma rises because higher RI materials provide better PBG performance but at the same time impose higher scattering rates and, hence, losses.

In this paper a review of recent developments in PCs will be given paying special attention to the materials aspects but giving a general overview of the field. The paper is structured as follows. This section serves as introduction of the basics, aims and applications. It will also deal with the main techniques of characterization, namely, optical spectroscopy and its fundamentals and with the interpretation of results in terms of photonic band structure. Second section is devoted to fabrication techniques reviewing different methods and leaving the self assembly approach, comprising synthesis of nano- and micro-particles, assembly thereof and posterior treatments to finish the sub-section regarding fabrication of 3D structures, focusing on it. The latter includes a summary of various composites that can be produced using opals as templates. Inverse structures resulting from the removal of the opal backbone are described with special emphasis on those materials with the highest dielectric constant.

1.1. Photonic Band Gap Materials

Different approaches to the subject can illuminate and give a broader perspective. For that reason several methods of tackling the problem of scattering by periodic media are described.

PBG materials, generally called photonic crystals, are a class of materials or structures in which the dielectric function suffers a spatially periodic variation. Of course the length scale in which the variation takes place (lattice parameter) determines the spectral range of functioning of the PC. As a rule of thumb, the wavelength at which the effects are felt corresponds approximately to the lattice parameter. This can be achieved by structuring a single compound or by constructing a composite made of homogeneous materials with different dielectric properties. A PC working in the optical range of the electromagnetic (EM) spectrum will present a modulation of the dielectric function with a period of the order of one micron; one designed for microwaves should be modulated with period of some centimeters; and a PC for X-rays should present a modulation of some Angstroms which is a solid state crystal. The most prominent feature of such structures is that they present iridescences as a result of diffraction. Some natural examples are the multilayered structure of pearls, the flashing wings of several insects [3] and natural opals [4]. The same effect appears in familiar manmade objects such as compact disks.

When modelling the diffraction of light by a periodic medium, as a layered structure, one computes the phase difference accumulated by rays that suffer scattering from different planes in the stack and adds up those with a multiple of 2π (constructive interference). With a little more algebra this leads to Bragg's law. This simple formulation as can be found in solid state physics text books [5] provides a description of the scattering properties of the periodic system in the form of sets of pairs (wavelength, direction) for which the light does not enter the structure but, rather, is Bragg diffracted. For any given direction several orders of diffraction may be considered inspiring the idea of various bands for a given wavevector. The previous description is fairly well suited to the X-ray scattering for which it was initially developed. Its validity relies on the fact that the RI negligibly departs from unity since the energy of the photons involved is immensely larger than the elementary excitations in the material (no absorption). If this approximation is to be adapted to the optical range a model in which something more is said about RI distribution is required.

The simplest such model might be a 1D periodic variation of dielectric function. Fourier expansions of periodic magnitudes become summations rather than integrals. These summations contain a discrete (though infinite) number of plane wave-like terms, each corresponding to a vector of the reciprocal lattice. The dielectric function can then be expanded in such a way, and so can all other periodic magnitudes involved. If introduced in a scalar wave equation this leads to an expression for the wavevector as a function of energy. The interesting result is that, for certain energy ranges – as marked stripes in Figure 1, no purely real solutions exist and the wavevector has a non vanishing imaginary part which means that the wave suffers attenuation and does not propagate (at least not through an infinite crystal). The wavevector at which this happens is that predicted by Bragg's law and the range of energies where k is no longer real (but complex) is called the stop band. One other interesting fact is that for this wavevector there are two solutions (energies): one above and one below the gap, both having the same periodicity but spatially shifted with respect to each other. The stop band width is mainly determined by the dielectric contrast: the greater the contrast the wider the gap. It also explains that even infinite crystals will present a finite diffraction peak width as opposed to X-ray diffraction where peak width ($\Delta\lambda/\lambda \sim 10^{-6}$) is mainly accounted for by crystal size broadening (Scherer's law). This very simple model can be further improved if we wish to consider the three dimensions [6]. If a single Fourier term in the expansion is assumed to dominate the sum close to a diffraction, simple expressions are obtained that fairly account for the width of diffraction peaks for common structures [7]. Of course the assumption that only one term in the summation is non-negligible limits the applicability to restricted energy ranges. The foregoing analysis can be summarized by saying that the propagation of waves in periodic media is peculiar in that it breaks the, otherwise linear, dispersion relation into bands (of propagating waves) and gaps (where propagation is forbidden).

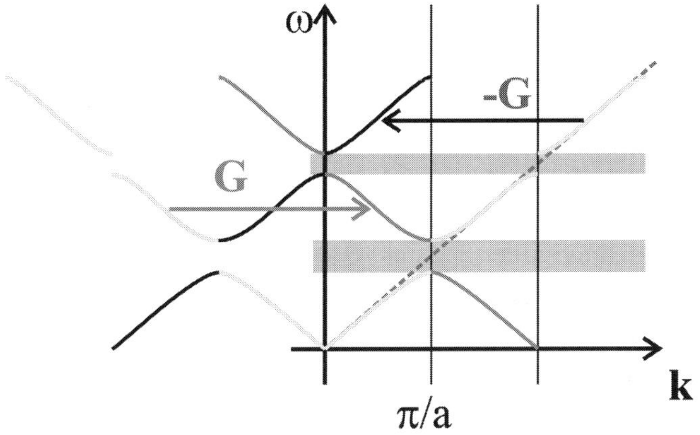

Figure 1. Dispersion relation for a homogeneous material (dashed straight line) and for a periodic material with the same average RI and period a.

Periodicity can be made to occur in one, two or three directions. Most of the properties discussed so far are independent of dimensionality and are well exemplified by sticking to one-dimensional systems. Such systems, extensively used in optical applications, are called Bragg reflectors and consist of stacks of alternating high and low RI transparent materials that base their superiority over metallic mirrors on their reflectance without dissipation.

There are however certain aspects that might require some inspection when more dimensions are considered. Two dimensional systems are more and more used for micro photonic applications and consist of periodic repetitions of objects in a two dimensional arrangement like, for instance, rods periodically arranged parallel to one another. In this case the dielectric function varies periodically in a plane perpendicular to the rods' axes but it is independent of position along the axis. Three dimensional systems present a modulation in the third direction as in a stack of spheres. Since the energy spectrum is direction dependent a gap found in one direction will not be a real gap if there are states in other directions within this energy.

Let us now turn to some aspects of the band structure that will help in interpreting optical spectra later on. For this, a good starting point is introductory solid state physics text books and specialized sources, such as the book by K. Sakoda [8], but we shall summarize the most important results.

A typical 3D dispersion diagram is shown in Figure 2. The X axis represents momentum and the Y axis represents energy measured in very convenient dimensional units: a/λ, where a is the lattice parameter. The different panels composing the diagram represent paths between important (high symmetry) points (labelled X, U, L, etc.) in reciprocal space which happen to be the midpoints of the segments joining the origin to its nearest neighbours in the reciprocal lattice and thus represent propagation directions.

Panels starting at Γ, e.g. ΓX, show the dispersion relation for states with increasing k, all pointing in the same direction (X = (001)) whereas in panels such as XW both direction and modulus of k vary continuously to sweep from X to W.

Among the many bands that are seen in the ΓL panel, some have the ordinary $\omega = c/n\ k$ form (n being an effective RI) or $\omega = c/n\ (k\text{-}2\pi/a)$ resulting from shifting such relation by one reciprocal lattice vector parallel to the vector involved. This is shown in Figure 1 as a translation of the red curve by $2\pi/a$ [9]. There are other curves that result from the folding back into the BZ by means of vectors not parallel to the ΓX direction as depicted in Figure 3. These bands have an expression like $\omega = c/n\ k' = c/n\ [(2\pi/a)^2 + k^2]^{1/2}$ which makes them less dispersive. This picture gets even more complicated for three dimensions, where many more vectors get involved, and is only simplified by symmetry properties making the bands degenerate. Of course this degeneracy may be lifted, for instance, by a slight deformation.

By looking at the bands in any panel in Figure 2 it can be seen that not all energies are covered by the bands. The regions where bands leave unfilled spaces are the gaps. By looking at the whole diagram we can see that none of these gaps is present for every wavevector: there are no complete gaps. As mentioned before, RI contrast enhances the width of the gaps. If we could continuously tune the contrast down to a homogeneous medium (of any RI) the bands would evolve to a situation where no gaps would be left, the bands covering the whole spectrum. This can be expressed by saying that larger Fourier coefficients lead to larger gaps and Fourier coefficients are synonymous to strong modulation. Gap widths are governed by refractive contrast while gap positions are governed by average RI. However, the existence of complete gaps very strongly depends also on topology. Gaps found in different directions occur at different energies. This can be roughly explained thus. Gaps open at the edges of the BZ located at k = π/d where d is the periodicity in that direction. The corresponding energy is ck/n so that, if a structure presents very different periodicities in different directions the gaps will open at very different energies and will hardly overlap. The lattice alone does not determine the band structure: the shape of the "atoms" also plays an important role: as suggested above the symmetry of the cell may lift degeneracies that eventually can open gaps. The best chances to find a structure with a complete gap are for lattices with as round as possible a BZ with a primitive cell with non spherical atoms inside. In 3D this points towards the fcc lattice [10], a particular case of which is the diamond lattice where each lattice site contains two atoms.

When bands depart form the "free photon" behaviour $\omega=ck/n$, the RI must be taken as $n=c/(\mathrm{d}\omega/\mathrm{d}k)$ rather than $n=ck/\omega$ and very peculiar behaviours can be obtained. The group (or energy transmission) velocity, the only meaningful one in these systems is now $v_g=\mathrm{d}\omega/\mathrm{d}k$ rather than $v=\omega/k$. Near the center and edges of the BZ bands flatten out and group velocities go to zero. The net effect is that the time taken by the EM field to traverse the sample is dramatically increased enhancing the interaction between light and matter. Since, for multidimensional systems, the derivative is actually a gradient, the correct expression for the velocity in the nth band is $\mathbf{v}_n(\mathbf{k}) = \nabla_n\omega_n(\mathbf{k})$ which means that propagation is (as the gradient) normal to equi-energy surfaces. Equi-energy surfaces cross Bragg planes at right angles. These and other concepts are best examined in a solid state physics text book [11].

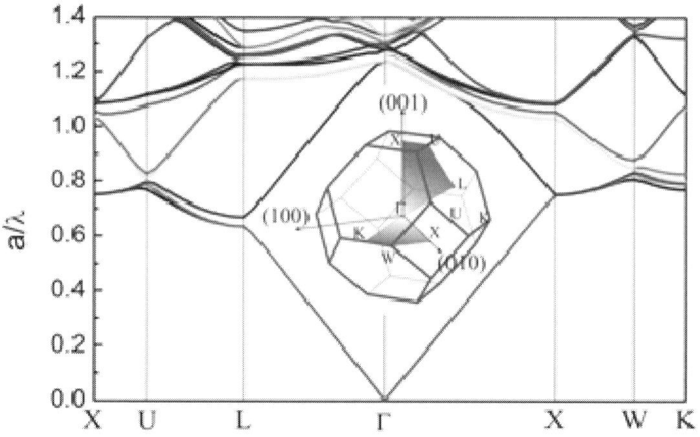

Figure 2. Dispersion relation for a PC made of silica spheres in a compact fcc. For these structures the lattice parameter is given by a=D√2 with D the sphere diameter. Owing to the scalability of the theory, the band diagram is independent of lattice parameter (and, thus, of sphere size) so long as the dielectric function is not dispersive. Therefore any such structure can be modelled by a single band diagram.

In the presence of a PBG many new phenomena can be expected. They are brought about by the two main features of the band structure: the suppression of the density of (photonic) states (DO(P)S) and the translational symmetry.

Control of spontaneous emission in photonic systems through the photonic DOS was the primary goal in the first proposal of PBG [1]. Spontaneous photon emission is genuinely quantum mechanical as vacuum fluctuations are at the heart of the phenomenon [12]. The rate of radiation of a dipole, according to the Fermi golden rule, is proportional to the density of final states. For an excited atom in vacuum that is the density of EM states at the energy of the excited state. So, if this DOS is suppressed, as in a PBG, no radiation is expected from the dipole, vacuum fluctuations are disabled, and the atom must remain in the excited state. The actual picture is that of a *dressed atom* [13] formed as the bound state between the atom and the emitted photon, the energy bouncing back and forth between the atom and the EM field and giving rise to a Rabi splitting. When the leading part of photon-atom interaction (spontaneous emission) is suppressed the next process (resonance dipole-dipole interaction) needs to be considered. If emission occurs near a band edge, where DOS jumps abruptly, and Fermi's golden rule fails to describe the process, the system shows a large differential gain.

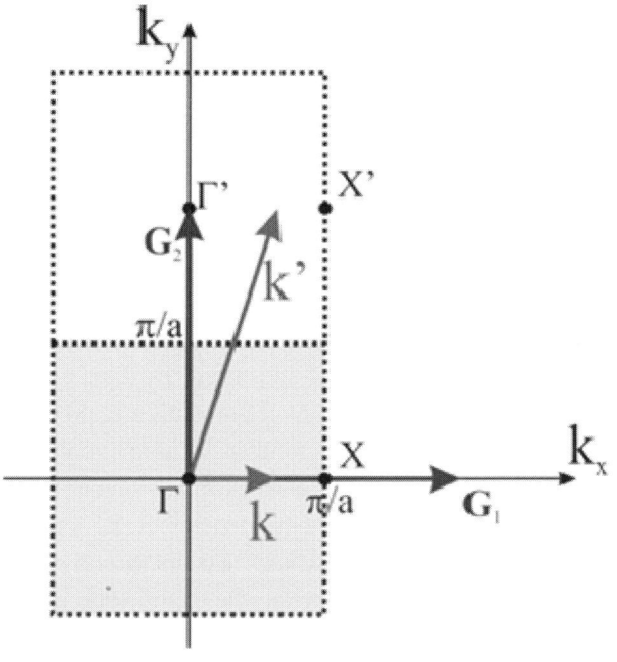

Figure 3. An arbitrary wave vector, k′, is folded back into the first BZ, k, by subtracting the closest reciprocal lattice vector (G_2).

The alternative approach to PBG was made bearing in mind localization [2]. As in semiconductor physics a defect in an otherwise perfect crystal introduces a state in the gap with a localized wave function. The state must be in the gap where no extended states exist. This is because if the energy of the localized mode coincided with that of a Block (extended) state they would interact and the mixed wave functions, being combinations of localized and the extended states, could not be localized. A mode of the EM field with its energy in the gap will remain confined in the vicinity of the defect since propagation is forbidden through the walls of the cavity. The magnitude describing the confining power of the defect is the quality, Q, that measures the rate of escape from the cavity, something like the number of oscillations of the mode before escape. It can also be expressed as the spectral width, $\Delta\omega/\omega$, of the state and directly relates to the decaying of the wave function into the PC: the more rapid the decay the less interaction (narrower state) and smaller coupling with extended states (smaller probability of tunnelling). High Q cavities are required to make low threshold lasers because they make better use of the pump energy. However, a compromise must be achieved to have an efficient power extraction.

The approximation that the polarization of a medium is proportional to the electric field breaks down for very intense fields and we enter the realm of non linear optics. For these processes to take place interaction light-matter is required, vacuum being linear. A typical example is second harmonic generation. These processes can be classified as parametric (when the final state of the atom involved is unaltered) or non-parametric (if a change in the level population is produced) and are given an order according to the number of terms considered in the polarization power expansion. Second harmonic generation is second order non linearity whereas intensity dependent RI is a third order process. Here PCs are useful for various reasons. Since very high fields are needed PC microcavities provide the right tool to confine EM radiation in small volumes. The control on group velocity (down to very small values) with the corresponding enhancement of light-matter interaction favours the onset of non linear interactions. As in any other class of process conservation of energy and momentum (phase matching) must be considered. However, in PCs momentum must be conserved to a reciprocal lattice vector which considerably relaxes the condition. Finally, symmetry considerations regarding the PC lattice must be taken into account that add to those relating to the atomic crystal symmetry of the host material.

1.2. Optical Characterization

The main technique for the characterization of PBG materials is, naturally, optical. Bragg diffraction is the primary feature of these systems and can be used for an initial characterization. An interpretation in terms of bands is most appropriate however.

When RI contrast is low it is a good approximation to use ordinary diffraction or dynamic diffraction theory to characterize PC. For this, small angle X-ray scattering is an appropriate technique if a two-dimensional detector is used [14].

Optical reflectance and transmission are the principal tools used to characterize 3D systems. In a reflectance experiment, a high symmetry facet of the PC is often chosen. This sets the normal on a high symmetry direction in reciprocal space. Light is shone with $\mathbf{k}_i$ forming a given angle with the normal, and detected in specular configuration, that is, $\mathbf{k}_o$ in the incidence plane and forming the same angle with the normal. Energy is then scanned. Typical examples are shown in Figure 4. The two conditions to be satisfied are energy conservation, which requires that incident and reflected wavevector lie on a circle centered at the origin; and momentum conservation, which is granted by requiring that the tips of $\mathbf{k}_i$ and $\mathbf{k}_o$ be joined by some reciprocal lattice vector, $\mathbf{G}$. The darker hexagon represents the first BZ, the normal is defined by the direction ΓX and the grey arrows are two reciprocal lattice vectors. In diagram a), reflection occurs at the X point (normal incidence) and the incident and reflected beams travel in opposite directions. Spectra can be taken varying the angle between $\mathbf{k}_i$ and the normal. If we slightly increase the energy of the incident photon the conditions can only be satisfied by tilting the incidence angle as in b). Further increase may give rise to a situation where two scattering mechanisms are possible as in c).

For transmission the detector is placed in line behind the sample and Bragg reflected photons are recorded as drops in transmitted intensity. At variance with the specular reflectance geometry the transmission experiment does not select a Bragg plane because it does not rely on detecting the scattered photon but its absence. Thus, any scattering event, regardless of the direction of $\mathbf{k}_o$, will be recorded as a dip in intensity. Every Bragg plane crossed by $\mathbf{k}_i$ gives rise to a reflection as marked in d) with small circles. These diagrams give not only an idea of when a reflection can be expected to occur, but also where the scattered photon should be observed; although in transmission this is overlooked.

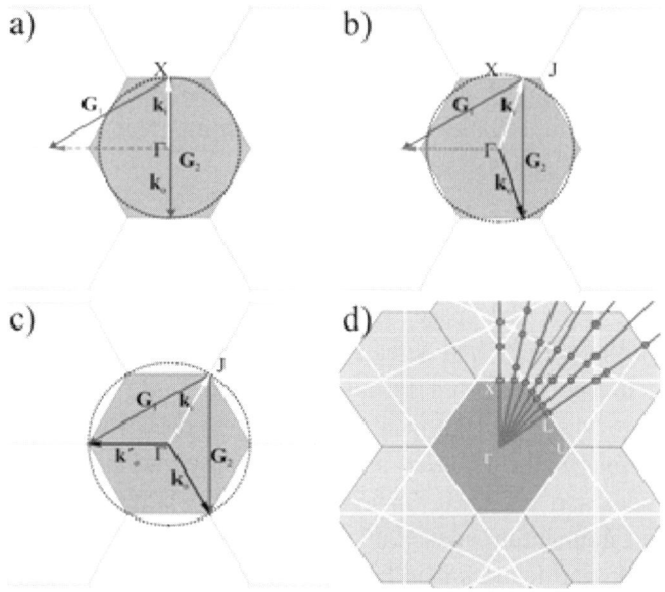

Figure 4. Scattering diagram in reciprocal space: incident (scattered) wavevectors represented by white (black) arrows. For nearly homogeneous medium, energy conservation is granted by staying on the circular equi-energy surface. For vectors in grey energy is not conserved and the process is not allowed. In d) Bragg planes other than those defining the first BZ are drawn and Bragg diffractions are marked with circles.

The preceding picture is valid for low RI contrast where the peaks are narrow, the interactions small, and the original character of the states preserved from those in the free photon picture. When RI is high equi-energy surfaces are no longer circular, this entire naïve picture breaks down and a rigorous band structure calculation is required.

1.2.1. Photonic Bands Interpretation

A full account of the interaction between PCs and light can only be given in terms of bands and gaps as described by a band structure diagram. As mentioned above the only meaningful wavevectors for states belonging to a PC are those contained in the first BZ. For a photon of energy ω, coming from outside, the wavevector, $\mathbf{k}'$, may have a modulus, k′, larger than the BZ boundaries. Yet, for any phenomenon involving states of the PC, this photon will behave entirely as if its wavevector was $\mathbf{k} = \mathbf{k}' - \mathbf{G}$ such that **k** fits in the BZ for some **G** of the reciprocal lattice (see Figure 3). These **G** vectors may or may not be parallel to **k** which will produce an interaction of external light with bands other than those originating from the folding of the nearly-homogeneous medium approximation (as in Figure 1). In principle, a wave that impinges on a PC with a given energy and a **k** vector in a given direction is transmitted through the PC if a band is available in the dispersion diagram for wavevector in the direction of **k**. In Figure 5 (left) the bands diagram for $\mathbf{k} \parallel (111)$ is plotted. According to it, photons incident in the (111) direction will be transmitted except for a narrow band (shaded area) around $a/\lambda \sim 0.65$ where no bands are available.

There may be circumstances when this picture is altered. Symmetry considerations [15] can impede coupling between the incoming EM field and the available state. If, for instance, the state has a mirror symmetry that the plane wave does not possess, overlap can vanish rendering the state useless for transmission. On the other hand even when a band is available and the symmetry allows the coupling to external waves the band may be such that the group velocity vanishes and transmission is virtually cancelled. This can be seen as a divergence of effective RI [16] with corresponding Fresnel reflectance approaching unity: $R = (n-1/n+1)^2 \sim 1$. The latter case usually occurs near the edges of the BZ where it is often masked by real gaps but it may also happen at intermediate **k**s as in the upper bands in Figure 5.

The preceding remarks account for some more peaks appearing without the need for gaps at energies around 1.2. The analysis is very complicated because many bands intervene [17] and diffraction phenomena, becoming important when $a/\lambda \sim 1$, can also contribute [18].

For any given frequency, propagation takes place by the contribution of many states that carry the energy in multiple directions. Only if the PC is a plane parallel slab, refractive phenomena at the entry facet will be reversed at the exit, restoring the fields to a collimated wave.

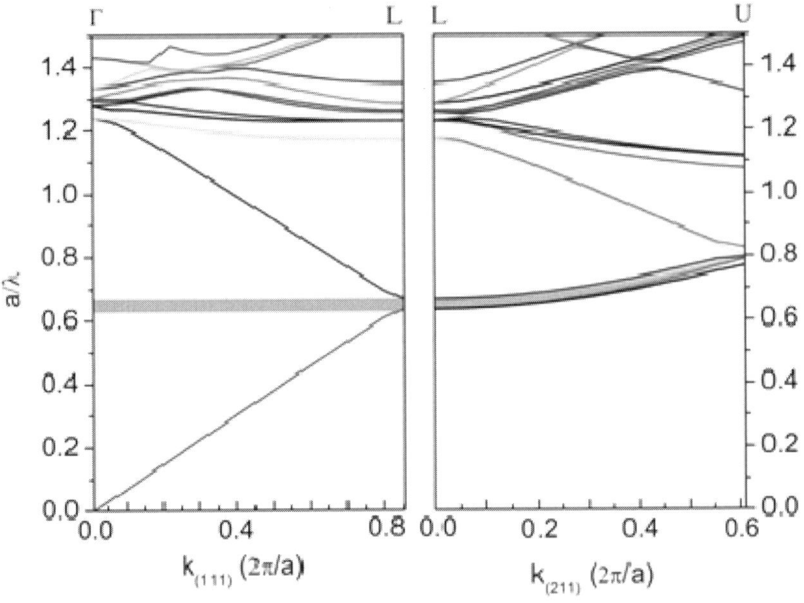

Figure 5. Typical band dispersion for a bare opal for k parallel to the (111) direction (left) and for directions comprised between the (111) and (411). A gap is present at a/λ~0.65 that shifts to higher energies upon changing direction. Bands with little or no dispersion can be seen at higher energies.

Searching the whole reciprocal space is required for a comprehensive characterization of the system and this entails scanning in all directions. It is often enough, however, to scan the principal corners defining the irreducible part of the BZ. The properties of lower symmetry points can be interpolated there from. The right panel in Figure 5 shows (shaded area) the evolution of the Bragg reflection highlighted on the left panel (grey rectangle) as the direction of propagation is changed. Here the abscissa is the wavevector component parallel to the LU segment. This is a typical example of an incomplete gap: the gap exists only for certain direction and, as the direction is changed so does the position (and possibly its width) [19]. If a gap, or at least some part of it, remained a gap for any direction of propagation a complete PBG would be found and reflectance and transmission would present a feature for the corresponding range of energies.

A detailed experimental study of the optical properties in the high-energy region (up to a/λ=1.8) of high-quality artificial opals can also be performed by means of reflection and transmission spectroscopy [20]. It can be observed that the main features of the optical response of artificial opals seem to be accounted for by bands originated by wave vectors parallel to the G*L* direction and their perturbations, such as gaps opening at the center and edges of the Brillouin zone and anticrossings elsewhere, introduced through the interaction with other bands of similar energy and symmetry.

1.3. Applications

Current and foreseen applications of PC can be divided according to their principle of functioning. Some rely on the existence or not of a complete gap, other rely on the peculiar properties of the bands and their dispersion. Applications that rely on a gap make use of the suppressed DOS. Thus the performance of solar cells or microelectronic devices can profit from the suppression of spontaneous emission. Antennas for microwave applications may be made to emit only in the desired direction by placing them on a substrate with a PBG structure [21]. Some applications are entirely new and spring from the new physics brought about by PBG systems.

The spectrum of black body radiation is very broad which results in an important loss of efficiency when thermal sources are used for light generation: most of the energy is emitted as heat. A mechanism to prevent this might have been found in the use of metallic band gap structures [22]. For 3D structures with the metal forming a continuous network (network topology), there is a cut-off frequency below which there are no propagating modes. If absorption can be neglected (thin structures) such an unusually large band gap can be used to recycle infrared blackbody radiation into the visible spectrum [23]. In the photon recycling process, an absolute 3D photonic band gap completely frustrates infrared thermal emission, and forces the radiation into a selective emission band. Energy is not wasted in heat generation, but channelled by thermal equilibrium into useful emission at the edge of the band gap. A prominent example is shown in Figure 6 where an all-metallic layer by layer structure is shown in two of the stages of its production. The tungsten 3D photonic crystal was made by selectively removing silicon from already fabricated polysilicon/SiO2 structures, and refilling the resulting mould with tungsten by CVD. Based on the same ideas, it is possible to enhance the efficiency of dye-sensitized solar cells [24] by using 3D colloidal photonic crystals [25]. By stacking different 3D structures with different lattice parameters one could cover the whole absorption spectrum of the dye, optimizing the light harvesting and therefore, enhancing the efficiency up to 60% [26].

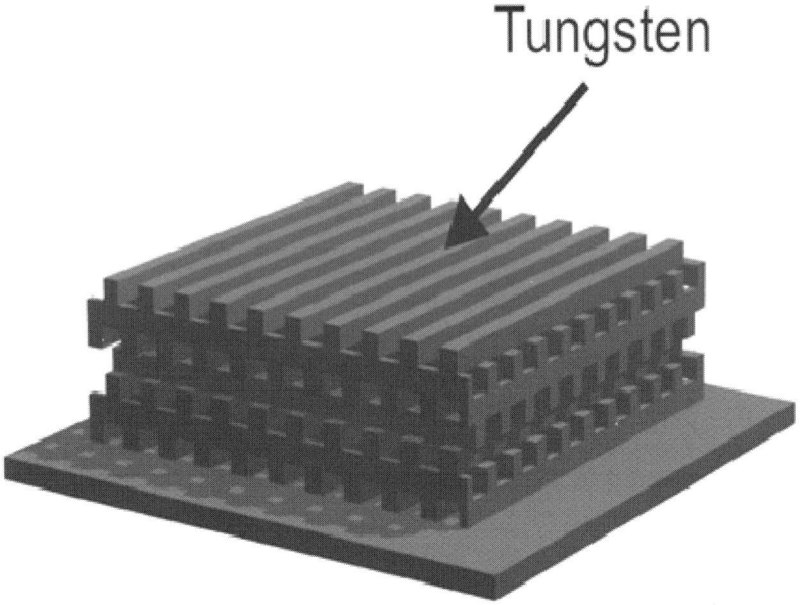

Figure 6. Image of a 3D model of tungsten photonic crystal woodpile structure. This metallic PBG structure has realized the channel of thermal emission into narrow discrete bands.

A laser is basically a medium where population inversion is attained whereby gain and amplification are achieved through stimulated emission. Spontaneous emission is the main foe to gain as it happens isotropically and cannot be prevented by (usually 1D) cavities. Cavities with a 3D complete gap present advantages in the sense that a single channel may be allowed for emission (through a defect with proper design) blocking other directions at the same time. Cavities in 2D systems are expected to present very high quality factors and, very recently, have been realized with Q values up to 500.000 by the group of S. Noda [27]. Lasers have been realized in 2D and the race for lowering threshold is on [28]. Using one such cavity laser action was demonstrated [29] and, very recently, it has been electrically driven [30]. By designing very carefully this kind of cavities one can even tailor at will the laser beam profiles [31]. Two-level atoms cannot be made to lase in vacuum because the rate of stimulated emission equals absorption. Thus, at most, population of the excited state reaches ½. Near a PBG edge, however, this situation changes, allowing for population inversion in two-level atoms. If a laser beam with intensity slightly below the threshold prepares such two-level atom system at population near ½ a second low intensity beam may place the system in gain condition and constitute an all optical transistor [32].

There is a wealth of proposed applications relating to the guiding of light to make real circuits for communications and processing. In this class fall linear waveguides (WG), bendings, crossings, splittings, junctions, add-drop filters, etc, all necessary to direct the flow of light as electrons are directed by wires. WGs are routinely made in 2D systems by placing linear defects in the desired configuration [33]. These are often done by removing rows of the elements constituting the PC. WGs can be made to bend through sharp corners with the enormous advantage in compaction as compared to ordinary fibers that can only be bent to a limited degree. The main workhorse in this field now is losses [34]. The optimization of guided mode properties and confinement in the third dimension and the design of bends, systems to input and extract signals from the systems, etc. account for most of the effort spent in this research. A WG can be bent on itself to form a ring that can be used as a laser cavity [35]. Modes in a WG may be designed to have linear dispersion and wide band width if many wavelengths are desired to be guided without pulse broadening – RI is the same for every wavelength – or they can be designed to have non linear dispersion in order that different wavelengths feel different RI and thus reshaping of propagating pulses can be achieved. This may be very useful in optical communications. A WG consisting of a chain of coupled defects can act as a phase compensator if the right group velocity dispersion is designed [36]. Losses due to coupling (extracting) light into (from) a PC are also very important and should be minimized in future devices. Very recently, E. Moreno *et al.* have proposed the use of crystal surface modes (modifying surface periodic corrugation) to enhance the propagation and coupling efficiencies in 2D PCs [37].

A 2D PC with a cavity in its center becomes a fiber when stretched in the 3rd dimension [38]. Photons localize in the core and can only propagate in the axial direction. Such "holey" fibers rely on the existence of a 2D PBG held by the cladding rather than on total internal reflection – the principle behind ordinary fibers. For that reason these fibers needn't actually have a material core and can guide light through an air pipe which makes them suitable for high power transmission: no undesirable non linear effects will be felt. On the other hand if the core is made of a material properly doped with, for instance, a non linear material they can be made to amplify the signal, compress pulses, etc.

Systems with a complete PBG are difficult to fabricate but some useful properties still exist when no such gaps are available. Dispersion properties can be taken advantage of for steering light within a PC. Very anisotropic equi-energy surfaces are used for superrefraction while the dispersive properties can be used for pulse reshaping.

Obviously a 1D PC (a Bragg reflector) cannot have a complete PBG: its gaps vary depending on direction such that a forbidden energy may be allowed to propagate just by changing incidence angle. It can, however, be used as an omnidirectional reflector by placing the pseudogap within the light cone. Thus, even though there exist states of either polarization for any given energy (no gaps) there are certain ranges for which the available states have **k**s which are inaccessible from outside the system [39]. These systems can be realized for any desired range and have a host of applications. The width of the omnidirectional reflection is mainly determined by the ratio of RIs of the layered materials and their ratio to the outer medium [40]. Using this principle hollow optical fibers are made where confinement of light in the hollow core is provided by the large PBG of an omnidirectional dielectric mirror [41].

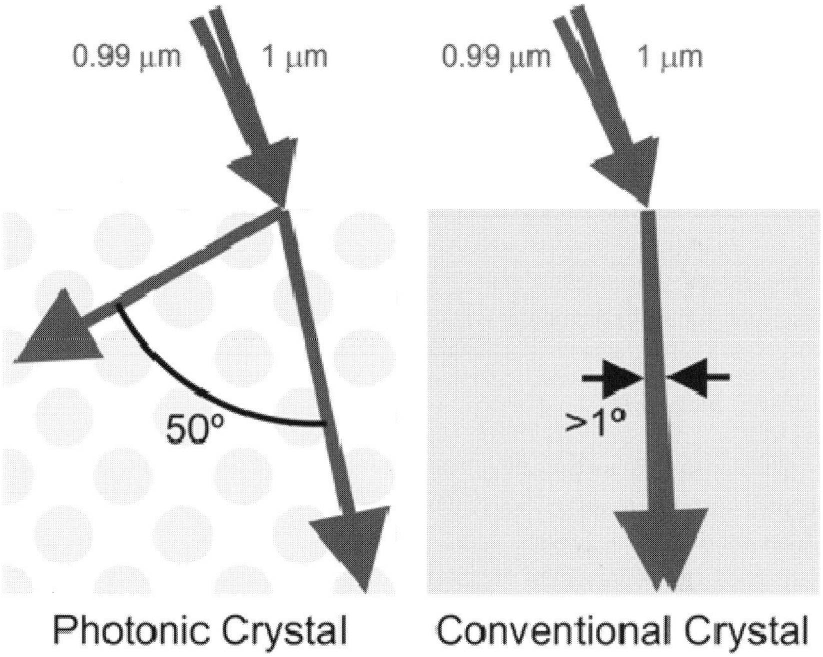

Figure 7. At variance with a conventional crystal a PC can present a strong chromatic separation based on superprism effect: beams of 0.99 μm and 1.0 μm suffer a very different refraction.

Proper design can allow for zero group velocity bands to contribute to laser action by enhancing the interaction between light and matter. Thus, a laser without a cavity can be realized [42] in which unit cell engineering can provide for the control of emission polarization [43].

The laws of refraction take on surprising forms in PCs. Propagation being normal to the equi-energy surfaces, regions of high curvature produce very strong beam swings for slight changes in wavevector (superlens). Similarly in regions where curvature changes sign (inflexion points) ample changes in **k** produce negligible changes in propagation direction (supercollimation). Additional control can be gained with a sound choice of bands; thus, using electron-like bands (ω grows with **k** – positive effective mass) yields positive effective RI, whereas the use of hole-like bands (negative effective mass) allows for negative effective RI [44]. Finally, in some regions of reciprocal space (usually where several Bragg planes meet) equi-energy bands show very strong energy dependence. The fact that a very slight change in energy results in a strong change of shape of the equi-energy surface can be applied for the design of a superprism [45]. The first demonstration is presented in Figure 7 where the deflection of two beams of very similar frequency is shown to behave very differently in a PC and in a homogeneous material. In 2D systems [46] this effect is being studied and experimentally implemented for multiplexers; more complex behaviour has been described for self assembled 3D systems [47].

1.4. Metamaterials

To close this section of introduction a few words on metamaterials should be said in order to better situate PCs in the materials word. Metamaterials is a general name for a larger class of materials whose properties (as represented by their electric permeability and magnetic permittivity) derive from their structure rather than from the materials composing them. PBG materials could be counted among them but the most striking examples are those with simultaneously negative ε and μ, called left handed materials where a negative RI is expected. Ordinary right-handed materials are such that **E**, **H** and **k** form a right handed trihedron meaning that **E**×**H** || **k** so that the direction of propagation of energy coincides with that of the wavevector. For left-handed materials the opposite occurs: **E**×**H** || -**k** and energy flows opposite to wavevector. One crucial consequence is that negative refraction, as long ago described by Veselago [48] may be realized [49] relying on which perfect lenses can be designed [50]. So far materials have been discussed in which the spectral position of optical features is close to the dimensions of their inner structure. In these, however, the periodicity of the structure occurs in a range much smaller than the wavelength thus accounting for the use of effective ε and μ. Strong spectral dispersion is inherent in these peculiar metamaterials where resonant phenomena involving some sort of plasma like excitation are usually invoked. Left-handed metamaterials working in the optical region have been recently fabricated by electron beam lithography [51,52]. Applications for these materials are expected to enhance the performance of Magnetic Resonance imaging systems [53]. Other applications involve emission from sources embedded in metamaterials built as stacks of metallic grids immersed in foam [54]. It has been shown that directive emission from a source can be confined to a narrow cone based on the fact that an effective RI can be made to take on very small values. Thus every ray leaving the material forms a right angle irrespective of the angle it formed within the metamaterial.

Frequency selective surfaces [55] and PBG structures operate on the principle of resonance based on multiple scattering by a series of periodically arranged metal or dielectric scatterers. Accordingly, the relevant frequencies where they function are closely related to the dimensions defining the periodicity. Fractals are mathematical objects with the unique property of being self-similar. This means that their general aspect is the same regardless of scale, that is, zooming in on the object's details gives a similar picture as zooming out for a panoramic view. This fact has a striking consequence on the resonances that can be sustained when such systems are built with scattering materials: the frequencies are detached from the scale. Such systems have been shown to present stop bands and pass bands like PCs over a broad range [56]. They can be built as 2D systems with design features well below the operating wavelength that can perform as 3D structures in terms of polarization- and incidence angle-independent reflection. Furthermore, the wavelengths of the lower gaps can be significantly longer than the size of the sample and can be driven by electrical signals.

Whether there can be PBG without periodicity and without a BZ is a fundamental question. 1D superlattices stacked as Fibonacci series were shown to posses optical stop bands; similar phenomena were observed in acoustics, all of which fostered the search for PBG in 2D quasiperiodic arrangements of dielectrics. These are structures arranged in orderly patterns that are not quite crystalline: the local symmetry is the same everywhere but there is no real overall periodicity. They are often built on forbidden rotational symmetries like 5-, 8- or 12-fold. Sizeable gaps that do not depend on incidence direction are predicted for these 2D systems [57] and have been experimentally found in the long wavelength regime (10 GHz) [58] (where WGs can be produced) and also in the far IR (100 μm) [59]. In 3D, Takeda *et al.* have recently fabricated a fractal structure with singular density of states presenting localized electromagnetic modes (fractal cavities) for microwaves [60].

2. Preparation of Photonic Crystals

A wave crystal can be designed for any interaction involving a wavy perturbation, be it light, sound, particles, spins etc. It only entails designing the appropriate periodic structure which, for EM interaction, is a PC. PCs may be built for any frequency range of the spectrum. Of course the realization greatly changes according to the spectral range of functioning, principally owing to the size imposed by the wavelength. Therefore, the fabrication methods vary depending on the length scale. The first PC shown to have a complete PBG [61] was machined into a low loss dielectric material with millimeter size drills. The size determined its working range to be in the microwave regime. A scaling down of the process, it was suggested, would lead to similar systems working in the visible spectrum. That challenge was actually faced and risen to [62] but the method showed rather ineffective in the number of layers that could be produced. An entirely new structure was later proposed with a sizable, robust band gap that could be built in a layer by layer fashion by stacking rods as a pile of logs [63]. This structure, initially realized in the GHz range, revealed as, probably, the best choice and has been scaled down to optical frequencies by several means.

In what follows a review of the methods employed to obtain systems functioning mainly in the optical range is made leaving the self assembly techniques for a section of their own.

2.1. One-Dimensional Systems

One-dimensional systems have been produced and used widely for a long time as, for instance, antireflective coatings, notch filters, or distributed Bragg reflectors. Typically evaporation techniques were used. A few recent developments are described next with specific applications in photonics.

Although dendritic nanowires can hardly be classified as 1D structures they can be described as a linear arrangement of light scatterers. These structures consist of grafted branches periodically grown on a straight WG acting as a stem. The synthesis exploits the self-organized dendritic crystal growth to assemble microscale periodic semiconductor single crystal nanowires in a vapour transport and condensation system. Laser action was demonstrated using this structure as resonator [64].

Cholesteric liquid crystals (LC) present a layered structure in whose every layer molecules are aligned in parallel but rotated with respect to the layer underneath. They can be viewed as 1D PC in which the dielectric anisotropy originates from the orientation of the molecular axes. The period is determined by the distance between the closest layers with parallel molecular axes and the interaction with circularly polarized light gives rise to bands and gaps for transmission. Laser action can be sustained at the edges of the gap when such systems are properly doped with a dye [65]. The mode closest to the edge has the lowest lasing threshold. Lately, by combining holographic photopolymerization with polymer dispersed LCs, electrically switchable 1D structures have been produced [66].

Layer by layer electrostatic assembly of polyelectrolytes is a technique that combined with the in situ synthesis of inorganic nanoparticles can lead to the formation of 1D structures with tuned morphologic (thickness) and dielectric properties (RI contrast) [67]. These systems have potential as sensors since they respond to environmental changes in pH or ionic strength by swelling or shrinking which changes their photonic signatures.

1D multilayers of controlled porosity can be also prepared by anodic etching of crystalline silicon [68]. Dielectric function is controlled through porosity which, in turn, is controlled through current density in the electrochemical cell. The layer thickness is dictated by the etching time. These stacks can be designed to act as 1D Bragg reflectors. Cavities can be inserted where luminescent ions can be included by doping through cathodic electromigration of Erbium into the porous silicon matrix, followed by high temperature oxidation. These structures can be used for biosensors after the inner surface of the porous Si is functionalized to bind some distinctive component in biological matter [69].

A photonic crystal fiber consisting of a pure silica core surrounded by a silica-air photonic crystal material with a hexagonal symmetry was originally made by drawing a bundle of hollow silica rods except for the center one that is solid. The drawing process thins the preform down to micrometer size, and fibers thus produced can support single robust low-loss guided modes over a very broad spectral range [70]. A fundamentally different type of optical WG structure is demonstrated, in which light is confined to the vicinity of a low RI region by a two-dimensional photonic band gap crystal [71]. The WG consists of an extra air hole in an otherwise regular honeycomb pattern of holes running down the length of the fiber. The core being air, guidance cannot occur through total internal reflection. Using the same principle, but with new materials, microstructured polymer optical fibers were also realized [72]. The low-cost manufacturability and the chemical flexibility of the polymers are here the main advantages.

A new concept in fiber fabrication was introduced by devising hollow fibers in which the wall is provided by an omnidirectional reflector [41]. These fibers comprise a hollow core surrounded by a multilayered cladding designed to reflect a certain range of frequencies that are guided.

2.2. Two-Dimensional Systems

The fabrication of 2D PBG systems has experienced the greatest development mainly boosted by the availability of mature techniques borrowed from many areas. Lithography (mainly optical and electronic) is the main contributor to this explosive expansion since microelectronics had, for a long time, demanded more and more precise processing.

Carbon nanotubes with controllable diameters, from tens to hundreds of nm, and lengths of tens of microns can be grown at large scale in a matter of minutes [73]. Plasma enhanced hot filament CVD is used with acetylene serving as source of carbon and often using ammonia as a catalyser. The nanotubes grow on thin metal films or nanoparticles previously deposited by various means. Nanoparticles can eventually be arranged in an orderly way leading to ordered arrangements of long, parallel, straight, carbon nanotubes [74]. In such circumstances the nanotubes act as scattering centers and produce a 2D PBG. In one step further using similar technique, Wang *et al.* have succeed to fabricate TiO_2 2D structures starting from ZnO nanorod arrays previously patterned on sapphire [75].

2D structure with a PBG in the NIR [76] can be fabricated using a technology developed for the production of microchannel plates. This consists basically in drawing a bundle of optical fibers in which core (SiO_2) and cladding (PbO) can be selectively dissolved. Thus the fibers are arranged in a hexagonal lattice and are drawn in order to reduce their diameter to the desired lattice parameter. The core is then dissolved leaving behind a network of air holes in a glass matrix. The filling fraction of the air holes can be adjusted through the initial core/cladding radii ratio. The technique can be extended to produce nanochannels with optical features in the VIS and UV [77].

Two dimensional PBG structures have acquired the widest diffusion due to the easy, albeit expensive, implementation of microelectronic technology techniques such as photolithography, electron beam lithography etc. Monorails and air bridges are examples of one-dimensional structures produced by these means [78]. Typically Si and III-V semiconductor technology is applied. Lattices composed of holes or pillars were often adopted to obtain different structures. The first such realization was in GaAs/GaAlAs [79] where a lattice of holes was etched through an e-beam lithographed mask. WGs can be defined in these structures by leaving rows of holes undone. Point defects produced near such multiple wavelength carrying WGs can be tuned for extraction of selected wavelengths [80]. Figure 8 shows an example of a membrane PC where two defect holes near a one-missing-row WG are tuned (by size) to extract selected wavelengths. Although these techniques are mostly applied to produce 2D systems (where they are most effective) they can also be used in conjunction with other processes to add a third dimension to the structures obtained. By using relatively standard microelectronic techniques in a layer by layer fashion, 3D crystals were soon fabricated for the IR region [81]. As an alternative, wafer bonding [82] was used to build 3D structures where WGs can be easily integrated during production [83].

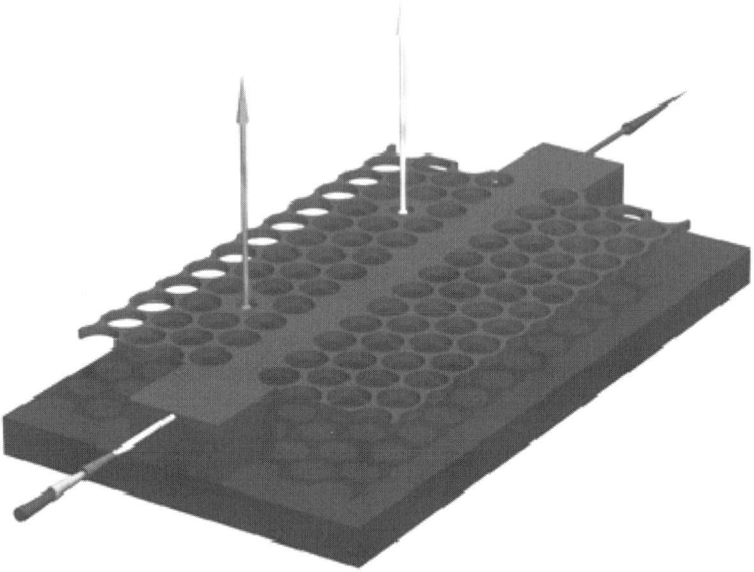

Figure 8. 3D model of a straight waveguide in an InGaAsP 2D PC with isolated defects formed to couple to selected transmitted wavelengths.

Soft lithography, extensively reviewed in Reference [84] by Xia *et al.*, is a non-photolithographic technique based on self-assembly and replica moulding for micro- and nanofabrication. In soft lithography, an elastomeric stamp with patterned relief structures on its surface is used to generate patterns and structures with feature sizes ranging from 30 nm to 100 microns. Several techniques have been derived: micro-contact printing, replica moulding, micro-transfer moulding, micro-moulding in capillaries, solvent-assisted micro-moulding, etc. In nanoimprinting a pattern recorded in a mould is press transferred to a thermoplastic polymer thin film where ridges and trenches with a minimum size of 25 nm and a depth of 100 nm are created [85]. Further processing may be applied to record the transferred pattern into the substrate for large scale replication. Although a 2D technique, it can be extended for building 3D structures by iterative processing.

2D structures can be obtained in porous silicon when a lithographical mask is transferred by means of electrochemical pore formation [86]. The preservation of the shape of the pattern through hundreds of microns in depth is guaranteed by the fact that electrochemical etching proceeds only upon provision of optically created electronic holes that promote the Si dissolution at the tip of the pore. The factor controlling the filling fraction is the illumination intensity. Therefore it can be used to shape the pores and confer on the structure a 3D character [87].

Holographic lithography is a new technique proposed by V. Berger *et al.* [88] and realized for the first time by S. Shoji *et al.* [89] for 3D structures. Interference of laser beams is able to produce large area high quality defects free 2D photonic crystals templates in a very fast and reproducible way. The interference of 3 laser beams produces a 2D light distribution pattern which is recorded in the bulk of a photosensitive resin. By this method 2D crystals [90] and quasicrystals [91] can be easily fabricated in a very large scale being also possible using other high refractive index photosensitive materials as chalcogenides [92].

2.3. Three-Dimensional Systems

The border line between two and three dimensional fabrication is sometimes rather diffuse because some 3D structures can be made with 2D techniques. Some can even range from one- to three-dimensional.

The transfer of corrugations from a patterned substrate can be made to act as an autocloning mechanism through which a 3D structure can be grown by sequential deposition of Si and SiO_2 [93]. The superprism effect was first demonstrated in these structures [45]. The technique initially developed for Si and SiO_2 was extended to new materials by direct growth [94] or by material exchange [95].

AsSeTe chalcogenide glasses are materials that are photosensitive and have a large RI. These properties make these glasses particularly suitable for the fabrication of photonic crystals. A grating is recorded in a thin layer of the chalcogenide by the interference of two laser beams. Since the material has a photoresist effect, it can be etched after exposure. The grooves are filled with a photoresist by spin coating and planarized. New material layers are evaporated on top and the process repeated rotating the orientation of the stripes. This technique is intrinsically self-aligned, providing a simple way to build layer-by-layer photonic crystals [92].

As seen above two interfering beams produce a diffraction grating that can be used in interference lithography to record a grating; three produce a 2D pattern. To go one step further, a fourth beam adds depth to the structure to yield a 3D periodic lattice. The period of the structure and its symmetry (lattice) are dictated by the laser wavelength, the relative phases, and incidence directions of the interfering beams whereas the shape of the repeated feature in the lattice (basis) results from the beams' polarizations. This set up applied to a photo sensitive polymer can lead to a 3D network if exposed polymer is crosslinked and unexposed polymer is dissolved [96] (Figure 9). Initially developed for UV sensitive polymers upon pulsed laser the technique has been extended to continuous wave VIS lasers [97] and with the required modifications real fcc [98] and woodpile [99] structures working in the VIS are possible. Exposed but undeveloped films are amenable to further processing by laser scanning to draw new structures for WG fabrication [100,101]. One unsatisfactory issue in this technique is that the available photo sensitive polymers (that have to be slightly absorbing to be sensitive but not in excess to allow exposure of the whole film) have low RIs. This makes a second process of templated infiltration necessary for the enhancement of dielectric contrast. Recent experiments have shown to be useful for infiltrating holographic 3D structures with TiO_2 by Atomic Layer Deposition (ALD) [102] and silicon by CVD [103], enhancing their photonic properties. Nevertheless, optical quality of this kind of 3D structures is still far from that presented by composites fabricated by other means (i.e. self-assembly).

Based on the same phenomenon, Direct Laser Writing (DLW) technique again takes advantage of light sensitive resins and with the help of a focused laser beam any feature can be written with a precision of 200 nm, as it involves a two photon absorption process which only takes place in the focus [104]. DLW offers many possibilities in fabrication, due to its versatility, only limited by sample thickness (about 10 μm) and lateral dimensions (around 100x100 μm^2). By this technique several different 3D photonic crystal templates have been fabricated in the last few years: woodpiles [105], tetragonal lattices of slanted pores [106], spirals [107], and even 3D/2D/3D heterostructures [108]. This technique can also be directly applied for writing 3D structures in chalcogenide glasses in which, due to their high refractive index, a full PBG is achievable [109]. When this is not possible, infiltration with other materials is needed, i.e. silicon [110].

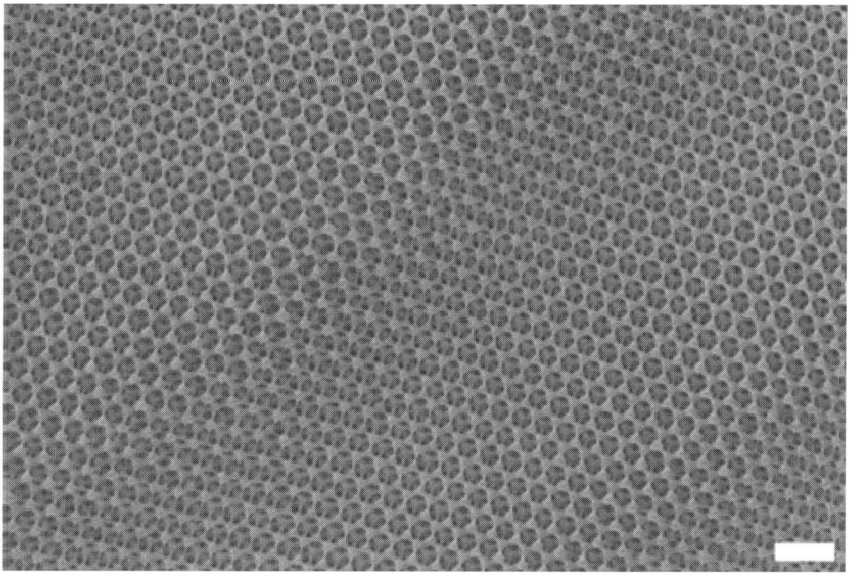

Figure 9. SEM picture of an fcc structures obtained by the holography technique (scale bar is 2 microns).

Deep X-ray lithography [111] allows an extra degree of precision and an extremely high depth of focus due to the reduced wavelength of the radiation used. A pattern is drawn on a resist and transferred to a metal sheet. This metal is then used as a mask whose shade can be projected in different directions tilted with respect to the normal to produce rod structures. Simple methods may be used to replicate yablonovite structures in other materials [112]. The pattern can then be transferred to a metal (by electrodeposition) or a high RI dielectric (by a sol-gel technique).

Glancing angle deposition [113] is a versatile technique that can be employed for the fabrication of large, 3D PBG crystals [114]. Physical vapour deposition with the vapour flux arriving at an oblique angle leads to the growth of inclined rods on the substrate. Rotation of the substrate by 90° every quarter period changes the direction of the rod to form a square spiral. The vapour flux angle determines the pitch in the vertical direction. Tetragonal square spiral crystals with a predicted PBG can be produced in the visible, NIR, and IR spectrum. A tetragonal lattice suitable for a large PBG (up to 15% for a silicon structure) can be synthesized by a regular array of square spiral structures grown from a simple, prepatterned substrate [115].

Copolymers are formed by linking together two or more classes of monomers to form a larger molecule. In a block copolymer, all of the monomers of each type are grouped together. A diblock copolymer can be thought of as two homopolymers joined together at the ends. Block copolymers self-assemble into one-, two-, and three-dimensional periodic equilibrium structures with length scales in the 10-100 nm range. The plethora of structures into which block copolymers can be assembled and the multiple degrees of freedom that they allow on the molecular level offer a large parameter space for tailoring new types of photonic crystals at optical length scales [116]. Very complex bicontinuous structures [117] with a 3D translation symmetry or composite systems with metallodielectric character are just two examples that can be realized with promising optical properties.

Similarly focused ion beam micromachining can be used in conjunction with macroporous etching to produce replicas of the Yablonovite structure [118]. A first set of holes is produced from a 2D mask by conventional photoassisted anodic dissolution. The focussed ion beam technique is then used to produce the other two set of holes. The structure thus produced is not exactly Yablonovite but a slightly distorted one.

Robot manipulation is a recently developed method that allows the arrangement of nano- and microscopic objects in 3D structures with a precision of nanometers and has allowed the first assembly of diamond opals [119]. The wood pile structure, that has been revealed as one of the best suited to sustain a complete PBG, is amenable to construction by this method. This has been recently demonstrated by stacking previously fabricated grids [120]. The grids were defined by lithography as wires held by a frame and they were stacked criss-crossed to form a 3D crystal. Crystals for infrared wavelengths of 3–4.5 μm of four to twenty layers (five periods) including one with a controlled defect, were integrated at predefined positions marked on a chip with an accuracy better than 50 nm. Microspheres are used to slot into holes defined in the frames to ensure correct alignment. Micromanipulation can also produce 3D structures using a nozzle as a paintbrush where polyelectrolyte inks can fluid. This new technique is similar to DLW and, in principle, any structure can be fabricated, in particular woodpiles [121]. These polymeric structures can also be infiltrated with silicon by CVD increasing RI contrast to develop 3D full PBGs [122].

Future PCs devices based on 3D structures will have to incorporate controlled defects, a technology already mature in 2D structures, but, due to its complexity, still young in the 3D case. Different strategies are being developed in order to produce, in a controlled manner, defects in 3D PCs being very recently reviewed in Reference [123].

2.3.1. Colloidal Crystals

Colloids are structures comprising small particles suspended in a liquid or a gas. In this context *small* refers to sizes, between nanometers and micrometers, much larger than atoms or molecules but much too small to be visible to the naked eye. Some times these particles self organize owing to their electrostatic or other interactions and give rise to crystalline structures. As such, they present the typical fingerprint of PCs: they are iridescent. Colloidal crystals can be used and studied as suspensions or can be further processed to obtain opals which, in turn, can be used to template other materials. Colloids of spherical particles, reviewed by Xia *et al.* in [124], are more useful when orderly assembly is required because their interactions are more isotropic and stacking is easier. Next, an update of fine particles synthesis and assembly methods is provided along with some materials aspects regarding infiltration and further processing.

2.3.2. Bare Opals

Opals are precious gems composed of sub-micron size silica spheres stacked together and cemented with a slightly different silica [125]. This structure confers photonic properties on the material and has inspired a whole branch of research in the PC field [126]. Fine spherical particles have been produced from numerous materials, both inorganic (see review by Matijevich in [127]) and organic such as polymers (find a comprehensive summary of synthesis and functionalization by Kawaguchi in [128]) aiming at replicating the natural gems. This technique allows the widest range of materials and procedures of synthesis. Only the range of topologies is narrow since natural assembly tends to produce thermodynamically stable structures that are restricted to a few lattices, mostly compact cubic and hexagonal. Furthermore, if special care is not given random stacking or disordered structures are obtained. This section is divided into two subsections reviewing particles synthesis and assembly and further processing and fabrication.

As will be explained later on, the assembly of opals can be seen not only as a goal in itself but as a means for posterior templated synthesis of other materials. In this sense the physicochemical properties of the material composing the initial opal must be taken in consideration. For this motive silica has acquired a supremacy on other materials because it is highly inert both chemically and thermally. Thus, if infiltration requiring high temperature (a few hundred ºC) is envisaged most organic polymers must be discarded in favour of silica. For polymeric opals, which in many aspects can be considered of better optical quality than those composed by silica spheres, new fabrication strategies are being developed, as we will see later on.

The best known method for monodispersed silica spheres was originally developed by Stöber *et al.* [129] and relies on the hydrolysis of a silicon alkoxide and posterior condensation of alcohol and water to form siloxane groups. Because of their chemical structure, these beads present a surface charge which is screened by the ions in the medium. Under appropriate conditions of temperature, pH, and concentrations this synthesis process yields spheres with diameters ranging from a few nanometers to a few microns. Although the reason why it is still controversial, very monodispersed colloids (a few percent standard deviation in diameter) can be obtained for sizes between 200 and 600 nm. For larger sizes a strategy based on the growth in the presence of small seeds has proven successful [130]. Everything seems to indicate that the final size of the particles produced responds to a correct balance between the nucleation process (by which new particles are created) and that of aggregation (through which those particles grow at the expense of available monomers). For that reason, the concentration of seeds in a regrowth process must correctly match the reaction condition in order that no nucleation occurs while an optimum use of the silica precursor is made. Although, most often the particles synthesized are spherical dimmers [131], and even ellipsoidal particles have also been produced by different methods such as ion irradiation [132] or inverse templating [133].

Among the materials used for opal assembly the widest spread are polymers, mainly polystyrene (PS) and poly(methyl methacrylate) PMMA. For the preparation of these beads there are a host of techniques all under the umbrella of polymerization but with different approaches such as precipitation polymerization, emulsion polymerization, dispersion polymerization, seeded polymerization, inversion emulsification, swelling polymerization, suspension polymerization, and modifications on them that can be found in the review by Arshady in [134]. Owing to a higher surface charge to density ratio, these particles tend to behave better than silica ones for the purpose of colloid crystallization. Therefore they are very appropriate for optical and crystallization studies.

In order to obtain a PC these beads must be arranged into a crystalline lattice. Depending on the properties of the particles used and the final product desired various methods have been developed that are summarized next.

Some colloidal particles, owing to their, mainly, electrostatic interaction, tend to organize into crystalline lattices when the volume fraction reaches certain value [135]. These crystals can be used for optical studies and find some applications but their liquid character prevents some uses since they must be kept confined in vials or the like. The process of crystallization has been extensively studied and these structures serve as models for liquids and phase transition investigations. The interaction has for instance been studied in absence of gravity where only electrostatic forces intervene [136]. These systems reveal useful for optical studies of the velocity of aggregation and make real space imaging possible [137].

Fields other than gravity can be used to induce order in the colloidal systems. For instance an electric field can polarize the beads which switches on a strong long range interaction that quickly orders structures [138]. With ac fields binary colloids present interesting behaviours depending on relative particles concentration and frequency [139]. Likewise, magnetic fields acting on colloids of magnetic beads (or non magnetic beads immersed in magnetic liquids) produce ordered arrangements [140]. Crossed electric and magnetic fields can induce a martensitic-like transition from the bct to the fcc structure in mesocrystallites consisting of multiple coated spheres. The multiple coatings serve to enhance the electrorheologic and magnetorheologic effect [141]. EM fields have been used by producing (through interference) 3D standing waves in whose potential minima ions are trapped. They can be similarly used to trap dielectric spheres. [142] Fields can also be used, after crystallization is achieved, as a tool to remove defects from the crystal [143]. An annealing process provides the activation energy needed to relax existing defects without generating new ones. Optical tweezers use forces exerted by intensity gradients in strongly focussed laser beams to trap and move colloidal particles. Rapidly translating optical tweezers can be used to introduce elastic strain locally within a thin colloidal crystal, thereby delivering kinetic energy to the lattice without heating the surrounding fluid.

Sedimentation is the natural way to obtain solid opals. This method produces thick opals and can be altered in different ways depending on the goal pursued. In this procedure a colloid is left to settle, a process that takes from days to months depending on the size of the spheres, at the end of which a sediment is obtained [144]. The sedimentation occurs driven by gravity and a clear sedimenting front can be seen in the vial separating clear water and colloid. When the concentration is low (~1 wt%) the beads behave as hard spheres and the velocity can be approximated by Stokes' law. As a matter of fact, this can be used to very accurately determine sphere size, if densities and viscosity are known. The supernatant liquid is removed and the sediment dried. At this stage the spheres are not in actual contact but kept together by water necks. The water content in the opal is about 10 wt% overall. A thermal treatment at 200ºC leaves the sample water-free and makes it truly compact fcc but, at the same time, turns it mechanically very weak and unmanageable. Besides, drying in these conditions invariably involves a crack formation process that can hardly be prevented when the lateral size of the sample is above tens of microns. Drying involves a contraction that does not occur in the supporting substrate that can only be accommodated by the creation of cracks as defects accommodate lattice mismatch in epitaxially grown materials. The final result is a compact of spheres arranged according to a face centered cubic structure [145] whose photonic properties scale with bead diameter revealing their origin [146]. A typical example is shown in Figure 10 where several facets are shown that belong to the fcc structure and are not found in other possible lattices like hexagonal close packed. The morphology includes the same kinds of defects that can be expected in ordinary solids. This is manifested in the optical spectra mostly as a broadening of the peaks, diffuse scattering and a decrease of the reflectance (increase in transmission) at the gap frequency [147]. It is worth mentioning that for this kind of structure the best estimate for effective RI is Maxwell Garnett approximation [148] rather than a weighed average.

When beads are too large or too small (and so is their deposition velocity) bad quality or no sedimentation is obtained in a reasonable time so a vertical electric field applied on the sedimenting vessel can help in achieving sedimentation. This is rendered possible by the beads surface charge that makes them respond to a macroscopical electric field [149]. By this means large diameter beads can be slowed in their sedimentation and small diameter spheres accelerated. The surface charge can be governed by pH which gives one extra handle on the process. Limits on the highest voltage applicable are mainly imposed by medium electrolysis potential for the electrodes used. When two kind of particles oppositely charged are used, new binary structures are possible, useful to study the phase behaviour of ionic species [150].

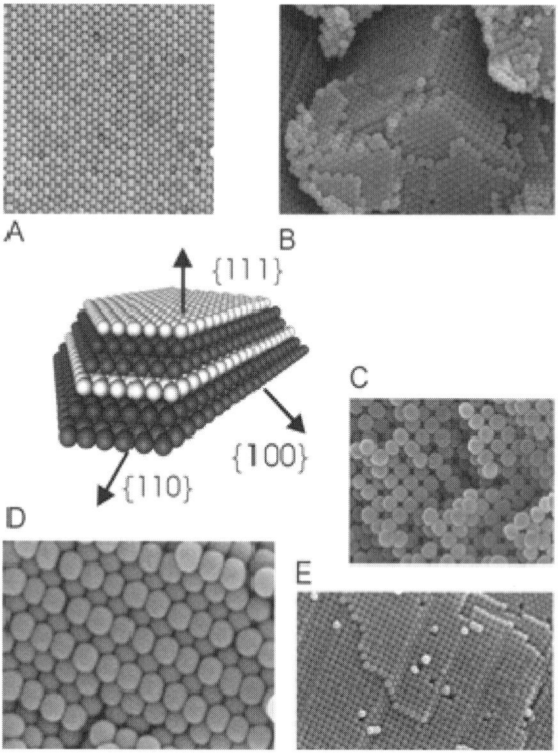

Figure 10. SEM images of four facets in a real opal to be compared with a model crystal (center). Images show A) outer (111), B) inner (111), C) inner (100), D) inner (110), and E) outer (100).

The Langmuir-Blodgett method of thin film preparation used for organic materials can be adapted to the growth of thin film layers of fine particles [151]. Here, a single layer of monodispersed hydrophobic latex spheres or silica spheres rendered hydrophobic by the use of a suitable reagent can be spread at the air-water interface in a colloidal solution. If sufficient care is employed in the purification of the material such films can be compressed and deposited by the Langmuir-Blodgett technique to form ordered multilayers, which show PBG properties. The compression causes the spheres to close pack on the water-air interface which is required to create a close packed layer on the substrate. The one layer process can be repeated several times to grow a 3D structure [152]. This method also allows a high degree of control in the synthesis of layered colloidal structures.

As opposed to the above mentioned method, where a single layer is transferred to the substrate in each upstroke, polymer particles can be made to crystallize in the air/water interface by a combined effect of capillarity and aggregation and then thick film transferred to a substrate [153]. A beaker containing a colloidal suspension of spheres is heated and the particles near the water/air surface start protruding due to evaporation. The capillary forces make the floating spheres assemble and convection aggregates new particles from the bath to build new layers from underneath. The crystal thus formed can be transferred to a glass substrate introduced at an angle in the beaker.

The capillary forces present in an advancing (owing to evaporation) meniscus can be taken advantage of to assemble ordered layers of colloidal particles [154]. This was initially exploited for monolayer films and later extended to virtually any number of layers [155]. The method, very similar to Langmuir-Blodgett for film deposition, yields ordered arrays on nearly any vertical surface. The set up is simple and only requires a vial containing a colloid where a flat substrate is inserted vertically. A meniscus is formed that draws particles to its vicinity by capillarity. Evaporation sweeps the meniscus along the substrate vertically feeding particles to the growth front. Colloid concentration and sphere diameter determine the thickness of the layer deposited. Successive growth processes can lead to the assembly of multilayers that can be constituted by similar or different particles [156]. Although the evaporation rate can be easily controlled through the vapor pressure in the surrounding atmosphere, controlling the sedimentation rate is not so easy as it is largely determined by sphere size. A major challenge is met when the size of the particles is too large (larger than about 700 nm) so that evaporation rate cannot match sedimentation rate: very soon particles abandon the meniscus region and the entire mechanism fails. This problem was overcome by the application of a thermal gradient to the colloid [157]. The combination of Langmuir-Blodgett to grow a single layer of spheres with convective assembly to grow two confining thicker layers has facilitated the build up of embedded defect structures [158].

Binary colloidal crystals can be grown in a layer-by-layer mode using a controlled drying process on vertical substrates [159]. A 2D hexagonal close-packed crystal of large (L) silica spheres is first grown on a clean glass substrate with a growth rate of 1 to 2 mm per day. The 2D crystal is then used as a template on which small (S) silica or polystyrene particles are deposited. The small particles arrange themselves in regular structures depending on their volume fraction and size ratio. Another layer of large particles is deposited, these steps being successively repeated to grow till a 3D structure of the desired thickness and composition is achieved. Structures with LS, LS_2, LS_3, formulas or kagomé lattices can be obtained depending on size ratio and concentration. If one of the two sizes is realized with latex a non compact structure can be obtained by calcination.

The method of growth in confinement cell [160] consists in the arrangement of particles from a colloid confined between two close up flat surfaces. The confinement cell consists of two glass slides separated by a lithographically defined resist in the shape of a rectangular frame with draining channels through which water leaks. One of the slides has a tube attached that serves to inject the colloid. The slides are kept together with binder clips. Sonication can be provided to help assembly. This ingenious method provides a route to grow thickness controlled opals by varying the thickness of the spacer so that the desired number of layers can be selected just by spinning the appropriate amount of resist. Although this technique requires lithographic facilities the spacer can be substituted for commercial Mylar foil that serves the purpose equally well although thicknesses are restricted to those foils available. Crystals formed simply by increasing the volume fraction beyond the liquid-solid phase transition tend to be small and predominantly random close packed. However, by forming these crystals while controlled shear to the glass plates bounding the crystal is applied, large-area single crystals can be made [161]. These crystals are much more regular and appear to be predominantly twinned face-centered cubic. The creation of single-orientation face-centered cubic crystals is possible when shear is applied in one direction although colloidal crystallization in confined spaces can present rather interesting facets orientations when no commensurable high symmetry layers are to be built [162].

In most of the methods described a layer is assembled on or grows in contact with one confining wall. Through the interaction with the wall this crystal facet tends to be the most compact possible and, as a consequence, the crystal naturally orients with the (111) planes parallel to the confining wall. It is sometimes desired to obtain other orientations (for instance to be able to do spectroscopy in other directions) or simply to extend the coherent growth length. To overcome this limitation the use of patterned substrates has proven to be advantageous. For this purpose a substrate is designed with a pattern such that settling or arrangement of the colloid particles is directed and crystallization takes place according to a predefined geometry. This technique can be combined with others to attain thickness and orientation control at the same time. Ordinary sedimentation combined with a patterned substrate may lead to a template directed sedimentation [163] that makes possible the growth in orientation other than the typical (111). This colloidal epitaxy not only allows growth in exotic directions but it gives a control on the size and shape of the grown crystal.

A system in which one of the slides of a confinement cell is properly patterned permits growth with control not only of thickness but also of orientation. This method relies on the dictation of structure and orientation by the pattern drawn on one of the confining walls. By means of ordinary or soft lithography a set of squares is recorded on a (100) silicon wafer that through anisotropic etching develops pyramidal pits. These pits can be designed such that their pitch is commensurate with the diameter of the spheres to be used. In such case if growth initiates in the apex and proceeds to fill the pyramids it can continue above the substrate to fill the whole confinement cell congruently [164]. By this means it has been possible to grow (100) oriented opals over areas as large as several square centimeters. This is not the only potential of the confinement cell with patterned floor as various other structures can be grown such as chiral chains with controlled handedness [165]. Clusters of a small number of beads in different arrangements can be assembled by adjusting the size and depth of the pits relative to the size of the spheres [166].

A versatile method for the growth of vectorial structures that combines self-assembly, micro-fluidics and soft lithography, can be used to control thickness, area, orientation and registry of silica colloidal crystals on silicon wafers [167]. This method relies on capillary forces to inject the suspension in V-grooves etched in a (100) Si wafer and on evaporation to assemble the spheres in a close packed structure. For geometrical reasons the opaline structure grows with the (100) planes parallel to the (100) of the templating wafer. The number of layers is easily controlled through the depth of the V-grooves, which, in turn, is proportional to the etching time.

Normally vertical deposition is performed on plain flat substrates, but new possibilities emerge when the substrate is previously patterned in micro channels with vertical walls [168]. This can be done by anisotropic etching. In such cases capillary forces at the surface relief act driving the assembly of compact cubic structures only in the concave trenches with well oriented lattice planes. The use of these principles and vast possibilities offered by soft lithography allows the building of structures with varying thickness or lattice parameters in combination with all sorts of shapes and architectures.

When opals are grown on solid substrates drying leads to crack formation because contraction has to be accommodated by fissures. One ingenious way to circumvent it is to perform the growth on a liquid surface [169]. As the sample is not firmly attached to a solid, undeformable surface, shrinkage is allowed without the need for a tension relief by breach opening. Eligible liquids are, for instance, gallium and mercury, both with a high density. Using this method, millimeter-wide, crack-free samples can be produced.

It was mentioned earlier that one of the best, if not the best, structures for complete PBG formation is the diamond structure. This is an fcc with a two-atom basis which confers on the structure all the *a priori* required characteristics for band gap opening. The lattice is as isotropic as it can be: the truncated octahedron that bounds its BZ is the closest to spherical of all possible. This implies that gaps opening at the boundaries of the BZ, whose energies are given by $\omega \sim ck_B/n$, will be very similar since all the boundaries of the BZ (k_B) are roughly the same magnitude in all directions. In this structure the gap opens between the second and third bands and is relatively wide and robust. The dielectric contrast threshold for this structure is rather low ($\varepsilon_2/\varepsilon_1 \geq 2$) which makes it very enticing. Its assembly, however, from spherical beads has been elusive so far and the structure has escaped researcher's hands and proved impossible. The reason is that the diamond lattice is very 'empty' and colloidal methods tend to produce only compact structures. A method was proposed [170] to build the structure by using nanorobot-assisted manipulation that was soon after realized [119]. This method relies on the fact that two diamond lattices properly interpenetrated form a body centered cubic (bcc) lattice. By building a mixed bcc lattice with two different materials, one of which is easily removable without affecting the other one, the goal can be realized. The nanorobot can place nano- to micrometric objects with nanometer precision and a plasma etching can remove latex without affecting silica. Direct growth (only one kind of spheres) was achieved too by making use of the contamination layer, inherent to the SEM observation, in order to glue particles together by their contact point. Thus, diamond lattices of silica spheres were obtained on silicon substrates, used as templates where the first layer was held in place.

2.3.3. Further Processing

Once the opal is assembled various treatments can be applied that enhance its mechanical and otherwise properties. Sintering, for instance, is a crucial treatment for thick freestanding opals for two reasons.

As-grown opals have a considerable amount of water both physically and chemically bound that can be removed by an appropriate thermal treatment. This treatment is seen as essential if close packing filling fraction is to be achieved since spheres in as-grown opals are not in close contact but separated by a thin water layer. The removal of water from the compact reduces the RI [171] and, as a consequence, blue shifts the Bragg wavelength. On the other hand, such thermal treatment may produce a matter viscous flow through which spheres are bound together conferring mechanical strength on the opal. The latter process (called sintering and routinely used in ceramic processing) can be applied to control not only mechanical and chemical properties but also optical properties. There is, however, a more important effect associated with the flow of matter (that leads to the formation of necks between the spheres) which is the increase in filling fraction and the lattice parameter contraction. All this causes the optical features to blue shift by stages as the baking temperature is raised, first owing to the loss of water and finally to the contraction of the lattice [172]. This phenomenon first observed in silica opals was then seen to occur also in latex opals [173]. By changing the sintering temperature or time the filling fraction can be controlled from close packing (ff=0.74) to glass state (ff=1.0) where no pores are left and photonic effects disappear.

In certain circumstances a control on the filling fraction or a strengthening of the opal without altering the lattice parameter may be desired. Here an alternative method based on the growth of a thin silica [174] or other oxide layer by CVD is useful. Typically the reaction involved is the hydrolysis of a metal chloride to produce the metal oxide and hydrogen chloride. The water used is hydrogen bonded to the surface of the silica spheres that can be recharged for further reactions by rehydrating through exposure to water vapor. This method allows the growth of several oxides by CVD at room temperature and atmospheric pressure which have no adverse effects on the structure.

Apart from obtaining opals in desired forms by controlled growth they can be used as raw material for further processing in order to fabricate devices. In particular, they must be cut and polished. Although there has been little progress in this direction some challenges are being explored.

When used in thin films, deposited opals may have to be lithographically or otherwise patterned. For instance, focused ion beams can be used to make trenches or vertical pits that can be used to connect different layers in a multiple stack. PMMA opals can be patterned through e-beam lithography [175,176] since this material is routinely used as a resist in microelectronic technology.

Recording a WG in the interior of an opal is a desirable goal if the opal can be made to sustain a PBG so that the WG could transmit light without loss. This could be done in steps by growing layers, some of which can be microlithographed. An ingenious way is to focus a writing beam in the interior of an opal whose pores are filled with a photopolymerizable resist [100]. By means of two-photon polymerization a sharp boundary between exposed and unexposed resist can be achieved since this is a largely non linear process that occurs only in those regions very close to the focus where intensity rises sharply to values high enough to produce the polymerization. Scanning the sample allows any desired shape to be drawn.

Silica opals can be carved by means of a micro lapping machine in order to expose inner facets and to create microprisms [177]. This can be done by attaching small pieces of opal at the end of a thinned glass drawn from a rod. A large single crystal domain has previously been marked by large current pulses of the electron microscope for optical identification. The opal, that can be rotated through different axes in order to properly orient it, is then polished with a microgrinder.

2.3.4. Composites

The mixing of different materials is used to provide structures with properties that a single component alone cannot offer. Thus luminescence, magnetism, electrical conductivity, etc. are properties a bare opal does not possess but can be supplied by the guest material. There are two main ways to make the composites: by synthesizing composite spheres prior to building the opal or by synthesizing new materials in or around the pores of an assembled opal. In either case the mixed composition can be homogeneously distributed or shaped in shells in or around the spheres. Different approaches to produce each morphology have been explored and are explained next.

Although core-shell beads are most often used spheres with homogeneously distributed nanoparticle inclusions have also been synthesized. For instance, nanocomposite monodispersed silica spheres containing homogeneously dispersed silver quantum dots can be produced by the photochemical reduction of silver ions during hydrolysis of tetraethoxysilane in a microemulsion [178]. Depending on the timing, Ag quantum dots can be directed to different annuli within the silica spheres, as well as onto the silica sphere surface. Alternatively, inclusions such as silicon can be incorporated in the spheres, by means of implantation techniques, that confer new properties on the opals [179].

Magnetically controllable photonic crystals can be formed through the self-assembly of monodispersed magnetic colloidal spheres [180]. These contain nanoscale ferrite particles synthesized through emulsion polymerization. The diffraction from these PCs can be controlled through a magnetic field, which readily alters the PC lattice constant. Magnetic fields can be also used to strain the fcc lattice of superparamagnetic colloids polymerized within hydrogels or to control the photonic crystal orientation.

Particles of, mainly but not only, silica with an inner core or an outer, or internal shell, of other material may present many new properties that can find very important applications [181]. The commonest materials used can be classified into organic dyes, metals and semiconductors.

Dyes, owing to their efficient luminescence, are often used to mark the spheres [182] by growing a thin internal shell. This opens a wide range of uses such as imaging the colloid formation [163] or studying defects in buried structures. They are also very useful in the study of the effects of a PBG on internal lucent sources. However impregnation of an already assembled opal is much easier than dye coating at the sphere synthesis stage and more widely employed.

Metallodielectric systems have received much attention for their photonic interest and practical realizations have been tackled both from the metal-core or metal-cap particles and metal infiltrated opal. The preparation and chemical reactivity of silver-core silica particles have been described [183]; similarly, gold colloids can be homogeneously coated with silica using a primer to render the gold surface vitreophilic [184]. After the formation of a thin silica layer in aqueous solution, the particles can be transferred into ethanol for further growth using the Stöber method. These spheres can be finally arranged into opal structures [185]. The opals thus built can be infiltrated with a polymer and silica can be subsequently dissolved. The result is an inverse polymer opal with the original metal cores confined in the holes left by the removed silica [186]. Conversely, metal shells can be grown on silica [187]. Reductive growth and coalescence of the small gold nanoclusters attached to the functionalized surface leads to the formation of a closed gold layer. It is also possible to produce hollow gold shells by dissolution of the silica core and to coat the silica core gold shell particles with an additional outer silica shell. Layer by layer growth of polyelectrolyte combined with their exposure to suspensions of metal nanoparticles, that bind themselves to the polyelectrolyte layers, gives rise to organic spheres with dense shells of metallic nanoparticles [188]. Latex cores can be coated with magnetic shells for instance by adding ferrite nanoparticles polyelectrolyte layers to produce magnetic colloids [189].

Nanometer-sized semiconductor particles such as CdS can be synthesized and subsequently surrounded by a homogeneous silica shell. The coating procedure uses a surface primer to deposit a thin silica shell in water which can be grown thicker in ethanol [190]. In a similar fashion CdTe, CdSe, and core-shell CdSe/CdS nanocrystals synthesized in aqueous solutions, may be homogeneously incorporated as multiple cores into silica spheres of 40-80 nm size [191]. These can be grown into larger silica spheres by the Stöber technique which gives semiconductor-doped silica globules the desirable size in the submicrometer range.

Even though opals cannot exhibit a full PBG (regardless of the RI of the constituting spheres) the infiltration with materials of high enough RI may turn them into interesting structures, and more so when the underlying silica is removed: inverse opals (IO) can sustain a complete PBG. Infiltration with substances of RI higher than 2.8 times that of the opal spheres provides the structure with a full PBG [192]. Considering that silica has a RI of 1.45, this can be hardy achieved without removing the silica. This can be appreciated in Figure 11 where the gap width relative to the center position is plotted as a function of the infiltrated material RI for inverse fcc opals. Silicon, germanium, the III-Vs and a few chalcogenides are the only materials complying with this requirement.

There are, however, many interesting applications that do not require a complete PBG such as sensors and superrefractive devices. IOs have even found applications as templates to grow spherical particles in their pore networks. When the goal is the build up of a complete PBG system the choice of material is not only determined by the RI but electronic gap must be taken into consideration. Of course, a high RI is required but the material must be transparent. As can be seen in Figure 12 the dielectric function decreases nearly exponentially with electronic gap making it hardly possible to find suitable materials transparent beyond 2.5 eV, that is, below 500 nm. Elemental materials are the easiest to synthesize but scarce. Compounds often require the reaction between two precursors, which is compromised by the special environment where reaction takes place not only in the way reactants are supplied but the ease with which products are eliminated. A good account of several periodic porous solids was given by Stein in [193].

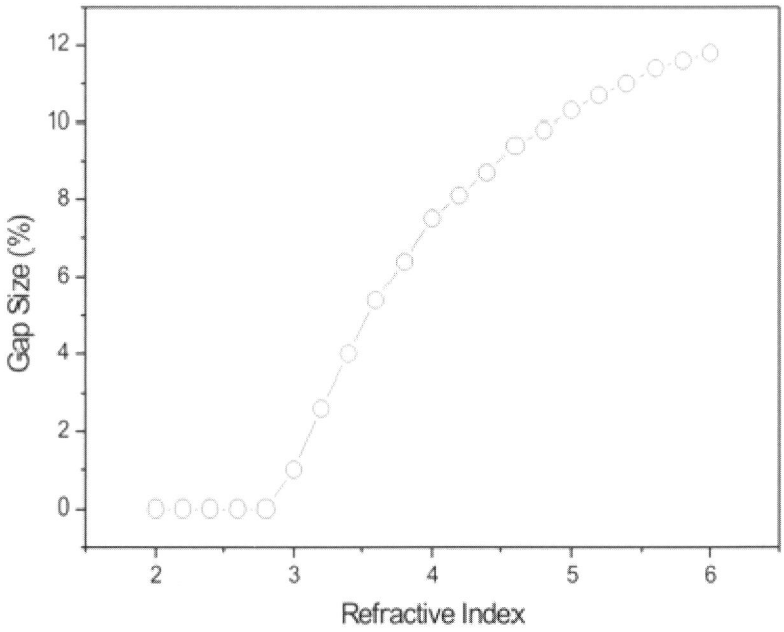

Figure 11. Gap width relative to center position in percent for composite opals as a function of the RI ratio between background and spheres. The filing fraction is that of compact fcc structure.

This section tries to contribute a summary of results regarding various materials of technological importance in photonics, along with the methods currently being used. Methods for the preparation of composites based on opals can be classified into three main categories. Wet methods are based on wet chemistry such as chemical bath deposition and electrodeposition whereas dry methods are mainly based on reactions in vapour phase. Injection of a molten substance (of low melting point) or infiltration with a suspension of nanoparticles can be classified as physical. One last consideration concerns the templating material itself. It must be born in mind that certain reaction conditions, such as high temperature, may not be withstood by organic materials as polymers. This can restrict their use as compared to silica – much more chemically inert and thermally stable – unless some tricks are applied, as we will see later on.

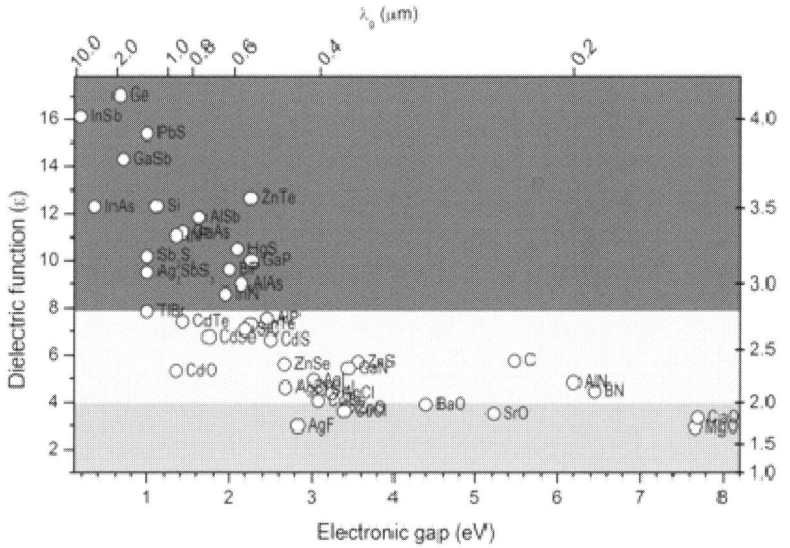

Figure 12. Some materials dielectric and electronic coordinates. As can be seen the greater the transparency range (larger electronic gap) the smaller the dielectric function. The threshold for complete PBG is indicated with shaded areas for fcc (white) and diamond (dark grey) opaline structures.

In the classification of compounds a division according to the electronic properties can be convenient.

Polymers can be easily synthesized within the pore network of opals just by impregnating with the monomer and then polymerizing it; inversion by dissolution of the original matrix leads to a perfect polymer replica [194]. Conducting polymers have a great interest and can be synthesized in the preformed opals by electro [195] or other polymerization and can show electrical conductivity and luminescent properties [196]. If the eletropolymerization is performed on a thick sedimented opal the thickness of the final infiltrated opal can be controlled by the reaction time [197]. Since polymerization occurs from the working electrode (substrate) and proceeds upwards increasing numbers of layers from the opal get cemented by the growing polymer. The reaction is stopped at any desired moment and excess layers (not bound by the polymer) can be washed. The original spheres can then be removed by dissolution. Polymers can be used for two-step templating in a "lost wax" mode whereby a replica from an opal (an IO) is used to template a second generation opal of nearly any desired material [133]. In this way it is possible to generate a wide variety of highly monodispersed inorganic, polymeric, and metallic solid and core-shell colloids, as well as hollow colloids with controllable shell thickness.

Hydrogels usually consist of randomly crosslinked polymer chains and contain a large amount of water filling interstitial spaces of the network, resulting in amorphous structures. Colloidal polystyrene spheres can be easily made to crystallize and if in their presence a hydrogel is polymerized that can be functionalized with a molecular recognition element to sense a given chemical agent, a specific sensor is created. In particular certain metal ions can be made to covalently attach to the hydrogel PC through complexes that cross-link the hydrogel and make it contract, blue-shifting its spectral features [198]. The same principle can be used to sense carbohydrates [199], redox state [200], pH or ionic strength [201]. Hydrogel encapsulated colloidal crystals can also find applications as mechano-chromic PCs [202]. Upon removal of water and reswelling, a completely water-free robust composite can be made. Glass transition temperature tunability (through monomer(s) composition) is granted that allows for a wide range of optical-mechanical properties.

Liquid crystals (LC) are a set of materials with tunable RI: their molecules are largely anisotropic so that their polarizability depends on the orientation of the molecules. The change of orientation varies the RI and, as a result, when included in a periodic dielectric the photonic band structure changes. They were shown to change the band structure of opal upon infiltration but, more interestingly, they could be used to enable a thermal tuning of the optical properties [203]. By controlling the temperature around the phase transition an abrupt change in RI can be attained. Thermal tuning is slow and electric control is more desirable. Based on the RI change caused by molecular reorientation in an electric field, a stop band can be driven by the field produced between the electrodes confining a thin opal film [204]. If an inverse, rather than a direct opal, is filled with LC a higher volume fraction is obtained which gives greater control over the optical properties [205]. In this case an abrupt change in the stop band is demonstrated. One further step can be made by infiltrating an IO with photochromic azo dye in LC. Illumination can induce the photochemical isomerization of the azo LC molecules from a rod-like shape into a bent one. This transition destabilizes the phase structure and randomizes the dielectric function distribution changing the band structure [206]. The most remarkable feature is that the isomerization is reversible by shining UV light on the structure which makes this a light driven switch.

Organic dyes have been long known for their efficient emission and used mostly as laser media. They are usually dissolved in organic solvents that facilitate impregnation and can be mixed with polymerizable compounds. All these properties have made them a perfect playground to probe the effects of PBGs on emitters embedded in PCs. Inhibition and stimulated emission were first observed in rhodamine dissolved in a polymer IO [207]. The effects of the PBG can also be observed in the luminescence lifetime. The stop band in these structures does not constitute a complete PBG and, as a consequence, shifts with angle with the corresponding effect on luminescence [208]. However, when the dielectric contrast is very strong the gaps widen and the angular dispersion may be negligible [209]. These structures may be made to lase [210] even in the absence of a complete PBG through a gap-enhanced distributed feed back process. Lasing can occur in different directions depending on the interplay between the dye emission band and the pseudogap [211]. Laser emission can even be obtained simultaneously in various directions if two dyes are infiltrated in the same piece of opal and properly pumped. Among the materials for this purpose, ZnO appears to be a very interesting one due to its remarkable properties as conductor and transparent in the visible range of the spectra. Recently, high quality ZnO inverted opals grown by CVD [212] and ALD [213] have been reported. The relative high ZnO refractive index provides enough dielectric contrast in the inverted structure to open additional pGs in the high energy regime [214] which has recently been used for enabling laser emission [215]. Its ample range of transparency allows it to host emitters in a broad spectral range from the infrared to the ultraviolet with no hindrance derived from absorption. All these reasons make ZnO an appropriate template material where quantum dots can be conformally self-assembled to modified their spontaneous emission making use of high energy photonic bands [216].

Dyes infiltrated in opals have served to demonstrate that intragap propagation is possible through excitations created by a highly polarizable medium with resonant frequency coupled to the PC [217]. These excitations are formed due to interaction between the (opal) Bragg gap and the (dye) polariton gap. As a result, the Bragg reflectivity plateau splits into two peaks, with a transmission spectral region in between. A similar effect was observed when an organic-inorganic layered perovskite [218] is introduced in the opal matrix such that the exciton couples to the opal stop band demonstrating a Rabi splitting.

The tuning or control of the photonic response with light is a major technological goal for whose realization several routes have been followed. A molecular aggregate is a collection of molecules, in which the individual molecules are closely coupled and respond in-phase to an external EM field. Collective coherent response in the molecular aggregate contributes a giant oscillation. Apparently, the giant oscillation can induce dramatic changes in the RI around the resonant frequency. Thus, a photochromic dye aggregate in which the resonance frequency of the aggregated state is close to the PBG, while that of the non aggregated state is far from the gap, would enable one to induce a large change in the stop-band by illumination. Photo-tunable photonic band gap crystals constituted from photochromic dyes and silica spheres have been fabricated by evaporating dye into the voids of opals. The dye undergoes photoisomerization upon illumination with UV light. The RI change induced by aggregate formation is quite large compared with the change induced by an orientation change of liquid crystal molecules or by the photochromic effect at the single-molecule level. The induction of such large changes of refractive indices provides a new approach toward the synthesis of materials with both large and small refractive indices [219]. Crystallization of a colloid depends on the balance of electrostatic forces between spheres. These forces can be strongly screened by free ions in the solution. Thus, dye molecules such as malachite green [220] that can be photoionized may permit a control on crystal order-disorder phase transitions induced by ultraviolet light irradiation.

Ferroelectric materials other than LCs have been synthesized within opals owing to the enormous increase in dielectric constant expected at the phase transition (divergent for materials with a second order transition). Thus a giant enhancement of the dielectric constant (up to $\sim 10^8$ at 100 Hz) in sodium nitrite infiltrated opals was found [221]. Barium [222], strontium, and lead titanate [223] in silica opals were also sol-gel synthesized with modest effects on the photonic properties. The same is true for barium titanate IOs [224]. This, however, is not surprising since divergence occurs only for static (or very low frequency) dielectric constant but not for the optical limit. Lithium niobate ($LiNbO_3$) is one of the most interesting materials for electrooptic and photorefractive applications [225]. $LiNbO_3$ IOs can be prepared by using polyelectrolyte multilayer coated polystyrene opals as templates for a sol-gel process and removing the initial opal by calcination.

Insulators have been explored in order to extend their use, from other fields where they have long standing applications, to opening areas in photonics. For instance, interesting behaviour should be obtainable for carbon-based structures with a dimensional scale between fullerenes or nanotubes and graphitic microfibers. Thus CVD produced carbon IO provide examples of both dielectric and metallic optical photonic crystals [226]. Most metal oxides, due to their insulator character, present a fairly low and non dispersive RI. They have been synthesized in the opal, mainly by electrochemical deposition or by hydrolysis of alkoxides. For instance zinc oxide [227], tungsten trioxide [228], and lead oxide [229] were obtained by electrochemical deposition and tin oxide from hydrolysis [230]. Among them, titanium dioxide has the highest dielectric constant. Its RI is not enough, however, to open a full PBG in the IO. More so considering that the phase resulting from the synthesis is the lower RI one: anatase. The high RI phase, rutile, may be obtained by thermal annealing of anatase but this process severely disrupts the structure. Nevertheless its ease of preparation (hydrolysis of titanium alkoxide, chloride or other precursors) and the range of studies it allows has made it very popular. When a diluted alkoxide is employed in several cycles shell structures grow around the templating spheres [231]. After infiltration with CdSe quantum dots, these systems have shown enhancement and inhibition of spontaneous emission provoking appreciable changes in the semiconductors emission life-time [232]. As an alternative, the use of highly concentrated alkoxides in a slow, one cycle process where a gel is initially formed in the pores, leads, upon calcination, to a titania skeleton structure [233]. This novel structure has very interesting properties as is the build up of two complete PBGs as opposed to common IOs where only one PBG occurring between the 8th and 9th bands is found. Physical methods have also been developed based on a dipping process whereby coassembly of monodispersed polystyrene spheres and nanoparticles of silica and titania from aqueous suspension is followed by a calcination treatment. It is possible to continuously adjust the position of the stop band of the IO by controlling the nanoparticle volume fractions [234]. Recently a method was described through which nanoparticles in the rutile phase were synthesized at room temperature from titanium tetrachloride [235]. These particles were infiltrated in the latex opal by centrifugation and the template removed by calcination without affecting the titania crystallographic phase.

Most materials introduced in opals as light sources have broad luminescence bands that cannot be completely blocked by the opal stop band. On the contrary, rare earth ions present narrow bands that may overcome this concern. These ions can be obtained from their salts (like europium from its nitrate) to outwardly impregnate assembled opals [236] or can be integrated in the opal spheres adding rare earth salt (like erbium from its chloride) at the time of synthesis [237]. Alternatively, fluorescent erbium-doped titania nanoparticles can be synthesized through a sol-gel process followed by peptization and hydrothermal treatment [238] that can, subsequently, be used to infiltrate previously assembled opals.

Magnetic materials have been introduced in opals not only but mostly for the magnetooptical properties they develop in this periodic environment [239]. By proper spatial arrangement of magnetic and dielectric components it should be possible to construct magnetic photonic crystal with strong spectral nonreciprocity: $\omega(\mathbf{k}) \neq \omega(-\mathbf{k})$. This spectral asymmetry, in turn, results in a number of interesting phenomena, in particular, one-way transparency: the magnetic photonic crystal, being transparent for a wave of frequency ω, becomes opaque for the same frequency propagating in the opposite direction. The practical realizations comprise magnetic nanoparticles precipitation from aqueous ferrofluid to form composite (then inverse) opals [240]; synthesis of polyferrocenysilane for magnetic ceramic IO [241]; or even ferromagnetic perovskite manganite, a colossal magnetoresistance material [242]. Potential applications for such a magnetoresistive ferromagnetic material are diverse, including a field tunable optical switch. A Faraday rotator has been fabricated in the form of a colloidal suspension with dysprosium nitrate dissolved [243]. An enhancement of the rotation is found in the stop band region due to multiple scattering.

The very interesting properties that arise when metals are involved (often associated with plasmons and negative dielectric function) make them of high technological interest. Originally infiltration of opals with metals such as gold was attained by filtration through a smooth membrane with pores that retained the latex and the gold particles but allowed flux of water [244]. Another physical method is the assembly of latex microspheres in the presence of concentrated gold nanoparticles. Thick multilayer latex crystals grew upon drying the films, probably due to the combination of the increasing latex volume fraction and convective assembly [245]. Periodic nanosphere arrays can be fabricated by a sequential templated infiltration. First a carbon IO is produced where molten metals are infiltrated at close to the melting point and pressures of 1-2 kbar resulting in nanosphere crystals for elemental metals and semiconductors (Pb, Bi, Sb, and Te) and thermoelectric alloys (Bi-Sb, Bi-Te, and Bi-Te-Se) [246].

Thiol surface-functionalized silica opals immersed in a toluene solution containing gold nanocrystals (that affix to the colloidal surfaces at the thiol sites) are partially covered with gold. These templates are then immersed in electroless deposition baths, and metal formation in the pores readily occurs [247].

Electrochemical deposition from a metal chloride in a typical three-electrode set up is, by far, the most widely used method. To grow gold in opals, the sample (assembled on an ITO substrate) is masked (to confine growth to the sample region) and suspended in the electrolyte solution with Pt counter electrode and a potentiostatical deposition is performed [248]. Nickel and gold electrodeposition can also be performed (on opals copper sputtered on the back) galvanostatically resulting in a dark opalescent metal 100 μm thick membrane [249]. Platinum, palladium and cobalt IOs are also prepared by electrochemical deposition on polystyrene latex spheres self-assembled on gold electrodes in a matter of minutes. In this case the thickness of the metal film is controlled through the charge passed. The polystyrene spheres are fully removed by washing in toluene [250]. Recently, Zn IOs have been prepared with a controlled morphology on semiconductor (silicon or ITO) substrates using similar techniques [251]. Metal IOs can be used for sequential templating. In the first step an opal slab is electrochemically infiltrated with nickel. After removal of the opal template the nickel mesh is slowly oxidized. The resulting poorly conducting nickel oxide mesh is then used as a nanomold for the electrochemical growth of a gold nanosphere array. Finally, the nickel oxide template is removed in dilute sulfuric acid to produce an array of gold nanospheres [246].

Mechanical processing of pre-formed three-dimensional metal IO has lead to lower dimensional structures [252]. One-dimensional photonic crystals (corrugated nanowires), and two-dimensional metal-mesh photonic crystals of between one and fifty sheets may be made by this process. Possible applications that could be enabled include fibrous photonic crystal colorants for plastics and 1D photonic crystal wires.

Semiconductors have been introduced in opal structures owing to their optical and electronic properties. The reduced electronic band gap makes them interesting for electronic and optoelectronic applications but, at the same time, implies (recall the correlation in Figure 12) an elevated RI which makes them good candidates for PBG applications. Very different synthesis methods have been used to produce composites and IOs depending on their elemental or compound nature [253]. Different application have been proposed depending on their optoelectronic characteristics.

Silicon is a very interesting material with current microelectronics technology in mind. It would be desirable to develop PCs that could be directly integratable with current technology. Besides its RI is well beyond the threshold for complete PBG opening in IO configuration and it is transparent in the communications window (*ca.* 1.5 μm). Considering that opal templating competes favourably with most other methods of PC fabrication, as far as cost, ease and large scale production is concerned, it is not strange that this has been one of the first systems where a complete PBG has been realized in the optical spectrum. The first silicon IO was synthesized by CVD in silica opaline matrices obtained by sedimentation in ethylene glycol aqueous solution [254]. The precursor used was disilane which decomposes at elevated temperature (*ca.* 300 °C) but not so elevated that the underlying opal could not withstand it. The reaction produces amorphous silicon shell around the spheres and, if desired, it can be crystallized by thermal annealing. A typical example is shown in Figure 13 where thin silicon shells can be seen after dissolving the underlying silica from the templating opal. Sintering of the initial opal provides the connectivity required for the etchant to be able to dissolve the silica when inverting the structure. The opal was designed in order that the complete PBG of the final structure fell in the optical communication wavelength. Following this approach, but using a different method for the assembly of the opal, thin film IO were produced [255]. Similarly IO fibers were produced by growing silicon within opals assembled in confined environments such as microchannels where the opals assembled have well controlled orientation [256]. The existence of a complete PBG in these systems was further supported by measuring the reflectivity in the principal directions of propagation [257]. It was found that a frequency band existed common to all the propagation directions where transmission is prevented.

The mentioned CVD process is made at temperatures (around 350-400°C) which limits the template materials, silica being usually employed. However, 3D polymeric structures (usually PS or PMMA in the case of opals, but others like SU-8 in holographic photonic crystals or

DLW structures) present some advantages with respect to silica templates: different symmetry, more versatility, and, often, superior optical quality when opals are grown by vertical deposition. This was initially dealt with by infiltrating the air cavities with silica through a sol-gel process at room temperature to produce resilient silica replica for further processing. This technique allows growing silicon in the inner surface of the silica inverse opals (MISO technique) [258]. This sol-gel infiltration, however, lends little or no control over infiltration.

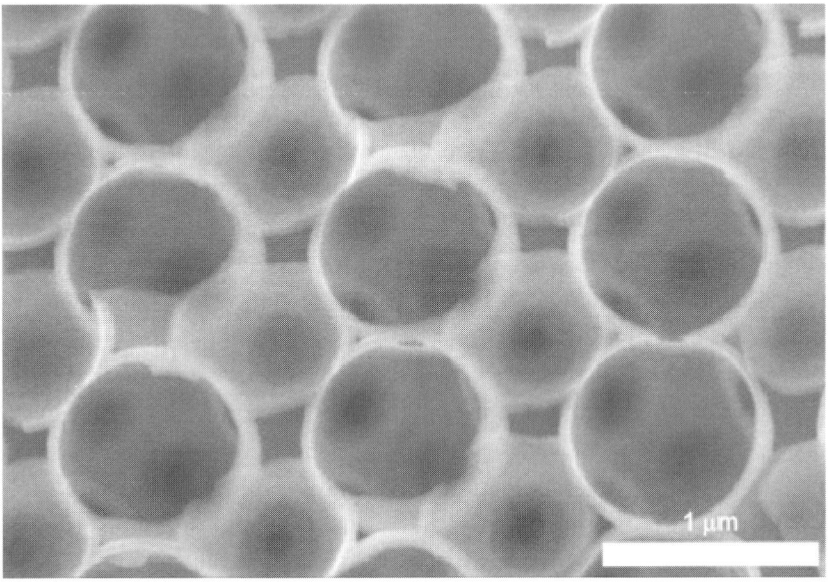

Figure 13. SEM image of a silicon inverse opal with a very low filing fraction.

A different strategy was recently demonstrated that allows not only silicon infiltration in polymer 3D templates but a fine tuning of their photonic crystal properties [259]. The use of silica CVD permits a fine control over the degree of infiltration in polymer opals which can be used to generate a novel topology consisting of two silicon concentric leaves. These 3D macroporous silicon structures may be of importance due to their versatile photonic properties tailored through design (Figure 14).

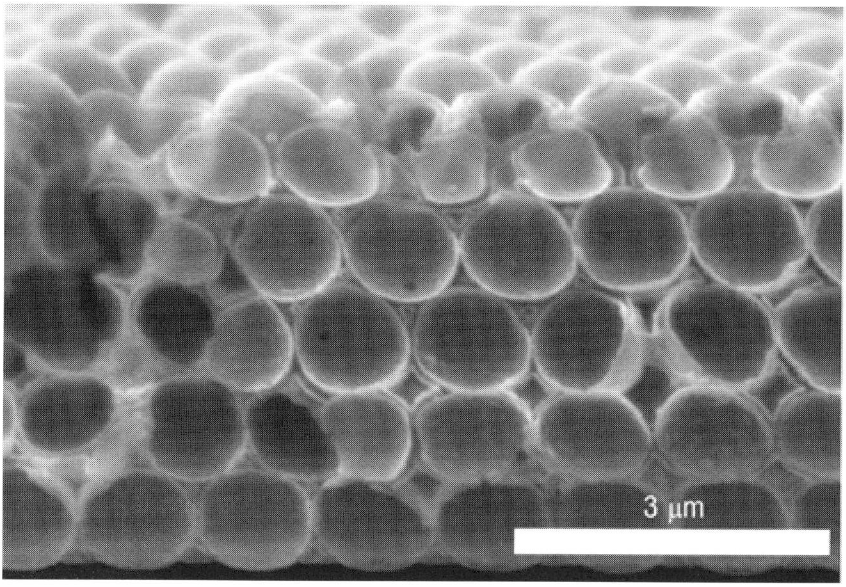

Figure 14. SEM image of an inverse opal composed by two concentric silicon layers separated by an air gap.

The material with the highest RI in the optical spectrum is germanium. Its absorption edge is small, however, which restricts its use to wavelengths greater than about 1.8 μm. This, in turn, sets the periodicity of the opals where useful structures can be built to large values, so that the spheres in a complying fcc lattice must be larger than 1.1 μm in diameter. Although germanium can be synthesized through a two step process of hydrolysis of an alkoxide (to form GeO_2) and reduction to metallic Ge; this procedure yields granular material and is laborious [260]. At variance, the CVD method grows layer by layer high quality material and is quicker and simpler [261]. The precursor can be germane that has a convenient decomposition temperature (270 ºC) and the infiltration can be controlled by the precursor pressure or baking time and temperature. Electrodeposition has also proven to work for this synthesis. Germanium tetrachloride is used and amorphous Ge is obtained that can be crystallized by thermal annealing [262]. Layered Si/Ge structures can be grown by CVD in opals. The control of semiconductor shell thickness and sequence, along with the selective oxidation, added to ability to etch the oxides opens a vast range of possibilities in PBG engineering. Thus, structures with mixed layers of silicon, germanium, both oxides and air can be easily and precisely created where double complete PBGs can be hosted [263].

II-V semiconductors have extremely interesting optoelectronic properties. They have luminescence, large dielectric functions and non linear coefficients and a mature technological backup. They are commonly grown by molecular beam epitaxy or CVD but their synthesis is difficult in the opal environment. This is so since two molecules are involved and the reaction must happen in a very intricate network of interstices; single source precursors have not obtained much success. Indium phosphide [264] and gallium phosphide [265] have been produced by atmospheric pressure metal organic CVD, showing, when inverted, an anomalous optical response due to their peculiar morphology [266,267]. Only possible by electrochemical deposition [268], very recently gallium arsenide IOs have been prepared by CVD [269]. The depositions of III-V semiconductor films on silica opals was much more difficult to achieve than those of II-VIs mainly because of the more covalent (less ionic) character of the former.

Gallium nitride has become a very interesting material for its wide band gap and thermal stability. Its introduction in opals has been realized by direct synthesis from metal gallium or its oxide (liquid at room temperature) annealed in the presence of nitrogen hydrides [270].

Selenium IOs can be prepared by imbibing a direct silica opal with molten selenium under pressure. Pressure is required because selenium does not wet silica [271]. Chalcogenides such as cadmium sulphide [272] or selenide [273] were first physically infiltrated in opals by soaking the porous structures with suspensions of quantum dots that presented optical gain near the PC stop band edge [274]. Evaporation of the solvent results in precipitation and formation of the composite. Subsequently, the template is removed and a three-dimensionally patterned material consisting solely of densely packed nanocrystals is obtained. The nanocrystals can be sintered (nanocrystals melt at temperatures much lower than the bulk) to form a macroporous bulk semiconductor. Chemical bath deposition methods, where two precursors supply the ions in successive impregnations, can be used to control the degree of infiltration allowing to produce cadmium sulphide [275] or antimony sulphide (Sb_2S_3) inverse opals [276]. However, electrochemical synthesis produces better quality material both of CdS and CdSe [277]. Furthermore, the method has proved itself versatile enough to produce virtually any selenide, telluride or sulphide. Only detailed studies of deposition parameters (current density, deposition time, concentrations of electrolytes, solvents, and temperatures) must be carried out first [268]. Gas-solid reaction techniques where a gas is passed through a pretreated opal has facilitated the synthesis of tin sulphide [278] and, lately antimony sulphide, the only semiconductor with silicon and germanium for which a complete PBG has been obtained [279]. The latter is a very interesting material with one of the highest RI available (enough to open a complete PBG) and a broad electronic gap. With these conditions a complete PBG has been demonstrated in visible that might set a record as the highest energy gap in opal based structures.

3. Summary

The field of photonic crystals is in continuous expansion and a review ages swiftly. We have tried to give a comprehensive overview of the state of the art with emphasis in the use of colloidal systems for photonics applications. Many groups in the world are researching in this field in different EM ranges, with different aims and employing different approaches. The use of self assembly techniques for the optical range is very widespread and the relative materials science involved is very high. New materials are being explored every day and old materials are being found useful for emerging needs. Giving a classification has become very difficult because metals, semiconductors, insulators can be found among organic, inorganic, and hybrid materials that present themselves in multiple forms, phases and flavours and can be viewed differently according to which of their properties they are chosen for. The bottom line is that new ways will emerge in which research will profit from very different sources of knowledge.

Some profound questions remain unanswered such as will the ultimate PBG device be all-optical or will it be electronic + optical? Most probably the answer will be the latter because the boundaries of the device will themselves be diffuse. More so if we consider the very likely integration in the bio and molecular worlds, towards which the field will inevitably drift. This will surely provide new questions for those that think that there is no new science left to unveil, only development. There is also the question whether 2D will be enough or 3D will be needed. It is undeniable that, if complete photon confinement is required for some application, 2D techniques will surely be not enough. 3D systems other than self assembled structures entail rather expensive production strategies and cost is a very strong constraint. Almost certainly we shall see the merging of techniques coming from the 2D and microelectronics background and techniques from self assembly.

Acknowledgements

This work was partially funded by the Spanish Ministry of Science and Education under contract MAT2003-01237 and NAN2004-08843, the EU under contract IST-511616 NoE PhOREMOST. A.B. also acknowledges *Ramón y Cajal* Programme.

References

1. E. Yablonovitch, Phys. Rev. Lett. 58, 2059 (1987).
2. S. John, Phys. Rev. Lett. 58, 2486 (1987).
3. H.E. Hinton, D. F. Gibbs. J. Insect. Physiol. 15, 959 (1969).
4. J.V. Sanders, Nature 204, 1151 (1964).
5. C. Kittel, "Introduction to Solid State Physics", Third Edition, John Wiley, New York, (1968).
6. I. I. Tarhan, G. H. Watson, Phys. Rev. B 54, 7593 (1996).
7. K. W. K. Shung, Y. C. Tsai, Phys. Rev. B 48, 11265 (1993).
8. K. Sakoda, "Optical properties of Photonic Crystals" Springer Series in Optical Sciences Vol. 80. Berlin (2001).
9. It is worth noticing that the point X ((001) direction) does not show such a gap at the edge of the BZ which is in agreement with a simple prediction based on structure factor as in X-ray scattering. However, if dielectric contrast is strongly increased a small gap opens signaling the failure of that theoretical approach.

10. H.S. Sözüer, J.W. Haus, and R. Inguva. (1993). Phys. Rev. B 45, 13962.
11. N. W. Ashcroft, N. D. Mermin, (1976) "Solid State Physics", Rinehart & Wilson, Philadelphia.
12. Fluctuations of the EM field may induce the emission from an excited atom and the transfer of its energy into the EM field. The opposite is not possible since fluctuations can not excite a ground state atom if there is no energy in the field.
13. S. John, J. Wang, Phys. Rev. Lett. 64, 2418 (1990).
14. W. L. Vos, M. Megens, C. M. vanKats, P. Bosecke, Langmuir 13, 6004 (1997).
15. W. M. Robertson, G. Arjavalingam, R. D. Meade, K. D. Brommer, A. M. Rappe, J. D. Joannopoulos, Phys. Rev. Lett. 68, 2023 (1992).
16. A. Imhof, W. L. Vos, R. Sprik, A. Lagendijk, Phys. Rev. Lett. 83, 2942 (1999).
17. W. L. Vos, H. M. van Driel, Phys. Lett. A 272, 101 (2000).
18. F. Garcia-Santamaria, J. F. Galisteo-Lopez, P. V. Braun, C. Lopez, Phys. Rev. B 71, 195112 (2005).
19. A remarkable feature of this gap is that its position is determined by very linear dispersion bands; as a consequence its position can be fairly approximated by an effective RI. This fact permits the use of Snell law to determine the wavevector inside the PC and the mapping of the bands.
20. J. F. Galisteo-Lopez, C. Lopez, Phys. Rev. B 70, 035108 (2004).
21. E. R. Brown, C. D. Parker, E. Yablonovitch, J. Opt. Soc. Am. B-Opt. Phys. 10, 404,(1993).
22. J. G. Fleming, S. Y. Lin, I. El-Kady, R. Biswas, K. M. Ho, Nature 417, 52 (2002).
23. C. F. Lin, C. H. Chao, L. A. Wang, W. C. Cheng, J. Opt. Soc. Am. B-Opt. Phys. 22, 1517 (2005).
24. B. Oregan, M. Gratzel, Nature 353, 737 1991).
25. S. Nishimura, N. Abrams, B. A. Lewis, L. I. Halaoui, T. E. Mallouk, K. D. Benkstein, J. van de Lagemaat, A. J. Frank, J. Am. Chem. Soc. 125, 6306 (2003).
26. A. Mihi, F. J. Lopez-Alcaraz, H. Miguez, Appl. Phys. Lett. 88, 193110 (2006).
27. B. S. Song, S. Noda, T. Asano, Y. Akahane, Nat. Mater. 4, 207 (2005).
28. M. Loncar, T. Yoshie, A. Scherer, P. Gogna, Y. M. Qiu, Appl. Phys. Lett. 81, 2680 (2002).
29. O. Painter, R. K. Lee, A. Scherer, A. Yariv, J. D. O'Brien, P. D. Dapkus, I. Kim, Science 284, 1819 (1999).
30. H. G. Park, S. H. Kim, S. H. Kwon, Y. G. Ju, J. K. Yang, J. H. Baek, S. B. Kim, Y. H. Lee, Science 305, 1444 (2004).
31. E. Miyai, K. Sakai, T. Okano, W. Kunishi, D. Ohnishi, S. Noda, Nature 441, 946 (2006).
32. S. John, K. Busch, J. Lightwave Technol. 17, 1931 (1999).

33. S.-Y. Lin, E. Chow, V. Hietala, P. R. Villeneuve, J. D. Joannopoulos, Science 282, 274 (1998).
34. H. Benisty, D. Labilloy, C. Weisbuch, C. J. M. Smith, T. F. Krauss, D. Cassagne, A. Beraud, C. Jouanin, Appl. Phys. Lett. 76, 532 (2000).
35. S. H. Kim, H. Y. Ryu, H. G. Park, G. H. Kim, Y. S. Choi, Y. H. Lee, J. S. Kim, Appl. Phys. Lett. 81, 2499 (2002).
36. K. Hosomi, T. Katsuyama, IEEE J. Quantum Electron. 38, 825, (2002).
37. E. Moreno, F. J. Garcia-Vidal, L. Martin-Moreno, Phys. Rev. B 69, 121402 (2004).
38. P. Russell, Science 299, 358 (2003).
39. J. N. Winn, Y. Fink, S. H. Fan, J. D. Joannopoulos, Opt. Lett. 23, 1573 (1998).
40. Y. Fink, J. N. Winn, S. H. Fan, C. P. Chen, J. Michel, J. D. Joannopoulos, E. L. Thomas, Science 282, 1679 (1998).
41. B. Temelkuran, S. D. Hart, G. Benoit, J. D. Joannopoulos, Y. Fink, Nature 420, 650 (2002).
42. M. Meier, A. Mekis, A. Dodabalapur, A. Timko, R. E. Slusher, J. D. Joannopoulos, O. Nalamasu, Appl. Phys. Lett. 74, 7 (1999).
43. S. Noda, M. Yokoyama, M. Imada, A. Chutinan, M. Mochizuki, Science 293, 1123 (2001).
44. M. Notomi, Phys. Rev. B 62, 10696 (2000).
45. H. Kosaka, T. Kawashima, A. Tomita, M. Notomi, T. Tamamura, T. Sato, S. Kawakami, Appl. Phys. Lett. 74, 1370 (1999).
46. L. J. Wu, M. Mazilu, T. Karle, T. F. Krauss, IEEE J. Quantum Electron. 38, 915 (2002).
47. T. Ochiai, J. Sanchez-Dehesa, Phys. Rev. B, 64, 245113 (2001).
48. V. G. Veselago, Soviet Physics Uspekhi-Ussr 10, 509 (1968).
49. R. A. Shelby, D. R. Smith, S. Schultz, Science 292, 77 (2001).
50. J. B. Pendry, Phys. Rev. Lett. 85, 3966 (2000).
51. S. Linden, C. Enkrich, M. Wegener, J. F. Zhou, T. Koschny, C. M. Soukoulis, Science 306, 1351 (2004).
52. G. Dolling, C. Enkrich, M. Wegener, C. M. Soukoulis, S. Linden, Science 312, 892 (2006).
53. M. C. K. Wiltshire, J. B. Pendry, I. R. Young, D. J. Larkman, D. J. Gilderdale, J. V. Hajnal, Science 291, 849 (2001).
54. S. Enoch, G. Tayeb, P. Sabouroux, N. Guerin, P. Vincent, Phys. Rev. Lett. 89, 213902 (2002).
55. B. A. Munk, Frequency Selective Surfaces, Theory and Design, Wiley, New York (2000).

56. W. J. Wen, L. Zhou, J. S. Li, W. K. Ge, C. T. Chan, P. Sheng, Phys. Rev. Lett. 89, 223901 (2002).
57. Y. S. Chan, C. T. Chan, Z. Y. Liu, Phys. Rev. Lett. 80, 956 (1998).
58. C. J. Jin, B. Y. Cheng, B. Y. Man, Z. L. Li, D. Z. Zhang, S. Z. Ban, B. Sun, Appl. Phys. Lett. 75, 1848 (1999).
59. M. Hase, H. Miyazaki, M. Egashira, N. Shinya, K. M. Kojima, S. Uchida, Phys. Rev. B 66, 214205 (2002).
60. M. W. Takeda, S. Kirihara, Y. Miyamoto, K. Sakoda, K. Honda, Phys. Rev. Lett. 92, 093902 (2004).
61. E. Yablonovitch, T. J. Gmitter, K. M. Leung, Phys. Rev. Lett. 67, 2295 (1991).
62. C. C. Cheng, A. Scherer, J. Vac. Sci. Technol. B 13, 2696 (1995).
63. K. M. Ho, C. T. Chan, C. M. Soukoulis, R. Biswas, M. Sigalas, Solid State Commun. 89, 413 (1994).
64. H. Yan, R. He, J. Johnson, M. Law, R. J. Saykally, P. Yang, J. Am. Chem. Soc. 125, 4728 (2003).
65. V. I. Kopp, B. Fan, H. K. M. Vithana, A. Z. Genack, Opt. Lett. 23, 1707 (1998).
66. R. Jakubiak, T. J. Bunning, R. A. Vaia, L. V. Natarajan, V. P. Tondiglia, Adv. Mater. 15, 241 (2003).
67. T. C. Wang, R. E. Cohen, M. F. Rubner, Adv. Mater. 14, 1534 (2002).
68. H. A. Lopez, P. M. Fauchet, Appl. Phys. Lett. 77, 3704 (2000).
69. S. Chan, S. R. Horner, P. M. Fauchet, B. L. Miller, J. Am. Chem. Soc. 123, 11797 (2001).
70. J. C. Knight, T. A. Birks, P. S. Russell, D. M. Atkin, Opt. Lett. 21, 1547 (1996).
71. J. C. Knight, J. Broeng, T. A. Birks, P. S. J. Russel, Science 282, 1476 (1998).
72. M. A. van Eijkelenborg, M. C. J. Large, A. Argyros, J. Zagari, S. Manos, N. A. Issa, I. Bassett, S. Fleming, R. C. McPhedran, C. M. de Sterke, N. A. P. Nicorovici, Opt. Express 9, 319 (2001).
73. Z. F. Ren, Z. P. Huang, J. W. Xu, J. H. Wang, P. Bush, M. P. Siegal, P. N. Provencio, Science 282, 1105 (1998).
74. K. Kempa, B. Kimball, J. Rybczynski, Z. P. Huang, P. F. Wu, D. Steeves, M. Sennett, M. Giersig, D. Rao, D. L. Carnahan, D. Z. Wang, J. Y. Lao, W. Z. Li, Z. F. Ren, Nano Lett. 3, 13 (2003).
75. X. D. Wang, C. Neff, E. Graugnard, Y. Ding, J. S. King, L. A. Pranger, R. Tannenbaum, Z. L. Wang, C. J. Summers, Adv. Mater. 17, 2103 (2005).
76. K. Inoue, M. Wada, K. Sakoda, A. Yamanaka, M. Hayashi, J. W. Haus, Jpn. J. Appl. Phys. Part 2 - Lett. 33, L1463 (1994).
77. A. Rosenberg, R. J. Tonucci, E. A. Bolden, Appl. Phys. Lett. 69, 2638 (1996).

78. K. Y. Lim, D. J. Ripin, G. S. Petrich, L. A. Kolodziejski, E. P. Ippen, M. Mondol, H. I. Smith, P. R. Villeneuve, S. Fan, J. D. Joannopoulos, J. Vac. Sci. Technol. B 17, 1171 1999).
79. J. R. Wendt, G. A. Vawter, P. L. Gourley, T. M. Brennan, B. E. Hammons, J. Vac. Sci. Technol. B 11, 2637 (1993).
80. S. Noda, A. Chutinan, M. Imada, Nature 407, 608 (2000).
81. S. Y. Lin, J. G. Fleming, D. L. Hetherington, B. K. Smith, R. Biswas, K. M. Ho, M. M. Sigalas, W. Zubrzycki, S. R. Kurtz, J. Bur, Nature 394, 251 (1998).
82. S. Noda, N. Yamamoto, A. Sasaki, Jpn. J. Appl. Phys. Part 2 - Lett. 35, L909 (1996).
83. S. Noda, K. Tomoda, N. Yamamoto, A. Chutinan, Science 289, 604 (2000).
84. Y. N. Xia, G. M. Whitesides, Annu. Rev. Mater. Sci. 28, 153 (1998).
85. S. Y. Chou, P. R. Krauss, P. J. Renstrom, Science 272, 85 (1996).
86. U. Gruning, V. Lehmann, C. M. Engelhardt, Appl. Phys. Lett. 66, 3254 (1995).
87. F. Müller, A. Birner, J. Schilling, U. Gösele, C. Kettner, P. Hänggi, Phys. Status Solidi A-Appl. Res. 182, 585 (2000).
88. V. Berger, O. GauthierLafaye, E. Costard, J. Appl. Phys. 82, 5300 (1997).
89. S. Shoji, S. Kawata, Appl. Phys. Lett. 76, 2668 (2000).
90. H. B. Sun, A. Nakamura, S. Shoji, X. M. Duan, S. Kawata, Adv. Mater. 15, 2011 (2003).
91. X. Wang, C. Y. Ng, W. Y. Tam, C. T. Chan, P. Sheng, Adv. Mater. 15, 1526 (2003).
92. A. Feigel, Z. Kotler, B. Sfez, A. Arsh, M. Klebanov, V. Lyubin, Appl. Phys. Lett. 77, 3221 (2000).
93. S. Kawakami, T. Kawashima, T. Sato, Appl. Phys. Lett. 74, 463 (1999).
94. T. Sato, K. Miura, N. Ishino, Y. Ohtera, T. Tamamura, S. Kawakami, Opt. Quantum Electron. 34, 63 (2002).
95. O. Hanaizumi, M. Saito, Y. Ohtera, S. Kawakami, S. Yano, Y. Segawa, E. Kuramochi, T. Tamamura, S. Oku, A. Ozawa, Opt. Quantum Electron. 34, 71 (2002).
96. M. Campbell, D.N. Sharp, M.T. Harrison, R.G. Denning, A.J. Turberfield, Nature 404, 53 (2001).
97. S. Yang, M. Megens, J. Aizenberg, P. Wiltzius, P. M. Chaikin, W. B. Russel, Chem. Mat. 14, 2831 (2002).
98. Y. V. Miklyaev, D. C. Meisel, A. Blanco, G. von Freymann, K. Busch, W. Koch, C. Enkrich, M. Deubel, M. Wegener, Appl. Phys. Lett. 82, 1284 (2003).
99. Y. K. Pang, J. C. W. Lee, C. T. Ho, W. Y. Tam, Optics Express 14, 9113 (2006).
100. W. M. Lee, S. A. Pruzinsky, P. V. Braun, Adv. Mater. 14, 271 (2002).

101. J. Scrimgeour, D. N. Sharp, C. F. Blanford, O. M. Roche, R. G. Denning, A. J. Turberfield, Adv. Mater. 18, 1557 (2006).
102. J. S. King, E. Graugnard, O. M. Roche, D. N. Sharp, J. Scrimgeour, R. G. Denning, A. J. Turberfield, C. J. Summers, Adv. Mater. 18, 1561 (2006).
103. J. H. Moon, S. Yang, W. T. Dong, J. W. Perry, A. Adibi, S. M. Yang, Optics Express 14, 6297 (2006).
104. S. Kawata, H. B. Sun, T. Tanaka, K. Takada, Nature, 412, 697 (2001).
105. M. Deubel, G. Von Freymann, M. Wegener, S. Pereira, K. Busch, C. M. Soukoulis, Nat. Mater. 3, 444 (2004).
106. M. Deubel, M. Wegener, A. Kaso, S. John, Appl. Phys. Lett. 85, 1895 (2004).
107. K. K. Seet, V. Mizeikis, S. Matsuo, S. Juodkazis, H. Misawa, Adv. Mater. 17, 541 (2005).
108. M. Deubel, M. Wegener, S. Linden, G. von Freymann, S. John, Opt. Lett. 31, 805 (2006).
109. S. Wong, M. Deubel, F. Perez-Willard, S. John, G. A. Ozin, M. Wegener, G. von Freymann, Adv. Mater. 18, 265 (2006).
110. N. Tetreault, G. von Freymann, M. Deubel, M. Hermatschweiler, F. Perez-Willard, S. John, M. Wegener, G. A. Ozin, Adv. Mater. 18, 457 (2006).
111. G. Feiertag, W. Ehrfeld, H. Freimuth, H. Kolle, H. Lehr, M. Schmidt, M. M. Sigalas, C. M. Soukoulis, G. Kiriakidis, T. Pedersen, J. Kuhl, W. Koenig, Appl. Phys. Lett. 71, 1441 (1997).
112. C. Cuisin, A. Chelnokov, D. Decanini, D. Peyrade, Y. Chen, J. M. Lourtioz, Opt. Quantum Electron. 34, 13 (2002).
113. K. Robbie, M. J. Brett, J. Vac. Sci. Technol. A-Vac. Surf. Films 15, 1460 (1997).
114. S. R. Kennedy, M. J. Brett, O. Toader, S. John, Nano Lett. 2, 59 (2002).
115. O. Toader, S. John, Science 292, 1133 (2001).
116. Y. Fink, A. M. Urbas, M. G. Bawendi, J. D. Joannopoulos, E. L. Thomas, J. Lightwave Technol. 17, 1963 (1999).
117. A. M. Urbas, M. Maldovan, P. DeRege, E. L. Thomas, Adv. Mater. 14, 1850 (2002).
118. A. Chelnokov, K. Wang, S. Rowson, P. Garoche, J. M. Lourtioz, Appl. Phys. Lett. 77, 2943 (2000).
119. F. Garcia-Santamaria, H. T. Miyazaki, A. Urquia, M. Ibisate, M. Belmonte, N. Shinya, F. Meseguer, C. Lopez, Adv. Mater. 14, 1144 (2002).
120. K. Aoki, H. T. Miyazaki, H. Hirayama, K. Inoshita, T. Baba, K. Sakoda, N. Shinya, Y. Aoyagi, Nat. Mater. 2, 117 (2003).
121. G. M. Gratson, M. J. Xu, J. A. Lewis, Nature 428, 386 (2004).
122. G. M. Gratson, F. Garcia-Santamaria, V. Lousse, M. J. Xu, S. H. Fan, J. A. Lewis, P. V. Braun, Adv. Mater. 18, 461 (2006).

123. P. V. Braun, S. A. Rinne, F. Garcia-Santamaria, Adv. Mater. 18, 2665 (2006).
124. Y. N. Xia, B. Gates, Y. D. Yin, Y. Lu, Adv. Mater. 12, 693 (2000).
125. J. V. Sanders, Nature 204, 1151 (1964).
126. C. Lopez, Adv Mater 15, 1679 (2003).
127. E. Matijevic, Chem. Mat. 5, 412 (1993).
128. H. Kawaguchi, Prog. Polym. Sci. 25, 1171 (2000).
129. W. Stöber, A. Fink, and E. Bohn, J. Colloid Interface Sci. 26, 62 (1968).
130. G. H. Bogush, M. A. Tracy, C. F. Zukoski, J. Non-Cryst. Solids 104, 95 (1988).
131. M. Ibisate, Z. Q. Zou, Y. N. Xia, Advanced Functional Materials 16, 1627 (2006).
132. E. Snoeks, A. van Blaaderen, T. van Dillen, C. M. van Kats, M. L. Brongersma, A. Polman, Adv. Mater. 12, 1511 (2000).
133. P. Jiang, J. F. Bertone, V. L. Colvin, Science 291, 453 (2001).
134. R. Arshady, Colloid Polym. Sci. 270, 717 (1992).
135. P. N. Pusey, W. van Megen, Nature 320, 340 (1986).
136. J. X. Zhu, M. Li, R. Rogers, W. Meyer, R. H. Ottewill, W. B. Russell, P. M. Chaikin, Nature 387, 883 (1997).
137. U. Gasser, E. R. Weeks, A. Schofield, P. N. Pusey, D. A. Weitz, Science 292, 258 (2001).
138. M. Trau, D. A. Saville, I. A. Aksay, Science 272, 706 (1996).
139. W. D. Ristenpart, I. A. Aksay, D. A. Saville, Phys. Rev. Lett. 90, 128303 (2003).
140. A. T. Skjeltorp, Phys. Rev. Lett. 51, 2306 (1983).
141. P. Sheng, W. Wen, N. Wang, H. Ma, Z. Lin, W. Y. Zhang, X. Y. Lei, Z. L. Wang, D. G. Zheng, W. Y. Tam, C. T. Chan, Pure Appl. Chem. 72, 309 (2000).
142. M. M. Burns, J. M. Fournier, J. A. Golovchenko, Science 249, 749 (1990).
143. P. T. Korda, D. G. Grier, J. Chem. Phys. 114, 7570 (2001).
144. R. Mayoral, J. Requena, J. S. Moya, C. Lopez, A. Cintas, H. Miguez, F. Meseguer, L. Vazquez, M. Holgado, A. Blanco, Adv. Mater. 9, 25 (1997).
145. H. Miguez, F. Meseguer, C. Lopez, A. Mifsud, J. S. Moya, L. Vazquez, Langmuir 13, 6009 (1997).
146. H. Miguez, C. Lopez, F. Meseguer, A. Blanco, L. Vazquez, R. Mayoral, M. Ocaña, V. Fornes, A. Mifsud, Appl. Phys. Lett. 71, 1148 (1997).
147. V. N. Astratov, A. M. Adawi, S. Fricker, M. S. Skolnick, D. M. Whittaker, P. N. Pusey, Phys. Rev. B 66, 165215 (2002).
148. S. Datta, C. T. Chan, K. M. Ho, C. M. Soukoulis, Phys. Rev. B 48, 14936 (1993).
149. M. Holgado, F. Garcia-Santamaria, A. Blanco, M. Ibisate, A. Cintas, H. Miguez, C. J. Serna, C. Molpeceres, J. Requena, A. Mifsud, F. Meseguer, C. Lopez, Langmuir 15, 4701 (1999).
150. M. E. Leunissen, C. G. Christova, A. P. Hynninen, C. P. Royall, A. I. Campbell, A. Imhof, M. Dijkstra, R. van Roij, A. van Blaaderen, Nature 437, 235 (2005).

151. K. U. Fulda, B. Tieke, Adv. Mater. 6, 288 (1994).
152. M. Bardosova, P. Hodge, L. Pach, M. E. Pemble, V. Smatko, R. H. Tredgold, D. Whitehead, Thin Solid Films 437, 276 (2003).
153. S. H. Im, Y. T. Lim, D. J. Suh, O. O. Park, Adv. Mater. 14, 1367 (2002).
154. N. D. Denkov, O. D. Velev, P. A. Kralchevsky, I. B. Ivanov, H. Yoshimura, K. Nagayama, Langmuir 8, 3183 (1992).
155. P. Jiang, J. F. Bertone, K. S. Hwang, V. L. Colvin, Chem. Mat. 11, 2132 (1999).
156. P. Jiang, G. N. Ostojic, R. Narat, D. M. Mittleman, V. L. Colvin, Adv. Mater. 13, 389 (2001).
157. Y. A. Vlasov, X. Z. Bo, J. C. Sturm, D. J. Norris, Nature 414, 289 (2001).
158. K. Wostyn, Y. Zhao. G. Schaetzen, L. Hellemans, N. Matsuda, K. Clays, A. Persoons, Langmuir 19, 4465 (2003).
159. K. P. Velikov, C. G. Christova, R. P. A. Dullens, A. van Blaaderen, Science 296, 106 (2002).
160. S. H. Park, D. Qin, Y. Xia, Adv. Mater. 10, 1028 (1998).
161. R. M. Amos, J. G. Rarity, P. R. Tapster, T. J. Shepherd, S. C. Kitson, Phys. Rev. E 61, 2929 (2000).
162. F. Ramiro-Manzano, F. Meseguer, E. Bonet, I. Rodriguez, Phys. Rev. Lett. 97, 028304 (2006).
163. A. van Blaaderen, R. Ruel, P. Wiltzius, Nature 385, 321 (1997).
164. Y. D. Yin, Y. N. Xia, Adv. Mater. 14, 605 (2002).
165. Y. D. Yin, Y. N. Xia, J. Am. Chem. Soc. 125, 2048 (2003).
166. Y. D. Yin, Y. N. Xia, Adv. Mater. 13, 267 (2001).
167. S. M. Yang, G. A. Ozin, Chem. Commun. 2507 (2000).
168. S. M. Yang, H. Miguez, G. A. Ozin, Adv. Funct. Mater. 12, 425 (2002).
169. B. Griesebock, M. Egen, R. Zentel, Chem. Mat. 14, 4023 (2002).
170. F. Garcia-Santamaria, C. Lopez, F. Meseguer, F. Lopez-Tejeira, J. Sanchez-Dehesa, H. T. Miyazaki, Appl. Phys. Lett. 79, 2309 (2001).
171. F. Garcia-Santamaria, H. Miguez, M. Ibisate, F. Meseguer, C. Lopez, Langmuir 18, 1942 (2002).
172. H. Miguez, F. Meseguer, C. Lopez, A. Blanco, J. S. Moya, J. Requena, A. Mifsud, V. Fornes, Adv. Mater. 10, 480 (1998).
173. B. Gates, S. H. Park, Y. N. Xia, Adv. Mater. 12, 653 (2000).
174. H. Míguez, N. Tétreault, B. Hatton, S. M. Yang, D. Perovic, G. A. Ozin, Chem. Comm. 2377, 2736 (2002).
175. M. Muller, R. Zentel, T. Maka, S. G. Romanov, C. M. S. Torres, Chem. Mat. 12, 2508 (2000).
176. B. H. Juarez, D. Golmayo, P. A. Postigo, C. Lopez, Adv Mater 16, 1732 (2004).

177. R. Fenollosa, M. Ibisate, S. Rubio, C. Lopez, F. Meseguer, J. Sanchez-Dehesa, J. Appl. Phys. 93, 671 (2003).
178. W. Wang, S. A. Asher, J. Am. Chem. Soc. 123, 12528 (2001).
179. J. Valenta, J. Linnros, R. Juhasz, J. L. Rehspringer, F. Huber, C. Hirlimann, S. Cheylan, R. G. Elliman, J. Appl. Phys. 93, 4471 (2003).
180. X. L. Xu, G. Friedman, K. D. Humfeld, S. A. Majetich, S. A. Asher, Adv. Mater. 13, 1681 (2001).
181. F. Caruso, Adv. Mater. 13, 11 (2001).
182. A. Vanblaaderen, A. Vrij, Langmuir 8, 2921 (1992).
183. T. Ung, L. M. Liz-Marzan, P. Mulvaney, Langmuir 14, 3740 (1998).
184. L. M. Liz-Marzan, M. Giersig, P. Mulvaney, Langmuir 12, 4329 (1996).
185. F. Garcia-Santamaria, V. Salgueirino-Maceira, C. Lopez, L. M. Liz-Marzan, Langmuir 18, 4519 (2002).
186. B. Rodriguez-Gonzalez, V. Salgueirino-Maceira, F. Garcia-Santamaria, L. M. Liz-Marzan, Nano Lett. 2, 471 (2002).
187. C. Graf, A. van Blaaderen, Langmuir 18, 524 (2002).
188. Z. J. Liang, A. S. Susha, F. Caruso, Adv. Mater. 14, 1160 (2002).
189. F. Caruso, A. S. Susha, M. Giersig, H. Mohwald, Adv. Mater. 11, 950 (1999).
190. M. A. Correa-Duarte, M. Giersig, L. M. Liz-Marzan, Chem. Phys. Lett. 286, 497 (1998).
191. A. L. Rogach, D. Nagesha, J. W. Ostrander, M. Giersig, N. A. Kotov, Chem. Mat. 12, 2676 (2000).
192. K. Busch, S. John, Phys. Rev. E 58, 3896 (1998).
193. A. Stein, Microporous Mesoporous Mat. 44, 227 (2001).
194. H. Miguez, F. Meseguer, C. Lopez, F. Lopez-Tejeira, J. Sanchez-Dehesa, Adv. Mater. 13, 393 (2001).
195. K. Yoshino, S. Satoh, Y. Shimoda, Y. Kawagishi, K. Nakayama, M. Ozaki, Jpn. J. Appl. Phys. Part 2 - Lett. 38, L961 (1999).
196. M. Deutsch, Y. A. Vlasov, D. J. Norris, Adv. Mater. 12, 1176 (2000).
197. T. Cassagneau, F. Caruso, Adv. Mater. 14, 34 (2002).
198. S. A. Asher, A. C. Sharma, A. V. Goponenko, M. M. Ward, Anal. Chem. 75, 1676 (2003).
199. S. A. Asher, V. L. Alexeev, A. V. Goponenko, A. C. Sharma, I. K. Lednev, C. S. Wilcox, D. N. Finegold, J. Am. Chem. Soc. 125, 3322 (2003).
200. A. C. Arsenault, H. Miguez, V. Kitaev, G. A. Ozin, I. Manners, Adv. Mater. 15, 503 (2003).
201. K. Lee, S. A. Asher, J. Am. Chem. Soc. 122, 9534 (2000).
202. S. H. Foulger, P. Jiang, A. Lattam, D. W. Smith, J. Ballato, D. E. Dausch, S. Grego, B. R. Stoner, , Adv. Mater. 15, 685 (2003).

203. K. Yoshino, Y. Shimoda, Y. Kawagishi, K. Nakayama, M. Ozaki, Appl. Phys. Lett. 75, 932 (1999).
204. Y. Shimoda, M. Ozaki, K. Yoshino, Appl. Phys. Lett. 79, 3627 (2001).
205. M. Ozaki, Y. Shimoda, M. Kasano, K. Yoshino, Adv. Mater. 14, 514 (2002).
206. S. Kubo, Z. Z. Gu, K. Takahashi, Y. Ohko, O. Sato, A. Fujishima, J. Am. Chem. Soc. 124, 10950 (2002).
207. K. Yoshino, S. B. Lee, S. Tatsuhara, Y. Kawagishi, M. Ozaki, A. A. Zakhidov, Appl. Phys. Lett. 73, 3506 (1998).
208. S. G. Romanov, T. Maka, C. M. S. Torres, M. Muller, R. Zentel, Appl. Phys. Lett. 75, 1057 (1999).
209. A. F. Koenderink, L. Bechger, H. P. Schriemer, A. Lagendijk, W. L. Vos, Phys. Rev. Lett. 88 (2002).
210. S. V. Frolov, Z. V. Vardeny, A. A. Zakhidov, R. H. Baughman, Opt. Commun. 162, 241 (1999).
211. M. N. Shkunov, Z. V. Vardeny, M. C. DeLong, R. C. Polson, A. A. Zakhidov, R. H. Baughman, Adv. Funct. Mater. 12, 21 (2002).
212. B. H. Juárez, P. D. García, D. Golmayo, A. Blanco, C. López. Adv. Mater. 17, 2761 (2005).
213. M. Scharrer, X. Wu, A. Yamilov, H. Yamilov, H. Cao, R. P. H. Chang, Appl. Phys. Lett. 86, 151113 (2005).
214. P. D. García, C. López, J. Appl. Phys. 99, 046103 (2006).
215. M. Scharrer, A. Yamilov, X. Wu, H. Cao, R. P.H. Chang, Appl. Phys. Lett. 88, 201103 (2006).
216. P. D. García, A. Blanco, A. Shavel, N. Gaponik, A. Eychmüller, B. Rodríguez-González, L. M. Liz-Marzán and C. López, Adv. Mater. 18, 2768 (2006).
217. N. Eradat, A. Y. Sivachenko, M. E. Raikh, Z. V. Vardeny, A. A. Zakhidov, R. H. Baughman, Appl. Phys. Lett. 80, 3491 (2002).
218. K. Sumioka, H. Nagahama, T. Tsutsui, Appl. Phys. Lett. 78, 1328 (2001).
219. Z. Z. Gu, S. Hayami, Q. B. Meng, T. Iyoda, A. Fujishima, O. Sato, J. Am. Chem. Soc. 122, 10730 (2000).
220. Z. Z. Gu, A. Fujishima, O. Sato, J. Am. Chem. Soc. 122, 12387,(2000).
221. S. V. Pankova, V. V. Poborchii, V. G. Solovev, J. Phys.-Condes. Matter 8, L203 (1996).
222. J. Zhou, C. Q. Sun, K. Pita, Y. L. Lam, Y. Zhou, S. L. Ng, C. H. Kam, L. T. Li, Z. L. Gui, Appl. Phys. Lett. 78, 661 (2001).
223. B. G. Kim, K. S. Parikh, G. Ussery, A. Zakhidov, R. H. Baughman, E. Yablonobitch, B. S. Dunn, Appl. Phys. Lett. 81, 4440 (2002).
224. I. Soten, H. Miguez, S. M. Yang, S. Petrov, N. Coombs, N. Tetreault, N. Matsuura, H. E. Ruda, G. A. Ozin, Adv. Funct. Mater. 12, 71 (2002).

225. D. Y. Wang, F. Caruso, Adv. Mater. 15, 205 (2003).
226. A. A. Zakhidov, R. H. Baughman, Z. Iqbal, C. X. Cui, I. Khayrullin, S. O. Dantas, I. Marti, V. G. Ralchenko, Science 282, 897 (1998).
227. T. Sumida, Y. Wada, T. Kitamura, S. Yanagida, Chem. Lett. 31, 38 (2001).
228. T. Sumida, Y. Wada, T. Kitamura, S. Yanagida, Chem. Lett. 2, 180 (2002).
229. P. N. Bartlett, T. Dunford, M. A. Ghanem, J. Mater. Chem. 12, 3130 (2002).
230. R. W. J. Scott, S. M. Yang, G. Chabanis, N. Coombs, D. E. Williams, G. A. Ozin, Adv. Mater. 13, 1468 (2001).
231. J. E. G. J. Wijnhoven, W. L. Vos, Science 281, 802 (1998).
232. P. Lodahl, A. F. van Driel, I. S. Nikolaev, A. Irman, K. Overgaag, D. L. Vanmaekelbergh, W. L. Vos, Nature 430, 654 (2004).
233. W. T. Dong, H. Bongard, B. Tesche, F. Marlow, Adv. Mater. 14, 1457 (2002).
234. Z. Z. Gu, S. Kubo, W. P. Qian, Y. Einaga, D. A. Tryk, A. Fujishima, O. Sato, Langmuir 17, 6751 (2001).
235. D. B. Kuang, A. W. Xu, J. Y. Zhu, H. Q. Liu, B. S. Kang, New J. Chem. 26, 819 (2002).
236. S. G. Romanov, A. V. Fokin, R. M. De La Rue, Appl. Phys. Lett. 76, 1656 (2000).
237. M. J. A. de Dood, B. Berkhout, C. M. van Kats, A. Polman, A. van Blaaderen, Chem. Mat. 14, 2849 (2002).
238. S. Jeon, P. V. Braun, Chem. Mat. 15, 1256 (2003).
239. A. Figotin, I. Vitebsky, Phys. Rev. E, 6306, 066609 (2001).
240. B. Gates, Y. N. Xia, Adv. Mater. 13, 1605 (2001).
241. J. Galloro, M. Ginzburg, H. Miguez, S. M. Yang, N. Coombs, A. Safa-Sefat, J. E. Greedan, I. Manners, G. A. Ozin, Adv. Funct. Mater. 12, 382 (2002).
242. E. O. Chi, Y. N. Kim, J. C. Kim, N. H. Hur, Chem. Mater. 15, 1929 (2003).
243. C. Koerdt, G. Rikken, E. P. Petrov, Appl. Phys. Lett. 82, 1538 (2003).
244. O. D. Velev, P. M. Tessier, A. M. Lenhoff, E. W. Kaler, Nature 401, 548 (1999).
245. P. M. Tessier, O. D. Velev, A. T. Kalambur, J. F. Rabolt, A. M. Lenhoff, E. W. Kaler, J. Am. Chem. Soc. 122, 9554 (2000).
246. L. B. Xu, W. L. Zhou, M. E. Kozlov, Khayrullin, II, I. Udod, A. A. Zakhidov, R. H. Baughman, J. B. Wiley, J. Am. Chem. Soc. 123, 763 (2001).
247. P. Jiang, J. Cizeron, J. F. Bertone, V. L. Colvin, J. Am. Chem. Soc. 121, 7957 (1999).
248. J. Wijnhoven, S. J. M. Zevenhuizen, M. A. Hendriks, D. Vanmaekelbergh, J. J. Kelly, W. L. Vos, Adv. Mater. 12, 888 (2000).
249. L. B. Xu, W. L. L. Zhou, C. Frommen, R. H. Baughman, A. A. Zakhidov, L. Malkinski, J. Q. Wang, J. B. Wiley, Chem. Commun. 997 (2000).
250. P. N. Bartlett, P. R. Birkin, M. A. Ghanem, Chem. Commun. 1671 (2000).

251. B. H. Juarez, C. Lopez, C. Alonso, J Phys Chem B 108, 16708 (2004).
252. F. Li, L. B. Xu, W. L. L. Zhou, J. B. He, R. H. Baughman, A. A. Zakhidov, J. B. Wiley, Adv. Mater. 14, 1528 (2002).
253. D. J. Norris, Y. A. Vlasov, Adv. Mater. 13, 371 (2001).
254. A. Blanco, E. Chomski, S. Grabtchak, M. Ibisate, S. John, S. W. Leonard, C. Lopez, F. Meseguer, H. Miguez, J. P. Mondia, G. A. Ozin, O. Toader, H. M. van Driel, Nature 405, 437 (2000).
255. Y. A. Vlasov, X. Z. Bo, J. C. Sturm, D. J. Norris, Nature 414, 289 (2001).
256. H. Miguez, S. M. Yang, N. Tetreault, G. A. Ozin, Adv. Mater. 14, 1805 (2002).
257. E. Palacios-Lidon, A. Blanco, M. Ibisate, F. Meseguer, C. Lopez, J. Sanchez-Dehesa, Appl. Phys. Lett. 81, 4925 (2002).
258. H. Miguez, N. Tetreault, S. M. Yang, V. Kitaev, G. A. Ozin, Adv. Mater. 15, 597 (2003).
259. A. Blanco, and C. López, Adv Mater. 18, 1593 (2006).
260. H. Miguez, F. Meseguer, C. Lopez, M. Holgado, G. Andreasen, A. Mifsud, V. Fornes, Langmuir, 16, 4405 (2000).
261. H. Miguez, E. Chomski, F. Garcia-Santamaria, M. Ibisate, S. John, C. Lopez, F. Meseguer, J. P. Mondia, G. A. Ozin, O. Toader, H. M. van Driel, Adv. Mater. 13, 1634 (2001).
262. L. K. van Vugt, A. F. van Driel, R. W. Tjerkstra, L. Bechger, W. L. Vos, D. Vanmaekelbergh, J. J. Kelly, Chem. Commun. 2054 (2002).
263. F. Garcia-Santamaría, M. Ibisate, I. Rodriguez, F. Meseguer, C. Lopez, Adv. Mater. 15, 788 (2003).
264. H. Miguez, A. Blanco, F. Meseguer, C. Lopez, H. M. Yates, M. E. Pemble, V. Fornes, A. Mifsud, Phys. Rev. B 59, 1563 (1999).
265. S. G. Romanov, R. M. De la Rue, H. M. Yates, M. E. Pemble, J. Phys.-Condes. Matter 12, 339 (2000).
266. E. Palacios-Lidon, H. M. Yates, M. E. Pemble, C. Lopez, Appl Phys B-Lasers O 81, 205 (2005).
267. H. M. Yates, M. E. Pemble, E. Palacios-Lidon, F. Garcia-Santamaria, I. Rodriguez, F. Meseguer, C. Lopez, Adv Funct Mater 15, 411 (2005).
268. Y. C. Lee, T. J. Kuo, C. J. Hsu, Y. W. Su, C. C. Chen, Langmuir 18, 9942 (2002).
269. I. M. Povey, D. Whitehead, K. Thomas, M. E. Pemble, M. Bardosova, J. Renard, Applied Physics Letters 89 (2006).
270. V. Y. Davydov, R. E. Dunin-Borkovski, V. G. Golubev, J. L. Hutchison, N. F. Kartenko, D. A. Kurdyukov, A. B. Pevtsov, N. V. Sharenkova, J. Sloan, L. M. Sorokin, Semicond. Sci. Technol. 16, L5 (2001).
271. P. V. Braun, R. W. Zehner, C. A. White, M. K. Weldon, C. Kloc, S. S. Patel, P. Wiltzius, Adv. Mater. 13, 721 (2001).

272. V. N. Astratov, V. N. Bogomolov, A. A. Kaplyanskii, A. V. Prokofiev, L. A. Samoilovich, S. M. Samoilovich, Y. A. Vlasov, Nuovo Cimento D 17, 1349 (1995).
273. Y. A. Vlasov, N. Yao, D. J. Norris, Adv. Mater. 11, 165 (1999).
274. Y. A. Vlasov, K. Luterova, I. Pelant, B. Honerlage, V. N. Astratov, Appl. Phys. Lett. 71, 1616 (1997).
275. A. Blanco, C. Lopez, R. Mayoral, H. Miguez, F. Meseguer, A. Mifsud, J. Herrero, Appl. Phys. Lett. 73, 1781 (1998).
276. B. H. Juarez, S. Rubio, J. Sanchez-Dehesa, C. Lopez, Adv. Mater. 14, 1486 (2002).
277. P. V. Braun, P. Wiltzius, Nature 402, 603 (1999).
278. S. G. Romanov, T. Maka, C. M. S. Torres, M. Muller, R. Zentel, Appl. Phys. Lett. 79, 731 (2001).
279. B. H. Juarez, M. Ibisate, J. M. Palacios, C. Lopez, Adv. Mater. 15, 319 (2003).

CHAPTER 3

NANOPARTICLE-MICELLE: A NEW BUILDING BLOCK FOR FACILE SELF-ASSEMBLY AND INTEGRATION OF 2-, 3-DIMENSIONAL FUNCTIONAL NANOSTRUCTURES

Hongyou Fan* and C. Jeffrey Brinker

Sandia National Laboratories, Advanced Materials Laboratory, 1001 University Blvd. SE, Albuquerque, New Mexico 87106, The United States of America; The University of New Mexico/NSF Center for Micro-Engineered Materials, Department of Chemical and Nuclear Engineering, Albuquerque, New Mexico 87131, The United States of America

**Tel: (505) 2727128, Fax: (505) 2727336, Email:hfan@sandia.gov*

In this chapter, we discuss the synthesis of a new building block – nanoparticle (NP)-micelle, and its further self-assembly and integration into 2-, 3-dimensional (D) functional nanostructures. The NPs are encapsulated inside the core of surfactant micelles in an interfacial micro-emulsion process. The flexible surface chemistry of the NP-micelles causes them to be water-soluble and allows further self-assembly and derivation. We discuss the synthesis of a new NP/silica mesophase through self-assembly of water-soluble NP-micelles with soluble silica. The mesophase comprises gold NPs arranged within a silica matrix in a face-centered-cubic (*fcc*) lattice with cell dimensions that are adjustable through control of the NP diameter and/or the alkane chain lengths of the primary alkanethiol stabilizing ligands or the surrounding secondary surfactants. Under kinetically controlled silica polymerization conditions and self-assembly process, ordered NP/silica mesophase films, mesocrystal arrays, and mesostructure powders are formed respectively. The NP-micelles are water-soluble and biocompatible, of interest for bio-labeling. The robust, 3-D NP mesophase solids are of interest for development of collective optical and electronic phenomena, and, importantly, for the integration of NP arrays into device architectures. Initial experiments on a MOS capacitor fabricated with an ordered gold NP/silica 'oxide' demonstrated charge storage on the gold NPs and discharge behavior dominated by electron transport within the ordered gold NP array.

1. Introduction

Nanometer-sized crystallites of metals, semiconductors, and oxides exhibit optical, electronic and chemical properties often different from those of the corresponding isolated molecules or macroscopic solids. The ability to adjust properties through control of size, shape, composition, crystallinity, and structure [1,2] has led to a wide range of potential applications for NPs in areas like optics, electronics, catalysis, magnetic storage, and biological labeling [1,3-7]. Furthermore NP assembly into 2- and 3-D arrays is of interest for development of synthetic solids with collective optical and electronic properties that can be further tuned by the NP spacing and arrangement. For example, using uniformly-sized CdSe NPs passivated with a close-packed monolayer of organic coordinating ligands (trioctylphosphine oxide), Murray *et al.* exploited the inherent tendency for monodisperse lyophobic colloids to self-assemble to create periodic, three-dimensional quantum dot superlattices [8]. Following this approach, Klimov *et al.* prepared ordered CdSe quantum dot films and demonstrated optical gain and stimulated emission [4,9]. In 2-D quantum dot monolayers formed in a Langmuir trough, Collier *et al.* reported quantum mechanical tunneling between adjacent silver NPs at an interparticle spacing below 1.2-nm and a reversible insulator-to-metal transition below 0.5-nm [10]. Even richer transport and collective phenomena are expected for 3-D NP arrays [11].

Despite recent advances in the synthesis and characterization of NPs and NP arrays, there remain numerous challenges that limit their practical utilization [5,6]. For example, synthesis procedures generally used for metallic and semiconducting NPs employ organic passivating ligands, making the NPs water insoluble. This is very problematic for biological imaging and for incorporation of NPs in hydrophilic sol-gel matrices like silica or titania needed for the fabrication of robust, functional lasers [12,13]. Furthermore, while steric stabilization of NPs with organic passivating layers suppresses strong, attractive particle-particle interactions, thereby facilitating self-assembly of NP arrays, it necessarily causes the arrays to be mechanically weak and often thermally and chemically unstable. Although evaporation of NP dispersion has been used to prepare quasi-3-D NP islands [11,14], there exist no procedures to reliably fabricate 3-D NP arrays as uniform thin film. These combined factors ultimately limit routine integration of NPs into 3D artificial solid devices, in which electronic, magnetic, and optical properties could be tuned through electron charging and quantum confinement of individual NPs mediated by coupling interactions with neighboring NPs [15,16]. Building block self-assembly is one of the important methods in materials self-assembly. From controlling surface chemistry, shape, and composition of building blocks, one can readily fabricate nanomaterials with controlled orientation, morphology, and function. In this chapter, we review the synthesis of a new building block – NP-micelle and its utilization in formation of 2-, 3-D ordered NP arrays and fabrication of nanodevices.

2. Synthesis of NP-Micelles

NPs have been often used as building blocks to form ordered 2-, 3-D superlattice films. Most of previous work has focused on the synthesis and self-assembly of NPs that are stabilized with alkane ligands ($CH_3(CH_2)_nR$, R=SH, NH_2, PO, etc) [1,8,11]. Such NPs can be made at fairly high quality (narrow size distribution, size deviation $< 10\%$, preferred shapes such as rod, cube, etc, and large production). These NPs are hydrophobic, and their self-assembly is limited to organic solvents. However, there are many applications requiring hydrophilic or aqueous environments, such as biolabeling and SERS-based NP films or arrays for biosensing [3,5,17]. In addition, water-soluble NPs and their ordered arrays/films provide great opportunity for further integration into inorganic ceramic frameworks that offer the chemical and mechanical robustness needed for enhanced device functionality [12,13]. Although some NP superlattices have been fabricated through aqueous media, the methods are limited to spherical gold NPs.

Recently We have developed an encapsulation method to synthesize NP-micelles [18-21]. The synthetic approach is general and avoids the complicated multi-step procedures reported previously [22]. The concept is to consider monosized, organically-passivated NPs as large hydrophobic molecules that, if incorporated individually into the hydrophobic interiors of surfactant micelles, would result in the formation of monosized NP micelles composed of a metallic (or other) NP core and a hybrid bilayer shell with precisely defined primary and secondary layer thicknesses (see Figure 1H). The hydrophilic NP micelle surfaces would cause them to be water-soluble and allow further assembly or derivatization as depicted in Figure 1.

To individually incorporate NPs in surfactant micelles and realize this concept, a micro-emulsion procedure was developed first (see Figure 2). A concentrated NP solution, prepared in organic solvent (chloroform, hexane, etc.), was added to an aqueous solution of surfactant with a volume ratio of 1:10 under vigorous stirring to create an oil-in-water micro-emulsion. Organic solvent evaporation (aided optionally by vacuum or heat treatments) transferred the NPs into the aqueous phase by an interfacial process driven by the hydrophobic van der Waals interactions between the primary alkane of the stabilizing ligand and the secondary alkane of the surfactant, resulting in thermodynamically defined interdigitated bilayer structures (Figure 1H). For single-tailed surfactants, an alkane chain of eight or more carbons is required to form micelles with gold NPs stabilized by C_{12} alkanethiols (dodecanethiol). Cationic, anionic, and non-ionic surfactants can all form NP-micelles, allowing facile control of micelle surface charge and functionality. In addition, fluorescent semiconducting CdSe NPs (stabilized by trioctylphosphine oxide) have been formed into NP-micelles with maintenance of optical properties, further supporting the general nature and flexibility of this approach.

NP-micelle solutions remained the same color of original hydrophobic nanoparticles and indefinitely stable (> 2-years). Generally, individual gold NP-micelles (as opposed to aggregated dimers, trimers etc.) were often formed. For spherical gold NP-micelles, the monodispersity was confirmed by UV/visible spectroscopy (Figure 1D), where no difference was observed in the position or width of the plasmon resonance band (~510-nm) of the C_{12}-alkanethiol stabilized gold NPs in chloroform and the resulting water soluble NP-micelles.

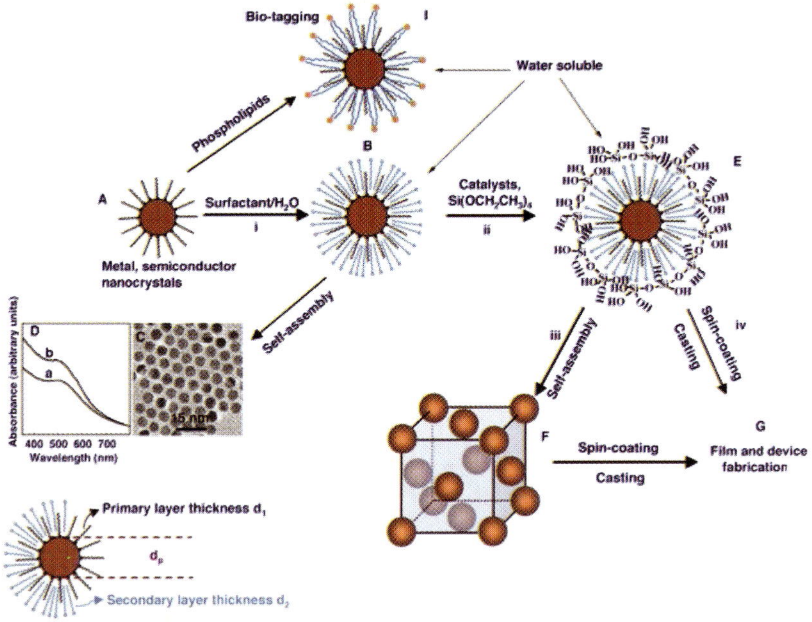

Figure 1. Processing diagram for the synthesis of water-soluble gold NP-micelles and periodically ordered gold NP/silica mesophases. Gold NPs (A) were prepared according to the method of Brust *et al.*[49], using 1-dodecanethiol (DT) as a stabilizing agent. Heat-treatments were employed to further narrow the particle size distributions. Thiol-stabilized NPs (B) are encapsulated in surfactants to form water-soluble NP micelles that, upon evaporation, self-assemble to form hexagonally-ordered NP arrays as shown in the TEM image (C). UV-visible spectra (D) of (a) gold NPs in chloroform and (b) gold NP-micelles in water both exhibit plasmon resonaNPe bands at 510-cm^{-1}. Silicic acid moieties (E) formed by hydrolysis of TEOS are organized at the hydrophilic surfactant-water interface of NP micelles, leading, under basic conditions, to a gold NP/silica mesophase (F) composed of NPs organized in a periodic *fcc* lattice within a dense silica matrix. Under acidic conditions (G) that suppress siloxane condensation, spin-coating or casting result in ordered thin film NP/silica mesophases that are readily integrated into devices. The lattice constant (H) of the NP/silica mesophase is controlled by the NP size (d_p), the primary layer thickness of the alkanethiol, d_1, and/or the secondary layer thickness of the surfactant, d_2 (see Figure 3). Polyethylene glycol-surfactants or lipids (I) can be used to prepare biocompatible water-soluble NP micelles for biolabeling. Reprinted with permission from [Ref. 21], H. Y. Fan, K. Yang, D. Boye, T. Sigmon, K. Malloy, H. Xu, G. P. Lopez, C. Brinker, Science 304, 567 (2004), Copyright @ AAAS.

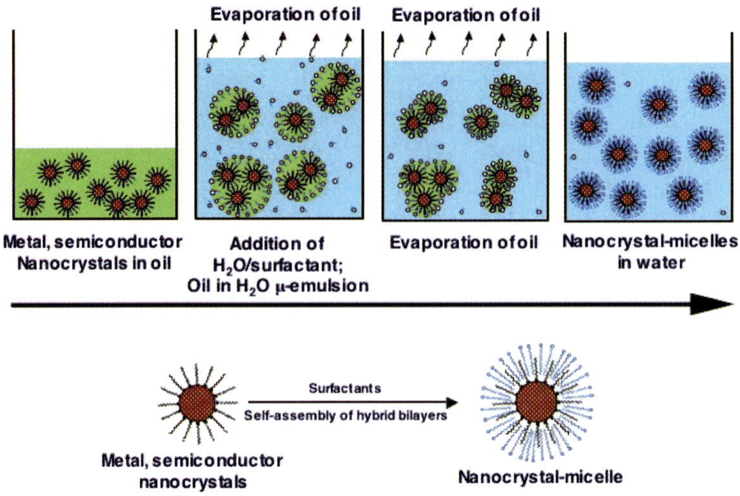

Figure 2. Process scheme for the synthesis of water-soluble NP micelles. Addition of oil containing gold NPs into a surfactant containing aqueous solution during vigorous stirring leads to the formation of an oil in H_2O μ-emulsion. Subsequent evaporation of oil transfers the NPs into the aqueous phase by an interfacial process driven by the hydrophobic van der Waals interactions between the primary alkane of the stabilizing ligand and the secondary alkane of the surfactant, resulting in interdigitated bilayer structures that encapsulate NPs. Reprinted with permission from [Ref. 21], H. Y. Fan, K. Yang, D. Boye, T. Sigmon, K. Malloy, H. Xu, G. P. Lopez, C. Brinker, Science 304, 567 (2004), Copyright @ AAAS.

In addition to spherical NPs, the concept has been successfully extended to stabilize and self-assemble NPs with other shapes such as rods and cubes. CdTe nanorods were prepared according to Peng *et al.* [23] using trioctylphosphine oxide as a stabilizing agent. The encapsulation was conducted using CTAB. The TEM image (Figure 3d) revealed that surfactant "bilayer" stabilized CdTe nanorods formed uniform arrays. Fe_2MnO_4 nanocubes were prepared according to Zeng *et al.* [24] using oleylamine as a stabilizing agent. Formation of Fe_2MnO_4 nanocube-micelles was conducted using CTAB. The TEM image (Figure 3e) shows that the Fe_2MnO_4 NP-micelles self-organize into ordered arrays with cubic symmetry.

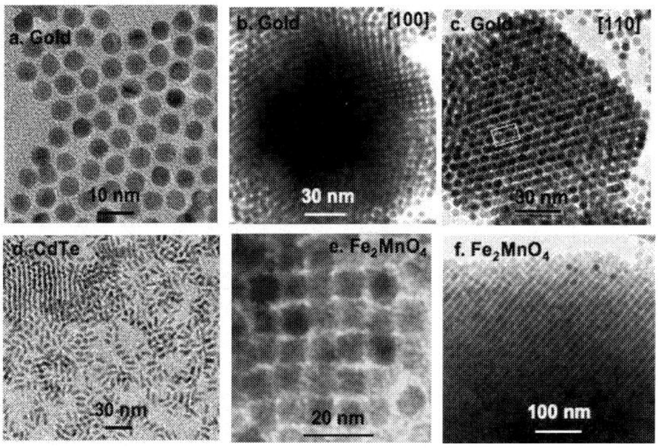

Fgure 3. TEM of NP-micelle arrays. Hexagonal gold NP-micelle arrays are shown in (a); (b) & (c) show [100] and [110] orientation of 3-D gold NP-micelle arrays; (d) shows ordered CdTe nanorod-micelle arrays; (e) shows 2-D Fe_2MnO_4 nanocube-micelle arrays; and (f) shows [100] orientation of 3-D ordered Fe_2MnO_4 nanocube-micelle arrays. Reprinted with permission from [Ref. 19], H. Y. Fan, E. Leve, J. Gabaldon, A. Wright, R. Haddad, C. Brinker, Adv. Mater. 17, 2587 (2005), Copyright @ Wiley InterScience.

Depending on the surfactants, the final NP-micelle surface possesses varied groups such as positive charges, negative charges, or neural groups (e.g., -OH), etc. Through electrostatic interactions and/or hydrogen bonding, 2-, 3-D NPs arrays can be formed [18,19]. Figure 3b & 3c show some representative TEM images of 3-D gold NP superlattices. Based on the [100]-orientation in Figure 3b, the measured unit cell is ~10.6-nm. Figure 4a shows a representative low-angle x-ray diffraction (XRD) pattern of a 3-D gold NP superlattice film prepared by drying an aqueous solution of NP-micelle on Si wafer. On the basis of *fcc* symmetry, the primary peaks are assigned as (111), (220), and (311). The calculated unit cell, a = ~11-nm, is close to TEM results. Through using concentrated $FeMnO_4$ nanocube-micelles, 3-D ordered magnetic NP arrays with cubic symmetry were also formed (Figure 3f).

Semiconductor NPs with superior optical properties are a promising alternative to organic dyes for fluorescence-based bio-applications. The potential benefits and use of conjugated NP probes to monitor cellular functions has prompted extensive efforts to develop methods to synthesize water-soluble and biocompatible NPs that can be widely used for fluorescent-based bio-imaging [3,22,25-31]. In general, the key to develop NPs as a tool in biological systems is to achieve water-solubility, biocompatibility and photo-stability, and importantly to provide flexible NP surface chemistry/functionality that will enable efficient coupling of these fluorescent inorganic probes to reagents capable of targeting and/or sensing ongoing biological processes. We have extended the encapsulation method to the synthesis of water-soluble, biocompatible semiconductor NP-micelles [20]. Monodisperse CdSe and CdSe/CdS core/shell NPs were synthesized according to previous methods using trioctylphosphine (TO) and oleylamine (OA) as stabilizing agents [23,32]. In a typical NP-micelle synthesis procedure, a concentrated suspension of NPs in chloroform is added to an aqueous solution containing a mixture of surfactants or phospholipids with different functional head groups such as ethylene glycol (-PEG) and amine (-NH_2). PEG is used to improve biocompatibility and amine groups provide sites for bioconjugation. Addition of the NP chloroform suspension into the surfactant/lipid aqueous solution under vigorous stirring results in the formation of an oil-in-water microemulsion.

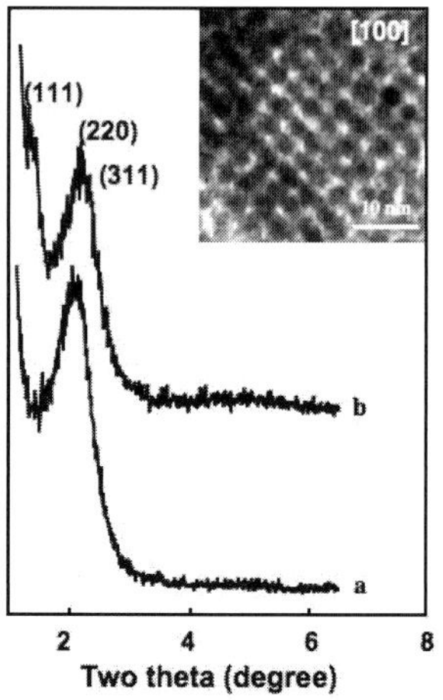

Figure 4. XRD patterns of *fcc* structured 3-D gold NP-micelle arrays (a) with and (b) without spacer group (benzene dicarboxylate). Inset: TEM image of [100] orientation of gold NP-micelles arrays after adding spacer molecules. Reprinted with permission from [Ref. 19], H. Y. Fan, E. Leve, J. Gabaldon, A. Wright, R. Haddad, C. Brinker, Adv. Mater. 17, 2587 (2005), Copyright @ Wiley InterScience.

Evaporation of chloroform during heating (40-80°C, ~10 minutes) transfers the NPs into the aqueous phase by an interfacial process driven by the hydrophobic van der Waals interactions between the primary alkane of the stabilizing ligand and the secondary alkane of the surfactant, resulting in thermodynamically defined interdigitated bilayer structures. Based on evaluation of water-solubility and stability using ultraviolet-visible spectroscopy (UV-vis), transmission electron micrograph (TEM), and fluorescent spectra, optimized lipids were those with at least 6-8 carbons in their hydrophobic alkane chains.

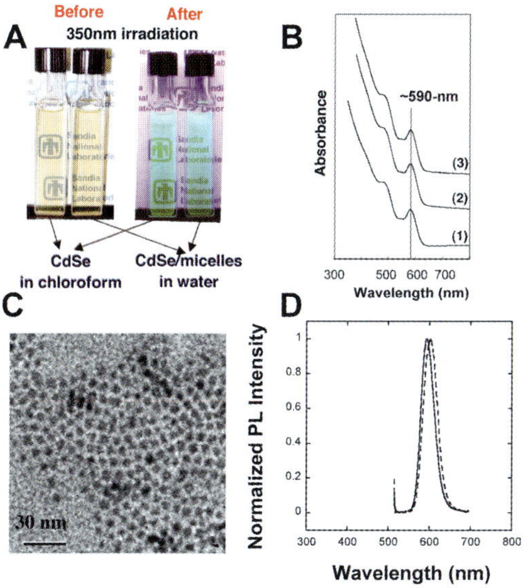

Figure 5. An optical micrograph (A) of CdSe NPs in chloroform and CdSe-micelles in water prepared using cetyltrimethylammonium bromide and imaged under room light and 350-nm UV light. UV-vis spectra (B) of (1) CdSe/CdS in hexane, (2) CdSe/CdS-micelles prepared in phosphate buffer solution by using 1,2-Dioctanoyl-*sn*-Glycero-3-Phosphocholine (C8-lipid) and hexadecylamine (HDA), and (3) 1,2-Distearoyl-*sn*-Glycero-3-Phosphoethanolamine-N-Amino (Polyethylene Glycol) (aPEG) and dipalmitoyl phosphatidylcholine (DPPC). TEM image (C) of CdSe/CdS-micelles deposited on TEM grid from a drop of CdSe/CdS-micelle water solution. Photoluminescence spectra (D) of CdSe/CdS in toluene (solid line) and C8-lipid encapsulated CdSe/CdS-micelles in water (dashed line). Reprinted with permission from [Ref. 20], H. Y. Fan, E. W. Leve, C. Scullin, J. Gabaldon, D. Tallant, S. Bunge, T. Boyle, M. C. Wilson, C. J. Brinker, Nano Lett 5, 645 (2005), Copyright @ American Chemical Society.

NP-micelle solutions exhibited the same visible and emission colors as hydrophobic NPs, as shown in the optical micrograph (Figure 5A). The formation of individual CdSe/CdS NP-micelles was further characterized by UV-vis spectroscopy shown in Figure 5B. No difference was observed in the position or width of the absorbance bands at ~480-nm and ~590-nm from hydrophobic NPs and NP-micelles, which suggests that NP-micelles maintain their optical properties in water. Formation of ordered hexagonal close packing shown in TEM (Figure 5C), as expected for individual monosized NPs, further confirmed the monodispersity of NP-micelles. Photoluminescence (PL) spectra of OA-stabilized hydrophobic NP and NP-micelles prepared using 1,2-Dioctanoyl-*sn*-Glycero-3-Phosphocholine (C8-lipids) are shown in Figure 5D. PL of NP-micelles exhibited no obvious shift in emission wavelength. Studies of photo-stability of these water-soluble NP-micelles under long-time laser irradiation showed that these NP-micelles exhibited no loss of PL intensity in water (Figure 6A inset). The stability of NP-micelles in a cellular aqueous environment was confirmed by incubation with rat hippocampal neurons (see below).

The PL persistence (lifetime) from OA-stabilized CdSe/CdS in toluene, CdSe/CdS-micelle water solutions, and Rhodamine B (RB) in ethanol were measured and calculated according to Donega *et al.* [33] and are shown in Figure 6A. The persistence was measured during pulsed, 450-nm excitation by monitoring emission at the peak of the PL profile. The RB persistence is only slightly longer than the instrumental bandwidth. The $\log_e$ plots of the persistence curves from the CdSe NPs in toluene and H_2O/C8-lipid solutions were similar and nearly linear, with a (1/e) relaxation constant of ~30 nanoseconds, which was much longer than the lifetime of RB (~5 nanoseconds).

The advantage of our method to provide water-soluble NP-micelles is its capability to utilize a wide variety of surfactants/lipids with different functionally-terminated groups, which can be used to improve biocompatibility and to enable bioconjugation. The encapsulation is accomplished in a single, rapid (10-minute) step of solvent evaporation. In our method, the stabilization of hydrophobic NPs in water relies on the thermodynamically favorable interdigitation of surfactants and NP stabilizing ligands formed after solvent evaporation, enabling rapid efficient transfer of the NPs into the aqueous phase. The NP-micelles preserve the optical properties of the original source NPs, such as PL intensity and lifetime etc., as the encapsulation process involves no chemical substitution reactions. In contrast, in other methods, such as the ligand exchange method, chemical substitution occurs at the NP interface. This usually results in NP aggregation and changes in NP physical properties [34,35]. Additionally, our method can effectively prevent NP-micelles from aggregation due to the interdigitated "bilayer."

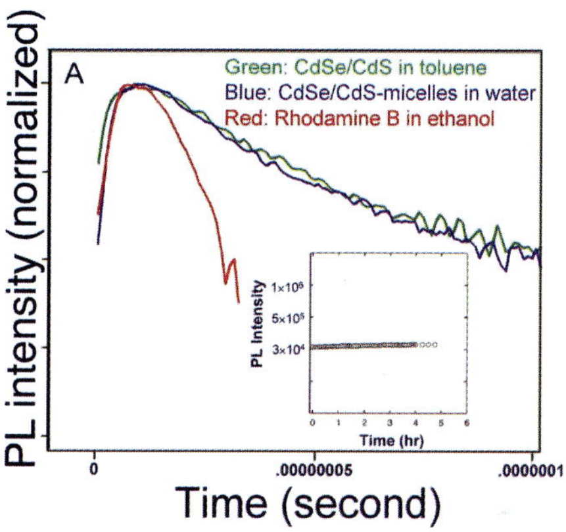

Figure 6. A. Persistence curves from CdSe/CdSe in toluene, RB in ethanol, and CdSe/CdS-micelles in water solution prepared by using C8-lipid. Inset. PL intensity CdSe/CdS-micelles in water solution vs laser irradiation time. Reprinted with permission from [Ref. 20], H. Y. Fan, E. W. Leve, C. Scullin, J. Gabaldon, D. Tallant, S. Bunge, T. Boyle, M. C. Wilson, C. J. Brinker, Nano Lett 5, 645 (2005), Copyright @ American Chemical Society.

To evaluate the biocompatibility of CdSe/CdS NP-micelles, an initial study was performed to examine their uptake in cultured rat hippocampal neurons. While the route through which NPs are taken up by cells has yet to be resolved fully, they do appear to accumulate in intracellular vesicular compartments [36,38] in several cell types suggesting uptake may be mediated, at least in part, though endocytosis. Because synaptic vesicle recycling is tightly coupled to neurotransmitter release [39], endocytosis can be manipulated by depolarization of the plasma membrane to excite neurons, evoking exocytosis for neurotransmitter release followed by endocytotic retrieval of synaptic vesicle membrane from nerve terminals. As shown in Figure 7A, NP-micelles (emission 592 nm, red) accumulate in neurons, distinguished by immmunohistochemical counterstain for NeuN (green), after exposure for 16 hours in basal serum-free growth media. A three-dimensional reconstruction of confocal images confirmed the intracellular localization of NP fluorescence after long-term exposure. The observation that NPs retain their fluorescence character after a fairly long incubation suggests that NPs can accumulate without aggregation and a general compatibility of the micelle coating within a cellular environment.

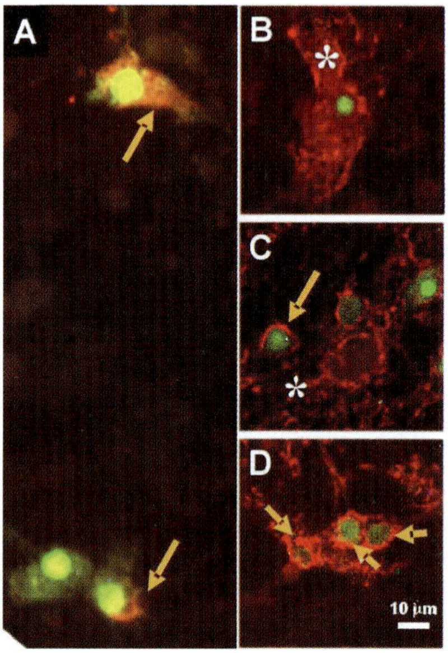

Figure 7. Images of CdSe/CdS-micelles, prepared using modified phospholipids including aPEG, DPPC, and HDA, in cultured rat hippocampal neurons. Panel (A) shows the fluorescent image of NP accumulation (arrows) after 16 hours of incubation (~ 1mg/ml NP-micelles) in neurons identified by nuclear staining for NeuN (green). Panels (B-D) show confocal images of pulse-chase labeling with NP-micelles. Incubation with NP-micelles during depolarization with 55 mM KCl for 10 min (panel B) shows dispersed, punctate staining (indicated by asterisks) characteristic of neurite processes; whereas after chase in basal media without NP-micelles for 1 hour (panel C) or 5 hours (panel D) shows increasing perinuclear localization (arrows) with a corresponding decrease in the punctate stain seen in the 10 min labeling. Reprinted with permission from [Ref. 20], H. Y. Fan, E. W. Leve, C. Scullin, J. Gabaldon, D. Tallant, S. Bunge, T. Boyle, M. C. Wilson, C. J. Brinker, Nano Lett 5, 645 (2005), Copyright @ American Chemical Society.

In order to obtain a more detailed view of NP uptake, pulse-chase labeling was used to expose hippocampal cultures to NP-micelles during membrane depolarization (induced by addition of 55 mM KCl) to trigger neurotransmitter release and coupled exo/endocytosis, followed by a return to basal media. After a 10 min exposure to NPs during depolarization, robust NP fluorescence was detected by confocal microscopy in wide-spread, yet clustered pattern surrounding NeuN-positive neurons Figure 7, panel B), consistent with a diffuse, punctate localization and uptake in neuronal processes and synapses. Interestingly, following exchange for basal media without NPs, there was an apparent time-dependent increase in the perinuclear accumulation of NP fluorescence accompanied by a decrease in the more wide-spread distribution seen immediately after KCl depolarization (Figure 7, Panels C, D). By 5 hours, the majority of NP fluorescence appeared to locate to a perinuclear compartment adjacent to NeuN stained neuronal nuclei. The results indicate that under these conditions NPs were effectively taken up by an endocytotic mechanism that was promoted by increased synaptic vesicle recycling and potentially followed by transport to lysosomes. Trafficking of NPs to lysosomal compartments may reflect autophagy, a vesicular-mediated process that underlies catabolic recycling and degradation of macromolecules, including both cellular and xenotypic agents, such as bacteria and viral pathogens [40]. Autophagy may be a particularly critical mechanism for maintaining the cellular integrity of neurons, which are long-lived and generally show little or no turnover in most regions of the nervous system, and could play a role in neuronal cell death under certain conditions.

3. Synthesis of Ordered NP Arrays

3.1. Synthesis of Hierarchically Ordered Mesostructured NP Arrays

Through a hydrothermal self-assembly process, the NP-micelles have been successfully utilized in synthesis of new, supported, and hierarchically ordered inorganic mesophase crystals [41]. Unlike the alkane chain capped NPs that are hydrophobic and chemically inert, NP-micelles provide a positively charged and hydrophilic interface for further self-assembly with metal oxides through charge interactions and hydrogen bonding, which is analog to the self-assembly of surfactant mesophase with metal oxides [42]. Depending on the substrates, these crystals consist of either individual or subunit crystals. Inside these mesophase crystals, the gold NPs were assembled as an *fcc* mesostructure with orientated feature.

The process (shown in Figure 8) involves two steps: (1) Preparation of a building block solution and (2) Hydrothermal nucleation and growth of mesophase crystals. In the first step, the NP-micelles were synthesized through interfacially mediated micro-emulsion encapsulation process using n-dodecanethiol (DT) capped gold NPs and cetyltrimethyl ammonium bromide (CTAB) as a secondary surfactant (Figure 1). The second step involves the nucleation and growth of mesophase crystals in a hydrothermal self-assembly process (Figure 7B). Bis(triethoxylsilyl)ethane (BTEE) was used as the precursor. In a typical synthesis, BTEE and sodium hydroxide are mixed with NP-micelle aqueous solution in a small vial. After 1 hour stirring at room temperature, a substrate such as microscope slide, thermal oxide coated silicon wafer, or fresh-peeled mica slide was vertically placed inside the small vial. The vials were then sealed and placed inside an oven at 100°C for crystal growth.

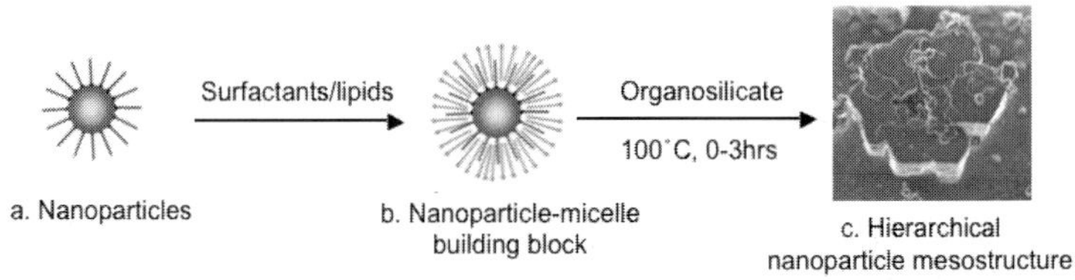

Figure 8. Formation of water-soluble NP-micelle and hierarchical NP mesostructure arrays through hydrothermal self-assembly of water-soluble NP-micelle with organosilicate. (a) shows an organic monolayer capped, hydrophobic gold NPs, (b) shows NP-micelle building, and (c) shows hierarchical NP mesostructure arrays, which are formed via a hydrothermal self-assembly process. Reprinted with permission from [Ref. 41], A. Wright, J. Gabaldon, D. B. Burckel, Y. B. Jiang, Z. R. Tian, J. Liu, C. J. Brinker, H. Y. Fan, Chem. Mater. 18, 3034 (2006), Copyright @ American Chemical Society.

Well-shaped and oriented NP/silsesquioxane mesophase crystals began to grow over several hours. These large mesophase crystals covered the whole substrate. Figure 9A shows a typical scanning electron microscopy (SEM) micrograph of large area arrays of the mesophase crystals grown on glass slides. These crystals are fairly uniform in shape and size (~15μm, Figure 9A). The high-resolution SEM image (Figure 9B) shows that these crystals exhibit flower-like hierarchical structure, which consists of three to five subunit crystals with hexagonal shape and size of ~ 7μm. A cross-sectional view of the crystal (Figure 9B inset) shows that the mesophase crystal is about ~8μm high. The transmission electron microscopy (TEM) image (Figure 9 C&D) and low-angle X-ray diffraction patterns (see below) reveal that the gold NPs self-assemble as an *fcc* mesophase inside these crystals. Figure 9 C&D show representative TEM images of [110] and [100] orientations of NP mesophase along with its corresponding selected electron diffraction pattern. The TEM images are consistent with a unit cell of ~10.6-nm and a uniform, minimum interparticle spacing of ~3-nm. To study the formation mechanism, samples at different growth times were examined by SEM (Figure 10) and X-ray diffraction patterns (XRD) (Figure 11).

Figure 10A through D show the mesophase crystals after growth for 10, 20, 30, and 40 minutes. After the initial 10 minutes, the macroscopic crystals (size <0.5μm) had just begun to nucleate with no recognizable shape. In the XRD (Figure 11A), the appearance of a small hump between two and three degrees suggests that at this stage, the NPs have started to self-assemble, exhibiting short range order. After 20 minutes, the crystals begin to form into triangle-shaped crystals with size between 0.5 to 1μm. On the XRD patterns, two peaks are observed at low angle corresponding to cubic symmetry. The first sharp peak is attributed to (111) reflection and the second broad hump can be assigned to (220) and/or (311). Between 30 and 40 minutes, hexagon-shaped subunit crystals with size of 7μm are developed. They are rather uniform in shape and size. Some of them have merged into hierarchical structures. XRD shows that *fcc* mesophase began to form as evidenced by the emergence of (311) and (222) peaks. After 60 minutes, the hierarchical mesophase crystals have completely developed, each of which consists of three to five uniformly merged subunit crystals. The XRD (Figure

11A curve e) suggests that the gold NP cubic mesophase has been fully established by full appearance of (111), (220), (311), and (222) reflections. The enhanced intensity from XRD reflections of (111) and (222) during the crystal growth course (after 40 minutes) suggests an orientated growth with (111) planes parallel the substrate. This is consistent with the shapes of the mesophase crystal changing from triangular to hexagonal shape during the growth phase.

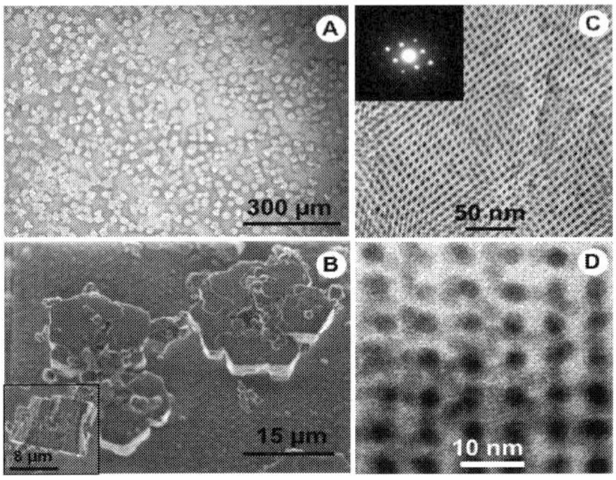

Figure 9. SEM and TEM images of hierarchical NP mesostructure arrays on glass slides. Sample (A) was prepared using BTEE as the precursor under basic conditions at 100°C for 60 minutes. A magnified SEM image (B) of (A) showing that the hierarchical mesostructure consists of 3 to 5 subunits. Inset shows a cross-sectional image of a main unit of the hierarchical mesostructure indicating that the crystal height is ~8μm. Representative TEM images from [110] (C) orientation and [100] orientation (D). Inset in (C) shows the selected area electron diffraction pattern from the image in (C) and suggests the 3-dimesionaly ordered features. Reprinted with permission from [Ref. 41], A. Wright, J. Gabaldon, D. B. Burckel, Y. B. Jiang, Z. R. Tian, J. Liu, C. J. Brinker, H. Y. Fan, Chem. Mater. 18, 3034 (2006), Copyright @ American Chemical Society.

Substrate surface properties strongly influence the nucleation and growth of inorganic materials [43]. In addition to amorphous glasses, other substrates such as mica, thermal oxide coated silicon wafer, and photo-addressable polymer films have been used to grow mesophase crystals. In a controlled experiment, fresh-peeled mica was used instead of glass, resulting in a distinct crystal shape. Figure 12 shows a typical SEM image of the large mesophase crystals formed on a mica surface.

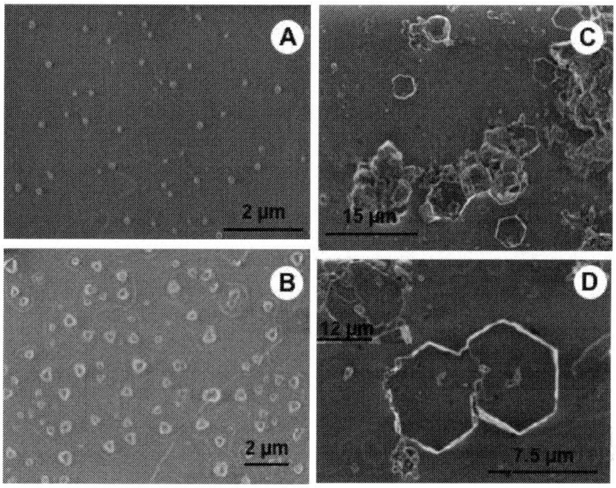

Figure 10. SEM images of samples at different growth time. At 10 minutes (A), ~0.3μm particles were observed. At 20 minutes (B), crystals with triangular shape were observed. At 40 minutes (C and D), a single subunit and some of the intermediate hierarchical structures have been formed. Reprinted with permission from [Ref. 41], A. Wright, J. Gabaldon, D. B. Burckel, Y. B. Jiang, Z. R. Tian, J. Liu, C. J. Brinker, H. Y. Fan, Chem. Mater. 18, 3034 (2006), Copyright @ American Chemical Society.

As with those grown on glass, these crystals are fairly uniform in shape and size (~ $12\mu m$), but the crystal possesses circular rather than flower-like structure, and has only one single unit instead of several merged subunit crystals. Another remarkable difference is that the orientated NP mesophase began to form at 30 minutes on mica, earlier than on amorphous glass. This could be due to the fact that mica has a crystalline surface with hexagonal patterns that promote preferential growth along (111) planes. Similar to the case of amorphous glass, the XRD patterns (Figure 11B, curve b&c), show enhanced reflections of the (111) and (222) planes suggesting the (111) planes of NP cubic mesophase are parallel to the substrate. It is noteworthy that prior work in self-assembly of silica and surfactants indicated that only a 1-dimensional hexagonal mesophase was formed. It was proposed that surfactants form hemi-rod micelles on the mica surface, leading to a 1-dimensional surfactant/silica mesophase [44]. In the current system, NP-micelles are pre-formed in a homogeneous aqueous solution. Self-assembly with organo-bridged silsesquioxane starts with "hard sphere" NP-micelle building blocks instead of "soft" pure surfactant micelles or liquid crystals. Thus, the cubic mesophase is preferentially formed.

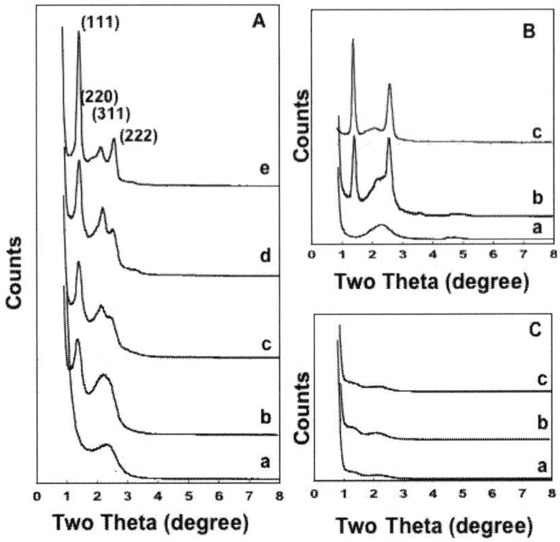

Figure 11. XRD patterns (A) of hierarchical NP mesostructure arrays on glass grown (a) through (e) for 10, 20, 30, 40, 60 minutes, respectively. The primary peaks for sample e can be assigned as (111), (211), (220), and (222) reflections of *fcc* with lattice constant a = 109.1Å. XRD patterns (B) of hierarchical NP mesostructure arrays on mica grown (a) through (c) for 10, 30, and 60minutes, respectively. XRD patterns (C) of powder samples precipitated in solutions at (a) through (c) 40, 60, and 120 minutes, respectively. The samples were prepared using ~2-nm gold NPs, cetyltrimethyl ammonium bromide surfactant, and sodium hydroxide. Reprinted with permission from [Ref. 41], A. Wright, J. Gabaldon, D. B. Burckel, Y. B. Jiang, Z. R. Tian, J. Liu, C. J. Brinker, H. Y. Fan, Chem. Mater. 18, 3034 (2006), Copyright @ American Chemical Society.

Following growths longer than 40 minutes, precipitate was observed in the bottom of the vial. To further explore the formation mechanism, we filtered the growth solution and performed XRD studies on the powders. Figure 11C shows the XRD patterns from the precipitated powders after growth for 40, 60 and 120 minutes. The irregular patterns suggest much less ordered/organized NP/silsesquioxane arrays. This unambiguously establishes that the mesophase crystals are grown from the solution rather than deposited via precipitation. Any gold NP agglomeration occurring during growth could be expected to have significant impact on the self-assembly process. A UV-vis spectrometer was used to measure the wavelength of the characteristic surface plasmon resonance, providing a measure of the size-dispersity of gold NP-micelles in the solution during the growth process: Relative to the original n-DT capped gold NPs, the gold surface plasmon band exhibited no shift, both for as-prepared solutions and after growth for 40, 60, and 120 minutes. This suggests that the gold NP-micelles remain monodisperse during growth and thus that the formation of the mesophase crystals was initiated from NP-micelle building blocks.

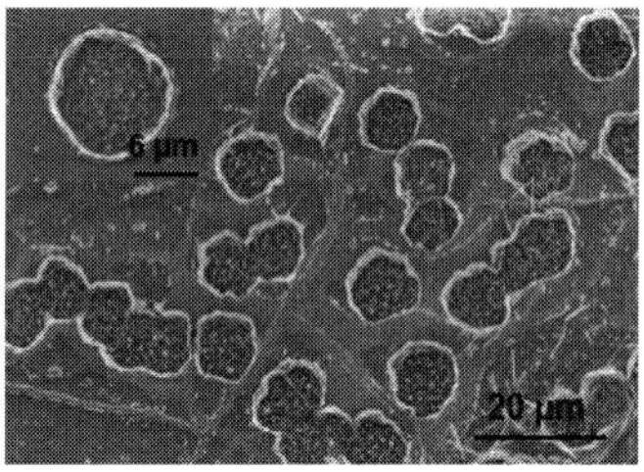

Figure 12. SEM image of hierarchical NP mesostructure arrays grew on mica surface using BTEE. Inset: a magnified SEM image showing an individual mesophase crystal. Reprinted with permission from [Ref. 41], A. Wright, J. Gabaldon, D. B. Burckel, Y. B. Jiang, Z. R. Tian, J. Liu, C. J. Brinker, H. Y. Fan, Chem. Mater. 18, 3034 (2006), Copyright @ American Chemical Society.

Thermal and mechanical stability is critical for NP arrays to be applied in practical nanodevices. Heat treatment over time usually broadens the NP size distribution [45]. We expect the organo-bridged silsesquioxane framework will provide extensive protection from thermally induced particle agglomeration. After crystal growth at 100°C for up to 3 hours, the gold NPs within the hierarchical mesophase crystals remain monodisperse, as shown by TEM (Figure 9C&D). UV-vis spectra further verified that the NP monodispersity was preserved inside the mesophase crystals after thermal annealing, showing no changes in surface plasmon resonance band of gold NP. This implies that the inorganic framework provides enhanced thermal stability. Additionally, in comparison with ordered NP superlattice films that exhibits cracking, these hierarchical NP mesophase crystals show no cracking, unit cell distortion or shrinking even after heating to 100°C for several hours, further confirming the mechanical stability.

3.2. Synthesis of Ordered NP Arrays in Thin Films

In aqueous media, NP-micelles organize hydrophilic components/precursors at the surfactant/water interface through electrostatic and hydrogen-bonding interactions in a mechanism analogous to that of surfactant-directed self-assembly of silica/surfactant mesophases (so-called 'mesoporous silicas') [42]. For example, addition of tetraethyl orthosilicate under basic conditions results in the formation of hydrophilic oligo-silicic acid species that organize with NP-micelles to form a new type of ordered gold NP/silica mesophase with face-centered-cubic symmetry (space group Fm3m). Figure 13B shows a representative low angle x-ray diffraction pattern of a NP mesophase powder prepared according to pathway i-ii-iii (Figure 1) using 2-nm diameter C_{12}-thiol stabilized gold NPs, cetyltrimethylammonium bromide ($C_{16}TAB$) surfactants, and sodium hydroxide catalyst. Based on *fcc* symmetry the primary peaks are assigned as 111, 220, and 311 reflections. Figures 14A and B show representative TEM images of [001]- and [012]-oriented NP mesophases (prepared as for Figure 13B) along with their corresponding electron diffraction patterns. The TEM images are consistent with an *fcc* unit cell with a = ~10.2 nm and a uniform, minimum (silica/surfactant) spacing between NPs of ~6 nm. This appears to be the first example of an ordered *fcc* NP array formed spontaneously by self-assembly in aqueous media (rather than by solvent evaporation [46,47]). Compared to other ordered NP arrays, the embedding silica matrix provides for greater chemical, mechanical, and thermal robustness and, compared to other connected NP systems (for example, those prepared by DNA hybridization [5,48]), thermodynamically controlled self-assembly provides greater order and control of NP spacing.

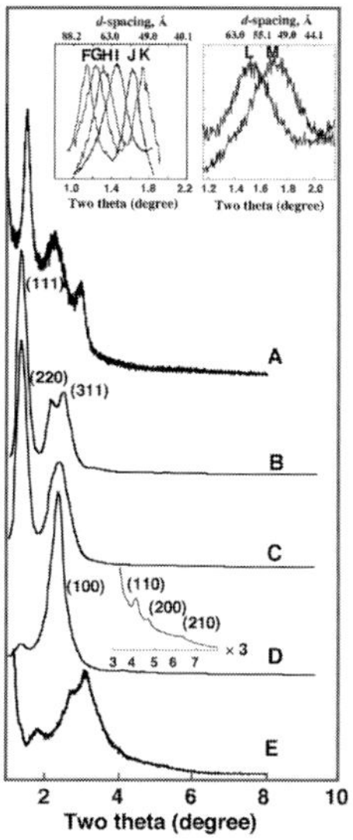

Figure 13. Representative x-ray diffraction patterns of gold NP/silica mesophases. Gold NP/silica thin film mesophase (A) formed by spin-coating. The pattern can be indexed as an *fcc* cubic mesostructure with lattice constant a = 95.5Å. XRD patterns (B –D) of bulk gold NP/silica samples prepared according to pathway i-ii-iii (Figure 1) by addition of NaOH to an aqueous solution containing TEOS and gold NP-micelles. The weight percent of gold NP micelles was progressively reduced from B to D, while the TEOS/surfactant molar ratios were kept constant. Sample (B) was prepared by using 0.36g 1-DT derivatized gold NPs. The pattern is consistent with an *fcc* cubic mesostructure with lattice constant a = 102Å. Sample (C) was prepared using 0.23g 1-DT derivatized gold NPs. The pattern is similar to that of sample B. However, the (220) and (311) reflections are not resolved. Sample (D) was prepared by using 0.12g 1-DT derivatized gold NPs. The XRD pattern (inset magnified 3x) is indexed as a hexagonal silica mesophase with lattice constant a = 43.4Å. A small (111) peak attributable to the NP/silica mesophase is still observed. Sample (E) was prepared using 0.36g 1-DT derivatized gold NPs. These NPs were synthesized directly by the method of Brust *et al.* without narrowing the particle size distribution. The amount of TEOS and surfactant were the same as B-D. The broader NP size distribution reduces considerably the extent of order. Insets: traces (F) through (K) are magnifications of the (111) reflections plotted linearly for samples prepared as in (B) using gold NPs with successively smaller diameters (F~33Å, G~27Å, H~23Å, I~17Å, J~12Å, K~11Å). Traces (L) and (M) are magnifications of the (111) reflections plotted linearly for samples prepared as in (B) with two different secondary alkane chain lengths of the surfactant ($CH_3(CH_2)_n(NCH_3)_3^+Br$-, n_L=15, n_M=11). Reprinted with permission from [Ref. 21], H. Y. Fan, K. Yang, D. Boye, T. Sigmon, K. Malloy, H. Xu, G. P. Lopez, C. Brinker, Science 304, 567 (2004), Copyright @ AAAS.

It is interesting to note that reducing the concentration of the gold NPs, while maintaining a constant surfactant/silica molar ratio, caused the progressive transformation of the cubic, gold NP/silica mesophase to a 2-D hexagonal silica/ surfactant mesophase (Figure 13 XRD patterns C and D). This is evident from the greatly diminished (111) reflection in XRD pattern D and the appearance of (100), (110), (200), and (210) reflections (see XRD pattern and D inset). This emphasizes that the highly ordered *fcc* gold NP/silica mesophase forms within a limited NP/surfactant/silica compositional range in which all the surfactant and gold is incorporated into NP micelles, allowing the host framework to be essentially pure, dense silica (polysilicic acid). Highly ordered NP mesophases also require very monosized NPs as is demonstrated by the broad, poorly-defined XRD pattern (Figure 13E) obtained for a gold NP/silica powder sample prepared as in Figure 13B but using gold NPs synthesized according to Brust *et al.* [49] and known to have a rather broad size distribution (*ca* 50% compared to *ca* 5% in the present study).

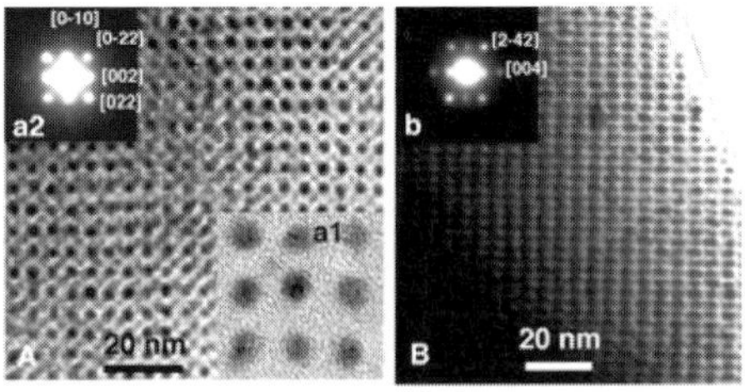

Figure 14. TEM images of gold NP/silica mesophases. [100] and [210] orientations (A and B) of bulk samples prepared according to pathway i-ii-iii (Figure 1) and corresponding to Figure 13, trace B. Inset (a1): high resolution TEM of sample (A) showing gold NP lattice fringes. Inset (a2): selected area diffraction pattern from the image in (A). Inset (b) selected area electron diffraction from the image in (B). Reprinted with permission from [Ref. 21], H. Y. Fan, K. Yang, D. Boye, T. Sigmon, K. Malloy, H. Xu, G. P. Lopez, C. Brinker, Science 304, 567 (2004), Copyright @ AAAS.

As suggested by Figure 1H, changing the primary NP particle size d_p, the primary layer thickness d_1, or the secondary layer thickness d_2 allows adjustment of the NP mesophase lattice constant. For example, Insets F through K in Figure 13 show the (111) *d*- spacing to change linearly from *ca* 5.0 to 7.2-nm through variation of d_p from 1.0 to 3.3-nm. Insets L and M show that changing the secondary layer thickness d_2 by 4 carbon units ($CH_3(CH_2)_n(NCH_3)_3^+Br^-$, n_L=11 to n_K=15) results in a 1.11-nm change in (111) *d*- spacing (1.38-Å/C-C bond) consistent with model predictions and structural studies.

For device fabrication, thin film is more desirable than powder. Using acidic conditions designed to minimize the siloxane condensation rate (pH~2), pathway i-ii-iv (Figure 1) leads to the formation of thin films using standard techniques like dip-coating, spin-coating, or micro-molding. By suppressing siloxane condensation and, thereby, gel formation, solvent evaporation accompanying coating induces self-assembly of NP-micelles into *fcc* NP thin film mesophases (Figure 13D) in a manner similar to the evaporation-induced self-assembly of cubic or hexagonal silica/surfactant thin film mesophases [46,50]. Silica condensation played important role in formation of ordered 3-D gold NP/silica arrays [51]. Oligo silica species in homogeneous coating solution is essential to form ordered, transparent gold/silica thin films without cracking. Extensive silica condensation leads to less ordered gold/silica mesophase. To this end, the precursor solution was prepared in acidic conditions (pH~2) by addition of aqueous hydrogen chloride to maximize silica gel time, facilitating the self-assembly and formation of ordered NP/silica thin films. By using organo-silsesquioxane, we were able to tune the framework chemistry and dielectrics. The final film consists of monodispersed gold NPs arranged within silica or organo-silsesquioxane host matrices in an *fcc* mesostructure with precisely controlled interparticle spacing.

The gold/silica superlattice films have an average refractive index of ~1.7 and thicknesses of 100 to 300-nm depending on the spin rate. The film thickness can reach up to 25μm when casting. Figure 15d shows an optical micrograph of a film that exhibits reddish color and very good optical transparency. Optical properties of the superlattice film have been characterized using UV-vis spectroscopy as shown in Figure 15a-c. The superlattice film exhibits a characteristic surface plasmon resonance band at ~518-nm, as expected from gold NPs. In comparison with the spectra of monodisperse DT-stabilized gold NPs, the position and width of the surface plasmon band from gold superlattice film stay unchanged, suggesting that gold NPs inside the film remain monodisperse. This is further confirmed by TEM results (see below). Low-magnification scanning electron micrograph (SEM) imaging (Figure 16A) shows that the film has a uniform and continuous surface without macroscopic granularity or cracks. Figure 16B shows a high-resolution SEM image of as-prepared film surface. Ordered gold NPs are distributed uniformly on the film surface.

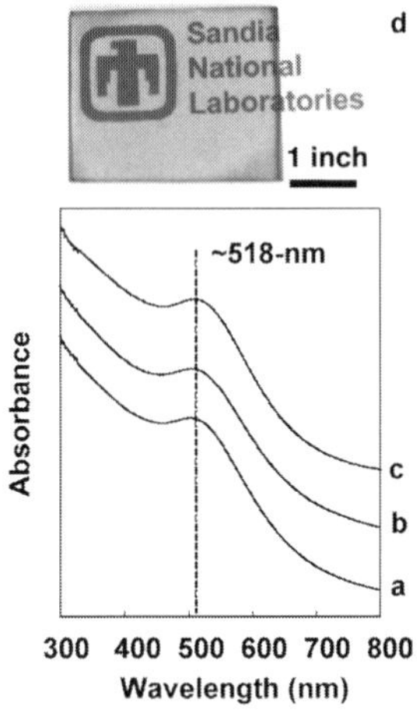

Figure 15. UV-vis spectra of DT-stabilized gold NPs (a), gold NP/silica superlattice films prepared by using TEOS (b) and (c) bis(triethoxylsilyl) ethane. An optical micrograph (d) of as-prepared ordered gold/silica film on glass substrate. Reprinted with permission from [Ref. 51], H. Y. Fan, A. Wright, J. Gabaldon, A. Rodriguez, C. J. Brinker, Y. Jiang, Adv. Func. Mater. 16, 891 (2006), Copyright @ Wiley InterScience.

Figure 16-C3 shows a representative low angle x-ray diffraction (XRD) pattern of a superlattice thin film prepared according to pathway i-ii-iii (Figure 1), using ~3-nm DM-stabilized gold NPs, CTAB, and TEOS. The patterns can be indexed as an *fcc* structure with unit cell a = ~10.5-nm. The primary peaks are assigned as (111), (220)/(311), and (222) reflections. Figure 16-C1 shows the XRD pattern of a film prepared using a spin solution aged at room temperature for 24 hours. The film exhibits lower degree of ordering with disappearance of (220)/(311) and intensity decrease of (222). Figure 17A shows a representative cross-sectional TEM image of an ordered superlattice film. Periodically ordered regions are observed throughout the whole film thickness. The sharp, continuous, and uniform air-film and film-substrate interfaces are consistent with SEM results showing no steps, kinks, or cracking that appear in superlattice films prepared by evaporation of NP solution [1,8,14]. A representative plan-view TEM image is shown in Figure 17B. The TEM images are consistent with an *fcc* structure (Fm3m space group) with a measured unit cell a = ~10.8-nm and a minimum average interparticle spacing is ~2.3-nm. Note that the gold NP/silica superlattice thin film has a larger unit cell than that of previous superlattice films and solids [14]. This is due to that here the gold NPs are arranged within a silica matrix. There exists a thin silica layer between each individual gold NP. The silica-insulating layer was further confirmed by high-resolution transmission electron micrograph (Figure 17B inset). As shown in plan-view image (Figure 17B), regions of ordered gold NP arrays inside silica exhibit no preferred orientation with seamless transition between ordered domains of gold NPs. Within the film, the gold NPs remain monodisperse.

Formation of ordered gold NP/silica thin films is analog to that of self-assembly of surfactant and silica [42,46,50]. Charge interaction and hydrogen bonding between hydrolyzed silica and surfactant head groups on NP-micelle surface drive the formation of ordered gold NP/silica mesophase [52]. However, the two systems exhibit distinct tendency to form mesostructures. Prior work on self-assembly of pure surfactant and silica indicated that a series of mesostructures can be formed including lamellar, 1-d hexagonal, cubic, and 3-d hexagonal periodic symmetries. In the case of self-assembly of NP-micelles and silica, only *fcc* mesostructure are preferentially formed regardless of basic and acidic catalytic conditions. This is probably due to the fact that the gold NP-micelles are pre-formed in a homogeneous solution and behave rather like a "hard" sphere tending to form *fcc* close packing than "soft" pure surfactant micelles that incline to undergo phase transformation. Vital to the formation of transparent, ordered gold NP/silica superlattice films is the use of stable and homogeneous spinning or casting solution that upon evaporation of water undergoes self-assembly of NP-micelles and soluble silica. For this purpose, we prepared oligomeric silica sols in NP-micelle aqueous solution at a low hydronium ion concentration (pH~2) designed to minimize the siloxane condensation rate, thereby enabling facile silica and NP-micelle self-assembly during spin-coating or casting

[46]. The aging experiments (Figure 16, C1 to C5) unambiguously demonstrate that extensive silica condensation, that results in polymeric silica species, does not favor the self-assembly, leading to a less ordered film. The method is flexible and allows tuning of framework composition, thus dielectrics, by using different sol-gel precursors. Organo-bridged silsesquioxane is an ideal low *k* host, exhibiting chemical and mechanical robustness [53,54]. In addition to silica, we have demonstrated the synthesis of ordered gold NP arrays inside organo-silsesquioxane framework. The ordered gold NP/ silsesquioxane was prepared by using ~3-nm DM-stabilized gold NPs, CTAB, and BTEE. The corresponding XRD patterns (Figure 16-C4&5) and TEM image (Figure 17C) reveal that films exhibit ordered *fcc* mesostructure.

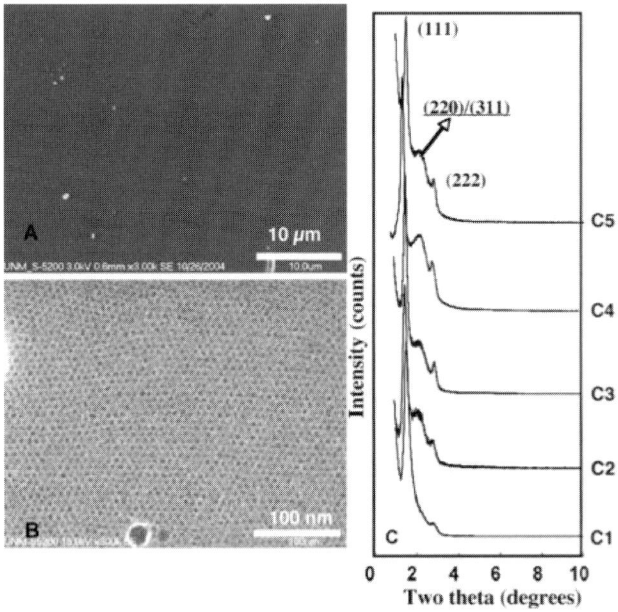

Figure 16. LR-SEM image (A) of ordered gold NP/silica superlattice film. HR-SEM (B) from same specimen in (A). XRD patterns (C) of gold NP/silica superlattice films. The film (C1) was prepared using a coating solution that was aged at ambient condition for 24 hours, and (C2) for 5 hours. Ordered gold/silica film (C3) prepared using a coating solution without aging. The film (C4) was prepared using a solution that was aged at ambient condition for 24 hours, and without aging (C5). Reprinted with permission from [Ref. 51], H. Y. Fan, A. Wright, J. Gabaldon, A. Rodriguez, C. J. Brinker, Y. Jiang, Adv. Func. Mater. 16, 891 (2006), Copyright @ Wiley InterScience.

In addition, we observed from XRD results that the self-assembly when using BTEE is not strongly affected by solution aging unlike that when using TEOS. This is due to that organo-bridged precursor has relatively slower hydrolysis and condensation rate than TEOS [55]. The ability to form patterned films is essential for device fabrication. The homogeneous solution of gold NP-micelle allows using several soft lithographic techniques such as μ-molding, pen writing, and ink-jet printing, etc [47,56] to pattern the ordered gold NP/silica superlattice films. We have demonstrated the formation of patterned gold NP/silica superlattice films based on our previous work on patterning surfactant templated silica mesophases. Figure 18D shows the patterned stripes and dots containing ordered gold NP/silica superlattice fabricated using μ-molding techniques [57]. The pattern sizes are determined by the feature sizes of the PDMS stamps.

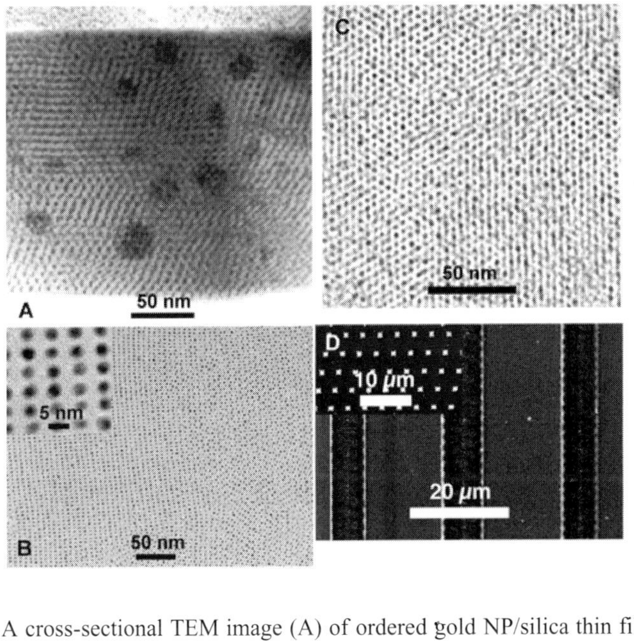

Figure 17. A cross-sectional TEM image (A) of ordered gold NP/silica thin film. A plan view TEM micrograph (B) of ordered gold NP/silica films through complete film thickness showing seamless transition among three ordered domains [211] and [111]. Inset is an HR-TEM image. A plan view TEM image (C) of ordered gold NP/silsesquioxane superlattice films prepared by using BTEE. Patterned gold NP/silica is shown in (D). Reprinted with permission from [Ref. 51], H. Y. Fan, A. Wright, J. Gabaldon, A. Rodriguez, C. J. Brinker, Y. Jiang, Adv. Func. Mater. 16, 891 (2006), Copyright @ Wiley InterScience.

In comparison to the previous methods to assemble NP superlattice through evaporation of a colloidal solution of NPs, our method provides several advantageous features. First, unlike previous superlattice films formed through evaporation of organic flammable solutions, the water-soluble gold NP-micelles allow to make superlattice films in water, resulting in enhanced safety and better compatibility with current semiconductor fabrication processing. Second, by using different sol-gel precursors, our method allows simple tuning of framework composition, thus dielectrics, between gold NPs. This is essential to achieve enhanced collective properties of such three-dimensional superlattice films [13,58]. Furthermore, the inorganic framework provides chemical and mechanical robustness, prevents films from cracking, which is important for device fabrication. It is believed that formation of NP superlattice through organic solvent evaporation is an entropy-driven process in which NPs organize in a way to achieve the highest packing density or maximum entropy. In our method, silica condensation affects ordering during formation of gold NP/silica superlattice film. It is interesting to note that previous methods to assemble NPs during solvent evaporation leads to two-level preferential orientations. First, the ordered NPs pack as *fcc* structure with (111) planes parallel to substrates [8,59]. Second, the crystal structure of each individual NP is also orientated relative to substrates [8]. In our system, the films consist of ordered "domains" randomly distributed throughout the film. From wide-angle XRD, we observe no preferred orientation of gold crystal structure relative to substrate.

4. Integration of NP Arrays for Charge Transport Study

The ordered arrays of metallic NPs in silica are potential implementations of several types of model systems. Electrostatically, such an ordered array of metallic particles is identical to the composite model systems used in textbooks to describe the properties of polarizable media [60]. They also may be a means of implementing the NP memory devices described by Tiwari and others [61,62]. In those applications, the NPs are present in a single plane and serve to store charge and modify the channel conductance. However, the three-dimensional array of NPs presented here offers a variety of opportunities to study the physics of transport. Thus our structures could permit experiments and encourage modeling of new systems.

As an initial investigation of charge transport in ordered three-dimensional NP arrays, planar metal-oxide-semiconductor (MOS) devices were fabricated (see schematic in Figure 18) with a gold NP/silica mesophase 'oxide' prepared according to pathway i-ii-iv (Figure 1). Charge storage and decay was assessed by measuring the time and voltage behavior of the capacitance. Silicon (100) *p*-type wafers of approximately 10^{15} cm^{-3} doping were cleaned using standard procedures [63], and backside contacts were deposited using ~ 420 nm of *e*-beam evaporated Al, followed by a 450 ºC forming gas anneal for 25 minutes. Gold NP/silica films, approximately 90-nm thick, were deposited by spin-coating and annealed in *UV* at room temperature for 1.5 hours to remove organics and promote additional siloxane condensation [64]. Capacitor structures were formed by *e*-beam evaporation of Al (~ 300 nm) through a shadow mask. Control samples were fabricated by spin-coating gold-free silica sols designed to form silica layers comparable to the silica matrices of the NP/silica films and by using a conventional thermal silicon dioxide insulator. High frequency capacitance-voltage (*C-V*), current-voltage (*I-V*) and charge storage measurements were performed on all films at room temperature.

Figure 18 presents the results of electric field-aided transport measurements performed on both types of MOS devices. For this measurement the sample is initially biased at V_1 for 50 seconds, then rapidly switched (<1μs) to V_2 and the time evolution of the capacitor charge monitored. Samples without gold exhibit an exponential discharge (with a time constant ~100 μs) consistent with normal *RC* discharge behavior. For the samples with gold, an ogive (S-shaped) profile with a ~7.5 second discharge time is observed. The total excess charge contained in this capacitor is approximately 2.8×10^{-11} C. Assuming the capacity of each NP to be one electron based on electrostatic energy considerations, this corresponds to 10^{18} cm^{-3} charged NPs. This is attributed to the charging of the gold in the oxide near the gate electrode when the gate is negatively biased. A diagram depicting the charges within the oxide and *p*-doped Si substrate is shown in Figure

19C for the gate bias sequence studied. After reversing the gate voltage to V_2, the electrons on the gold NPs are swept out of the oxide into the gate sequentially from the gate side first. No change in substrate capacitance occurs until the gold NPs in the oxide are discharged as they effectively pin the Si-surface in accumulation.

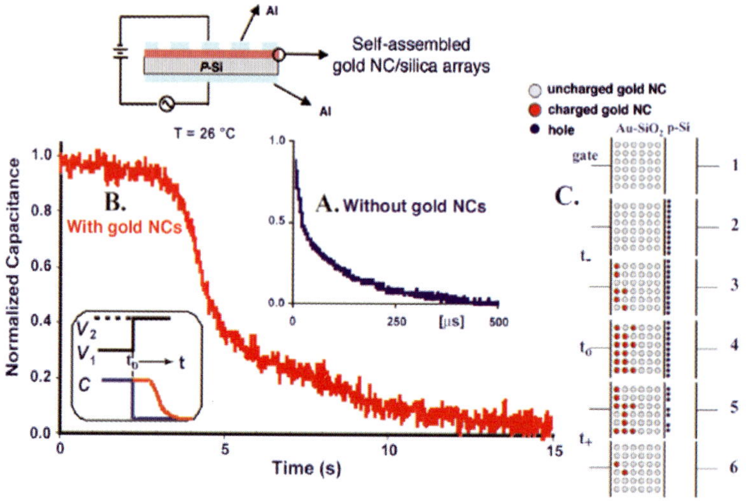

Figure 18. Time dependence of normalized capacitance for MOS capacitors prepared using silica or ordered gold NC/silica as the 'oxide' dielectrics. The charge is measured when the capacitors are switched from V_1 to V_2. The discharge time for capacitors without gold NPs (A) is exponential with a characteristic time constant of ~100 μs. For gold NC-containing capacitors (B), the decay is no longer exponential and the discharge time (10 to 90%) is about 7.5 seconds. The time behavior of the discharge curve may be explained by following the biasing sequence in (C). Diagram 1) depicts an unbiased neutral p-type Si substrate, and 2) shows that once a negative bias V_1 is placed on the gate, an accumulation of holes is quickly established at the Si-oxide interface. In 3) the capacitance of the device with gold NPs is higher than that of an equivalent gold NP-free device due to the excess stored charge in the gold levels. Experiments indicate that this process continues until all of the gold NP's in the first 25% of the oxide film near the gate are charged. Upon reduction of the gate bias to V_2 in 4), the accumulation layer begins to disappear. However, in 5) the accumulation layer is partially sustained in the presence of the charged gold NPs. Only as the electrons on the gold NPs move into the gate does the accumulation layer dissipate completely as shown in 6). The minimum capacitance for these structures occurs when the p-type semiconductor surface is biased into inversion. This value is the same for both gold NP and NP-free capacitors. Reprinted with permission from [Ref. 21], H. Y. Fan, K. Yang, D. Boye, T. Sigmon, K. Malloy, H. Xu, G. P. Lopez, C. Brinker, Science 304, 567 (2004), Copyright @ AAAS.

Using the *fcc* lattice constant, a uniform gold NP concentration was estimated in the oxide of ~$4x10^{18}$ cm^{-3}. However, only those NPs located near the gate electrode can respond to the high frequency signal used to measure the capacitance, resulting in roughly *all* of the NPs in the first 25% of the oxide film being occupied by electrons. Given the spacing for these dots, coulomb blockade effects are expected to control transport amongst the gold NPs. However, disorder and trapping within the silica matrix could prevent collective effects, so transport in this situation is probably also influenced by a combination of local kinetic and diffusive factors, particularly at room temperature where these experiments were performed. Still, the charge storage and transport behavior is completely different from that of the corresponding MOS capacitor prepared *sans* gold with silica identical to the host matrix of the NP array, it is evident that charge is stored on the gold NPs and that the discharge characteristics are dominated by electron transport involving the NPs.

5. Conclusions and Outlook

The synthesis of NP-micelles represents a new synthetic methodology toward the mass-production (grams) of water-soluble NPs. The method is easy, rapid (10 minutes), and avoids the complicated multi-step procedures reported previously. The robustness of this approach should allow formation of a variety of other water-soluble nanomaterials such as C60, carbon nanotubes, porphorins that currently require complex procedures to achieve water solubility. Switching from normal surfactants to phospholipids enable the NP-micelles to be biocompatible, which is essential for them to be used for bioapplications. The dense- packed surfactant/lipid layers help to prevent NP s from oxidation and preserve their physical properties, such as luminescence, etc. depending on the surfactants used, the NP-micelles provide varied surface chemistry and ability for further reactions and cross-linking with other ligands.

With improvement, these NP arrays could be the ideal media for the study of the Hubbard Hamiltonian and the variety of transport and collective phenomena predicted to occur for such systems. Key to such studies is the ability to control the array lattice constant and hence the spacing between individual NPs as we demonstrate. Briefly summarizing the implications for transport studies, wider spacing between the NP s leads to weaker coupling and creates the situation for observing the collective coulombic blockade behavior described for a one-dimensional array [65]. Stronger coupling arising in more closely spaced dots permits efficient tunneling between the dots and observation of such phenomena as quantum Hall behavior and intricate energy gap dependence on magnetic field as described for a two dimensional array of quantum dots in [66]. The ability to modulate this behavior with an external electric field may allow a controllable Mott transition such as discussed in [67].

Beyond transport, the robust, highly ordered NP arrays we described could be useful in photonic devices such as lasers and the water-soluble NP micelle intermediates are promising for biological labeling. Finally, due to their uniform gold spacings, NP arrays may serve as an optimal platform for sensors and molecular electronic 'nanocell logic gates' formed by 'wiring' the NPs with conjugated dithiol oligomers for chemical sensors [68].

Acknowledgements

This work was partially supported by the U.S. Department of Energy (DOE) Basic Energy Sciences Program, Sandia National Laboratory's Laboratory Directed R&D program, and Center for Integrated Nanotechnologies (CINT). TEM investigations were performed in the Department of Earth and Planetary Sciences at the University of New Mexico. We acknowledge the use of the SEM facility supported by the NSF EPSCOR and NNIN grants. Sandia is a multiprogram laboratory operated by Sandia Corporation, a Lockheed Martin Company, for DOE under contract DE-AC04-94ALB5000.

References

1. A. P. Alivisatos, V. F. Puntes, K. M. Krishnan, M. Bruchez, M. Moronne, P. Gin, S. Weiss, Science 933 (1996).
2. M. A. El-Sayed, Accounts of Chemical Research 34, 257 (2001).
3. B. Dubertret, P. Skourides, D. J. Norris, V. Noireaux, A. H. Brivanlou, A. Libchaber, Science v.298, p.1759 (2002).

4. V. I. Klimov, A. A. Mikhailovsky, S. Xu, A. Malko, J. A. Hollingsworth, C. A. Leatherdale, H. J. Eisler, M. G. Bawendi, Science 290, 314 (2000).
5. C. A. Mirkin, R. L. Letsinger, R. C. Mucic, J. J. Storhoff, Nature v.382, p.607 (1996).
6. C. Murray, C. Kagan, M. Bawendi, Ann. Rev. Mater. Sci. 30, 545 (2000).
7. S.H. Sun,C.B. Murray,D. Weller,L. Folks,A. Moser, Science 287, 1989 (2000).
8. C. B. Murray, C. R. Kagan, M. G. Bawendi, Science 270, 1335 (1995).
9. V. I. Klimov, A. A. Mikhailovsky, D. W. McBranch, C. A. Leatherdale, M. G. Bawendi, Science 287, 1011 (2000).
10. C. P. Collier, R. J. Saykally, J. J. Shiang, S. E. Henrichs, J. R. Heath, Science 277, 1978 (1997).
11. M. P. Pileni, J. Phys. Chem. B 105, 3358 (2001).
12. M. A. Petruska, A. V. Malko, P. M. Voyles, V. I. Klimov, Adv. Mater. 15, 610 (2003).
13. V. C. Sundar, H. J. Eisler, M. G. Bawendi, Adv. Mater. 14, 739 (2002).
14. Z. L. Wang, Adv. Mater. v.10, p.13 (1998).
15. A. A. Middleton, N. S. Winggreen, Phys. Rev. Lett. 71, 3198 (1993).
16. R. Parthasarathy, X. Lin, K. Elteto, T. F. Rosenbaum, H. M. Jaeger, Phys. Rev. Lett. 92, 076801 (2004).
17. Y. W. C. Cao, R. C. Jin, C. A. Mirkin, Science 297, 1536 (2002).
18. H. Y. Fan, Z. Chen, C. Brinker, J. Clawson, T. Alam, J. Am. Chem. Soc. 127, 13746 (2005).
19. H. Y. Fan, E. Leve, J. Gabaldon, A. Wright, R. Haddad, C. Brinker, Adv. Mater. 17, 2587 (2005).
20. H. Y. Fan, E. W. Leve, C. Scullin, J. Gabaldon, D. Tallant, S. Bunge, T. Boyle, M. C. Wilson, C. J. Brinker, Nano Lett 5, 645 (2005).
21. H. Y. Fan, K. Yang, D. Boye, T. Sigmon, K. Malloy, H. Xu, G. P. Lopez, C. Brinker, Science 304, 567 (2004).
22. M. Bruchez, M. Moronne, P. Gin, S. Weiss, A. P. Alivisatos, Science 281, 2013 (1998).
23. Z. A. Peng, X. G. Peng, J Am. Chem. Soc.124, 3343 (2002).
24. H. Zeng, J. Li, J. P. Liu, Z. L. Wang, S. H. Sun, Nature 420, 395 (2002).
25. W. C. W. Chan, S. M. Nie, Science 281, 2016 (1998).
26. D. Gerion, F. Pinaud, R. C. Williams, W. J. Parak, D. Zanchet, S. Weiss, A. P. Alivisatos, J. Phys. Chem. B 105, 8861 (2001).
27. D. R. Larson, W. R. Zipfel, R. M. Williams, S. W. Clark, M. P. Bruchez, F. W. Wise, W. W. Webb, Science 300, 1434 (2003).
28. I. L. Medintz, A. R. Clapp, H. Mattoussi, E. R. Goldman, B. Fisher, J. M. Mauro, Nature Mater. 2, 630 (2003).
29. T. Pellegrino, L. Manna, S. Kudera, T. Liedl, D. Koktysh, A. L. Rogach, S. Keller, J. Radler, G. Natile, W. J. Parak, Nano Lett 4, 703 (2004).

30. J. O. Winter, T. Y. Liu, B. A. Korgel, C. E. Schmidt, Adv. Mater. 13, 1673, (2001).
31. X. Y. Wu, H. J. Liu, J. Q. Liu, K. N. Haley, J. A. Treadway, J. P. Larson, N. F. Ge, F. Peale, M. P. Bruchez, Nature Biotech. 21, 41 (2003).
32. X. G. Peng, L. Manna, W. D. Yang, J. Wickham, E. Scher, A. Kadavanich, A. P. Alivisatos, Nature 404, 59 (2000).
33. C. D. Donega, S. G. Hickey, S. F. Wuister, D. Vanmaekelbergh, A. Meijerink, J. Phys. Chem. B 107, 489 (2003).
34. J. Aldana, Y. A. Wang, X. G. Peng, J. Am. Chem. Soc. 123, 8844 (2001).
35. S. F. Wuister, C. D. M. Donega, A. Meijerink, J. Am. Chem. Soc. 126, 10397 (2004).
36. K. Hanaki, A. Momo, T. Oku, A. Komoto, S. Maensono, Y. Yamaguchi, K. Yamaguchi, Biochem. Biophys. Res. Commun. 302, 496 (2003).
37. J. K. Jaiswal, H. Mattoussi, J. M. Mauro, S. M. Simon, Nature Biotechnol. 21, 47 (2003).
38. D. S. Lidke, P. Nagy, R. Heintzmann, D. J. Arndt-Jovin, J. N. Post, H. E. Grecco, E. A. Jares-Erijman, T. M. Jovin, Nature Biotech. 22, 198 (2004).
39. P. De Camilli, V. I. Slepnev, O. Shupliakov, L. Brodin, Synaptic Vesicle Endocytosis, The Johns Hopkins Press, Baltimore, 2001.
40. T. Shitani, D. J. Klionsky, Science 306, 990 (2004).
41. A. Wright, J. Gabaldon, D. B. Burckel, Y. B. Jiang, Z. R. Tian, J. Liu, C. J. Brinker, H. Y. Fan, Chem. Mater. 18, 3034 (2006).
42. C. Kresge,M. Leonowicz,W. Roth,C. Vartuli,J. Beck, Nature 359, 710, (1992).
43. J. Aizenberg, A. J. Black, G. M. Whitesides, Nature 398, 495 (1999).
44. H. Yang, A. Kuperman, N. Coombs, S. MamicheAfara, G. A. Ozin, Nature 379, 703 (1996).
45. M. M. Maye, W. X. Zheng, F. L. Leibowitz, N. K. Ly, C. J. Zhong, Langmuir 16, 490 (2000).
46. C. J. Brinker, Y. F. Lu, A. Sellinger, H. Y. Fan, Adv. Mater.11, 579, (1999).
47. H. Y. Fan, Y. F. Lu, A. Stump, S. T. Reed, T. Baer, R. Schunk, V. PerezLuna, G. P. Lopez, C. J. Brinker, Nature 405, 56 (2000).
48. A. P. Alivisatos, K. P. Johnsson, X. G. Peng, T. E. Wilson, C. J. Loweth, M. P. Bruchez, P. G. Schultz, Nature v.382, p.609 (1996).
49. M. Brust, M. Walker, D. Bethell, D. J. Schiffrin, R. Whyman, Chem. Commun. 801 (1994).
50. H. Y. Fan, Y. F. Lu, A. Stump, S. T. Reed, T. Baer, R. Schunk, V. Perez-Luna, G. P. Lopez, C. J. Brinker, Nature 405, 56 (2000).
51. H. Y. Fan, A. Wright, J. Gabaldon, A. Rodriguez, C. J. Brinker, Y. Jiang, Adv. Func. Mater. 16, 891 (2006).
52. Q. S. Huo, D. I. Margolese, U. Ciesla, P. Y. Feng, T. E. Gier, P. Sieger, R. Leon, P. M. Petroff, F. Schuth, G. D. Stucky, Nature 368, 317 (1994).

53. H. Y. Fan, H. R. Bentley, K. R. Kathan, P. Clem, Y. F. Lu, C. J. Brinker, J. Non-Crystalline Solids 285, 79 (2001).
54. Y. F. Lu, H. Y. Fan, N. Doke, D. A. Loy, R. A. Assink, D. A. LaVan, C. J. Brinker, J. Am. Chem. Soc. 122, 5258 (2000).
55. C. J. Brinker, G. W. Scherer, Sol-gel science: the physics and chemistry of sol-gel processing, Academic Press INC,, San Diego, CA, 1990.
56. P. D. Yang, G. Wirnsberger, H. C. Huang, S. R. Cordero, M. D. McGehee, B. Scott, T. Deng, G. M. Whitesides, B. F. Chmelka, S. K. Buratto, G. D. Stucky, Science 287, 465 (2000).
57. Y.N. Xia,J.A. Rogers,K.E. Paul,G.M. Whitesides, Chem Rev. 99, 1823, (1999).
58. J. Lee, V. C. Sundar, J. R. Heine, M. G. Bawendi, K. F. Jensen, Adv. Mater. 12, 1102 (2000).
59. M. B. Sigman, A. E. Saunders, B. A. Korgel, Langmuir 20, 978 (2004).
60. R. P. Feynman, R. B. Leightton, M. L. Sands, The Feynman Lectures on Physics, Vol. 2, California Institute of Technology, Pasadena, Calif:, 1963.
61. W. K. Choi, W. K. Chim, C. L. Heng, L. W. Teo, V. Ho, V. Ng, D. A. Antoniadis, E. A. Fitzgerald, App. Phy. Lett. 80, 2014 (2002).
62. S. Tiwari, F. Rana, H. Hanafi, A. Hartstein, E. F. Crabbe, K. Chan, App. Phy. Lett. 68, 1377 (1996).
63. W. Kern, Handbook of Semiconductor Cleaning Technology, Vol. chapter 2, Noyes Publishing, Park Ridge, NJ, 1992.
64. T. Clark, J. D. Ruiz, H. Y. Fan, C. J. Brinker, B. I. Swanson, A. N. Parikh, Chem. Mater. 12, 3879 (2000).
65. C. A. Stafford, S. DasSarma, App. Phy. Lett. 72, 3590 (1994).
66. G. Kirczenow, Phys. Rev. B: Condensed Matter 46, 1439 (1992).
67. R. Ugajin, J. Appl. Phys. 76, 2833 (1994).
68. J. M. Tour, W. L. Van Zandt, C. P. Husband, S. M. Husband, L. S. Wilson, P. D. Franzon, D. P. Nackshi, IEEE Tansactions on Nanotech. 1, 1 (2002).

CHAPTER 4

ELECTROSPINNING NANOFIBERS WITH CONTROLLED STRUCTURES AND COMPLEX ARCHITECTURES

Dan Li[1,2], Jesse T. McCann[1], Manuel Marquez[3-5], Younan Xia[1,*]

*[1]Department of Chemistry, University of Washington, Seattle, WA 98195-170; [2]INEST Group, Philip Morris USA, Richmond, VA 2323; [3]NIST Center for Theoretical and Computational Nanosciences, Gaithersburg, MD 2089; [4]Harrington Department of Bioengineering, Arizona State University, Tempe, AZ 85287; [5]Research Center, Philip Morris USA, 4201 Commerce Road, Richmond, VA 23234; *Corresponding author, Email: xia@chem.washington.edu*

Electrospinning is a simple and versatile technique for generating nanofibers made of various materials. This article presents an overview of some recent advances in this field and the development of new variants of electrospinning that allow for the fabrication of nanofibers with complex secondary structures and nanofiber assemblies with controllable architectures.

1. Introduction

Although electrospinning of fibers from polymer solutions and melts has been known since the 1930s, this technique has received great attention in recent years due to its potential use in the manufacture of nanostructured materials. As in a conventional fiber spinning process, electrospinning involves the ejection of a viscous solution (or melt) from an orifice and the subsequent drawing and solidification of the jet to form thin fibers. Instead of the mechanical stretching or pulling that is involved in a conventional spinning process, electrostatic repulsion are the driving force in electrospinning. The most attractive advantage of electrospinning is that it is capable of generating fibers with diameters down to the nanometer scale. Due to the simplicity and versatility of this technique, electrospinning has gained growing interest in the past several years. Uniform nanofibers comprised of a variety of materials have been successfully fabricated using this technique. Applications of electrospun fibers in many fields are also emerging, including tissue engineering, catalysis, sensing, as well as fabrication of composite electrodes for supercapacitors and batteries [1].

Electrospinning has been employed primarily for producing polymer fibers with simple compositions or structures. Recent efforts on this technique is focused on the development of electrospinning for generating nanofibers with intriguing functionalities and/or complex secondary structures, as well as the exploration of new techniques to assemble electrospun nanofibers into ordered assemblies. This article will concentrate on how conventional electrospinning can be modified to produce nanofibers with multiple components and/or having special secondary structures such as core/sheath, hollow, or porous structures as well as nanofiber assemblies with controllable architectures. The research activities before 2004 in this field have been summarized in several review articles [1-4]. In this chapter, we would like to limit the scope to the progress that was achieved in the past two years (2004-2006). The applications of electrospun fibers have been discussed in other recent reviews [1-5] and will not be covered in this chapter.

2. Experimental Setup for Electrospinning

The equipment required for electrospinning is remarkably simple and readily available. A basic electrospinning setup consists of three major components: a high voltage power supply, a spinneret, and an electrically conductive collector (Figure 1). The solution for spinning is usually loaded in a plastic syringe that is connected to the spinneret. In order to control the quality of the fibers, the syringe is placed in a syringe pump so that one can maintain a constant and adjustable feeding rate for the solution. An ordinary hypodermic metallic needle works well as the spinneret for many experiments. However, the conventional setup often needs to be modified to better control the electrospinning process or to fabricate more complex nanofibers. For this purpose, microfabricated sharp tips and spinnerets containing more than one capillary have been developed [1].

As for the collector, it can be constructed from various materials such as metallic plates and wires, silicon wafers, and even liquids. The material of the collector may have a considerable effect on the packing of collected fibers. It was found that the packing density was dependent on the conductivity of the collector. Less conductive collectors yielded a more porous structure due to the difficulty of dissipating the remaining charges on the fibers. In addition, porous collectors, such as paper and copper mesh, produced a less-packed structure as compared to fibers collected on a solid conductive substrate [6]. The geometry of the collector also has a significant effect on the distribution and orientation of electrospun fibers (see the discussion in Section 8).

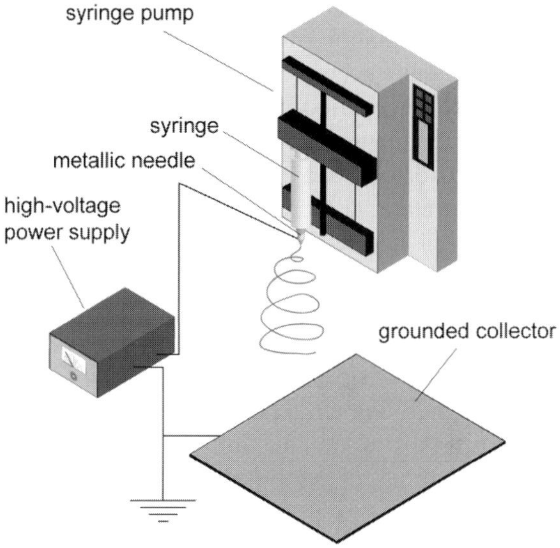

Figure 1. Schematic of a typical setup for electrospinning.

3. History and Mechanism of Electrospinning

The beginning of electrospinning can be traced back to the early studies on the behavior of electrified droplets in the nineteenth century. In 1882, Lord Rayleigh predicted that a spherical droplet would become unstable in shape once the density of surface charges exceeded a threshold (now known as Rayleigh limit), beyond which the surface tension could be overcome by the electrostatic repulsion. Such instability would lead to elongation of the droplet and emission of fine charged jets [7]. In 1917, John Zeleny described the observation of liquid jets created by electrical forces in a paper entitled "Instability of Electrified Liquid Surfaces" [8]. In 1960s, G. I. Taylor studied and explained the shape change when a high voltage was applied to a suspended droplet [9]. It was observed that the droplet at the nozzle initially formed a hemispherical surface. As the electric field increased, the surface underwent shape change from hemispherical to spherical and eventually to conical. These morphological changes were determined by the competition of the increasing surface charges on the droplet with its surface tension, and the final conical shape has come to be known as the Taylor cone. When the applied voltage was sufficient to induce enough charge to overcome surface tension, a stream was ejected from the tip of the Taylor cone. The formation of the jet from electrified droplets due to Rayleigh instability was recently confirmed by an observation of the shape evolution of an electrified microdroplet using high-speed optical microscopy [10]. As shown in Figure 2, once the charge density of an ethylene glycol microdroplet was increased beyond the Rayleigh limit due to solvent evaporation, the droplets first deformed into an ellipsoid and then into two thin jets which were subsequently emitted from the droplet (Figure 2C).

Further experiments have shown that for low viscosity liquids, the ejected jet would disintegrate into droplets due to varicose instability, resulting in the spraying of droplets from the initial droplet (Figure 2D). This phenomenon is now known as electrospray, which has found many commercial applications such as ink-jet printing, spray painting, production and patterning of nanoparticles, encapsulation, as well as ionization of macromolecules in mass spectrometry. However, if the liquid is sufficiently viscous to prevent the jet from disintegrating, subsequent solidification of the jet via solvent evaporation or cooling will lead to the formation of fibers. This process is now referred to as electrostatic spinning or electrospinning.

The first report on electrospinning of fibers was disclosed as a patent by A. Formhals in the 1930s for a process that produced fine fibers from a cellulose acetate solution [11]. In the 1970s, Baumgarten electrospun acrylic fibers with diameters ranging from 50 to 1,100 nm [12]. Larrondo and Manley published several papers on the electrospinning of polymer melts [13].

Electrospinning received little attention before the 1990s, probably due to the low productivity of this technique. However, as interest in nanostructured materials and nanotechnology increased tremendously in the last decade, there has been a great deal of renewed interest in this simple and versatile technique. Significant advances have been achieved in the last few years. In the theoretical aspects, a number of groups have studied the complex electrohydrodynamics associated with the electrospinning process [1]. It has been established that the formation of ultrathin fibers from the electrically ejected jets primarily originates from the elongation and stretching of the jets during a bending or whipping instability process, instead of splaying of the jets. This discovery has enabled further development and refinement of the electrospinning process, including coaxial electrospinning.

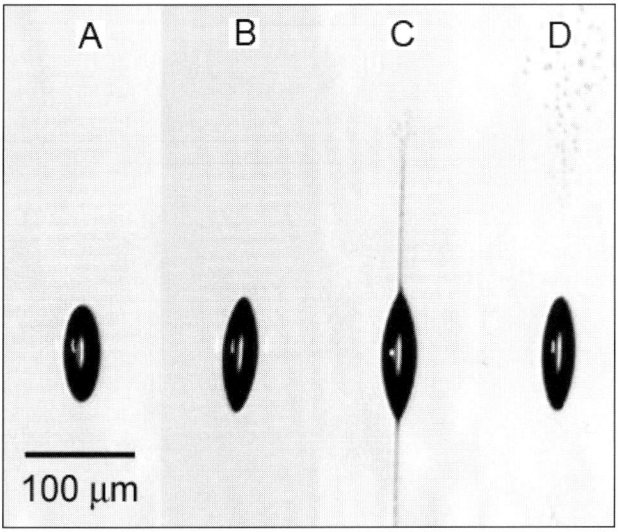

Figure 2. High-speed microscopic images of a levitated ethylene glycol droplet charged to the Rayleigh limit. The images were taken at different times (in μs): A) 140, B) 150, C) 155, and D) 160. The droplet changed from a sphere to an ellipsoid (A), tips appear at the poles (B) and a fine jet of liquid is ejected from each tip (C) and the jets disintegrate (D). This figure was adapted from [Ref. 10] with permission from Macmillan Publishers Ltd., © 2004.

In the experimental aspects, several groups have significantly advanced this technique by modifying the conventional setup, as well as by combining the traditional electrospinning process with appropriate chemical reactions (e.g., sol-gel reactions) and post-spinning treatments. These efforts have made it possible to directly electrospin fibers from a variety of materials such as ceramics, carbon, and composites. In particular, the secondary structures of individual fibers can now be tailored to form porous, hollow, and core/sheath fibers. In addition, electrospun fibers can be conveniently assembled into ordered arrays for certain applications. This chapter will only cover recent advances in the field of electrospinning technology.

4. Nanofibers Containing Nanoscale Fillers

Incorporation of nanoscale fillers into polymers has been extensively studied as an effective approach to enhance the mechanical, electrical, and magnetic properties of conventional polymers.

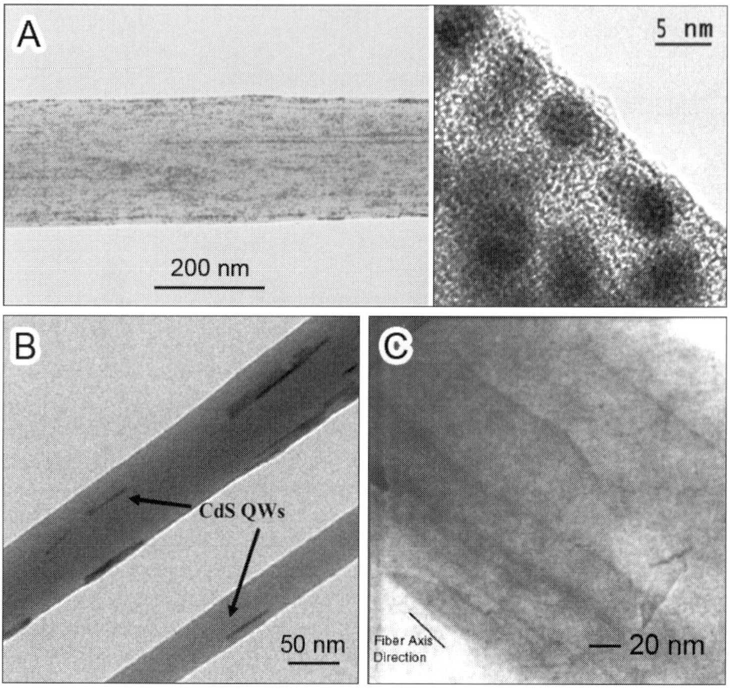

Figure 3. TEM images of electrospun nanofibers incorporated with nanostructures of different shapes, showing A) gold nanoparticles in PEO nanofibers [16], B) CdS nanowires in PEO nanofibers [22], and C) Montmorillonite nanosheets in PMMA-MAA nanofibers [27]. This figure was adapted from [Ref. 16] ©2005 American Chemical Society, [Ref. 22] ©2006 John Wiley and Sons, and [Ref. 27] ©2005 Elsevier Ltd. Used with permission.

To this end, electrospinning has been used to introduce functional nanostructures (Q-dots, nanowires, and nanotubes) into electrospun fibers with the hope that the introduction of such nanoscale fillers could endow nanofibers with new functionalities or enhanced properties.

Another motivation to these studies is that electrospun nanofibers can serve as nanoscale hosts to arrange the nanostructures into one-dimensionally ordered arrays, which is important for both fundamental studies and fabrication of nanodevices. In addition, the small diameters of electrospun fibers allow for convenient characterization of the internal microstructure of the resultant composite nanofibers using conventional microscopic tools without special sampling techniques. Figure 3 shows transmission electron microscopy (TEM) images of electrospun nanofibers containing Q-dots, nanowires, and nanosheets.

4.1. Nanoparticles as the Fillers

Electrospun nanofibers that contain metallic, magnetic, and semiconducting nanoparticles have all been fabricated [14-21]. In general, there are two ways to introduce nanoparticles into the fibers. The first one is to directly disperse pre-synthesized nanoparticles with unique and functional shapes into the polymer solution for electrospinning. Foro this approach, the surface of the nanoparticles must be modified in order to obtain well-dispersed, stable solutions for continuous spinning. An alternative approach that prevents the aggregation of particles is to electrospin polymer solutions containing reactive precursors, followed by chemical treatment of the nanofibers. For example, Wang and co-workers have prepared nanofibers loaded with Q-dots by electrospinning polymer nanofibers containing metal ions (Ag^{+}, Pb^{2+}) and *in situ* generation of nanoparticles by exposure of the as-spun nanofibers to H_2S gas [14,15].

The arrangement of the nanoparticles in electrospun nanofibers could greatly differ from that of bulk nanocomposites fabricated by solution casting or conventional spinning processes. The presence of nanoparticles can also change the conformation of the polymer chains in solution. Electrospinning involves the rapid stretching of an electrified jet and rapid evaporation of solvent. The polymer chains are expected to experience an extremely strong shear force during the electrospinning process, which could significantly affect the orientation of the entrained nanoparticles. Kim and co-workers observed that alkanethiol-capped Au nanoparticles incorporated into electrospun poly(ethylene oxide) nanofibers (Figure 3A) could form chain-like structures [16]. Thermal analysis and FTIR measurements showed that Au nanoparticles served as preferential nucleation sites during the crystallization of PEO and induced the PEO chains to transform from a helix to a *trans*-planar zigzag conformation. As a result, the PEO chains were preferentially aligned along the long axis of each nanofiber, resulting in a perpendicularly-arranged lamellar morphology. These results suggest that the semicrystalline PEO acts as a template to arrange the Au nanoparticles within the fibers during electrospinning.

Rutledge and co-workers have recently prepared superparamagnetic polymer nanofibers by electrospinning polymer-stabilized magnetite nanoparticle suspensions in PEO and poly(vinyl alcohol) (PVA) solutions [17]. It was also observed that the nanoparticles lined up within the fibers in columns parallel to the long axis of the fiber in both polymer matrixes, apparently induced by the strong stretching associated with the electrospinning process.

The introduction of functional nanoparticles into electrospun nanofibers not only adds new functionalities to the nanofibers, but also often makes the resulting fibers stronger. Rutledge and co-workers performed nanoindentation tests on the nanofibers containing magnetite nanoparticles and found that the nanoparticles reinforced the mechanical properties of the nanofibers [17]. Kim and co-workers added ferritin nanoparticles into PVA solutions. The resulting nanofibers also exhibited enhanced mechanical properties. The elastic modulus of the PVA/ferritin nanofibers was comparable to that of PVA composite films containing carbon nanotubes (CNTs) [21].

It is worth noting that the addition of nanoparticles into the spinning solution could change the properties of the pristine solutions such as conductivity and viscosity, therefore, their electrospinning behavior could be different from that of the nanoparticle-free solutions. For example, Rutledge and co-workers found that at low PEO concentrations (1% by wt.) and in the absence of magnetite nanoparticles, the PEO fibers adopted a bead-on-string morphology while fibers with uniform diameters were obtained when 3.52 wt% magnetite nanoparticles were added to the spinning solution [17]. Similar phenomena were also observed when other types of nanostructures were incorporated.

4.2. Nanowires and Nanotubes as the Fillers

Zussman and co-workers have prepared PEO nanofibers loaded with CdS nanowires by electrospinning PEO solutions containing solution-synthesized CdS nanowires [22]. TEM images clearly show that the CdS nanowires were unidirectionally aligned with their long axes parallel to the stretching direction of the fibers (Figure 3B). They further collected the composite nanofibers as uniaxially aligned arrays on a rotating wheel. The emission spectra of the resulting 1-D ropes shows linear polarization, which can be useful as polarized-light sources in high-resolution detection of polarized light or in other optical elements in the visible and near-IR spectral regime [22].

Stimulated by the unique properties of CNTs, their incorporation into nanofibers has received much attention. Several groups have investigated CNT-encapsulated polymer nanofibers [23-26]. Reneker, Cheng and co-workers found that surface-oxidized multi-wall CNTs could be well-dispersed in poly(acrylonitrile) solutions without the need for any surfactants or binding agents [23]. Continuous sheets made of such composite nanofibers were successfully electrospun with up to 20% (by wt.) CNTs incorporated. A high degree of orientation of the CNTs in the composite nanofibers was confirmed by both TEM and two-dimensional WAXD analysis. The degree of orientation for the surface-oxidized CNTs in the nanofibers was significantly greater than that of the PAN matrix crystals. The electrical conductivity, thermal stability/conductivity and tensile modulus were all enhanced as CNTs were incorporated in the nanofibers.

4.3. Nanosheets as the Fillers

Incorporation of layered compounds (e.g., montmorillonites) into polymer matrices has shown great promise for yielding nanocomposites with enhanced mechanical strength, chemical resistance, thermal stability, and self-extinguishing fire resistance when compared to pristine polymers. To further improve the properties of electrospun nanofibers, several groups have fabricated electrospun nanofibers containing exfoliated layered compounds (nanosheets) and have investigated the thermal and mechanical properties of the nanosheet-filled nanofibers [27-31]. Rutledge and co-workers synthesized copolymer dispersions consisting of methyl methacrylate (MMA), methacrylic acid (MAA), and montmorillonites (MMT) by *in situ* emulsion polymerization [27]. The clay layers were exfoliated and well-dispersed in the polymer matrix. The as-polymerized composites were then dispersed in *N,N*-dimethylformamide (DMF) from which the nanofibers were electrospun. It was found that the clays in the nanocomposite dispersions increased the zero-shear-rate viscosity over that of the pristine polymers. Dispersion of clays within the nanocomposites improved the electrospinnability of these materials through increased apparent extensional viscosity and strain hardening. TEM imaging showed that the exfoliated MMT nanosheets were well-distributed within the fiber volume and along the fiber axis (Figure 3C). Thermal analysis demonstrated that the nanofibers containing clay exhibited a higher glass transition temperature and enhanced thermal stability compared to the pure polymer nanofibers.

Youk and co-workers directly dispersed organically-modified MMT in poly(urethane) (PU) via solution intercalation [28]. The authors then studied the electrospinning behavior of these dispersions as well as the morphology and physical properties of the PU/MMT nanofiber mats. It was found that the conductivities of the PU/MMT solutions were linearly increased with increasing MMT, which caused a decrease in the average diameter of the PU/MMT nanofibers. The exfoliated MMT layers were well-dispersed in the composite nanofibers and oriented parallel to the long axes of the fibers. The elongation at the breaking point of the composite nanofiber mats was slightly lower than that of the clay-free nanofibers. However, their Young's modulus and tensile strength were improved by up to 200%.

Rafailovich and co-workers investigated the surface nanomechanical properties of electrospun poly(styrene)/clay fibers using shear modulation force microscopy [29]. Surface relative modulus measurements suggested that both the polymer chains and montmorillonite layers were oriented along the fiber axis. The existence of clay enhanced the shear modulus of the fibers and raised the glass transition temperature by nearly 20 °C.

In addition to clays, another class of layered compound, graphite, has also been tested for the reinforcement of electrospun nanofibers. Kaner and co-workers have synthesized graphite nanoplatelets using a specially designed exfoliation process. Electrospun nanofibers of poly(acrylonitrile) incorporated with carbon nanosheets exhibited significantly enhanced mechanical and thermal properties. The Young's modulus of the composite nanofibers could be doubled with the addition of 4% (by wt.) graphite nanoplatelets [30].

The orientation of the filler particles within the fibers during electrospinning allows for the possibility of manipulating nanoparticles through appropriate handling of the fibers in which they are embedded and oriented. For example, the assembly of CNTs in predetermined patterns is one of the most challenging tasks in the fabrication and mass-production of CNT-based electronic devices. As described in Section 8, it is relatively simple to control the alignment of electrospun nanofibers. By using electrospun nanofibers as carriers, Haddon and co-workers have demonstrated that large-scale aligned arrays of CNTs could be fabricated [32]. The local orientation of CNTs in the electrospun composite fibers can be conveniently transferred to macroscopically aligned structures by aligning the host nanofibers. By this method, high-strength oriented nanofiber-CNT composite arrays could be fabricated.

5. Electrospinning with a Dual-Capillary Spinneret

The conventional setup for electrospinning involves the use of a single capillary as the spinneret, and thus is only suitable for generating fibers with one particular composition. To make nanofibers with multiple compositions and more complex structures, spinnerets with two capillaries have been designed by several groups. These new designs allow one to co-spin two different solutions simultaneously, which provides additional control to the internal structure of electrospun nanofibers. Several groups have previously demonstrated that nanofibers with core/sheath and hollow or porous structures can be fabricated using a spinneret containing two coaxial capillaries [1]. These successes have stimulated more work in this direction in the last two years.

5.1. Core/Sheath Nanofibers

Core-sheath structured nanofibers hold a number of advantages over single-composition nanofibers for applications in biomedical areas. For example, nanofibers made of an unstable biological agent (core) can be protected by a sheath made of a stable material. To this end, Zhang and co-workers have demonstrated that poly(caprolactone)-gelatin composite nanofibers can be made by coaxial electrospinning of poly(caprolactone) and gelatin solutions [33]. TEM and XPS analysis showed that the resulting nanofibers had a core/shell structure (Figure 4A).

In addition to polymer solutions or viscous oils, nanoparticle dispersions can serve as the core material as well. Song and co-workers have coaxially electrospun a hexane solution of FePt nanoparticles and a solution of PCL in 2,2,2-trifluoroethanol as the core and shell fluids, respectively [34]. TEM imaging showed that FePt nanoparticles were encapsulated predominantly in the core of the nanofibers (Figure 4B).

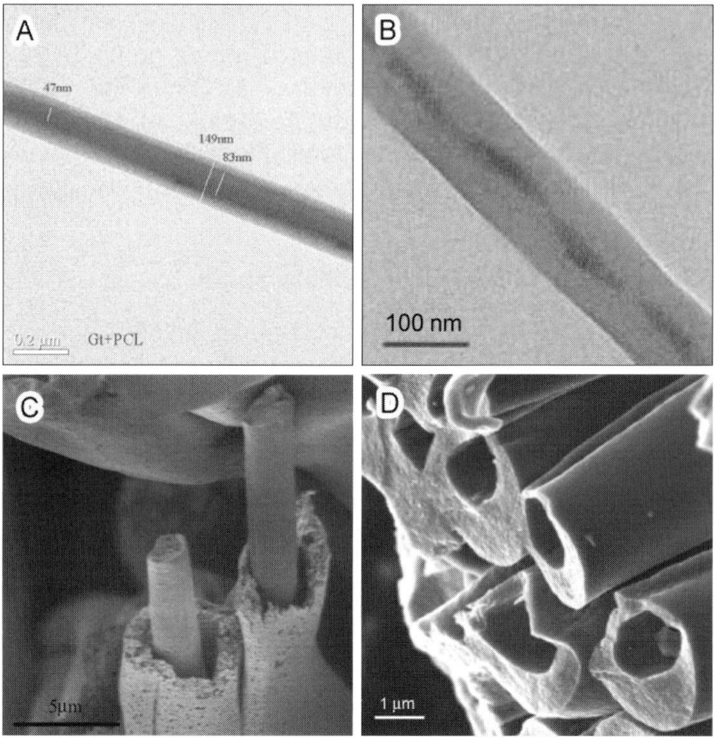

Figure 4. Examples of core/sheath and hollow nanofibers prepared by electrospinning with a coaxial setup and two different solutions for the inner/outer capillary: Images show A) PCL/gelatin [33], B) PCL/FePt nanoparticles [34], C) PAN/PMMA [35], and D) hollow carbon fibers that were obtained via carbonization of the sample in (C) at high temperatures [35]. Adapted from [Ref. 33] ©2004 American Chemical Society, [Ref. 34] ©2005 Elsevier Ltd., and [Ref. 35] ©2006 John Wiley and Sons. Used with permission.

Choosing the right solvents for both the inner and the outer liquids appears to be important for fabricating composite nanofibers with well-defined core/sheath structures using the co-electrospinning strategy. The diffusion between the outer and inner liquids during the spinning process can lead to overlap for the core and sheath materials. Zussman and co-workers have demonstrated that if poly(acrylonitrile) (PAN) and poly(methyl methacrylate) (PMMA) were both dissolved in DMF, core/sheath nanofibers with good interfaces were not formed by coaxial electrospinning of the two solutions [35]. However, if PMMA was dissolved in acetone and served as the inner solution, well-defined core/sheath nanofibers were successfully obtained. Interestingly, the PMMA core was narrower than the inner diameter of the PAN sheath. Note that acetone is not a solvent for PAN. It was postulated that during the co-spinning process, the PAN solution (in DMF) that was in contact with acetone was precipitated, resulting in a barrier layer to enable the formation of core/sheath structures (Figure 4C). Carbonization of the PAN sheath and decomposition of the PMMA core at high temperatures led to the formation of hollow carbon fibers (Figure 4D).

5.2. Hollow Nanofibers with Controlled Surface Structures

Xia and Li have previously demonstrated that hollow composite or ceramic nanofibers could be fabricated by co-electrospinning a sol-gel precursor solution with a mineral oil, followed by selective removal of the oil core [36]. Taking advantage of the immiscibility of the two liquids, they further demonstrated that the inner/outer surfaces of the electrospun hollow fibers could be easily functionalized with small molecules or nanoparticles [37,38]. By adding various precursors into the oil core, hollow nanofibers filled with different functional components could be fabricated. For example, hollow nanofibers of titania filled with fluorescent dyes or iron oxide nanoparticles could be readily obtained by adding an oil-soluble dye or oil-based ferrofluid to the core capillary, followed by octane extraction of the oil phase, resulting in the formation of optically or magnetically active nanotubes (Figure 5, A and B).

It was further demonstrated that by using oil-soluble, long-chain silanes (with 18 carbons) the inner and outer surfaces of titania nanotubes could be selectively modified. For example, the inner surface can be made hydrophobic by doping the oil with a methyl-terminated silane. This coating on the inner surface could be used to differentiate the inner and outer surfaces. For example, if the inner surface was first coated with methyl-terminated silane, it was protected when the sample was immersed in amino-functionalized silane. The inner surface would be inert when it was immersed in a solution of gold nanoparticles and the nanoparticles selectively adhered to the outer surface (Figure 5C). In comparison, if a methyl-terminated silane was not added into the oil, the surface chemistry of both surfaces of the resulting nanotubes would be the same. When the nanotubes were treated with amino-terminated silane and gold nanoparticles, gold nanoparticles adhered to both the inner and outer surfaces (Figure 5D) [37].

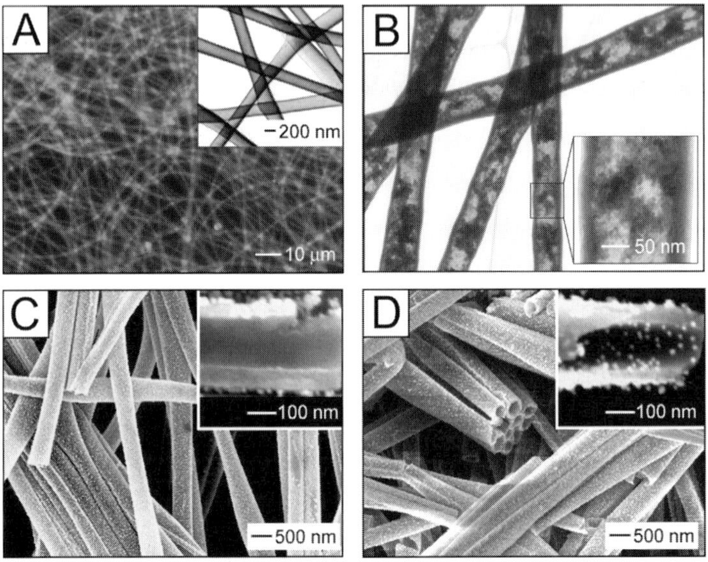

Figure 5. Fluorescence microscopy image (A) of dye-loaded hollow nanofibers made of a PVP/TiO_2 composite, with a TEM image in the insert. TEM image (B) of hollow titania nanofibers with iron oxide nanoparticle-coated inner surfaces. SEM images (C) of hollow titania nanofibers; the inner surfaces of which were functionalized with methyl-terminated silane, followed by functionalization of the outer surfaces with amino-terminated silane. SEM images (D) of hollow titania nanofibers; the inner and outer surfaces of which were both functionalized with amino-terminated silane. Reprinted from [Ref. 38] and ©2005 The Royal Society for Chemistry. Used with permission.

5.3. Improvement of Electrospinnability

It is worth noting that the use of a coaxial spinneret for electrospinning not only provides an additional route to the production of nanofibers with complex secondary structures, it also can greatly expand the capability of the conventional electrospinning technique for producing nanofibers that cannot be electrospun readily from conventional single-capillary spinnerets. It is known that not all polymer solutions are electrospinnable. The electrospinnability is limited by the viscosity, conductivity, and solvents of the solution as well as the conformation and molecular weight of the polymer. Many polymers (especially conjugated polymers) are not directly spinnable due to their limited solubility in suitable solvents. Although their electrospinnability can be greatly improved by blending the desired polymers with other spinnable polymers, the generality of this method is limited by the availability of a common solvent capable of dissolving both blend components. To overcome this problem, two groups have independently demonstrated that an unspinnable polymer can be fabricated as nanofibers by co-spinning with a spinnable polymer solution.

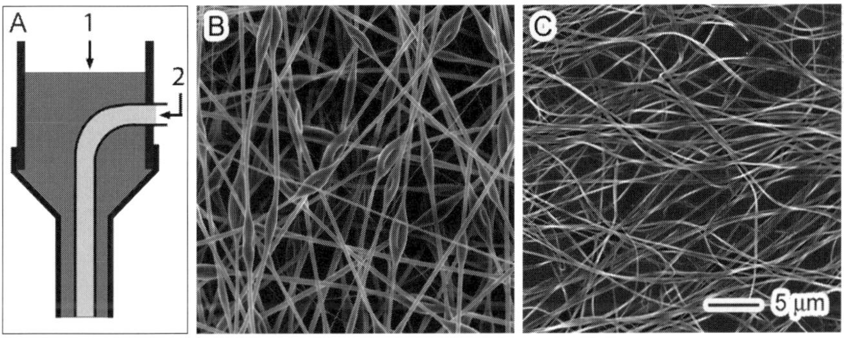

Figure 6. Schematic drawing (A) of the spinneret constructed from two coaxial capillaries. Two different solutions (1 and 2) can be fed to the orifice simultaneously. Composite nanofibers (B) of MEH-PPV/PVP that were prepared by co-electrospinning of an MEH-PPV solution in chloroform and a PVP solution in ethanol through the inner and outer capillaries, respectively. MEH-PPV nanofibers (C) that were prepared from sample B by selectively removing PVP using ethanol. Reprinted from [Ref. 39] and ©2004 John Wiley and Sons. Used with permission.

Xia and co-workers have shown that MEH-PPV/PVP composite nanofibers can be readily fabricated by co-spinning an MEH-PPV solution in chloroform (core fluid) and a PVP solution in ethanol (sheath fluid) through a two-capillary spinneret (Figure 6, A and B) [39,40]. Pure MEH-PPV nanofibers can be readily obtained by selective removal of PVP using ethanol (Figure 6C). Their diameters can be conveniently controlled by adjusting the feeding rate and concentration of the MEH-PPV solution. It is important to note that MEH-PPV nanofibers cannot be spun from solution using a single capillary setup. This polymer is insoluble in ethanol, making it impossible to make its blend fibers with PVP through the conventional blending strategy.

Using a similar approach, Rutledge and co-workers have fabricated a range of composite nanofibers such as poly(acrylonitrile)/ poly(acrylonitrile-*co*-styrene), poly(aniline sulfonic acid)/poly(vinyl alcohol) and silk/poly(ethylene oxide) [41,42]. After the removal of shell polymer, much thinner nanofibers of the core polymer can be made. For example, they have produced silk nanofibers with diameters as small as 170 nm, much thinner than natural silk fibers.

6. Porous Nanofibers

Porous nanofibers have a wide range of applications in fields such as tissue engineering, sensing, filtration, separation, catalysis, and superhydrophobic coating. In addition, controllable pores can be used to encapsulate active materials. To this end, a number of techniques have been developed for the generation of porous fibers. Electrospinning of blends followed by selective removal (leaching) of one component has been met with some success. For the formation of porous fibers with polymers that are immiscible (while the solvents are miscible), researchers in the Xia group have pioneered the use of the coaxial spinneret with miscible solvents in order to create blend fibers which can then be made porous via selective removal by calcination [36,38]. These methods rely on polymer-polymer phase separation. Porogen methods involving thermally-induced phase separation of polymer and solvent have also been of interest as they can be used to fabricate porous fibers with controllable surface chemistry in a single step. Evaporative cooling of highly volatile solvents has resulted in thermally-induced phase separation. Subsequently, techniques have been developed to induce polymer-solvent phase separation in order to extend the utility of this method. Leaching of salts from the fibers has also been a method generating porous polymer matrices for applications in tissue engineering, although these methods are becoming unattractive when compared with more recently developed methodologies that preserve the mechanical properties and structural integrity of the nanofiber assemblies.

6.1. Porous Nanofibers by Bicomponent Spinning

One major method for manufacturing porous fiber is the electrospinning of polymer blends, followed by selective removal of one of the components. Zhang and coworkers electrospun porous fibers of poly(ε-caprolactone) by co-spinning poly(ε-caprolactone) and gelatin with selective removal of the gelatin component by immersion in water at 37 °C [43]. Lyoo and coworkers created porous fibers by co-spinning poly(vinyl cinnamate) with poly(3-hydroxybutyrate-*co*-3-hydroxyvalerate), followed by photo-crosslinking of poly(vinyl cinnamate) and selective removal of the non-crosslinked component by leaching in chloroform [44]. This method is useful in that it can be applied to combinations of polymers that form a blend but do not share a solvent.

6.2. Porous Fibers by Polymer-Solvent Phase Separation

Another promising method for fabricating porous fibers by electrospinning is to utilize polymer-solvent phase separation. This has several advantages over co-spinning methods. First, there is no residual polymer left behind following selective dissolution, thus the surface chemistry of the fibers can be carefully regulated. Furthermore, since the pore generation occurs during the spinning process and not following fiber deposition, there are no interconnections formed between the fibers and thus the structural integrity of the resultant nanofiber mat is not compromised.

The formation of fibers with pores and surface architecture was observed when spinning certain polymers from highly-volatile solvents such as methylene chloride. Evaporative cooling during the spinning and attenuation process was believed to induce phase separation, and the resultant fibers were found to be porous. Pores were formed when spinning semi-crystalline materials such as poly(L-lactic acid) [45].

When spinning amorphous materials such as poly(carbonate) and poly(vinyl carbazole), some surface texture was visualized. Rabolt and coworkers also observed the formation of surface structure on electrospun nanofibers while using highly volatile solvents [46].

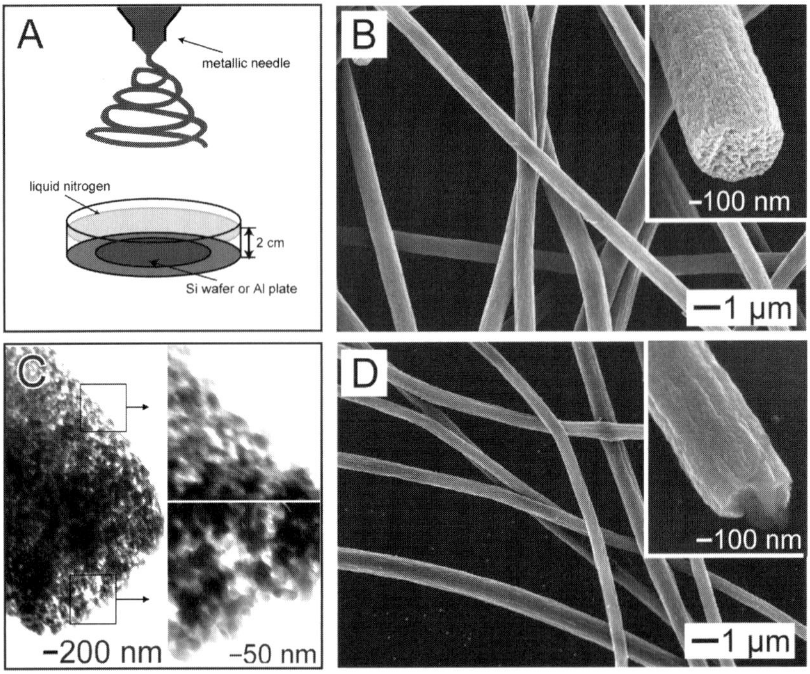

Figure 7. Schematic (A) of the setup used to electrospin porous fibers into liquid nitrogen. Porous polystyrene fibers (B) fabricated by electrospinning a dimethylformamide solution of poly(styrene into a bath of liquid nitrogen, followed by drying *in vacuo.* TEM (C) of the edge of the fiber shown in (B). SEM (D) of poly(styrene fibers electrospun into a bath of liquid nitrogen, followed by reheating in ambient air. Reprinted from [Ref. 47] with permission from the American Chemical Society, © 2006.

Xia and coworkers developed a method to induce polymer-solvent phase separation by electrospinning directly into a cryogenic liquid. In this procedure, various polymer solutions were electrospun directly into a bath of liquid nitrogen (see Figure 7A for a diagram of the setup) [47]. If the collection distance was short enough and the solvent was sufficiently non-volatile, the spinning jet reached the liquid nitrogen bath before the solvent fully evaporated, thereby inducing polymer-solvent phase separation in the remaining materials. The fibers were immersed in liquid nitrogen before the solvent evaporated completely. The rapid cooling causes phase-separation. The fibers were then dried *in vacuo* to sublime the solvent, resulting in highly porous fibers. Poly(styrene), poly(vinylidine fluoride), poly(acrylonitrile) and poly(ε-caprolactone) were all made into porous fibers using this method. Figure 7C shows a TEM image of the porous poly(styrene) fibers fabricated using this method. The fibers are unique in that they are porous throughout. By controlling the way the solvent is evaporated, the resultant morphology of the fibers can also be controlled. When poly(vinylidene fluoride) fibers were subjected to this treatment, the resultant fiber mats had a larger contact angle with water compared to untreated fibers. This method can be extended to a multitude of functional polymer-solvent combinations.

7. Complex Nanofibers via Post-Spinning Treatment

In addition to the use of specially-designed spinning setups to make complex nanofibers, post-spinning treatments of electrospun nanofibers can also be applied to make more complex nanostructures. For example, mono- or multiple-layers of nanoparticles or functional molecules can be deposited on the as-spun fibers via chemical or physical adsorption and even chemical reactions to fabricate core/sheath nanofibers. It has been well documented that electrospun nanofibers can be used as templates to produce a variety of nanotubes [1].

Complex nanofibers other than core/sheath configurations can also be made through appropriate post-spinning reactions. For example, Hou and Reneker have demonstrated that carbon nanotubes can be grown on carbon nanofibers that were derived from electrospun poly(acrylonitrile) nanofibers (Figure 8, A and B) [48]. In their method, an iron precursor was added into PAN solutions before electrospinning. Electrospun PAN nanofibers were carbonized at elevated temperatures while the iron compound was reduced to form iron nanoparticles on the surface of the carbonized fibers. These iron nanoparticles then served as catalysts for the growth of carbon nanotubes when a hexane vapor was flowed over the fibers at high temperature.

Figure 8. SEM (A) and TEM (B) images of carbon nanotubes on carbon nanofibers that were produced by electrospinning of PAN nanofibers, followed by carbonization of PAN and catalytic growth of carbon nanotubes [48]. SEM (C) and HRTEM (D) images of V_2O_5/TiO_2 nanofibers that were electrospun from an alcohol solution of $VO(OiPr)_3$, $Ti(OiPr)_4$ and PVP, followed by calcination at 425 °C for 6 h [49]. Adapted from [Ref. 48] ©2004 John Wiley and Sons, and [Ref. 49] ©2006 the American Chemical Society. Used with permission.

Xia and co-workers have demonstrated that nanorod-on-nanofiber structures of TiO_2/V_2O_5 composite nanofibers can be fabricated by controlled calcination of amorphous TiO_2/V_2O_5 fibers [49]. The as-spun TiO_2/V_2O_5 nanofibers were amorphous and had a smooth surface. When the fibers were calcined under appropriate conditions, single-crystalline V_2O_5 nanorods were simultaneously grown on the polycrystalline TiO_2 fiber (Figure 8, C and D). The length and diameter of these nanorods can be controlled by adjusting the compositions of the composite nanofibers and/or calcination conditions. No additional source of V_2O_5 was required in their work as all components were included in the nanofibers during the electrospinning process.

8. Ordered Architectures of Electrospun Nanofibers

Control of the spatial orientation of nanofibers is necessary for constructing hierarchical architectures of nanofibers and is important for many applications such as the reinforcement of composite materials, bone and blood vessel engineering, and the fabrication of electronic devices. Because of the bending instability associated with a spinning jet, electrospun fibers are often deposited on the surface of collector as randomly oriented, nonwoven mats. In the past several years, a number of approaches have been developed for the alignment of electrospun nanofibers. In the review article published in 2004 [1], we summarized several basic strategies to control the special orientation of electrospun nanofibers through mechanical motion or manipulation of the electrical field. In the last two years, there has been a great deal of work continued in this direction.

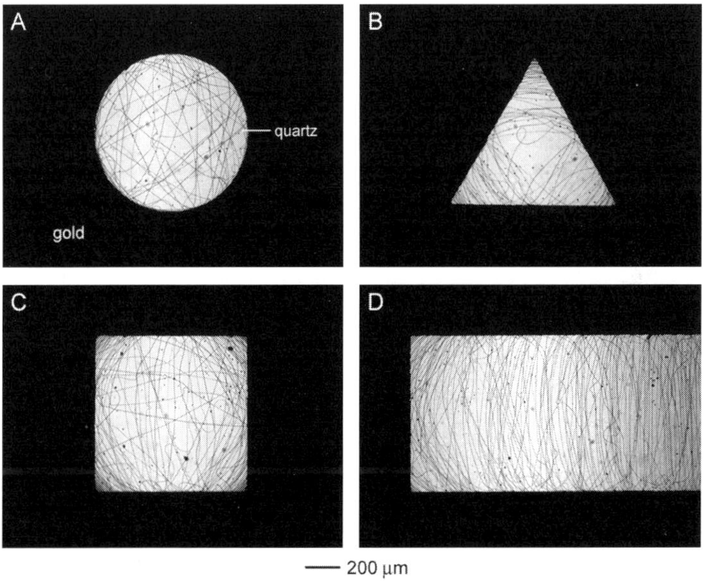

Figure 9. Transmission optical micrographs of electrospun nanofibers collected on gold electrodes (dark area) of various shapes patterned on insulating, quartz substrates (white area). Reprinted from [Ref. 52] with permission from the American Chemical Society, ©2005.

Xia and co-workers previously demonstrated that the nanofibers can be uniaxially aligned by introducing an insulating gap to the conductive collector [50,51]. To elucidate the mechanism of alignment, they have systematically studied the effect of the area and geometric shape of the insulating gap on the alignment of fibers [52]. It was found that the orientation of the fibers is strongly dependent on the design of the electrode. For the circular hole pattern (Figure 9A), the fibers were randomly oriented. For the triangular and square hole patterns (Figure 9, B and C), the fibers tended to be concentrated at the vertices and appeared to be aligned with their long axes perpendicular to the bisector of the vertex. Only a few fibers were collected in the middle of the insulating region. In the case of a rectangular hole pattern (Figure 9D), the orientation of fibers collected near the vertices is similar to the triangular and square patterns, while the fibers deposited outside of the diagonal corners were uniaxially aligned with their long axis perpendicular to the longer sides of the rectangle. By modeling the electrostatic forces acting on the fiber, it was established that the fibers tended to be oriented along a direction such that the net torque of electrostatic forces applied to the two ends of a discrete segment of the fiber were minimized. It was possible to control both alignment and assembly of the electrospun nanofibers by varying the design of the electrode pattern.

Several derivatives of the insulating-gap collectors such as two pointed collectors [53], two conducting rings with an air gap [54] and a wire-framed drum [55] have been developed to collect aligned nanofiber arrays. Twisted nanofiber yarns can be conveniently made from the dual-ring designs [54].

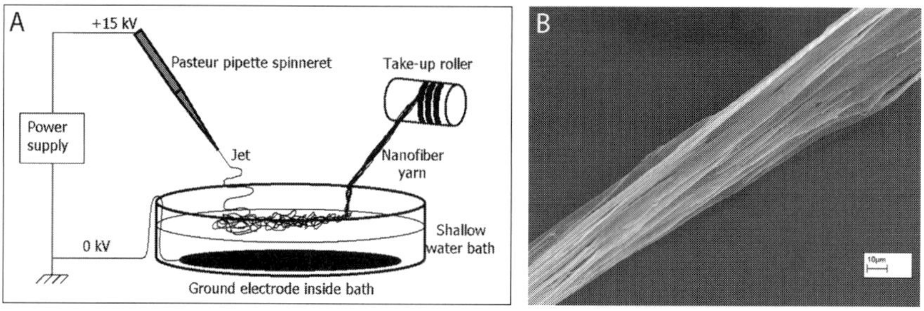

Figure 10. Schematic (A) showing a yarn-spinning setup with water bath as the collector. drawing yarn. SEM image (B) of an electrospun nanofiber yarn fabricated using this setup. Reprinted from [Ref. 56] with permission from Elsevier, Ltd. ©2005.

While conventional electrospinning collectors are usually made of solid materials, Sanderson and co-workers have demonstrated that a water bath can be used to collect the fibers [56]. Of particular interest is that aligned and continuous yarns of electrospun nanofibers can be obtained when the fibers floating on the water surface are drawn out. The fiber yarns can be wound on a take-up roller, enabling continuous producing of fiber yarns (Figure 10A). It was found that the nanofibers were uniaxially aligned along the drawing direction (Figure 10B).

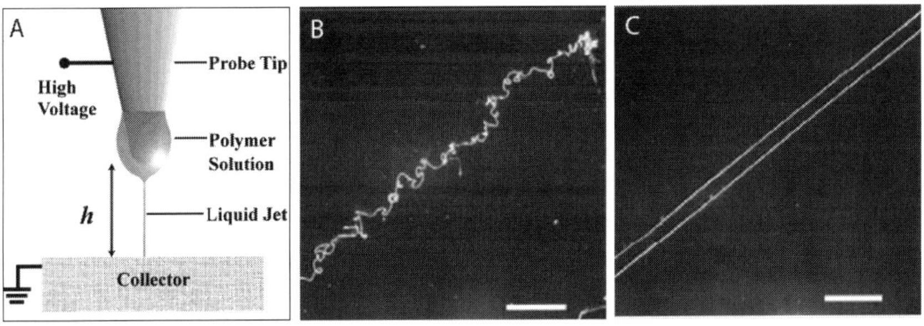

Figure 11. Schematic representation (A) of near-field electrospinning. The polymer solution is attached to the tip of a tungsten electrode in a manner analogous to that of a dip pen. SEM image (B and C) of nanofibers collected on a moving substrate. The collector was moving at B) 5 cm/s and C) 20 cm/s, respectively. Scale bar, 100 μm. Adapted from [Ref. 57] with permission from the American Chemical Society, ©2006.

In order to make the electrospinning more controllable, efforts have been made to redesign the conventional electrospinning setup. Lin and co-workers have demonstrated the so-called "near-field" electrospinning (NFES). Figure 11A illustrates the schematic setup of their design [57]. Differing from an ordinary design, the electrode-to-collector distance was greatly reduced to be in the range of 500 μm to 3 mm to utilize the stable liquid jets region for controllable deposition. The spinneret is a small solid tungsten tip (25 μm diameter) to position nanofibers with sub-100-nm resolution. Discrete droplets of polymer solutions were supplied in a manner analogous to that of a dip pen by immersing and pulling the tungsten electrode into and out of the polymer solution. The collector was put on an *x-y* stage and was moved during the collection process to make nanofiber patterns in a controllable fashion. It was shown that in order to collect straight nanofibers, the moving speed of the collector needed to be faster than the electrospinning speed (Figure 11C). Otherwise, spiral fibers were collected (Figure 11B). This demonstration suggests that electrospinning could serve as a simple method for addressable direct-write deposition of nanofibers comparable to those achieved with much more expensive and sophisticated lithography tools. It provides a useful tool for the heterogeneous integration of nanoscale materials to devices prepared by conventional lithographic and manufacturing methods.

9. Concluding Remarks

Electrospinning is an extremely simple and versatile technique for producing nanofibers from a broad range of materials including polymers, composites, and ceramics. Recent efforts have made the electrospinning technique capable of fabricating complex nanostructures with controllable hierarchical features. Nanofibers with core/sheath, hollow, or porous structures have been successfully produced by employing specially designed spinnerets or, adjusting the spinning parameters, and by performing post-spinning modifications. Electrospun nanofibers can also be collected as single fibers, nonwoven mats, uniaxially aligned arrays, or multilayered films by modifying the electrospinning setup. The ability to control both macroscopic organization of nanofibers and the secondary structures of individual fibers makes this technique a new platform for fabricating complex material systems for a variety of applications.

Acknowledgements

This work has been supported in part by an AFOSR-MURI grant on smart skin materials awarded to the UW, and a research fellowship from the David and Lucile Packard Foundation. Y.X. has been a Camille Dreyfus Teacher Scholar (2002-2007) and an Alfred P. Sloan Research Fellow (2000-2005). This work was performed in part at the Nanotech User Facility (NTUF) at the UW, a member of the National Nanotechnology Infrastructure Network (NNIN) funded by the NSF.

References

1. D. Li, Y. Xia, Adv. Mater. 16, 1151 (2004).
2. D. H. Reneker, I. Chun, Nanotechnology 7, 216 (1996).

3. Z. M. Huang, Y. Z. Zhang, M. Kotaki, S. Ramakrishna, Compos. Sci. Technol. 63, 2223 (2003).
4. A. Frenot, I. S. Chronakis, Curr. Opin. Colloid Interface Sci. 8, 64 (2003).
5. S. Ramakrishna, K. Fujihara, W.-E. Teo, T. Yong, Z. Ma, R. Ramaseshan, Mater. Today 9, 40 (2006).
6. H. Q. Liu, Y. L. Hsieh, J. Polym. Sci. Pt. B-Polym. Phys. 40, 2119 (2002).
7. L. Rayleigh, Phil. Mag. 14, 184 (1882).
8. J. Zeleny, Phys. Rev. Second Series, 10, 1 (1917).
9. G. I. Taylor, Proceedings of the Royal Society, A, 313, 453 (1969).
10. D. Duft, T. Achtzehn, R. Muller, B. A. Huber, T. Leisner, Nature 421, 128 (2003).
11. A. Formhals, US patent 1,975,504 (1934).
12. P. K. Baumgarten, J. Colloid Interface Sci. 36, 71 (1971).
13. L. Larrondo, R. St. J. Manley, J. Polym. Sci. Polym. Phys. 19, 909 (1981).
14. X. F. Lu, L. L. Li, W. J. Zhang, C. Wang, Nanotechnology 16, 2233 (2005).
15. X. F. Lu, Y. Y. Zhao, C. Wang, Adv. Mater. 17, 2485 (2005).
16. G. M. Kim, A. Wutzler, H. J. Radusch, G. H. Michler, P. Simon, R. A. Sperling, W. J. Parak, Chem. Mat. 17, 4949 (2005).
17. A. Wang, H. Singh, T. A. Hatton, G. C. Rutledge, Polymer 45, 5505 (2004).
18. M. K. Shin, S. I. Kim, S. J. Kim, S. K. Kim, H. Lee, Appl. Phys. Lett. 88(2006).
19. H. K. Lee, E. H. Jeong, C. K. Baek, J. H. Youk, Mater. Lett. 59, 2977 (2005).
20. Y. Z. Wang, Q. B. Yang, G. Y. Shan, C. Wang, J. S. Du, S. G. Wang, Y. X. Li, X. S. Chen, X. B. Jing, Y. Wei, Mater. Lett. 59, 3046 (2005).
21. P. Gupta, R. Asmatulu, R. Claus, G. Wilkes, J. Appl. Polym. Sci. 100, 4935 (2006).
22. M. Bashouti, W. Salalha, M. Brumer, E. Zussman, E. Lifshitz, ChemPhysChem 7, 102 (2006).
23. J. J. Ge, H. Q. Hou, Q. Li, M. J. Graham, A. Greiner, D. H. Reneker, F. W. Harris, S. Z. D. Cheng, J. Am. Chem. Soc. 126, 15754 (2004).
24. J. Liu, T. Wang, T. Uchida, S. Kumar, J. Appl. Polym. Sci. 96, 1992 (2005).
25. S. Kedem, J. Schmidt, Y. Paz, Y. Cohen, Langmuir 21, 5600 (2005).
26. H. Q. Hou, J. J. Ge, J. Zeng, Q. Li, D. H. Reneker, A. Greiner, S. Z. D. Cheng, Chem. Mat. 17, 967 (2005).
27. A. Wang, A. J. Hsieh, G. C. Rutledge, Polymer 46, 3407 (2005).
28. J. H. Hong, E. H. Jeong, H. S. Lee, D. H. Baik, S. W. Seo, J. H. Youk, J. Polym. Sci. Pt. B-Polym. Phys. 43, 3171 (2005).
29. Y. Ji, B. Q. Li, S. R. Ge, J. C. Sokolov, M. H. Rafailovich, Langmuir 22, 1321 (2006).
30. J. J. Mack, L. M. Viculis, A. Ali, R. Luoh, G. L. Yang, H. T. Hahn, F. K. Ko, R. B. Kaner, Adv. Mater. 17, 77 (2005).
31. Y. H. Lee, J. H. Lee, I. G. An, C. Kim, D. S. Lee, Y. K. Lee, J. D. Nam, Biomaterials 26, 3165 (2005).

32. J. B. Gao, A. P. Yu, M. E. Itkis, E. Bekyarova, B. Zhao, S. Niyogi, R. C. Haddon, J. Am. Chem. Soc. 126, 16698 (2004).
33. Loscertales, I.G., A. Barrero, M. Marquez, R. Spretz, R.Velarde-Ortiz, and G. Larsen, J. Am. Chem. Soc. 126, 5376 (2004).
34. J. E. Díaz, A. Barrero, M. Márquez, I. G. Loscertales, Adv. Funct. Mater. 16, 2110, (2006).
35. Y. Z. Zhang, Z. M. Huang, X. J. Xu, C. T. Lim, S. Ramakrishna, Chem. Mat. 16, 3406 (2004).
36. T. Song, Y. Z. Zhang, T. J. Zhou, C. T. Lim, S. Ramakrishna, B. Liu, Chem. Phys. Lett. 416, 317 (2005).
37. E. Zussman, A. L. Yarin, A. V. Bazilevsky, R. Avrahami, M. Feldman, Adv. Mater. 18, 348 (2006).
38. D. Li, Y. Xia, Nano Lett. 4, 933 (2004).
39. D. Li, J. T. McCann, Y. Xia, Small 1, 83 (2005).
40. J. T. McCann, D. Li, Y. Xia, J. Mater. Chem. 15, 735 (2005).
41. D. Li, A. Babel, S. A. Jenekhe, Y. Xia, Adv. Mater. 16, 2062 (2004).
42. A. Babel, D. Li, Y. Xia, S. A. Jenekhe, Macromolecules 38, 4705 (2005).
43. J. H. Yu, S. V. Fridrikh, G. C. Rutledge, Adv. Mater. 16, 1562 (2004).
44. M. Wang, J. H. Yu, D. L. Kaplan, G. C. Rutledge, Macromolecules 39, 1102 (2006).
45. Y. Z. Zhang, Y. Feng, Z.-M. Huang, S. Ramakrishna, C. T. Lim, Nanotechnology 17, 901 (2006).
46. W. S. Lyoo, J. H. Youk, S. W. Lee, W. H. Park, Mater. Lett. 59, 3558 (2005).
47. J. T. McCann, M. Marquez, Y. Xia, J. Am. Chem. Soc. 128, 1436 (2006).
48. H. Q. Hou, D. H. Reneker, Adv. Mater. 16, 69 (2004).
49. R. Ostermann, D. Li, Y. D. Yin, J. T. McCann, Y. Xia, Nano Lett. 6, 1297 (2006).
50. D. Li, Y. L. Wang, Y. Xia, Nano Lett. 3, 1167 (2003).
51. D. Li, Y. L. Wang, Y. Xia, Adv. Mater. 16, 361 (2004).
52. D. Li, G. Ouyang, J. T. McCann, Y. Xia, Nano Lett. 5, 913 (2005).
53. W. E. Teo, S. Ramakrishna, Nanotechnology 16, 1878 (2005).
54. P. D. Dalton, D. Klee, M. Moller, Polymer 46, 611 (2005).
55. P. Katta, M. Alessandro, R. D. Ramsier, G. G. Chase, Nano Lett. 4, 2215 (2004).
56. E. Smit, U. Buttner, R. D. Sanderson, Polymer 46, 2419 (2005).
57. D. H. Sun, C. Chang, S. Li, L. W. Lin, Nano Lett. 6, 839 (2006).

CHAPTER 5

STRUCTURE OF DOPED SINGLE WALL CARBON NANOTUBES

L. Duclaux [(a)], J-L. Bantignies [(b)], L. Alvarez [(b)], R Almairac [(b)], J-L. Sauvajol [(b)]

[(a)]Laboratoire de Chimie Moléculaire et Environnement, Polytech'Savoie, Université de Savoie, Campus de Savoie Technolac, 73376 Le Bourget du Lac Cedex, France; [(b)]Laboratoire des Colloides Verres et Nanomatériaux (UMR CNRS 5587), Université Montpellier II, 34095 Montpellier Cedex 5, France

A review of the structure of carbon nanotubes doped either by electron donors or electron acceptors is presented. The structure of doped carbon nanotubes probed by X-ray and neutron diffraction and the short range atomic arrangement by TEM and EXAFS spectroscopy were found to depend on the crystallinity and purification treatment of host SWCNTs. The modifications of 2D lattice of SWCNTs ropes induced by doping reactions are examined. The sites of insertion at the surface of bundles, in between the individual tubes or inside the tubes are discussed as a function of the type of dopant and the mode of doping, electrochemical, vapor or liquid phase. Finally, some Raman studies of alkali-doped bundled SWCNTs are reported. The striking dependence of the Raman modes under doping are explained on the basis of structural information.

1. Introduction

A single-walled carbon nanotube (SWCNT) can be described as a graphite sheet rolled up into a cylinder tube. Progress in the laser ablation production of SWCNTs has enabled to observe that they are self organised into 2D limited size crystals usually called nanobundles of SWCNTs [1].

The one dimensional trigonal channels between SWCNTs and their inner cylindrical cavities (central canal) have suggested the possibilities to use them as carbon host lattice, for insertion doping reactions. The idea of doping carbon nanotubes has been attractive to obtain a new class of quasi 1D synthetic metal.

Thus, in analogy to graphite or C_{60}, the electronic properties of SWCNTs have been tailored by doping either by electron donors or by acceptors [2-4]. The reactivity, i.e. the shift of their Fermi Levels is expected to depend upon the electronic properties of host SWCNTs, which gives either a metallic or a semiconducting behavior as a function of the wrapping angle and tube diameter [5].

Pure SWCNTs bundles of strictly one kind of well defined diameter (and wrapping angle) are at present time not available. In fact, the crystal structure of SWCNTs nanobundles is poorly defined: structural analysis (Transmission Electronic Microscopy {TEM} and diffraction analysis) has shown a dispersion of tube diameter, of bundle size [6,7] and of chiralities [8]. The size of structural units of doped SWCNTs in the section of the bundles remains close to 10 nm for bundles that usually contains of the order of 30 tubes [6]. The nanotubes can also be bended with small radii of curvature, so structural units along the tube axis are not even straight beyond the scale of 100 nm. Then, the bad crystallinity of the SWCNTs bundles hinders the interpretation of structural characterization data.

A lot of methods exist to produce SWCNTs: laser ablation [1], electric arc discharge [9], catalytic carbon vapour deposition using methane [10] or disproportion of CO [11], etc... Each method yields to various products with different diameter range, length, crystallinity, and purity, but as produced SWCNTs are always capped at their extremities. The SWCNTs extracted from soot represent a fraction of the total weight of the whole material that contains metallic impurities (10-20 %), turbostratic carbon usually called "amorphous" carbon and graphite particles. Though purification process of SWCNTs have progressed this last decade, the oxidative and acidic treatments, often applied in purification process, allow removing the metallic catalytic particles and the amorphous carbon [12-13]. These treatments induce also the partial opening of the side walls of the nanotubes, the removing of their end caps and shortening of the tubes [14, Monthioux *et al.*]. As a consequence, it appears clearly that the structure, the stoichiometry and the physical properties of doped nanotubes might depend on the origin and purification treatment of host materials.

Doping of nanotubes is intended to modify the electronic properties (i.e. the transport properties by conduction carriers). It can be carried out by substitution of carbon atoms by dopants directly during the synthesis of SWCNTs (example: with B, or N substitution) or by a reaction of the carbon host with a guest electron donor or acceptor (atoms or molecules). We will focus on this later type of doping accompanied by the accommodation or insertion of dopant in the host structure. The sites where the dopants are accommodated are similar to those of gas adsorption [15]. Hence, the closed SWCNTs exhibit three "insertion" sites (Figure 1): interstitial channels (IC sites), grooves (G) and surface sites at the curve outer surface of the bundles. In addition to these sites, the open-ended SWCNTs provide insertion sites (INT) inside the central canal of the empty nanotubes. These INT sites can be differentiated in wall sites (T), closed to the inner wall of the tubes and in axial sites (t) at the center of the tube.

Despite numerous studies on intercalated single-walled nanotubes, the exact localization of the doping agent within the bundles is still debatable. In particular, the detailed knowledge about the local structure of alkali and halogen-doped SWCNT, especially the local geometry around the doping ion is not achieved. Theoretical studies have investigated several models of intercalation with alkali metals (Na, K, Rb or Cs) within SWCNTs [16] or bundles of SWCNTs [17-21]. All the results suggest that the alkali cation is facing a carbon hexagon of the nanotubes. In addition, the locations of alkali ions either between tubes within the bundles [17] or in the hollow sites of a tube [18,22], are energetically stable. However, no experimental evidence has been reported on the precise localization of the guest ion within bundles of SWCNTs. Moreover, the determination of the occupied sites in doped SWCNTs is complicated by the poor crystallinity and heterogeneity of SWCNTs samples.

In this paper, the analysis of the structure of doped SWCNTs by X-ray and neutron diffraction will be reviewed, completed by the investigation of the local structure of dopant by Transmission Electron Microscopy (TEM) and Extended X-ray-Absorption Fine Structure

(EXAFS) spectroscopy. Based on structural data, the Raman investigations of alkali-doped bundled SWCNTs are reported and explained.

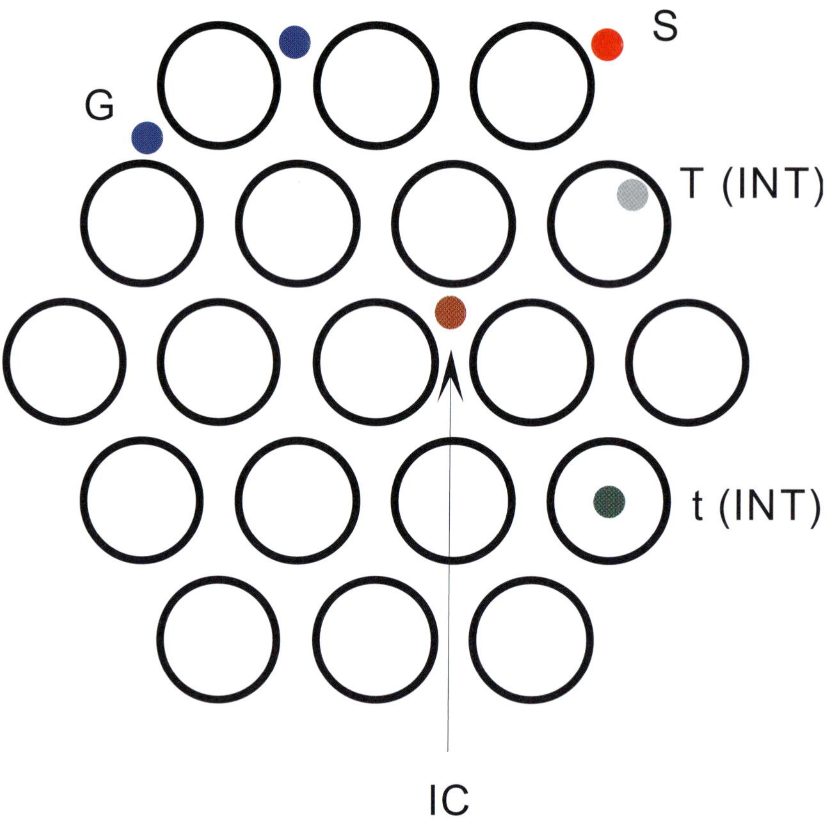

Figure 1. Schematic section of a nanotube bundle containing 19 tubes. The different insertion (intercalation) sites are shown. S is the surface (exterior surface) site (from [Ref. 15]).

2. Structure of Doped SWCNTs (X-Ray Diffraction and Neutron Diffraction Studies)

The diffraction techniques (X-ray and neutron) were used to demonstrate the triangular lattice of the bundles of SWCNTs. In the initial model developed by Thess *et al.* [1], the SWCNTs are assumed to be represented by uniformly charged cylinders. A zero order cylindrical Bessel function was used to approximate the scattering form factor of the nanotubes. The bundle size, the lattice parameter and the nanotube radius are varied until a reasonable agreement with the experimental spectrum is obtained. The Van der Waals inter–tube gap distance was estimated to be approximately 0.32 nm [1]. This model can take also into account a SWCNTs bundle composed of a mixture of different diameter nanotubes [23].

To reproduce the broadening of the lines and the intensity decrease at small Q, due to the diameter distribution of the SWCNTs, a diffraction equation derived from the general formula of diffraction was developed by Rols *et al.* [6-7]. This model can better take into account the diffraction by small crystals such as SWCNTs nanoscopic size bundles defining reduced coherent domains. It allows reproducing clearly disordered 2D lattice of SWCNTs bundles from different origin [6]. This calculation method of diffraction patterns was used successfully to characterize doped SWCNTs by electron donors or electron acceptors.

2.1. Electron Acceptors

A lot of work were devoted to the structural studies of acceptor doped SWCNTs by X-ray or neutron diffraction. Chemical oxidation of SWCNTs can occur by using nitric acid as a reactant. An expansion of 0.185 nm of the internanotube spacing and an increase in the amount of H were observed after immersion for 2 h into HNO_3 (70%) concentrated solution [24]. This expansion deduced from the position of the (*10*) line on the X-ray diffraction pattern, was explained by the insertion of HNO_3/NO_3^- molecules filling the (1/3, 2/3) interstitial voids (IC sites) in the expanded 2D lattice. However, the diffractograms of HNO_3 doped SWCNTs shows only the (*10*) line of the 2D structure so that the doping has lead to a more disorganised and defected bundled 2D structure. Exfoliation of SWCNTs was also reported by Jin *et al.* by electrochemical treatment of SWCNTs in nitric acid [25]. Moreover An *et al.* [26] have shown by microscopy observations that at shorter refluxing time in HNO_3 (30%), the SWCNTs bundles are partially disordered and exfoliated at preferably defective sites but with increasing refluxing, the distorted SWCNTs are transformed into new phases : multi-walled carbon nanotubes, cone-capped phases and onion-like phases.

Similarly to the reaction in the concentrated HNO_3, a charge transfer spontaneous reaction can arise by a direct immersion of SWCNTs in concentrated H_2SO_4. According to Sumanasekera *et al.* [27], the spontaneous electrochemical doping of SWCNTs by H_2SO_4 yields to the intercalation of H_2SO_4 and HSO_4^- molecules in the interstitial channels. Ramesh *et al.* [28] has observed that SWCNTs can be dissolved in anhydrous sulphuric acid where they induce a partial positional order in the surrounding solvent. The calorimetric and X-ray diffraction data have shown that intercalated ropes in superacid are formed of individual nanotubes surrounded by ordered layers of sulphuric acid [29]. It was proposed that the direct protonation of SWCNTs is responsible for the formation of acid layers and consequently their dissolution [30]. Zhou *et al.* [31] has studied by X-ray diffraction the super acid intercalated SWCNTs in the range $200\ K<T<300\ K$, and they proposed a model of disorder-order transition for the templated crystallisation of H_2SO_4 molecules through protonation interaction.

The iodine doping, was carried out by immersion in molten phase at 140°C leading to the average IC_{12} composition [32]. After heavy doping, the intensities of the diffraction peaks due to the 2D rope lattice are observed to fall below the detection limit. The authors interpret this absence of diffraction signal from the doped bundles by the fact that iodine acting as a chemical wedge enters the interstitial channels between the tubes overcoming the Van der Waals attraction between the tubes to force them farther apart [32]. In the deintercalated iodine sample, the positions of the diffraction lines were found to be similar to those before reaction but the overall intensity distribution was different from that of pristine sample. This was interpreted by a partial recover of the initial stacking of the tubes in the rope lattice. The analysis of these intercalated samples (Z contrast electronic microscopy) has also shown that iodine is incorporated inside SWCNTs. These studies on the iodine-doped SWCNTs were all performed on laser produced or arc produced SWCNTs mats [32,33].

In more recent studies, Bendiab *et al.* [34] has prepared from arc-electric raw materials, by the same method as Grigorian *et al.* [32] (simple immersion in iodine), iodine doped SWCNTs with also an IC_{12} approximate composition. Nevertheless, a diffraction pattern from the 2D lattice is observed by X-ray or neutron scattering after immersion in iodine. Moreover, this author has confirmed that iodine doping (generally as all dopants) is accompanied by disordering of the bundle array. The weak (*hk*) reflections well defined in the raw SWCNTs are no longer visible after doping and only the (*10*) line is observed. The contrast in the (*10*) line position was pointed out between X-ray and neutrons diffractograms (Figure 2). Indeed, the (*10*) peak is more shifted to the low angle in the case of X-ray than in the case of neutrons. This difference is notably due to a higher sensibility of carbon atoms than iodine atoms using neutrons compared to X-ray scattering. As an example, the contribution of iodine in IC_{12} to the neutron diffraction signal at Q=0 is only 0.4% (contribution of carbon: 88%, and contribution of the cross term both by iodine and carbon: 11.6%), instead of 18% by X-ray (contribution of carbon: 33%, and contribution of the cross term: 49%). The decrease in intensity of the (*10*) line after iodine intercalation is in fact attributed to the T inner sites occupancy. Though the arc-electric tubes are initially closed, the iodine acts as an oxidant and open the tubes enabling the insertion in the T sites. Similar evolution of the intensity of the (*10*) line was reported after gas adsorption in the T sites [15].

The simulations of the diffraction patterns [34] were performed, by using a model derived of the formalism of Rols *et al.* [6] as a function of the radius of iodine cylinders inside the tubes. The best agreement between simulation (X-ray and neutron) and experimental data was established for an IC_{12} composition in a model with 85% of iodine in T sites (radius of iodine cylinders: 0.48 nm) and 15% in IC sites [34]. The large Q shift of the (*10*) line, well reproduced by the simulations, have been demonstrated in the case of IC_{12} stoichiometry (Figure 3). This large Q shift of 15% is also strongly related to the large intrinsic diffraction signal explained by the small coherent domains of the bundles and the presence of a zero of the structure factor in the vicinity of the (*10*) reflection.

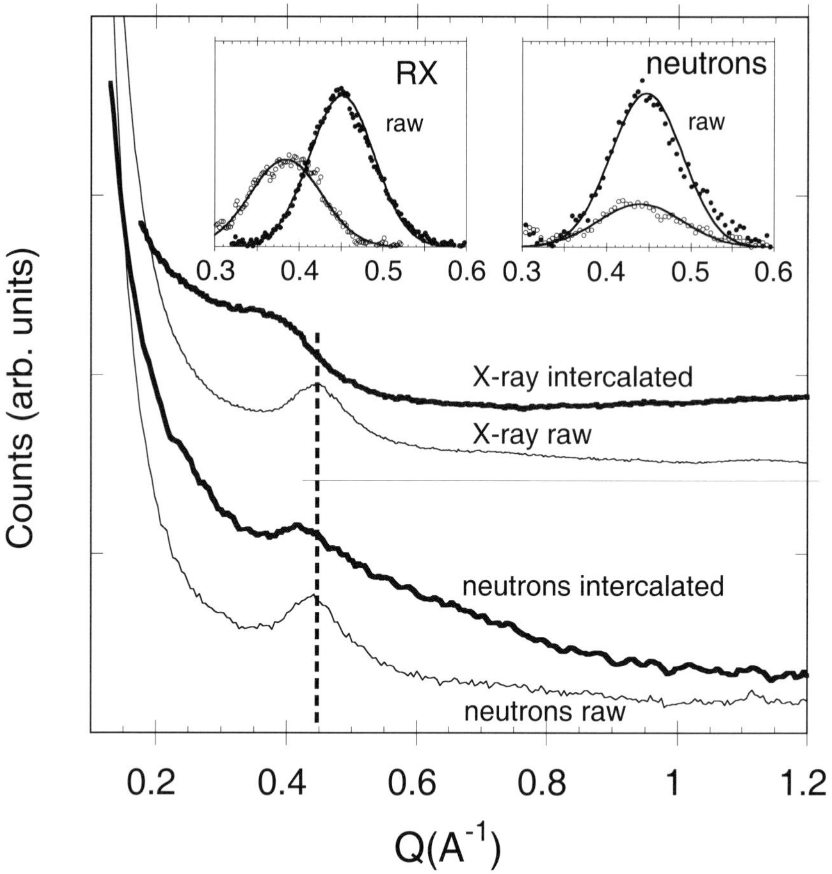

Figure 2. Diffraction diagrams in the small Q range (i.e. (10) peak) of pristine and iodine-inserted SWNT samples using X-ray and neutrons. The insets show the same data after subtraction of the background [Ref. 34]. Q is defined by $Q=4\pi \sin\theta/\lambda=2\pi S$ where $S=2 \sin\theta/\lambda$.

This fine combined neutron and X-ray diffraction simulations (on iodine doped compound) demonstrate that the shift to higher distance of the *(10)* line cannot be attributed systematically to expansion of the 2D lattice concomitant to intercalation in the IC sites. Similarly, neutron diffraction studies of the adsorption in the close-ended SWCNTs of two Ar isotopes [35], namely ^{36}Ar with large coherent section versus ^{40}Ar which is almost invisible for neutrons, has demonstrated that the experimentally observed shift peak and slight intensity growth of the (*10*) line are due to diffraction arising from adsorbate located on the outer part of the surface of bundles (Surface S and groove G sites). These authors have proved that adsorbed argon is in fact practically not inserted in the interstitial channels though the (*10*) peak shift toward higher distances.

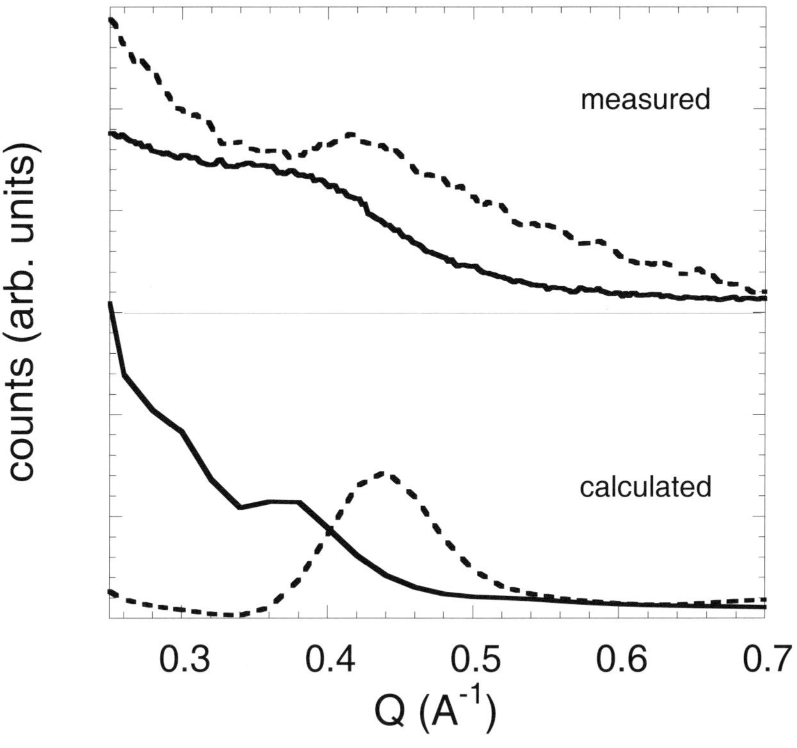

Figure 3. X-ray diffraction patterns of iodine-doped SWNT bundles in the small Q range (i.e. (10) peak). The top graph shows measured patterns, and the bottom graph shows calculated patterns. The dashed curve is for the pristine sample and the full line for the iodine-doped one [Ref. 34].

2.2. Electron Donors

2.2.1. Insertion of Li and Na

The SWCNT bundles can be doped by light alkali-metals such as Li or Na. An ideal maximum stoichiometry close to LiC_2 was predicted for open SWCNT, considerably higher than the maximum level of insertion in graphite (LiC_6) [19,36]. The possibility of site intercalation or insertion in the tubes was explored by energy calculations [37]. The authors have noticed that Li ions favour two positions: either inside the tube close to the wall or outside the tube. This former site is energetically slightly more favourable.

Li insertion can be achieved electrochemically [38-40]. The purified and as grown SWCNTs showed a reversible saturation composition of $Li_{1.23}C_6$ and $Li_{1.6}C_6$ respectively. This composition is significantly higher than the value for graphite, but the SWCNTs material displayed a very large irreversible capacity (x=3-4 in Li_xC_6). The capacity can be further improved by chemical etching [41] or ball milling the nanotubes [22] with a maximum capacity as high as $Li_{2.7}C_6$ (~1000 mAh/g).
Studies of Li electrochemical insertion by in situ X-ray diffraction have shown the irreversible loss of the triangular lattice [38] consistent with the disorder observed by doping with heavy metals in the vapour phase in some conditions and also with the electrochemical doping of K in THF solutions [39]. The use of THF molecules as a solvent, instead of electrolyte such as EC-DMC, are likely to solvate the Li ions and to co-intercalate it in host structure [39].

Petit *et al.* has [42-43] developed a chemical route to prepare LiC_x compound with a precise stoichiometry via Li^+ doping in tetrahydrofuran solution (THF) and the control of redox potential with several radical anions (naphtalene, benzophenone, fluorenone, anthraquinone, and benzoquinone).

The monotonous increase of conductivity with the dopant concentration has confirmed the absence of intercalation stages in SWCNTs contrary to GICs [42-43]. At saturation a composition close to LiC_6 was found. For example using naphthalene, $LiC_{5.88}(THF)_x$ sample can be prepared. The structure of this compound was studied by neutron diffraction [44]. Hydrogen-Deuterium substitution of THF in the ternary doped compound has allowed determining the structure of the doped bundles. The simulations of the neutron diffraction patterns were performed by assuming the shell of TDF molecules as a homogeneous density of scatterers while carbon nanotubes and the surrounding lithium shell were assumed to be uniform cylinders (Figure 4).

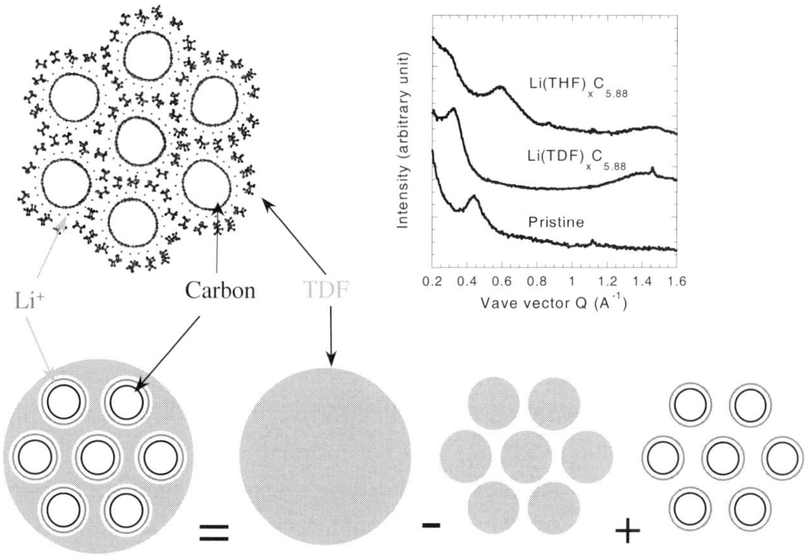

Figure 4. Structure of Li $(THF)_x$ $C_{5.88}$. The top left image is the model of the structure, and the top right image is the small Q measured neutrons diffraction patterns of pristine, deuterated, and hydrogenated compounds (i.e. (10) peak). The bottom image is the principle of the simulation. Holes are made in the large grey volume, which are replaced by the lithium and carbon curved sheets (from [Ref. 44]).

In these compounds, the alkali ions were found to form a shell surrounding the SWCNTs decorated by the THF molecules so that the 2D triangular lattice was preserved and strongly dilated (the lattice parameter value is 2.4 nm instead of 1.65 nm in raw ropes) by the intercalated guest molecules (alkali+THF) [44]. The structure of a $K_{5.88}C$-THF compound, investigated also by neutron diffraction, reveals that the spacing distance between the nanotubes is controlled by the size of the alkali cations. As a matter of fact, in THF solvent, with large radii Rb and Cs cations, the maximum of expansion is expected to be attained.

2.2.2. Heavy Alkali Metals (K, Rb ,Cs)

Most of electron donor's intercalation experimental works have been devoted to alkali metal. After reaction with heavy alkali metals (Cs and K) in the vapour phase, the saturation composition of doped SWCNTs was reported to be MC_8 by various authors similarly to the composition of first stage graphite intercalation compounds [45-47]. No evidence of "intercalation stage" or stable definite compounds was found as the alkali concentration increases from zero to its maximum value. The explanation given by Fischer [48] is that intercalation site energies are affected by the rotation of SWCNTs about their axes. Indeed, the site energy can be minimized by the local commensurability of the intercalated species with the hexagonal structure of carbons on the tube wall. But this commensurable ordered guest structure in closed packed 2D lattice can only occur if the three fold rotational symmetry is preserved. This symmetry, that can be found in bundles formed of tubes with wrapping indices (n,m) modulo 3 is rare and exceptional, so all other tubes are frustrated. Thus, the incommensurability prevents the formation of the "intercalation stage" and the alkali structure would most likely be of lattice gas type, with variable random filling as a function of guest concentration [48].

In many reports, crystalline 2D structure of nanotube bundles are found to be totally disorganized after alkali vapour reaction so that X-ray diffractograms of doped materials definitively show no lines of the 2D lattice [45-50]. Despite TEM showing no clear electron diffraction, Suzuki *et al.* conclude that alkali were intercalated in the interstitial channels of nanotube bundles causing the structural disorder of the 2D lattice [49]. Using in situ electron diffraction coupled with EEL spectroscopy, Pichler *et al.* has brought out the progressive shift of the (*10*) peak characteristic of the 2D arrangement of the tubes as the K doping ratio was increasing [51].

The purification treatment of SWNTs (based on oxidation treatment) [11-13] modifies considerably the structure of host nanotubes as it removes the amorphous carbon covering the bundles, open partially the nanotubes and creates defects. Alkali doping (K, Rb, and Cs) on various SWCNTs hosts has shown the effect of the nature of the host on the doped SWCNTs structure observed by X-ray diffraction [47]. These studies have brought out the difference of doped SWCNTs structure since host were purified or not.

The alkali doping at saturation of chemically purified bundles forms disordered structure as no clear diffraction (*hk*) lines from the 2D structure appears on the diffractograms [47] except a non indexed peak that appears at low angle (at d=0.89 nm marked by a star) (Figure 5) . The large diffusion band corresponding to first neighbor distances between alkali is observed on the diffractograms. The inter-alkali distances are not commensurate with the graphitic layers cylinders and shorter than in graphite intercalation compounds. Low temperature diffraction experiments in the range 77K-300K, showing no liquid-solid transitions have demonstrated that the alkali atoms are not arranged in a liquid-like structure as in MC_{24}-GIC [52] but in disordered structure with short range order. Annealing at 200°C in static vacuum is not effective to deintercalate totally purified SWCNTs mats doped with Cs but the initial diffraction pattern is completely recovered after immersion in water. Most of reports on purified SWCNTs doped by alkali metals (Li, Rb, Cs) conclude to the intercalation of SWCNTs in the interstitial channels accompanied by disorder [45-50]. Los *et al.* have also observed the absence of 2D lattice diffraction response by neutron scattering of purified SWCNTs sample doped by K at saturation [53]. The trial of D_2 sorption on this sample at $T<77$ K has remained unsuccessful indicating that all the sites (even the interstitial voids) are occupied by alkali atoms so that the doped purified SWCNTs materials is definitively not porous.

The structure of raw doped SWCNTs (collaret from electric-arc origin) was found less disordered than purified doped materials as the 2D triangular lattice probed by diffraction method is preserved after doping [2,47,54-55]. The X-ray diffraction patterns of pristine SWCNTs soots reacted with heavy alkali metals (K, Rb, or Cs) at saturation display a shift of the (*10*) line toward higher distance. The shifts are found to increase together with the size of alkali ions. These shifts have been interpreted by an expansion of ropes caused by the intercalation of alkali metals within the ropes in IC sites. Following this interpretation, the values of the expansion of the host lattice calculated from the position of the (10) peak are almost similar to those obtained in the same alkali metals GICs [2,47].

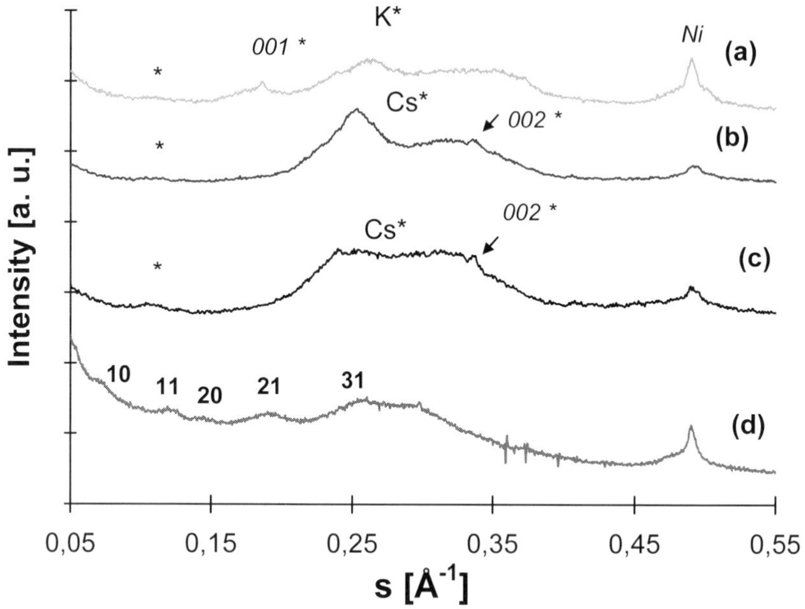

Figure 5: X-ray diffractogram of the pristine Rice carbon deposit SWNTs intercalated by (a) K, and (b) Cs at saturation. Diffractogram (c) is from the de-intercalated sample (first reacted at saturation with Cs) by heat treatment at 200°C in a 2 10^{-5} mbar static vacuum, and diffractogram (d) is from the intercalated sample washed in water and dried. The *001** and *002** lines belongs to the *00l* graphite intercalation compounds formed also by reaction with alkali metal. The alkali diffusion band and Ni impurities are indexed by M* (M=K, Cs), and Ni respectively. A non indexed peak that appears at low angle (at d=0.89 nm) is marked by a star.

However, the host SWCNTs materials used in these studies is known not to be pure and contains both amorphous carbon (turbostratic carbon) that covers the nanotube bundles, graphitic particles and residual

catalytic particles. The ends of the tubes are closed and some metallic particles are found at the extremities of the tubes. This means that the stoichiometry of the doped SWCNTs is definitively not MC_8 as some possibilities of insertion exist both in graphite and amorphous carbon.
As the diffractogram of K-intercalated collaret presents a few peaks attributed to the 2D lattice, the simulation of this diffraction pattern was performed by the calculation procedure of Thess *et al.* [1], using a model in which three K ions occupies the triangular cavities in respect with the threefold symmetry considering a distance between first neighbor K ions close to the distance in metallic state (d_{K-K}=0.49nm). Assuming that the distance between alkali atoms along the tube axis and the in-plane section is the same, the global stoichiometry of the K-doped SWCNTs deduced from structural models should be close to KC_{13} [47].

Gao *et al.* [20] has developed models obtained by molecular dynamics simulation based on a 2×2 lattice of K-ions in register with the graphitic lattice of the (10, 10) tubes that are distorted in order to accommodate such a commensurate lattice. In this model, the 2×2 lattice of K is decorating the outside or inside of the tubes. The optimum structure predicted by Gao *et al.* [20] shows a KC_{16} stoichiometry if K^+ ion cannot penetrate the (10, 10) tube and KC_{10} if the tubes are allowed to be inserted by alkali cations. However the theoretical predictions are in poor agreement with experimental works possibly because real SWCNTs materials are known to exhibit helicities, to present some diameter distribution and to possess a weak crystallinity. So real alkali doped bundles lattice is found to be disordered with only short range order.

Neutrons diffraction have also been performed on KC_{10} electric arc SWCNTs samples prepared by deintercalation at 200°C in sealed vacuum tube of a saturated doped SWCNTs (KC_8 stoichiometry) [56]. The sample was then studied by neutron diffraction for its properties of D_2 adsorption. KC_{10} doped neutron diffraction diffractograms display very small shift of the *(10)* line compared to raw SWCNTs corresponding to a calculated expansion of only 0.05 nm (expected shift from [Ref. 47] observed by X-ray diffraction is 0.18 nm in fully doped sample). Moreover, in situ studies of deuterium adsorption by neutron diffraction have shown that in this KC_{10} closed SWCNTs, the microscopic porosity that correspond to the triangular channels in between the tubes can be filled by adsorbed molecular D_2 at low temperature (T<77 K) [56]. By the way, the (*10*) diffraction intensity is progressively raised as the D_2/C atomic ratio of adsorbed deuterium increases (while the applied deuterium pressure is increased) (Figure 6).

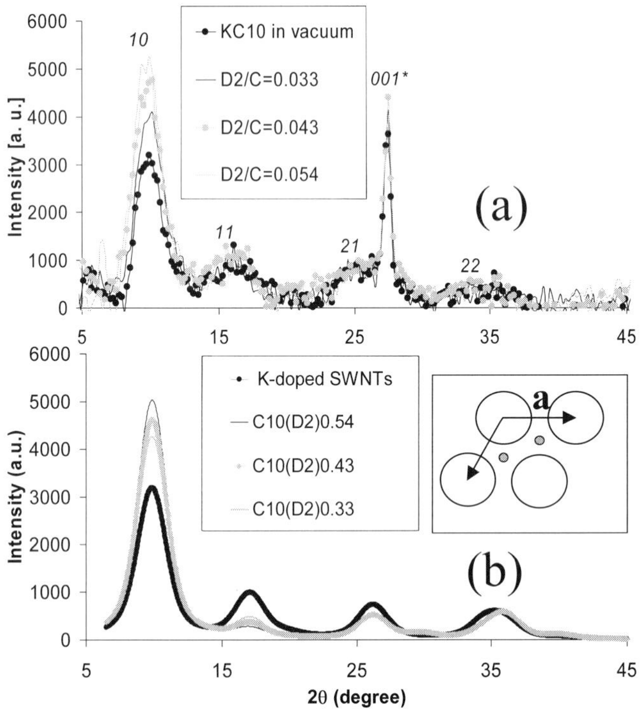

Figure 6. Schematic of the (a) background subtracted neutron diffraction patterns of KC10 doped SWNTs sample as a function of the D_2/C sorbed ratio at 22.6 K (λ=2.52 Å, acquisition time 30 min.). The 2D SWNTs structure (*hk*) peaks (10, 11, 21, and 22) are observed. The *001** line is due to the presence of KC_8 first stage GIC. Also shown is the schematic of the (b) simulated neutron diffractograms using the model showed in inset for various $KC_{10}(D_2)x$ compositions (x=0.33, 0.43 and 0.54). A good agreement is obtained between experimental and calculated patterns of K-doped SWNTs using a simple model of uniformly charged cylinders without K intercalated ions. Inset: Model of the 2D SWNTs lattice (tubes are shown by large circles) with D_2 molecules (grey filled circles) sorbed at the center of the triangle cavities in between the tubes (tube radius is 0.67 nm and lattice parameter a=1.7 nm).

The fact that the estimated pore volume (due to interstitial channels) is also consistent with the micropore volume (0.15 cm^3/g) of adsorbed D_2 deduced experimentally (from isotherms at 18 K) confirms the D_2 insertion into the free voids of this KC_{10} SWCNTs sample. This study shows that slightly doped raw SWCNTs (from electric arc origin) are not homogeneously intercalated and that the alkali atoms are mainly at the surface of the bundles (probably in groove and surface sites). The presence of amorphous carbon coating at the surface of nanotubes bundles in this raw SWCNTs material can also hinder the diffusion of alkali metals within the bundles compared to purified SWCNTs. It suggests that by contrast to purified SWCNTs, electric arc raw products are hardly intercalated due to impurities of amorphous carbon. The absence of intercalated-K ions in the triangular lattice is in disagreement with previous interpretation of X-Ray diffraction data [2] by the intercalation of dopants within the bundle in saturated compounds. In fact, one can wonder whether in alkali-doped SWCNTs from electric-arc origin, the X-ray diffraction downshift was related to an intercalation or not, as previous studies have shown that this shift can occur in the absence of dilatation of the triangular lattice [35]. To solve this question, more fine simulation of the diffraction signal comparing the same sample probed by the two techniques (neutron and X-ray- see [Ref. 34]) should be carried out.

The oxidative treatment of purification can led to an opening of the tubes [13], so that insertion was found possible in inner sites of the tubes (INT). For example, X-ray diagram of SWCNTs purified mats after potassium insertion-deinsertion cycle (deinsertion in vacuum) is typical of a filled tube pattern (Figure 7). Then, the diffraction response can be simulated using a model in which open tubes are partially filled by K atoms in a KC_{10} composition [47]. The tube might have been filled after deinsertion by remaining alkali metals. Others authors have also reported the accommodation of alkali metals to fill the inner cylindrical cavity after ion irradiation treatment [57] of SWCNTs immersed in a magnetised alkali metal plasma. This method of synthesis avoids the air instability inherent problem of alkali-doped SWCNTs, as alkali-encapsulated SWCNTs are created (for example Cs@SWCNTs). The authors have also indicated by ab initio band structures and density of states determination, the possibility of using Cs-doped SWCNTs as doped junctions, with potential application in nanoelectronics [58]. The fullerene peapods C60@SWCNTs doping can be also achieved by introduction of potassium vapour into the peapods [59-62]. Doping by K can lead to an insertion of the alkali metal not only between the tubes in a bundle but also into the interior of the peapod [61]. Sun *et al.* [63] has reported the synthesis of peapods encapsulating exohedral Cs metallofullerenes $(CsC_{60})_n$@SWCNTs by the direct vapour reaction of CsC_{60} fullerite with open-ended SWCNTs.

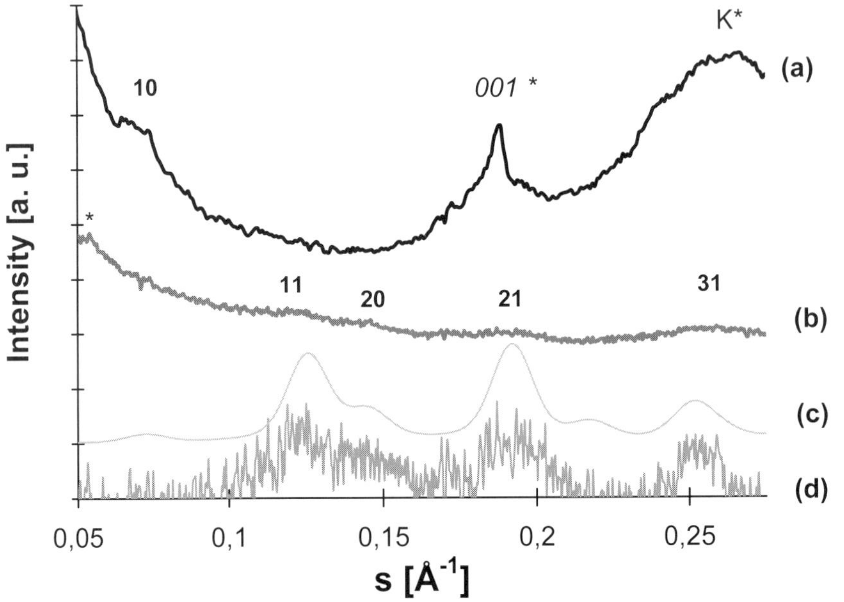

Figure 7. X-ray diffractogram of degassed Rice carbon deposit (at 370°C 10-6 mbar), when (a) reacted with K at saturation, and (b) intercalated, then de-inserted in a dynamic vacuum (10-6 mbar) at 200°C (simulated diffractogram). Diffractogram (c) represents the simulation using a model of triangular lattice of SWNTs filled by potassium atoms in their inner sites and empty interstitial voids (KC_{10} composition, lattice parameter a=1.59 nm, nanotube radius R= 0.629 nm , model described in [Ref. 47]). Diffractogram (d) is the background-subtracted b-diffraction pattern.

3. The Local Structure (EXAFS and TEM)

Imaging by means of TEM appears to be the most direct technique to visualize the intercalation sites within bundles of SWCNTs. For both iodine and cesium doping, Z-contrast scanning TEM experiments clearly identify the presence of dopants inside the tube [33,57-58,64]. For example, Z contrast TEM has revealed the incorporation of helical iodine chains inside single walled carbon nanotubes [33]. The observation of different periodicities of the iodine helix has suggested that the iodine chains adopt a preferred orientation with respect to the nanotube wall depending on the different chiralities of the tube. Furthermore, Raman spectroscopy shows the charge transfer to iodine typically corresponds to the I_3^- and I_5^- species. In the case of encapsulated Cs in the central channel of SWCNTs, three varieties were observed by field emission transmission electron microscopy and scanning transmission electron microscopy: linear configuration, spiral chains of Cs and crystallised Cs [57]. Even in case of potassium doped peapods, high resolution TEM gives direct proofs on the localization of potassium ions between fullerenes inside the tube [62]. Hence, from TEM experiments, it is proven that the site inside the tube is occupied upon doping since the tubes are opened, before or during insertion [33,57,62].

However, there is no direct evidence from TEM images about the presence of the dopant in between the tubes. Indeed, some articles on alkali doping report the vanishing of the lattice fringes upon intercalation [46,49]. These observations might be explained by the structural disorder induced by the intercalation process. Some experimental EXAFS works has been successfully used to investigate the local structural properties of Se confined in SWCNTs [65]. It was shown that confined selenium is arranged in well organized rings in SWCNTs and disordered trigonal structure in multiwall carbon nanotubes. Recently, the various intercalation sites of doped SWCNTs have been probed by X-ray absorption spectroscopy in Rb-doped SWCNT (Rb-SWCNT) [66] and I-doped SWCNT (I-SWCNT) samples [67]. The local arrangement around the doping species (iodine or rubidium) has been derived from the experimental EXAFS oscillations and simulation. In particular, data about the number of nearest neighbors of ions and the distances between doping ions and carbon atoms have been obtained.

3.1. Rubidium Doping

Rubidium doped SWCNTs prepared by reaction of electric-arc raw SWCNTs with vapor phase of rubidium have been studied by X-ray absorption spectroscopy at the Rb K-edge. Figure 8 displays the Rb K-edge absorption spectra, μ(E), of the saturated phases of RbC_8 (1st stage graphite intercalation compound), and Rb-doped SWCNT (RbC_9) obtained from electric-arc closed SWCNTs, respectively. By contrast with the strong and well-defined EXAFS oscillations of the Rb-doped graphite, in agreement with its high degree of crystallinity [68], the weak EXAFS oscillations in the Rb-doped SWCNT sign the presence of a strong local disorder around the rubidium ions.

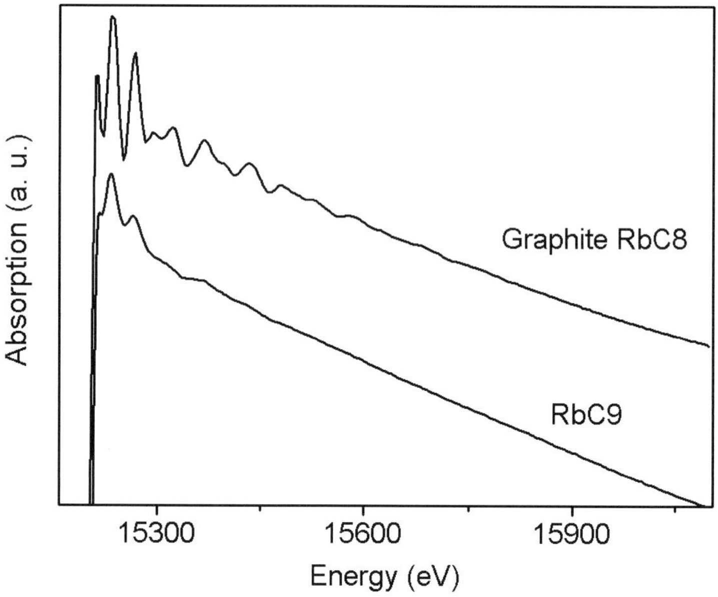

Figure 8. Rb K-edge absorption spectra of Rb-doped graphite RbC_8 and Rb-doped SWCNTs RbC_9. Spectra are vertically shifted for clarity.

The $\chi(k)$ EXAFS oscillations, measured on Rb-doped SWCNT samples with increasing doping time (NT2H doped two hours and RbC_9 doped one week) are very similar to each other (Figure 9a). The $\chi(k)$ profiles mainly exhibit a single oscillation assigned to a single coordination shell. The pseudo radial distributions, $FT(k^3\chi(k))$, derived from the Fourier transform of the EXAFS spectra (Figure 9b) exhibit one single neighbor shell centred between two an three angstroms.

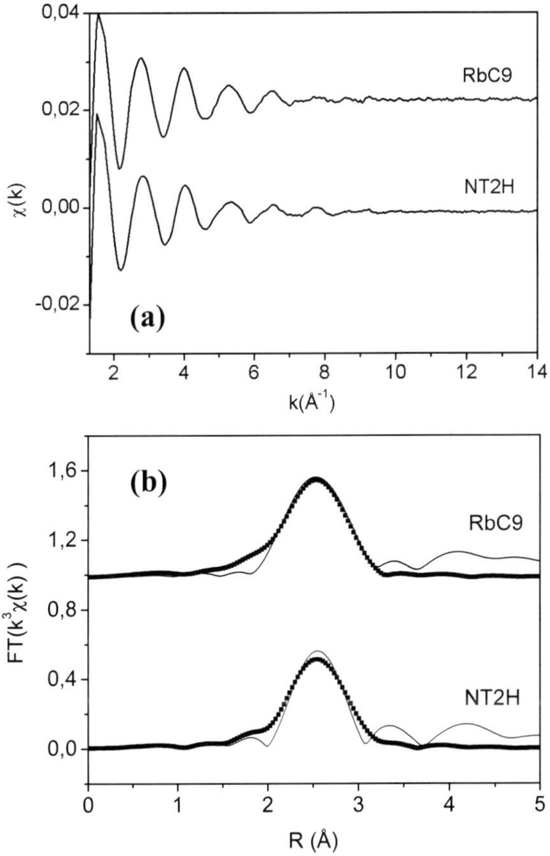

Figure 9. Plot of (a) the EXAFS oscillations $\chi(k)$ of two Rb-doped SWCNT samples, and (b) the pseudo radial functions, $FT(k^3\chi(k))$ of two different Rb-doped SWCNT : RbC9 et NT2H. The fits are performed using backscattering amplitude and total phase shift extracted from FEFF8 calculations (square). Note that the difference between the peak position (2.7 Å) and the real Rb-C distance (3.15 Å) is due to a phase term.

To determine the structural parameters of the Rb-doped SWCNT compounds: the N coordination number, the R first neighbor distance, the σ^2 Debye-Waller factor and the ΔE energy shift; this defined neighbor shell is fitted using standard theoretical amplitudes and phase functions extracted from FEFF8 simulations [69] by considering that the first neighbors correspond to carbon atoms. For this first coordination shell, a comparison of the results of the fit for NT2H and RbC_9, is reported in Table 1. The average Rb-C distance (called R in the tables) is found to be close to 3.15 Å, in good agreement with the distance calculated in RbC_8 first stage graphite intercalation compound where the Rb ions are centred above graphitic carbon hexagons [70]. The N number of first nearest neighbors is found to be clearly dependent of the doping level, increasing from 2.8 to 5.5 with the doping level (related to the time of doping). This point is discussed in detail for several doping rates in [Ref. 66].

Table 1. Structural parameters deduced from the least squares fit of the first shell of the Rb-doped SWNT samples. N is the coordination number, R the fitted Rb-C distance, σ^2 the Debye-Waller factor, ΔE the energy shift.

	RbC_9	NT2H
N	5.5	2.8
R (Å)	3.16	3.15
σ^2 (Å^2)	0.015	0.008
ΔE (eV)	-5.02	-5.4

In RbC_9 SWCNT saturated sample, the Rb-C distance value and the coordination number, suggest that a rubidium ion is facing an hexagonal ring of carbon atoms. Then, in this sample, the local order of rubidium ions is close to the one in the saturated phase of the Rb-doped graphite compounds [68]. For the low doping levels (NT2H), the preferential site is not clearly identified. However, as the number of coordination is significantly lower than 6 (Table 1), rubidium ions are assumed to be located off centre of the carbon hexagons.

In RbC_9 SWCNTs doped at saturation, the interstitial site, usually reported as the preferential site of insertion in alkali-doped SWCNT bundle [2], is ruled out from the EXAFS results since the number of carbon atoms at a Rb-C distance below 4 Å is found to be six. This number is significantly smaller than the one expected (around twenty-six for a (10,10) bundles) for rubidium ions localized in the interstitial space. In addition, the non expansion of the 2D triangular lattice after the rubidium doping of SWCNT bundles, recently stated from neutron diffraction experiments [71], prevents a position of rubidium ions in between two or three tubes.

Ab-initio FEFF8 simulations of the EXAFS oscillations, χ(k) have been performed to investigate the position of the rubidium ions within the bundles. In all calculations, the first Rb-C distance is kept close to the one derived from the EXAFS analysis of the first coordination shell. The main intercalation sites have been considered and only the surface sites with the Rb ions at the center of the hexagonal ring appear to be possible intercalation sites (see details in [66]).

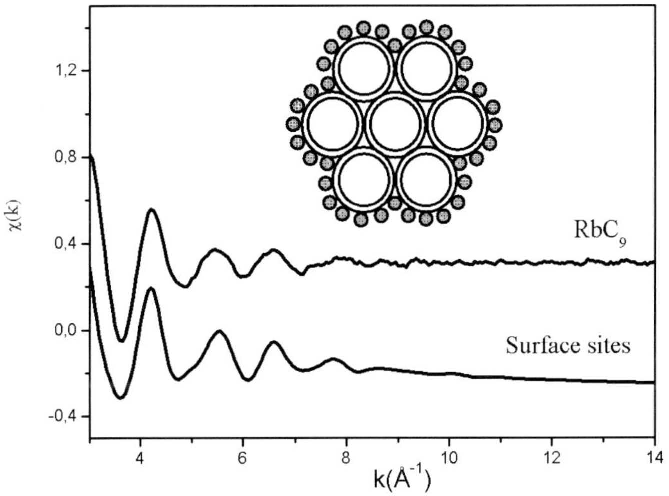

Figure 10. Comparison between the experimental Rb K-edge EXAFS oscillations, χ(k), measured in the saturation-phase of Rb-doped SWCNT (RbC_9), and FEFF8 calculations for surface bundle sites. The different local structures of the FEFF calculations are built for clusters of 4.7 Å around Rb and the curvature corresponds to the one of a (10,10) nanotube. For the outer surface site (Rb on outer SWCNT surface) six Rb-C distances are found in the 3.12-3.23 Å range, eight Rb-C distances between 4 and 4.7 Å. For the groove site (Rb on groove) four Rb-C distances are found between 3.11-3.28 Å, six Rb-C distances from 3.5 to 4 Å, and eighteen from 3.9 to 4.7 Å.

Though the EXAFS signals for Rb ions either at the inner surface or at the outer surface of SWCNTs are very similar [66], the hollow center of SWCNTs can be excluded as a possible site from EXAFS data because the electric-arc raw SWCNTs are not purified and thus close ended. Thus, in electric-arc pristine SWCNTs doped by heavy alkali metals, the absence of intercalation of the alkali metals in IC sites of bundles previously observed by neutron diffraction [56,71] is confirmed by EXAFS spectroscopy. The average EXAFS response of the rubidium ions at the surface of the bundle doped at saturation is a sum of all the sites available at the surface of the bundle (6 groove sites and 30 outer SWCNT surface sites for a (10,10) bundle). The simulation-experiment agreement is satisfactory (Figure 10), suggesting that the rubidium ions can be located only around the bundles.

3.2. Iodine Doping

Purified and raw SWCNTs deposits were intercalated by iodine in the molten phase as in [Ref. 32].The iodine doped SWCNT were washed with ethanol in order to get rid of non intercalated iodine present as I_2 crystal clusters in the samples [67]. The washed samples provide a very good EXAFS signal due to I-I contributions. The Fourier transform $FT(k^3\chi(k))$ of the EXAFS signal of I-SWCNT has given rise to only one peak located between to 2 and 3 Å (Inset Figure 11). To quantify the local arrangement around iodine, the experimental EXAFS signal is well fitted by considering a neighboring shell of iodine (Figure 11) and a second neighboring shell of carbon atoms. The structural parameters given by the fit (Table 2) indicate that Debye-Waller factor associated to the disorder in the structure is quite low in the case of iodine-iodine distance due to the formation of well organized structures by iodine species (I_3^- and/or I_5^- chains). From EXAFS results on iodine doped conjugated polymer, the I-I distance and the number of neighbors as a function of the chain length were determined [72]: I_3^- chain is attributed the I-I distance of 2.9 Å whereas the number of neighbors is 1.2; and I_5^- chain leads to an I-I distance of 2.73 Å and a coordination number close to 0.85. The neighbor number and I-I distance in iodine doped SWCNTs samples are estimated to be approximately 0.8 and 2.74 Å, respectively.

These values of the structural parameters have confirmed the dominant presence of I_5^- species in the sample doped at saturation in good agreement with the conclusion of Grigorian *et al.* [32].

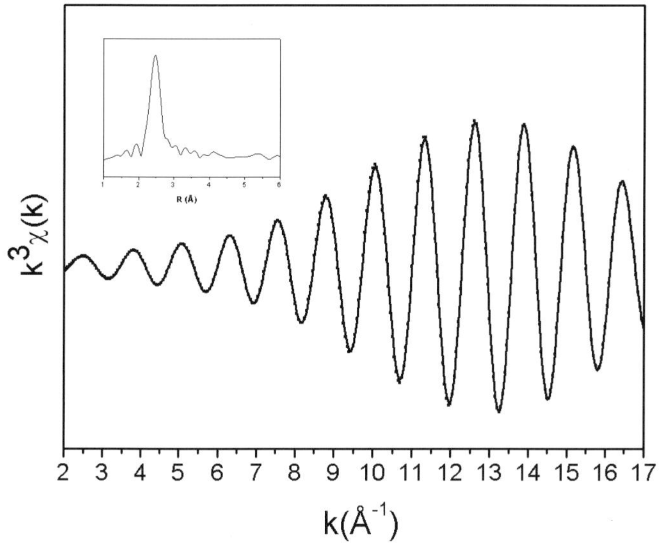

Figure 11. The EXAFS oscillations $\chi(k)$ of the I-doped SWCNT. The fits are performed using backscattering amplitude and total phase shift extracted from FEFF8 calculations (square). Inset: Pseudo radial function, FT($k^3\chi(k)$) of the I-doped SWCNT.

FEFF simulations is used to confirm the iodine chain structure (I_5^- chain) derived from the fit. Figure 12 (from [Ref. 67]) displays the EXAFS spectra simulations for three kinds of iodine chains. Due to the inter-atomic I-I distance (2.9 Å), the I_3^- chain EXAFS response is out of phase at high k values with respect to the experimental signal, so that an I_3^- structure is ruled out from simulation. Considering now an I_5^- chain, the agreement is still not satisfactory since both experimental and simulated data are not in phase and there is a discrepancy around 9 $Å^{-1}$. In fact, an I_5^- chain is assumed to be an I_3^- specie weakly interacting with an I_2 molecule [72,73]. Such a structure give rise to two I-I distances, respectively around 2.73 and 3.25 Å. This second I-I distance accounts for the discrepancies, the phase shift and the interference pattern observed around 9 $Å^{-1}$ in the simulated EXAFS signal but not in the experimental spectrum. However, considering the presence of the extra electron in such species, we can assume large fluctuations of the second distance, giving rise to a distorted I_5^- chain (D- I_5^-). Such fluctuations can kill out the response of the second distance. Moreover, in iodine-doped poly(octylthiophene) the corresponding signal was not observed in samples heavily doped [72]. Hence, by considering a D- I_5^- structure, the simulation really matches our experimental signal (Figure 12). Only slight discrepancies are observed in the low k region (k<5Å-1). These tiny differences are not really surprising since the carbon atoms environment around the iodine species is not simulated.

Table 2. Structural parameters deduced from the least squares fit of the first shell of the I-I doped SWNT samples. N is the coordination number, R the fitted I-I distance, σ^2 the Debye-Waller factor, ΔE the energy shift.

	Iodine Shell	Carbon Shell
N	0.8	2.2
R (Å)	2.74	3.02
σ^2 (Å^2)	0.0016	0.02
ΔE (eV)	12.11	2.8

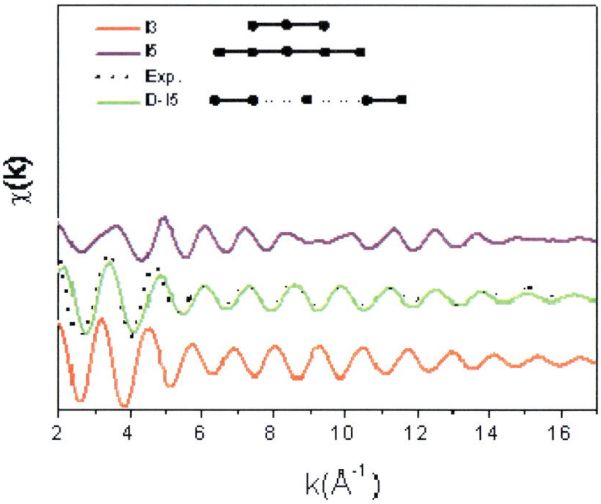

Figure 12. Experimental and simulated EXAFS spectra of several probable environments around the iodine absorbing atom. Inset: back scattering amplitudes of the iodine-iodine and iodine-carbon pair of atoms; drawings of the different I_n^- chains

Considering now the localization of iodine chains within bundles of SWCNTs, as for Rb doped SWCNTs, the interstitial site (between 2 or 3 tubes) is once again ruled out from steric considerations and the low number of carbon atoms surrounding the absorbing specie. Furthermore, ethanol washing removes the iodine species at the surface of SWCNTs bundles. Hence, from EXAFS results, the iodine D- I_5^- chains are mainly located in the hollow core of SWCNTs.

4. Raman Spectroscopy of Bundled SWCNT

Resonant Raman spectroscopy is known to be one of the most efficient tools for investigating the phonon modes in SWCNTs in relation to their structural and electronic properties [74]. In resonant Raman spectroscopy, the intensity of the spectrum goes through a maximum when the laser excitation energy matches an allowed optical transition of the material. The allowed optical transition energy of SWCNTs depends on tube diameter and chiral angle. The diameter dependence of the energy of the first allowed optical transitions (the so-called Kataura plot [75]) displays well separated curves for metallic and semiconducting tubes. This makes resonant Raman scattering a very powerful tool for studying selectively metallic and semiconducting SWCNTs [74].

The Raman spectrum of SWCNT bundle is dominated by two bunches of peaks. The first is specific to SWCNTs and appears in the range 100-400 cm^{-1}. It corresponds to in-phase radial modes of the carbon atom and it is therefore referred as radial breathing modes (RBM). The frequency of RBM is mainly related to the diameter of the tubes. The second is in the range 1400-1600 cm^{-1} and corresponds to the elongation of the carbon atoms at the surface of the tube. The modes of this bunch are therefore referred as tangential modes (TM) or G-modes because similar modes are observed in graphite. By contrast to RBM, the profile of G-modes strongly depends on the semiconducting/metallic character of the tubes in bundles. For semiconducting SWCNTs, the bunch of G-modes is composed of a series of narrow and symmetric lines. By contrast, for metallic SWCNTs, the bunch of G-modes is broader and the dominant low-frequency component is asymmetric. Then, the responses of metallic or semiconducting SWCNTs in bundles are easily identified from the profile of the G-modes. Spectra, in the G-modes range, measured at different excitation energies on a set of SWCNT bundles are given in Figure 13. It can be pointed out that the drastic enhancement of the Raman intensity when the excitation energy matches an allowed optical transition energy allows the performance of the experiment on individual SWCNT. Recently, by combining resonance Raman spectroscopy and electron diffraction on the same free-standing carbon nanotubes, the (n,m) indices and the RBM and TM frequencies of several semiconducting and metallic tubes were accurately and independently measured [76,77].

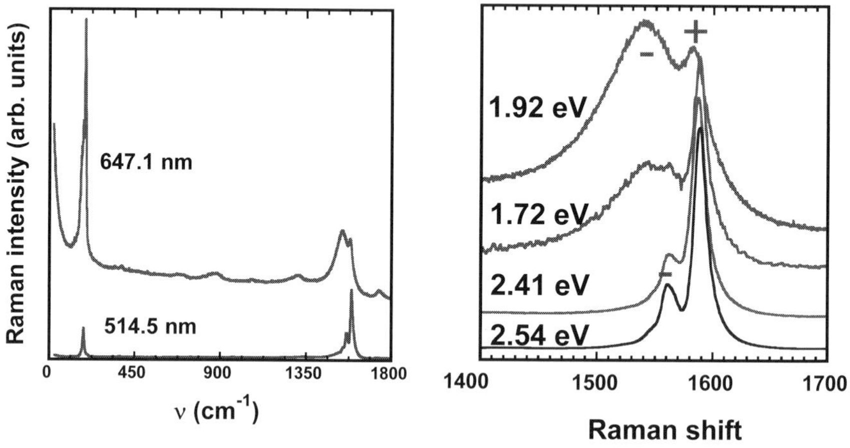

Figure 13. Raman spectra (left) of SWNT bundles measured at 647.1 nm (top) and 514.5 nm (bottom). G-modes range (right) at different excitation energies. The diameter distribution for this sample is between 1.2 nm and 1.5 nm.

SWCNT bundles can be doped and, like for graphite and fullerenes [78,79], charge transfer between the doping species and the tube can be probed in Raman scattering through the frequency of Raman modes involving the C-C stretching. Some reviews were published on the dependence of the Raman spectrum under doping [80,81,82]. Our goal here is not to duplicate these references. We will summarize the main results and focus on the most recent ones. We will discuss the Raman behavior under doping at the light of the recent information about the localization of the doping species obtained by diffraction and EXAFS on the same samples.

Several route have been used to doped bundled SWCNTs. Electrochemical doping of SWCNTs is a complex process that may involve the co-intercalation of large solvent molecules present in the electrolyte solution into the SWCNT bundles or side reactions at the electrode surface [83]. The chemical route based on the exposure of bundled SWCNTs to molecule of different redox potentials leads to a ternary compound: alkali metal, THF, SWCNT, (tetrahydrofuran (THF) is the solvent used in the doping procedure) [84]. As reported previously, neutron diffraction investigation indicates that alkali atoms form a monolayer surrounding each tube while the THF molecules intercalate between the decorated tubes and at the surface of the bundles [44]. Raman investigations performed on both electrochemically doped bundled SWCNTs [83,85-86] and ternary compound reveals interesting behavior [87]. However, their understanding is in part obscured by the presence of solvent molecules inside the bundles. Doping in vapor phase does not present this disadvantage. In the following, we review the results of Raman studies of the slow vapor phase doping of alkali metal into bundles of SWCNTs. Obviously, it is important to connect the Raman results obtained on vapor-phase, electrochemical and chemical doping to establish the common trends.

4.1. Raman Spectra of Alkali-Doped SWCNT Bundles

4.1.1. Doping at Saturation Level

After the pioneering Raman study performed by Rao *et al.* on K-doped SWCNTs [88], other Raman investigations have been performed on SWCNT bundles doped with rubidium (Rb) and cesium (Cs) [89]. All these studies have stated that, concomitant with the vanishing of the radial breathing modes, the dominant G-mode exhibits a large downshift with respect to its position in undoped SWCNT (typically from 1590 to 1550 cm^{-1}) and displays a broad and asymmetric profile. These features are independent of the excitation energy indicating a loss of the resonance character of the Raman signal due to the filling of electronic states and a shift of the Fermi level. Figure 14 compares the Raman spectra of pristine and Rb-doped sample at two excitations energies. The profile of the dominant G-mode is well described by a Breit-Wigner-Fano component in agreement with the expected metallic character of the alkali-doped SWCNT bundles at saturation level.

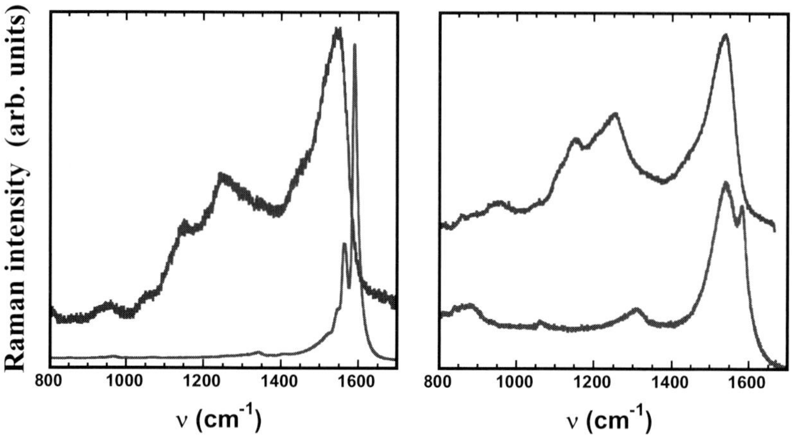

Figure 14. Comparison of Raman spectra of pristine (bottom) and Rb-SWNTs (top) doped at saturation level with excitation energy = 2.41 eV (left) and excitation energy = 1.92 eV (right).

4.1.2. Progressive Doping

Some Raman experiments have been performed in order to investigate the dependence of the Raman spectrum as a function of the doping level [90-92]. All these studies have revealed the same kind of behavior independently of the alkali metal: Rb in [90], K in [91], Cs in [92]. This one is summarized in Figure 15 for pristine semiconducting and metallic SWCNTs respectively. For semiconducting SWCNTs, in the first steps of doping, the dominant G-mode broadens and shifts to high frequency while the intensity of the RBM decreases. At a peculiar level of doping, (associated to a first minimum of the resistivity, see [Ref. 90]), the spectrum is featured by a single line, upshifted of about 4

cm^{-1} with respect to its position in undoped SWCNTs [90,92]. For the metallic SWCNTs, the broad and asymmetric low-frequency component of the G-bunch fastly vanishes upon doping, while the high-frequency component of the G-bunch upshifts [80]. At the peculiar doping level corresponding to a first minimum of the resistivity [90], the spectrum is featured by a single line close in shape and position with that observed for semiconducting tubes. At higher saturation doping level, the G-mode downshifts and take the line shape found in the spectrum measured at saturation doping level (see Figure 14). A detailed investigation of the progressive doping with Cs of semiconducting SWCNT bundles has been recently performed by Chen *et al.* [92]. These authors confirm all the features previously found in [Ref. 90], and they identify four distinct doping intervals after starting the doping: *Interval 1*, no evolution of the RBM and G-mode. This behavior is assigned to an induction period, *Interval 2*, decreasing of RBM and TM intensity, faster for RBM. The RBMs vanish at the end of this interval. *Interval 3*, vanishing of the low-frequency components of the G-modes; towards the end of this interval a single symmetric G-mode, upshifted of about 4 cm^{-1} with respect to its position in undoped sample, is observed. *Interval 4*, the G-mode broadens and progressively takes an asymmetric shape. The maximum of this band undergoes a gradual and substantial downshift in frequency. In interval 3 and 4, the intensity of the G-mode exhibits a huge loss of intensity. In graphite, the intercalation of M^+ (A^-) ions between the graphite layers adds electrons (holes) to the carbon sub lattice leading to a decrease (increase) of the C-C distance and to a softening (stiffening) of the E_{2g} G-mode [93]. In this picture, the well-identified striking behavior of the G-mode in alkali-doped SWCNT bundles can be understood as first an anomalous contraction of the C-C bond at intermediate doping level followed by an expansion of the C-C bond at high doping level. It can be pointed out that the theoretical studies of the dependence of the C-C- bond length in the honeycomb lattice of SWCNT upon electron doping always predict an expansion of the C-C-bond and then a softening of the G-mode [94]. Clearly, these predictions are not consistent with the well established experimental results [90-92].

Diffraction and EXAFS studies have shown that the alkali ions were localized mainly around the bundles, first in the grooves and at higher doping level at the surface of the tubes. Based on these results, we can propose the following scenario: assuming that the intermediate doping level (phase I in [90] and interval 3 in [92]) corresponds to most of ions inserted in the grooves, the upshift of the G-mode results from the competition between the effect of the charge transfer (downshift) and the pressure effect (upshift) due to the presence of ions in the grooves. This latter effect dominates at intermediate doping level. At high doping level (phase II in [90] and interval 4 in [92]), the ions are mainly localized at the surface of the tube. This organization is closer of that observed in graphite. Then, the effect of charge transfer is now dominant and the G-mode shows the same features than in graphite: downshifts, broadens and progressively takes an asymmetric shape.

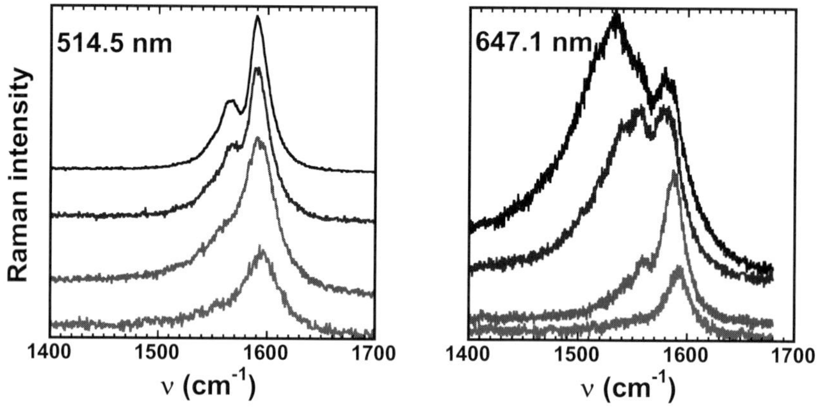

Figure 15. Dependence of the G-modes excited at 514.5 nm (left) and 647.1 nm (right) during in-situ Rb doping. The level of doping decreases from top to bottom. In this series of spectra, the highest doping level is weaker than the saturation doping level

As previously indicated the presence of additional molecules in electrochemical doped SWCNTs and ternary compounds adds some difficulties in the understanding of the results. However, an upshift is found in electrochemically K-doped SWCNT bundles at low level doping followed by a downshift at higher doping level [83]. On the other hand, the dependence of the G-modes measured in ternary compounds (Li_x-THF-C) is similar to that observed in vapor-phase doped bundled

SWCNTs, with an upshift of the G-mode of semiconducting or metallic tubes with the doping level [80,87]. This behavior is summarized in Figure 16. However, no downshift is observed for the $Li_{0.17}$-THF-C highest level doping. The structure of this last compound was recently determined. It was found that the alkali atoms form a monolayer surrounding each tube while the THF molecules intercalate between the decorated tubes and at the surface of the bundles [44]. In consequence, we can assume that the pressure effect due to the alkali ions and THF molecules surrounding the tubes acts again the effect of the charge transfer leading finally to a progressive upshift of the G-modes with the level doping. Finally, it can be emphasized that the upshift of the G-modes in vapor-phase doped SWCNT bundles and ternary compounds is concomitant to a loss of absorption bands in the optical spectra and to a monotonic increase of the conductivity related to the effective charge transfer on SWCNTs.

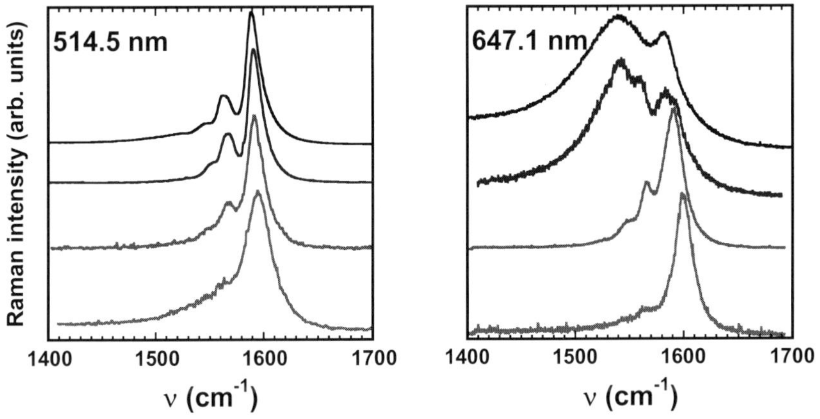

Figure 16. Raman spectra in the G-modes range excited at 514.5 nm (2.41 eV) (left) and 647.1 nm (1.92 eV) (right). From top to bottom: undoped, $Li_{0.04}C$, $Li_{0.14}C$ and $Li_{0.17}C$.

5. Conclusion

The difficulties in the determination of the structure of doped SWCNTs are due to host materials with low purity, heterogeneity and bad crystallinity. The SWCNTs are arranged in bundles but the structural characterizations (diffraction by neutron and X-ray) often indicate a dispersion of tube diameter, and of bundle size. Thus, predicting models based on theoretical energy calculation, of the structure of the guest dopant relatively to the nanotubes arrangement are rarely in agreement with the real structural data. Indeed, the simulations are mainly based on an ideal lattice of exclusively (10, 10) nanotubes, so that the models are rarely realistic and the real situation is much more complicated.

Up to now the different authors conclude that the intercalation can occur in the interstitial channels but all the experimental work has shown that insertion in these kinds of sites is rather obtained when the tubes are first purified and can be exempt of amorphous carbon. It was shown that no intercalation stage occurs in SWCNTs, and the structure of intercalated guest in the interstitial channels is always showing short range order in the case of small cations such as alkali metals. Thus , the alkali metals intercalated in the interstitial channels of purified SWCNTs are organised in short range order structure, globally seen as incommensurate with the graphitic structure, as the three fold symmetry is not preserved in the bundles made of tubes with various wrapping angles. The Rb ions environment in doped SWCNTs at saturation, probed by EXAFS spectroscopy, shows that rubidium ion is facing a carbon hexagon at a distance of 3.16 angstroms. This preferential site is also given for first stage intercalated graphite compounds where a transfer of carbon π electrons to the intercalated agent occurs. Thus at very local scale, the alkali ions appears to be commensurate with the closer carbon hexagons of first neighbor SWCNTs but inevitably incommensurate with others neighbor tubes as the arrangement of the host do not respect three fold symmetry.

In the case of the presence of polar solvent that is often used in electrochemical doping, the polarising alkali ions (i.e., Li or Na) are always solvated by the solvent molecules but still interact with the surface of the nanotubes, decorating their surface. In this case, the expansion of the 2D lattice is related to the size of alkali solvated ions.

This separation of the nanotubes is more favorable to the ordering of the guest intercalated molecules as their interactions with SWCNTs are weakened by increasing distance between the dopants and tubes and only local interactions occurs close to the surface of nanotubes. As an example, it was shown that the individual nanotubes can be separated by layers of superacid whom arrangement at low temperature is templated by the SWCNTs. Then, it is clear that the lack of crystallinity of the nanotubes bundles host structure strongly influence the structure of guest intercalated dopant. Moreover, strong interaction between dopant and the SWCNTs matrix induces disordered dopant arrangement, while in the case of weak interaction (for example, attenuated by the presence of solvent or a separation of the nanotubes) the dopant can be organized in less or more ordered structure, oriented at the surface of the individual nanotubes.

The reactivity with dopant is clearly dependent on the structure and purity of the host SWCNTs, related to the condition of preparation of the tubes and their further treatment. Since the tubes are opened through purification treatment, the guest ions or molecules are possibly inserted in the inner sites of nanotubes (i.e. central canal). While confined in the inside of nanotubes, the structures are strongly dependent on the interactions with the walls, and the diameter size and wrapping angle should orient the structure of the dopant. For example both iodine and cesium doping, can lead to the formation of spiral chains inside the tube, suggesting a preferred orientation with respect to the nanotube wall depending on the different chiralities of the tube. Selenium confined in carbon nanotubes can also give various structures as a function of the tube diameter.

The insertion of various dopants in the interstitial sites was also shown to induce important disorder of the 2D lattice. In the case of purified host, reaction with dopants such as alkali metals lead to the loss of the 2D triangular lattice of the ropes of nanotubes while a diffraction signal is obtained after doping raw SWCNTs. The doping of non–purified close-ended SWCNTs (i.e. electric arc raw SWCNTs) by heavy alkali metals has been first interpreted in terms of intercalation of the alkali-metals in the interstitial channels due to the shift of the X-ray diffraction (*10*) line concomitant to reaction with alkali. But the intercalation in interstitial channels can be hindered by the presence of amorphous carbon covering the nanotubes that prevent from the diffusion of the dopant within the bundles. Moreover, the EXAFS experiments show that in Rb-doped SWCNT bundle, the rubidium ions are not located in the IC. However, more fine simulation of the diffraction signal comparing the same alkali-doped sample probed by the two techniques (neutron and X-ray) is required to prove the absence of intercalation.

Finally, Raman investigations reveal a non-monotonic behavior of the Raman spectrum under doping. In agreement with the structural information, this behavior is explained by the progressive filling of the grooves followed by the adsorption of alkali ions at the surface of the bundles. The competition between the effect of charge transfer (a downshift of the G-modes is expected) and the one of tube deformation induced by the insertion (an upshift of the G-modes is expected) can explain the non monotonic behavior under doping observed in all the Raman experiments performed on alkali-doped SWCNT bundles.

References

1. A. Thess, R. Lee, P. Nikolaev, H. Dai, P. Petit, J. Robert, C. Xu, Y.H. Lee, S. G. Kim, A. G. Rinzler, D.T. Colbert, G. E. Scuseria, D. Tománek, J. E. Fischer, R. E. Smalley, Science 273, 483-487 (1996).
2. L. Duclaux, Carbon 40, 1751-1764 (2002).
3. J. E. Fisher, Acc. Chem. Res. 35, 1079-1086 (2002).
4. R-H. Xie, J. Zhao, Q. Rao, in Encyclopedia of Nanoscience and Nanotechnology edited by H. S. Nalwa, American Scientific Publishers, (2004), Vol. 10, p. 1-31.
5. J. W. Mintmire, B.I. Dunlop and C. T. White., Phys. Rev. Lett. 68, 631-634 (1992).
6. S. Rols, R. Almairac, L. Henrard, E. Anglaret and J. L. Sauvajol, Eur. Phys. J. B 10, 263-270 (1999).
7. S. Rols, R. Almairac, L. Henrard, E. Anglaret and J. L. Sauvajol, Synthetic Metals, 103, 2517-2518 (1999).
8. L. Henrard, A. Loiseau, C. Journet and P. Bernier, European Physical Journal B. 13, 661-669 (2000).
9. C. Journet, W. K. Maser, P. Bernier, A. Loiseau, M. Lamy de La Chapelle, S. Lefrant, P. Deniard, R. Lee, J.E. Fischer, Nature 388, 756-758 (1997).
10. J.-F. Colomer, C. Stephan, S. Lefrant, G. Van Tendeloo, I. Willems, Z. Konya, A. Fonseca, Ch. Laurent and J. B.Nagy, Chem. Phys. Lett. 317, 83-89 (2000).
11. P. Nikolaev, M. J. Bronikowski, R. K. Bradley, F. Rohmund, D. T. Colbert, K. A. Smith, R. E. Smalley, Chem. Phys. Lett. 313, 91-97(1999).

12. A.G. Rinzler, J. Liu, H. Dai, P. Nikolaev, C.B. Huffman, F.J. Rodríguez-Macías, P.J. Boul, A.H. Lu, D. Heymann, D.T. Colbert, R.S. Lee, J.E. Fischer, A.M. Rao, P.C. Eklund, R.E. Smalley, Appl. Phys. A 67, 29-37 (1998).
13. I. W. Chiang, B. E. Brinson, A. Y. Huang, P. A. Willis, M. J. Brownikowski, J. L. Margrave, R. E. Smalley and R. H. Hauge, Journal of Physical Chemistry B 105, 8297-8301 (2001).
14. M. Monthioux, B.W. Smith, B. Burteaux, A. Claye, J. E. Fisher and D. E. Luzzi, Carbon 39, 1251-1272 (2001).
15. S. Rols, M.R. Johnson, P. Zeppenfeld, M. Bienfait, O. E. Vilches and J. Schneble, Phys. Rev. B 71, 155411 (2005).
16. G. Mpourmpakis, E. Tylianakis, D. Papanikolaou and G. Froudakis, Rev. Adv. Mater. Sci., 11, 92, (2006).
17. B. Akdim, X. Duan, D. Shiffler and R. Patcher, Phys. Rev. B, 72, 121402R (2005).
18. Y. Miyamoto, A. Rubio, X. Blase, M. Cohen, S. Louie, Phys. Rev. Lett, 74, 2993 (2004).
19. X. Yang and J. Ni, Phys. Rev. B 69, 125419 (2004).
20. J. Zhao, A. Buldim, J. Han and J. P. Lu, Phys. Rev. Lett. 85, 1706 (2000)
21. G. Gao, T. Çağin, and W. A. Goddard III, Phys. Rev. Lett. 80, 5556-5559 (1998).
22. A. Farajian, K. Ohno, K. Esfarjani, Y. Maruyama, Y. Kawazoe, J. of Chem. Phys., 111, 2164 (1999).
23. B. Gao, C. Bower, J.D. Lorentzen, L. Fleming, A. Kleinhammes, X.P. Tang, L. E. McNeil, Y. Wu and O. Zhou, Chem. Phys. Lett.327, 69-75 (2000).
24. C. Bower, A. Kleinhammes, Y. Wu and O. Zhou, Chemical Physics Letters 288, 481-486 (1998).
25. H. J. Kim, K. K. Jeon, K. H. An, C. Kim, J. G. Heo, S. C. Lim, D. J. Bae, and Y. H. Lee, Adv. Mater. 15, 1757 (2003).
26. K. H. An, K. K. Jeon, J-M. Moon, S. J. Eum, C. W. Yang, G-S. Park, C. Y. Park and Y. H. Lee, Synthetic Metals, 140, 1-8 (2004).
27. G. U. Sumanasekera, J. L. Allen, S. L Fang., A. L. Loper, A. M. Rao.and P. C. Eklund, J. Phys. Chem. B 103, 4292-4297 (1999).
28. S. Ramesh, L. M. Ericson, V. A. Davis, R. K. Saini, C. Kittrell, M. Pasquali, W. E. Billups, W. W. Adams, R. H. Hauge and R. E. Smalley, Journal of Physical Chemistry B, 108, 8794-8798 (2004).
29. W. Zhou, J. E. Fischer, P. A. Heiney, H. Fan, V. A. Davis, M. Pasquali, and R. E. Smalley, Physical Review B 72, 054440 (2005).
30. L. M. Ericson, H. Fan, H. Peng, V. A. Davis, W. Zhou, J. Sulpizio, Y. H. Wang, R. Booker, J. Vavro, C. Guthy, A. N. G. Parra-Vasquez, M. J. Kim, S. Ramesh, R. Saini, C. Kittrell, G. Lavin, H. Schimdt, W. W. Adams, W. E. Billups, M. Pasquali, W. H. Hwang, R. H. Hauge, J. E. Fischer and R. E. Smalley, Science 305, 1447-1450 (2004).

31. W. Zhou, P. A. Heiney, H. Fan, R. E. Smalley and J. E. Fischer, J. Am. Chem. Soc.127, 1640-1641 (2005).
32. L. Grigorian, K. A. Williams, S. Fang, G. U. Sumanasekera, A. L. Loper, E. C. Dickey, S. J. Pennycook, and P. C. Eklund, Phys Rev Lett. 80, 60-5563 (1998).
33. X. Fan, E.C. Dickey, P.C. Eklund, K.A. Williams, L. Grigorian, R. Buczko, S.T. Pantelides and S.J. Pennycock, Phys. Rev. Lett. 84 4621-24 (2000).
34. N. Bendiab, R. Almairac, S. Rols, R. Aznar, J. L. Sauvajol and I. Mirebeau, Physical Review B, 69, 195415 (2004).
35. M. Bienfait , P. Zeppenfeld, N. Dupont-Pavlosky, J.-P. Palmari, M. R. Johnson, T. Wilson, M. DePies, and O. E. Vilches, Phys. Rev. Lett. 91, 035503-1 (2003).
36. M. Zhao, Y. Xia and L. Mei, Phys. Rev. B 71, 165413 (2005).
37. T. Kar, J. Pattanatak and S. Scheiner, J. Phys. Chem. A 105, 10397 (2001).
38. A. Claye and J. E. Fischer, Mol. Cryst. and Liq. Cryst, 340, 743-748 (2000).
39. A. S. Claye, N. M. Nemes, A. Jánossy and J. E. Fischer Phys. Rev. B 62, R4845-R4848 (2000).
40. A. S. Claye, J. E. Fischer, C. B. Huffman, A. G. Rinzler and R. E. Smalley, J. Electrochem. Soc. 147, 2845-2852 (2000).
41. H. Shimoda, B. Gao, X. P. Tang, A Kleinhammes L Fleming, T. Wu and O. Zhou, Phys. Rev. Lett. 88, 015502 (2002).
42. P. Petit, C. Mathis, C. Journet and P. Bernier, Chemical Physics Letters 305, 370-374 (1999).
43. E. Jouguelet, C. Mathis and P. Petit, Chemical Physics Letters; 318, 561-564 (2000).
44. J. Cambedouzou, S. Rols, N. Bendiab, R. Almairac, J. L. Sauvajol, P. Petit, C. Mathis, I. Mirebeau and M. Johnson, Physical Review B 72, 041404 (2005).
45. R.S. Lee, H.J. Kim, J.E. Fischer, A. Thess and R. E. Smalley, Nature 388, 255-256 (1997).
46. C. Bower C, Suzuki S, Tanigaki, K, O. Zhou, Appl. Phys. A, 67, 47-52 (1998).
47. L. Duclaux, J. P. Salvetat, P. Lauginie, T. Cacciaguera, A. M. Faugère, C. Goze-Bac, P. Bernier, Journal of Physics and Chemistry of Solids 64, 571-581 (2003).
48. J. E. Fischer, Acc. Chem. Res. 35, 1079-1086 (2002).
49. S. Suzuki, C. Bower, O. Zhou, Chemical Physics Letters 285, 230-234 (1998).
50. J. E. Fischer, A. Claye, R. Lee, Mol. Cryst. Liq. Cryst. 340, 737-742 (2000).
51. T. Pichler, M. Sing, M. Knupfer, M.S. Golden and J. Fink, Solid State Communications 109, 721-726 (1999).
52. F. Rousseaux, R. Moret, D. Guérard and P. Lagrange, Phys. Rev. B 42, 725 (1990).
53. S. Los, P. Azaïs, R. Pellenq, Y. Breton, O. Isnard, L. Duclaux, Ann. Chim. Sci. Mat. 30, 393-400 (2005).
54. L. Duclaux, K. Metenier, J. P. Salvetat, P. Lauginie, S. Bonnamy, F. Beguin, Mol. Liq. Cryst. 340, 749-751 (2000).

55. N.M. Nemes, J.E. Fischer, K. Kamarás, D.B. Tanner, and A.G. Rinzler, in Structural and Electronic Properties of Molecular Nanostructures, edited by H. Kuzmany, J. Fink, M. Mehring, and S. Roth, American Institute of Physics, Melville (2002) pp. 259-262.
56. S. Challet, P. Azaïs , R.J-M. Pellenq, L. Duclaux, Chemical Physics Letters, 377, 544-550 (2003).
57. G.-H. Jeong, R. Hatakeyama, T. Hirata, K. Tohji, K. Motomiya, T. Yaguchi and Y. Kawazoe, Chem. Commun. 152; (2003).
58. G.-H. Jeong, A. A. Farajian, R. Hatakeyama, T. Hirata, T. Yaguchi, K. Tohji, H. Mizuseki and Y. Kawazoe Phys. Rev. B 68, 075410 (2003).
59. X. Liu, T. Pichler, M. Knupfer, J. Fink, and H. Kataura, Phys. Rev. B 69, 075417 (2004).
60. T. Pichler, H. Kuzmany, H. Kataura, and Y. Achiba, Phys. Rev. Lett. 87, 267401 (2001).
61. M. Kalbac , L. Kavan, M. Zukalova and L. Dunsch, J. Phys. Chem. B 108, 6275-6280 (2004).
62. L. Guan, K. Suenaga, Z. Shi, Z. Gu and S. Ijima, Phys. Rev. Lett. 94, 045502 (2005).
63. B-Y. Sun, Y. Sato, K. Suenaga, T. Okazaki, N. Kishi, T. Sugai, S. Bandow, S. Ijima and H. Shinohara, J. Am. Chem. Soc. 127, 17972-17973 (2005).
64. G-H. Jeong, A. A. Farajian, T. Hirata, R. Hatakeyama, K. Tohji, T. M. Briere, H. Mizuseki, and Y. Kawazoe, Thin Solid Films, volume 435,307-311 (2003).
65. J. Chancolon, F. Archaimbault, S. Bonnamy, A. Traverse, L. Olivi, G. Vlaic, J. of Non Crys. Solids, 352, 99 (2006).
66. J. L. Bantiginies, L. Alvarez, R. Aznar, R. Almairac, J. L. Sauvajol, L. Duclaux, F. Villain, Phys. Rev. B, 71, 195419 (2005).
67. T. Michel, L. Alvarez, J-L. Sauvajol, R. Almairac, R. Aznar, J-L. Bantignies and O. Mathon, Phys. Rev. B, 73, 195419 (2006).
68. J. Bouat, D. Bonnin, L. Facchini, and F. Beguin, Synthetic metals, 7, 233 (1983).
69. A. L. Ankudinov, J.J. Rehr, and S.D. Conradson, Physical Review B 58, 7565-7576 (1998).
70. D. Guérard and P. Lagrange, Journal de Chimie Physique, 81, 853-856 (1984).
71. N. Bendiab, Ph. D Thesis, University of Montpellier II (2003).
72. K.E. Aasmundtveit, E. J. Samuelsen, C. Steinsland, C. Meneghini and A. Filipponi, Synthetic metals 101, 363-364 (1999).
73. R.C. Teitelbaum, S. L. Ruby, and T. J. Marks, J. Am. Chem. Soc., 101,25 (1979).
74. M. S. Dresselhaus and P.C. Eklund, Adv. Phys. 49 , 405 (2000).
75. H. Kataura, Y. Kumazawa, Y. Maniwa, I Umezu, S. Suzuki, Y. Ohtsuka and Y. Achiba, Sythetic Metals 103, 2555-2558 (1999).
76. J. Meyer, M. Paillet, T. Michel, A. Moreac, A. Neumann, G. Duesberg, S. Roth, and J.-L. Sauvajol, Phys. Rev. Lett. 95, 217401 (2005).

77. M. Paillet, T. Michel, C. Meyer, S. Roth, V. Popov, L. Henrard and J.-L. Sauvajol, Phys. Rev. Lett. 96, 257401-1-4 (2006).
78. M. S. Dresselhaus and G. Dresselhaus, Adv. Phys. 30, 139 (1981).
79. H. Kuzmany, M. Matus, B. Burger, and J. Winter, Adv. Mat. 6, 731 (1994).
80. J.-L. Sauvajol, N. Bendiab, E. Anglaret, and P. Petit, C.R. Physique 4,1035 (2003).
81. R.-H. Xi, J. Zhao, and Q. Rao, in Encyclopedia of Nanoscience and Nanotechnology, Edited by H. S. Nalwa, Vol. X, pages 1 (2004).
82. J.-L. Sauvajol, E. Anglaret, S. Rols, and O. Stephan, in Understanding Carbon Nanotubes: from Basics to Application. Eds: A. Loiseau, P. Launois, P. Petit, S. Roche, J.-P. Salvetat. Springer. Chap. 5, (2006) p. 277-335.
83. A. Claye, S. Rahman, J. E. Fischer, A. Sirenko, G. U. Sumanasekera, and P. C. Eklund, Chem. Phys. Lett. 333, 16 (2001).
84. P. Petit, C. Mathis, C. Journet, and P. Bernier, Chem. Phys. Lett. 305, 370 (1999).
85. L. Kavan, M. Kalbac, M. Zukalova and L. Dunsch, J. Phys. Chem. B 105, 10764 (2001).
86. L. Kavan, M. Kalbac, M. Zukalova and L. Dunsch, J. Phys. Chem. B 109, 19613 (2005).
87. N. Bendiab,, E. Anglaret, J.-L. Bantignies, A. Zahab, J.-L. Sauvajol, P. Petit, C. Mathis, S. Lefrant., Phys. Rev. B 64, 245424 (2001).
88. A. M. Rao, P. C. Eklund, S. Bandow, A. Thess, and R. E. Smalley, Nature (London) 388, 257 (2001).
89. N. Bendiab, A. Righi, E. Anglaret, J.-L. Sauvajol, L. Duclaux and F. Beguin, Chem. Phys. Lett. 339, 305 (2001).
90. N. Bendiab, L. Spina, A. Zahab, P. Poncharal, C. Marlière, J.-L. Bantignies, E. Anglaret, and J.-L. Sauvajol, Phys. Rev. B 63, 153407 (2001).
91. Y. Iwasa, H. Fudo, Y. Yatsu, T. Mitani, H. Kataura, and Y. Achiba, Synth. Met. 121, 1203 (2001).
92. G. Chen, C. A. Furtado, S. Bandow, S. Iijima, and P. C. Eklund, Phys. Rev. B 71, 045408 (2005).
93. M. S. Dresselhaus, in "Light Scattering in Solids", Topics in Applied Physics. Eds: M. Cardona and G. Güntherodt. Vol. 8 (1983).
94. M. Verissimo-Alves, B. Koiler, H. Chacham, and R. B. Capaz, Phys. Rev. B 67, 161401 (2003).
95. A. Kukovecz, T. Pichler, R. Pfeiffer, and H. Kuzmany, Chem. Comm. 1730 (2002).

CHAPTER 6

ELECTRON TRANSPORT IN MOLECULAR ELECTRONIC DEVICES

Shimin Hou, Zekan Qian, Rui Li

Key Laboratory for the Physics and Chemistry of Nanodevices, Department of Electronics, Peking University, Beijing 100871, China

We review molecular electronic devices in light of recent experimental advances. The implementation of a fully self-consistent approach for calculating electron transport through molecular devices is presented. This new approach combines the non-equilibrium Green's function (NEGF) formalism with density functional theory (DFT) calculations (the NEGF+DFT approach). This approach introduces into the DFT portion of the calculation a Hamiltonian matrix of the extended molecule derived from its density matrix, the electrostatic correction induced by the charge distribution in the electrodes and the exchange-correlation correction due to the spatial diffuseness of localized basis functions. Thus the effects of the electrodes are completely included at least for devices at equilibrium. We conclude by discussing future improvements to the NEGF+DFT approach.

1. Introduction

Following the continuous miniaturization of microelectronic devices, a great deal of attention has been devoted to molecular electronic devices in recent years [1,2]. A basic question that needs to be addressed before the fabrication of functional molecular electronic devices is: how we can construct, measure and understand the conductance of a molecule connected to two metallic electrodes? A more challenging task in the development of molecular electronics is how to control electron transport through a single molecule with a third electrode (gate), an analog of the field-effect-transistor (FET) in microelectronics. Such a molecular transistor is highly desired, because it can provide power gain that is an essential requirement for large-scale integrated circuits. In this review, we will present recent advances in molecular electronics from both experimental and theoretical aspects.

2. Experimental Progress in Molecular Electronic Devices

When a molecule is coupled to electrodes, the electronic structure of the molecule is modified significantly. Its discrete molecular levels are broadened, shifted and attain a finite lifetime. This is due to the interaction with the continuous electronic states of the electrodes. The modification of the electronic structure is strongly dependent on both the interaction and the band structure of the electrodes. Therefore, it is meaningless to discuss the conductance of a molecule since only the conductance of the electrode-molecule-electrode junction is well defined.

Gold is perhaps the most popular electrode material due to its good conductivity and chemical inertness. Recently, single-walled carbon nanotubes (SWCNTs) have been used as quasi-one-dimensional electrodes [3,4]. In order to connect a molecule to electrodes reliably, two appropriate terminal groups are always attached to both sides of the molecule. For gold electrodes, the most commonly used groups are thiol, pyridine and amine groups, thus various molecules can bind to two gold electrodes through the Au-S or Au-N bonds [5-12]. For SWCNT electrodes, the amide group provides a robust linkage [4].

It is well known that the steady-state current through a two-terminal molecular device is influenced by the quantum nature of the molecule, the geometry-dependent molecule-electrode coupling and the electronic properties of the electrodes near the Fermi level. Especially, the microscopic details of the molecule-electrode contact play a vital role in the device conductance. For the Au-S contact, these details include whether the molecule is bonded properly to both electrodes, that is, have both thiol hydrogen atoms been removed? What is the binding site of the thiol group on the gold electrode (for example, the top, bridge, hollow or adatom position on the Au(111) surface)? How does the molecule orient itself with respect to the gold electrodes, vertically or at a slight angle?

Unfortunately, these geometrical details cannot be determined or controlled in current experiments, which leads to large variations of the conductance value measured even for the combination of the same molecule and electrodes. Therefore, in order to obtain a complete picture of the junction conductance, a statistical analysis of a large number of molecular junctions is necessary. This approach was first introduced in 2001 by Cui and coworkers and further developed by other groups [13,6,11]. In this method, even pronounced peaks appear in the histogram constructed from thousands of conductance curves, one must also find a signature for identifying that the measured conductance is due to not only the sample molecules but also a single sample molecule.

Tao and coworkers have developed an efficient experimental approach that can rapidly produce a large number of experiments on different molecular junctions, which makes it possible to study the junction conductance through a systematic statistical analysis [6]. A modified scanning tunneling microscope (STM) is used to move repeatedly an Au tip (source electrode) in and out of contact with an Au substrate (drain electrode) in a solution containing the sample molecules. The individual molecular junctions are then created during the separation of the tip and the substrate electrode. This approach is referred to as the STM break junction method. During the initial stage of pulling the tip out of contact with the substrate, the conductance decreases in discrete steps of near integer multiples of conductance quantum $G_0=2e^2/h$ (Figure 1a). A histogram constructed from ~1000 conductance curves shows pronounced peaks at $1G_0$, $2G_0$ and $3G_0$, indicating that the cross section of the contact is reduced down to that of a few and eventually a single atomic chain (Figure 1b). When pulling the tip away farther, the atomic chain is broken and a new sequence of steps in a lower conductance regime appears in the presence of a certain molecule. The statistical analysis of the conductance values represented by these steps reveals clear peaks at the integer multiples of the fundamental conductance value characteristic for that molecule, which provides an unambiguous determination of a single molecule conductance averaged over a large number of measurements. For the 4,4' bipyridine molecule, the corresponding histogram shows pronounced peaks near 0.01 G_0, 0.02 G_0 and 0.03 G_0 (Figure 1c). These conductance steps are directly related to the formation of stable molecular junctions, and the corresponding peaks at 1×, 2× and 3×0.01 G_0 are ascribed to one, two and three molecules in the junction, respectively. In other words, the conductance near zero bias is determined to be about 0.01 G_0 for a single 4,4' bipyridine molecule.

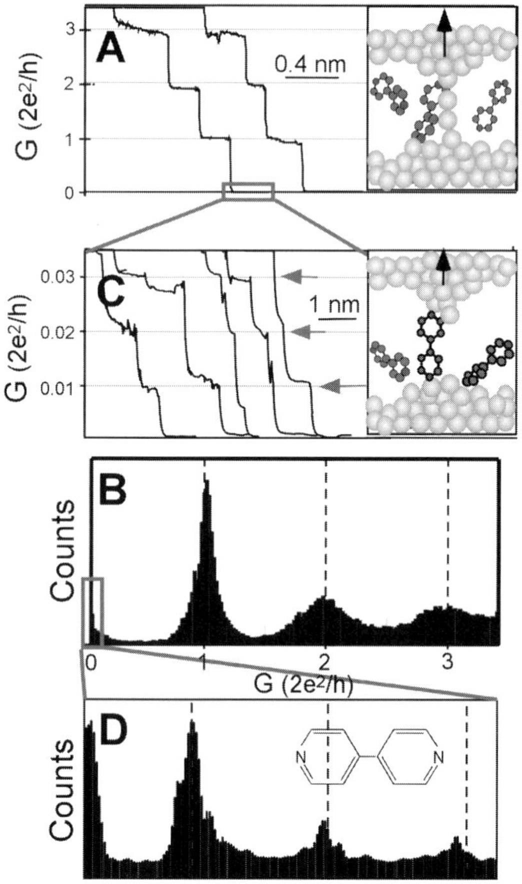

Figure 1. Transient conductance trace and the corresponding conductance histogram of gold point contact (A and B) and 4,4' bipyridine junctions (C and D) [reprinted by permission from Ref. 6].

Using methods like the STM break junction, two kinds of molecules have been investigated extensively. One is saturated molecules like alkane chains, which are considered highly insulating due to their large HOMO-LUMO gap. For the conductance of a series of single alkanes sandwiched between two gold electrodes, the general agreement is that the conducting mechanism is through-bond tunneling and the junction resistance can be described by R=A exp(βN), where A is the contact resistance, β is the tunneling decay constant and N is the number of methylene units in the alkane chain. Most of the experimentally measured values of β are comparable (1.0±0.05 [6], 0.83 [14], 0.91±0.03 [11]), regardless of the linker groups (dithiol or diamine groups). The small difference could be ascribed to the difference in the alignment of the molecule HOMO and LUMO relative to the electrode Fermi level. However, it should be mentioned that a much smaller value of β (0.52 ± 0.05) was also reported [15]. For the junction resistance that sensitively depends on the microscopic details of the contact, large variations exist in the literature. For example, Xu and Tao reported a junction resistance of 10.5±0.5 MOhm for the hexanedithiol molecule [6], while a much larger value of ~385 MOhm was reported by Wierzbinski and Slowinski [14]. These large disparities reflect the difficulty of forming identical molecular junctions even for the simple alkanedithiols.

The second is conjugated molecules, such as oligophenyl [8,16], oligophenylene-ethynylene [17,18], oligothiophene [19] and oligoaniline [4,20], which possess delocalized π-electrons. The smaller HOMO-LUMO gap and the possibility of functionalizing these molecules make them more attractive for potential molecular electronic devices. A consensus about the conductance of such conjugated molecule is that the electrical conduction depends critically on the delocalization of the molecular electronic orbitals and their connection to the metallic contacts. Dadosh *et al.* compared the electrical conduction through three short organic molecules: 4,4' –biphenyldithiol (BPD), a fully conjugated molecule; bis-(4-mercaptopheyl)-ether (BPE), in which the conjugation is broken at the center by an oxygen atom; and 1,4-benzendimethanethiol (BDMT), in which the conjugation is broken near the contacts by a methylene group [16]. They found that the oxygen in BPE and the methylene groups in BDMT both suppress the electrical conduction relative to that in BPD. Xiao and coworkers also found that the conductance of a benzenedithiol (BDT) molecule ($0.011G_0$) is much larger than that of the BDMT (0.0006 G_0) [8]. However, when we talk about the conductance value of a specific conjugate molecule, the published figures vary significantly. The conductance of benzene, the most typical aromatic molecule, has always been at the heart of research in molecular electronics. In 1997, Reed *et al.* reported for the first time the conductance of BDT between two gold electrodes, which was determined to be about $4\times10^{-4}G_0$ at 1V bias [5]. This pioneering work stimulated a large theoretical and experimental effort to investigate the conductance of these molecules. Recently, the average conductance of the gold-BPD-gold junction was measured to be 0.011 G_0 using the STM break junction method and a statistical analysis [8], much larger than that reported by Reed; whereas Ulrich and coworkers claim that they cannot assign a value to the conductance of the gold-BDT-gold junction [10], because the corresponding histogram constructed from a large number of conductance traces does not show clear peaks though individual conductance traces measured as a function of increasing gold electrode displacement show clear steps below the quantum conductance steps of the gold contact. The discrepancy of experimentally determined conductance values originates from the different atomic-scale details of the contact in the gold-dithiol molecule-gold junction, suggesting that the Au-S bond is not a well-defined contact. In order to improve the quality of contacts, Venkataraman and coworkers adopt the amine group as linkers to gold electrodes, and find that the variability of the observed conductance for the diamine molecule-Au junctions is less than the variability for dithiol molecule-Au junctions [11]. For example, they determined the most prevalent conductance value for the 1,4-benzenediamine molecule to be 0.0064 ± 0.0004 G_0. It can be seen that one of the greatest challenges to overcome in molecular electronics for future studies is the ability to measure and manipulate the atomic structure of the molecule-electrode contacts precisely.

Besides these difficulties encountered in the construction of two-terminal molecular junctions, more challenges have appeared in the fabrication of an efficient molecular FET. First, to achieve a gate field that is large enough to effectively control electron transport through the molecule, the distance between the gate electrode and the molecule must be extremely small. It has been shown that, to get good gate control for the conventional silicon FET, the gate oxide thickness needs to be less than 3% of the channel length [21]. For a molecular FET, the source-drain distance is determined by the length of the molecule, which is usually less than 10 nm. Therefore, the gate electrode must be placed within a few angstroms distance from the molecule. Second, the screening of the gate field by the source and drain electrodes must be minimized. This screening effect becomes serious when the distance between source and drain electrodes is small in comparison to the distance between the molecule and the gate electrode. At present, one possible method to realize a molecular FET is to use electrochemical gate [9,20,22], in which the molecular junction is immersed in an electrolyte and the source and drain potentials are biased with respect to a reference electrode inserted in the electrolyte. This reference electrode is called the electrochemical gate. The effective gate–molecule distance is determined by the double layer thickness at the electrode–electrolyte interfaces, which is of the order of the size of a few solvated ions. This allows the electrochemical gate to create a rather large field in the molecule. The STM break junction method is easily generalized to the electrochemical gating approach, which can be realized in two different ways. The first one is to perform statistical analysis at different gate voltages. The peak position in conductance histograms at different gate voltages with a fixed bias voltage is used to identify the average gate effect. In the second method, once the average conductance of a single molecule is determined, the source and drain electrodes are pulled apart until the conductance drops to the lowest step, which corresponds to the value of a single molecule conductance. Then, the electrodes are frozen and the current (I_{sd}) are recorded while sweeping the source–drain voltage (V_{sd}) or gate voltage (V_g). Even for the electrochemical gating approach, these are requirements for the molecule: 1) the molecule must be much longer than the thickness of the double layer. One obvious solution to this problem is to choose relatively long molecules. However, since the conductance of a molecule often decreases exponentially with its length, highly conductive conjugated molecules are clearly among the best choice. 2) Because there exists a window in which the electrochemical gate voltage can be varied without electrochemical reactions, the molecule must have an orbital (usually the HOMO or

LUMO) close enough to the Fermi level of the source or drain electrode so that the applied gate voltage can shift it between the Fermi levels of both electrodes and thus increase conductivity.

Tao and coworkers investigated the electron transport properties of single molecules covalently bonded to two gold electrodes in electrolytes using the electrochemical gating approach [8,9,23]. They find that the electrochemical gating approach can effectively control the current through the molecules, depending on the electronic properties of the molecules. To measure the gate effect of benzenedithiol, conductance histograms are constructed from a large number of conductance curves of BDT junctions formed in 0.1 M $NaClO_4$ supporting electrolyte at different electrochemical gate voltages with a fixed bias voltage (0.1 V) between the source and drain electrodes. Within the electrochemical gate voltage window from 0.4V to –1.1V, the peaks in the conductance histograms are still pronounced; however, the average conductance of the molecule does not depend on the electrochemical gate voltage within the experimental uncertainty [8]. For an individual BDT junction, the source-drain current changes only slightly (~5%) with the gate voltage [23]. The electrochemical gate effect on the conductance of 4,4 bipyridine has also been studied in a similar way, there is a only ~25% increase in one typical source-drain current as the gate voltage is swept from 0.6V to -0.2V [23]. The same measurement on different molecular junctions prepared under the same conditions is also performed to examine the reproducibility of this observation. The histogram of the gate-induced current change shows that the small gate effect is reproducibly observed despite of the run-to-run variations.

A possible reason for the lack of significant gate effect on the conductance of 4,4'-bipyridine and benzenedithiol is that both HOMOs and LUMOs of the two molecules are far away from the Fermi levels of the Au electrodes, and the applied gate voltage is not large enough to shift the HOMOs or LUMOs close to the Fermi levels. At present, there are still no direct experimental data about the alignment of the frontier orbitals of these two molecules with the Fermi level of gold electrodes available for the gold-molecule-gold junction structure. The energy difference between the HOMO of benzenethiol and the Fermi level of gold electrode is 2.0 eV [24], and the HOMO-LUMO gap of benzenedithiol is more than 4.0 eV, so that this argument should be correct for the benzenedithiol molecule. This argument is also consistent with the lack of electrochemical activity of these molecules. Another possible reason is the screening effect. In order to achieve effective gate control with a single gate electrode, the gate thickness has to be much smaller than the molecular length. In the case of electrochemical gating, the gate thickness is given by the double layer thickness, which is comparable with the lengths of 4,4'-bipyridine and benzenedithiol (0.7 nm). This may result in substantial screening of the gate field.

To overcome the difficulties caused by the screening effect and the large energy difference between the Fermi level and the closest molecular orbital, an appropriate molecule should be found first. Perylene tetracarboxylic diimide (PTCDI) just meets these demands. Its nominal length is about 2.3 nm, much greater than the gate thickness (0.7 nm). Therefore, compared to the cases of 4,4'-bipyridine and benzenedithiol, the screening of gate field due to the proximity of source and drain electrodes should be significantly reduced. The HOMO-LUMO gap of PTCDI is about 2.37 eV [25], and the LUMO of a PTCDI-Spent molecule is determined to be only 0.4 eV above the Fermi level of the gold substrate for the PTCDI-Spent monolayer film [26]. Furthermore, the cyclic voltammogram of one PTCDI derivative terminated with two thiol groups on both sides (PTCDI-dithiol), which can bind to two gold electrodes simultaneously, shows two pairs of peaks at -0.55 V and -0.8 V, respectively, corresponding to two reversible reduction processes [9]. Hence, the levels of the empty molecular states are close to the Fermi level. These attributes make PTCDI an excellent candidate for molecular FET.

The average conductance of a single PTCDI-dithiol molecule is determined to be 1.2×10^{-5} G_0 using the STM break junction method [9] (Figure 2b). The charge transport through this PTCDI-dithiol molecule shows strong dependence on the gate voltage, a typical curve of the source–drain current vs. the gate voltage (I_{sd}-V_g) is given in Figure 2c. As we can see, the conductance increases rapidly with decreasing V_g and reaches a peak at about -0.65 V. For a fixed source-drain bias voltage V_{sd}=0.1 V, the peak conductance is about 500 times greater than that at V_g = 0 V. Because the source-drain current depends on the source-drain voltage nonlinearly, I_{sd} can be controlled over a larger range at higher bias. Figure 2d shows the I_{sd}-V_{sd} curves obtained at various V_g. At V_{sd} = 0.4 V, the current can be controlled over ~1000 times. The current peak observed here is located at Vg≈-0.65 V, close to the first reduction potential (-0.55 V) of PTCDI adsorbed on gold electrodes. Therefore, the current enhancement should be due to an empty molecular state-mediated electron transport process. One possible mechanism is resonant tunneling, in which the current is predicted to reach a peak when an empty state is shifted to the Fermi level by the gate [27].

It is a great progress for molecular electronics; to control and vary the current reversibly through the PTCDI over three orders of magnitude using the gate demonstrates that FET-like behaviors can also be realized at single molecule level. At present this molecular transistor works only in an electrolyte and more work is required to produce a solid-state molecular transistor, which is more desirable for future applications in integrated circuits.

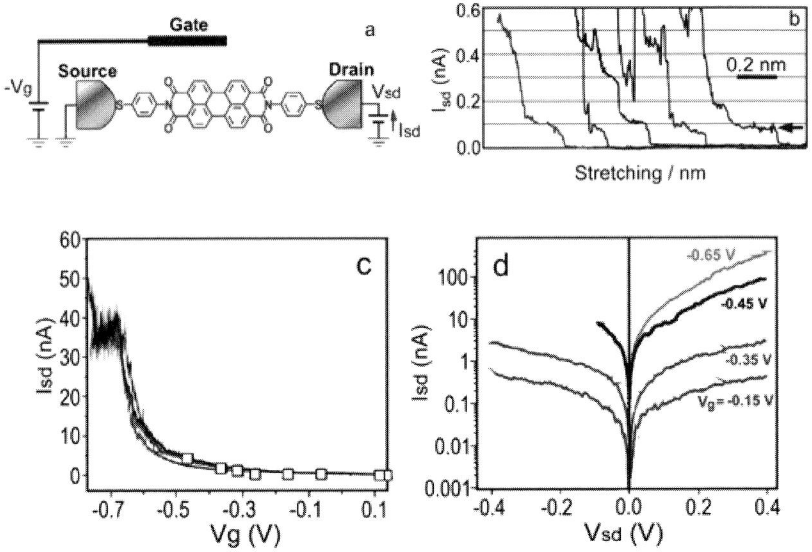

Figure 2. Schematic illustration of a) a single molecule transistor with an electrochemical gate, b) transient conductance curves of PTCDI junctions, c) source-drain current (I_{sd}) versus gate voltage (V_g) for a single PTCDI molecule transistor, and d) I_{sd} versus bias voltage (V_{sd}) characteristic curves at various gate voltages [reprinted by permission from Ref. 9].

3. The NEGF+DFT Approach

Experimentally, it is still a great challenge to precisely determine the atomic structure of the molecule-electrode contacts. Therefore, neither the influence of atomic structure on charge transport through the devices nor the method to improve the device performance has been known yet. As a result, first-principles theoretical studies on the atomic and electronic structure as well as the transport properties of molecular devices are important and useful, which can supplement and guide the experiments. To calculate charge transport across an electrode-molecule-electrode junction, it is necessary to specify the geometry, the methodology for conductance computation, the model Hamiltonian and how the bias voltage on the junction will be treated. For molecular transistors, consideration about the gate voltage is also needed. For the calculation of the junction conductance, the non-equilibrium Green's function (NEGF) formalism has proved to be a powerful and formally rigorous approach [28,29], in which the molecular devices are characterized as a central region connected to electron reservoirs via non-interacting leads. Certainly, arbitrary interactions in the central region can be included [30]. To apply the NEGF formalism to practical calculations, the Hamiltonian of electrode-molecule-electrode systems is needed to calculate the single-particle Green's functions. In molecular devices, the scattering of an electron in the central region is determined by the potential of a few particular atoms and the potential of other electrons. Hence, it's necessary to explicitly investigate the electronic structure of the molecular device from first principles, especially the central region. Though the electronic structure of isolated molecules and periodic solids can be calculated accurately and efficiently with density functional theory (DFT) under the Kohn-Sham (KS) ansatz [31-33], conventional DFT methods cannot rigorously handle molecular electronic devices, an open system that is infinite, nonperiodic and out of equilibrium, because the theory of DFT is based on the study of systems in their ground state or in thermodynamic equilibrium [34-36]. Since a truly first-principles treatment of the electronic process in such non-equilibrium system does not exist yet, DFT is only the best choice currently available. Therefore, the so-called NEGF+DFT approach that combines the NEGF formalism with DFT employing a finite set of local orbitals should be treated as a practical method [37-45], rather than a rigorously exact theory. In order to use the NEGF+DFT approach properly and to know where further improvements are needed in future studies, we should know various approximations involved. Hence, first we will outline the basic formalism of NEGF; then, we will give some details of the implementation of the NEGF+DFT approach. Finally, we will discuss the application and challenge of the NEGF+DFT approach.

3.1. Current Formula for an Electrode-Molecule-Electrode Junction

In a seminal work [28], using the non-equilibrium Keldysh formalism, Meir and Wingreen derived an exact formula for the current through a region of interacting electrons coupled to two multichannel leads where the electrons are not interacting. The system was partitioned into three parts (two leads plus the central region), and the basis of the single-particle Hilbert space was taken as an orthonormal set of functions each belonging to one of the three regions. Recently, Thygesen has generalized the current formula to nonorthogonal basis sets [29], which lays a firm foundation for the commonly used NEGF+DFT approach in which local basis functions are often employed. However, for simplicity, we still follow the derivation of the current formula given by Meir and Wingreen.

When a bias voltage is applied, the molecular device is driven out of equilibrium. For systems out of equilibrium, we can define the contour-ordered Green's function:

$$G(1,1') = -i < T_c[\psi_H(1)\psi_H^+(1')] > , \quad (1)$$

where the average given by $< >$ is defined on a suitable non-equilibrium statistics, T_C is the contour-ordering operator, ψ_H and ψ_H^+ are the Fermi field operators in the Heisenberg picture, and the shorthand notation $(1) \equiv (\vec{x}_1, \tau_1)$. The contour-ordered Green's function plays an analogous role in non-equilibrium theory as the causal Green's function plays in equilibrium theory. For more comprehensive introductions to the general Green's function theory the readers are referred to Ref 46. For future use, we also define the retarded, advanced and lesser Green's functions:

$$G^r(1,1') = -i\theta(t_1 - t_1') < \{\psi_H(1), \psi_H^+(1')\} > ,$$

$$G^a(1,1') = i\theta(t_1' - t_1) < \{\psi_H(1), \psi_H^+(1')\} > ,$$

$$G^{<}(1,1') = i < \psi_H^{+}(1')\psi_H(1) > .$$

Actually all of these three Green's functions can be derived from the contour-ordered Green's function. At equilibrium, these Green's functions are linked via the dissipation-fluctuation theorem. However, in non-equilibrium situations the retarded Green's function and the lesser Green's function are independent, and they are both very important. The retarded Green's function has a nice analytic structure (poles in one half-plane) and is suitable for calculating a physical response. The lesser Green's function is directly linked to physical observables, for example, the electron density is determined as follows:

$$n(\vec{r},t) = < \psi^{+}(\vec{r},t)\psi(\vec{r},t) > = -iG^{<}(\vec{r},t;\vec{r},t) .$$

For steady states, the Green's functions only depend on the time difference $t-t'$, for which we can use Fourier transform on the energy.

We consider the transport of electrons through a system that can be divided into three regions: the left lead (L), the right lead (R) and the central region (C), the Hamiltonian of which is written as:

$$\begin{aligned} H &= H_{L,R} + H_C + H_T \\ &= \sum_{k\alpha \in L,R} \varepsilon_{k\alpha} c_{k\alpha}^{+} c_{k\alpha} + H_C(\{d_n^{+}\};\{d_n\}) + \sum_{\substack{k\alpha \in L,R \\ n}} (V_{k\alpha,n} c_{k\alpha}^{+} d_n + V_{k\alpha,n}^{*} d_n^{+} c_{k\alpha}), \quad (2) \end{aligned}$$

where $c_{k\alpha}$ ($c_{k\alpha}^{+}$) and d_n (d_n^{+}) are annihilation (creation) operators in the leads and the central region, respectively. In this picture the description of the leads is at the single-particle level, while only the coupling between the leads and the central region makes the problem to be fully interacting. Before the coupling between the three regions is established, the left and right leads maintain its own thermal equilibrium with the associated Fermi levels μ_L and μ_R, respectively. The corresponding Green's functions in the leads for the uncoupled systems are $g_{k\alpha}^{<}(E) = 2\pi i f(\varepsilon_{k\alpha} - \mu_{L(R)})\delta(E - \varepsilon_{k\alpha})$ and $g_{k\alpha}^{r,a}(E) = \dfrac{1}{E - \varepsilon_{k\alpha} \pm i\eta}$, where $f(E) = 1/(1 + \exp(E/K_B T))$ is the Fermi distribution function and η is a small positive number ensuring proper convergence of the Fourier integral. As the leads are coupled to the central region, a current will start to flow. After some time the system achieves a steady state. In the following we consider the steady current at a certain time t. The steady current from the left lead into the central region can be defined as [46]:

$$I_L(t) = -e < \dot{N}_L(t) > = \frac{ie}{\hbar} < [N_L, H](t) > , \quad (3)$$

where $N_L = \sum_{k\alpha\in L} c_{k\alpha}^+ c_{k\alpha}$ is the number operator of the left lead. Doing the algebraic calculations for the commutator, the current turns out to be expressed as:

$$I_L(t) = \frac{ie}{\hbar}\sum_{\substack{k\alpha\in L\\ n}} (V_{k\alpha,n} < c_{k\alpha}^+(t)d_n(t) > -V_{k\alpha,n}^* < d_n^+(t)c_{k\alpha}(t) >) \,. \quad (4)$$

We introduce two Keldysh Green's functions:

$G_{n,k\alpha}^<(t,t') = i < c_{k\alpha}^+(t')d_n(t) >$ and $G_{k\alpha,n}^<(t,t') = i < d_n^+(t')c_{k\alpha}(t) >$.

It can be seen that the current is given by the time-diagonal components of these two Green's functions:

$$I_L(t) = \frac{e}{\hbar}\sum_{\substack{k\alpha\in L\\ n}} [V_{k\alpha,n} G_{n,k\alpha}^<(t,t) - V_{k\alpha,n}^* G_{k\alpha,n}^<(t,t)] = \frac{2e}{\hbar}\sum_{\substack{k\alpha\in L\\ n}} \mathrm{Re}\{V_{k\alpha,n} G_{n,k\alpha}^<(t,t)\} \,. \quad (5)$$

In order to obtain an expression for $G_{n,k\alpha}^<(t,t')$, we first obtain a general relation for the contour-ordered Green's function $G_{n,k\alpha}(\tau,\tau')$ using the structure equivalence between the non-equilibrium theory and the equilibrium theory [46]:

$$G_{n,k\alpha}(\tau,\tau') = \sum_m \int d\tau_1 G_{nm}(\tau,\tau_1) V_{k\alpha,m}^* g_{k\alpha}(\tau_1,\tau') \,. \quad (6)$$

Here, $G_{nm}(\tau,\tau') = -i < T_c[d_n(\tau)d_m^+(\tau')] >$ is the central region contour-ordered Green's function. Applying the analytic continuation rules, we find

$$G_{n,k\alpha}^<(t,t') = \sum_m \int dt_1 V_{k\alpha,m}^* [G_{nm}^r(t,t_1) g_{k\alpha}^<(t_1,t') + G_{nm}^<(t,t_1) g_{k\alpha}^a(t_1,t')] \,. \quad (7)$$

The Fourier transform of Eq. (7) is

$$G_{n,k\alpha}^<(E) = \sum_m V_{k\alpha,m}^* [G_{nm}^r(E) g_{k\alpha}^<(E) + G_{nm}^<(E) g_{k\alpha}^a(E)] \,. \quad (8)$$

Thus, the current becomes

$$I_L = \frac{2e}{\hbar}\int \frac{dE}{2\pi} \mathrm{Re}\{\sum_{\substack{k\alpha\in L\\ n,m}} \{V_{k\alpha,n} V_{k\alpha,m}^* [G_{nm}^r(E) g_{k\alpha}^<(E) + G_{nm}^<(E) g_{k\alpha}^a(E)]\} \,. \quad (9)$$

In the following, explicit reference to the E dependence will sometimes be omitted to simplify the notations. It is useful to convert the momentum summation into energy integration. Introducing the broadening function of the left lead $\Gamma_{mn}^L(E) = 2\pi \sum_{k\alpha\in L} \delta(E-\varepsilon_{k\alpha}) V_{k\alpha,n} V_{k\alpha,m}^*$, then the current from the left leads into the central region becomes

$$I_L = \frac{ie}{\hbar}\int \frac{dE}{2\pi} Tr[\Gamma^L G^< + f_L \Gamma^L (G^r - G^a)] \,. \quad (10)$$

Here, $f_L(E)=f(E-\mu_L)$ is the Fermi function of the left lead, the broadening function Γ and the central-region Green's functions $G^{<,r,a}$ are matrices in the central region indices n and m.

Likewise, the current from the right lead to the central region takes the form:

$$I_R = \frac{ie}{\hbar}\int \frac{dE}{2\pi} Tr[\Gamma^R G^< + f_R \Gamma^R (G^r - G^a)]\,. \quad (11)$$

In steady state, the current will be uniform, so that the general expression for the current through a two-terminal electrode-molecule-electrode device is

$$\begin{aligned} I = I_L = -I_R = \frac{I_L - I_R}{2} &= \frac{ie}{\hbar}\int \frac{dE}{2\pi} Tr[(\Gamma^L - \Gamma^R)G^< + (f_L\Gamma^L - f_R\Gamma^R)(G^r - G^a)] \\ &= \frac{e}{\hbar}\int \frac{dE}{2\pi} Tr[(\Gamma^L - \Gamma^R)iG^< + (f_L\Gamma^L - f_R\Gamma^R)i(G^r - G^a)] \end{aligned}\,. \quad (12)$$

Clearly, the current is determined by three factors: (1) the coupling between the central region and the leads, accounted for by the broadening functions Γ^L and Γ^R; (2) the occupation of the energy levels, given by the Fermi functions f_L and f_R for the leads and by $iG^<$ for the central region; (3) the energy levels in the central region, given by $i(G^r - G^a)$. This current formula is a very powerful result, which is valid for two-terminal devices with any arbitrary structure. The practical problem of its application is how to calculate these Green's functions, which is not an easy task. Both the lesser Green's function and the retarded Green's function of the central region must be calculated in the presence of tunneling, which can be realized by simultaneously solving the Keldysh equation for $G^<$ and the non-equilibrium Dyson equation for $G^{r,a}$:

$$G^<(E) = G^r(E)\Sigma^<(E)G^a(E) = G^r(E)[\Sigma_L^<(E) + \Sigma_R^<(E) + \Sigma_{\text{int}}^<(E)]G^a(E)\,, \quad (13)$$

$$G^{r,a}(E) = g^{r,a}(E) + g^{r,a}(E)\Sigma^{r,a}(E)G^{r,a}(E) = g^{r,a}(E) + g^{r,a}(E)[\Sigma_L^{r,a}(E) + \Sigma_R^{r,a}(E) + \Sigma_{\text{int}}^{r,a}(E)]G^{r,a}(E)\,. \quad (14)$$

Here, $g^{r,a}(E)$ are the reference non-interacting retarded and advanced Green's function for the central region. The lesser, retarded and advanced self-energies $\Sigma^{<,r,a}$ for the central region include two parts, one is due to the coupling to the left and right leads, the other is due to the interactions limited in the central region that should be determined by suitable approximations. It should be noted that the exact many-body self-energy operators are usually non-Hermitian and energy-dependent.

For the non-interacting case, that is, the interaction in the central region is also treated with a mean-field approximation like the Hartree-Fock (HF) approximation or the Kohn-Sham DFT, the Hamiltonian of the central region can be written as $H_C = \sum_n \varepsilon_n d_n^+ d_n$ and the current formula can be further simplified. In this case, one part of the retarded and advanced self-energies that originates from the electron-electron interaction in the central region is approximated by some suitable potentials which can be adsorbed in the Hamiltonian for the central region. For example, in the Hartree-Fock approximation, they are approximated as the Hartree potential and the exchange potential [47]. Thus, the retarded and advanced self-energies only include the part due to the coupling to the left and right leads, which take the following form:

$$\Sigma_{nm}^{r,a}(E)=\sum_{k\alpha\in L,R} V_{k\alpha,n}^{*} g_{k\alpha}^{r,a}(E) V_{k\alpha,m}=\sum_{k\alpha\in L,R}\frac{V_{k\alpha,n}^{*}V_{k\alpha,m}}{E-\varepsilon_{k\alpha}\pm i\eta}\,. \quad (15)$$

Comparing with the definition of the broadening function, it can be seen that the broadening function is equal to twice of the imaginary part of the retarded self-energy, i.e.,

$$\Gamma^{L}=i(\Sigma_{L}^{r}-\Sigma_{L}^{a})\,,\ \Gamma^{R}=i(\Sigma_{R}^{r}-\Sigma_{R}^{a})\,. \quad (16)$$

Similarly, the lesser self-energy also includes only the part due to the coupling to the leads:

$$\begin{aligned}\Sigma_{nm}^{<}(E)&=\sum_{k\alpha\in L,R} V_{k\alpha,n}^{*} g_{k\alpha}^{<}(E) V_{k\alpha,m}=\sum_{k\alpha\in L} V_{k\alpha,n}^{*}V_{k\alpha,m} g_{k\alpha}^{<}(E)+\sum_{k\alpha\in R} V_{k\alpha,n}^{*}V_{k\alpha,m} g_{k\alpha}^{<}(E)\\ &=\sum_{k\alpha\in L} V_{k\alpha,n}^{*}V_{k\alpha,m} 2\pi i f_{L}(\varepsilon_{k\alpha})\delta(E-\varepsilon_{k\alpha})+\sum_{k\alpha\in R} V_{k\alpha,n}^{*}V_{k\alpha,m} 2\pi i f_{R}(\varepsilon_{k\alpha})\delta(E-\varepsilon_{k\alpha}) \quad .\ (17)\\ &=if_{L}(E)\Gamma_{nm}^{L}(E)+if_{R}(E)\Gamma_{nm}^{R}(E)\end{aligned}$$

Thus, the lesser Green's function for the non-interacting central region is

$$G^{<}=if_{L}G^{r}\Gamma^{L}G^{a}+if_{R}G^{r}\Gamma^{R}G^{a}\,. \quad (18)$$

Putting Eq. (18) into Eq. (12), and using the relation $G^{r}-G^{a}=-iG^{r}(\Gamma^{L}+\Gamma^{R})G^{a}$, we get the current formula for the non-interacting case:

$$I=\frac{e}{h}\int dE(f_{L}-f_{R})Tr[\Gamma^{L}G^{r}\Gamma^{R}G^{a}]=\frac{e}{h}\int dE(f_{L}-f_{R})T(E)\,, \quad (19)$$

where $T(E)=Tr[\Gamma^{L}G^{r}\Gamma^{R}G^{a}]$ is the transmission function. If we assume that the two spin contributions are degenerate, then the current formula becomes

$$I=\frac{2e}{h}\int dE(f_{L}-f_{R})T(E)\,. \quad (20)$$

This is the celebrated Landauer formula, which connects the current to the transmittance across a scattering region. It should be stressed that this formalism and its interpretation break down in the presence of interactions in the central region.

3.2. Implementation of the NEGF+DFT Approach

Though the current formula (Eq. (12)) derived by Meir and Wingreen is an exact result for a central region of interacting electrons coupled to two non-interacting leads, two approximations are always introduced in the NEGF+DFT approach [41]: the first one is that the exact, non-Hermitian, energy-dependent, many-body self-energy operator due to the electron-electron interaction in the central region is approximated by the energy-independent, real DFT exchange-correlation potential; the second one is that the Green's functions are expanded in terms of a finite, incomplete basis set. The first approximation indicates that the original interacting electrons in the central region are replaced by the non-interacting Kohn-Sham quasi-electrons, so that the current formula (Eq. (20)) for the non-interacting case is actually employed to calculate the current-voltage characteristics of molecular devices in the NEGF+DFT approach. In other words, only the device characteristics within the coherent transport regime are calculated in the NEGF+DFT approach, and the calculated transmission functions have resonances at the non-interacting Kohn-Sham excitation energies which in general do not coincide with the true excitation energies of the original interacting systems. The second approximation indicates that the choice of basis sets also plays a critical role in determining the coupling strength and the electronic structures of the molecules and metal electrodes [48].

An important consideration on molecular electronic devices is that charge transfer and atomic relaxation will occur around the two molecule-electrode contact regions when a molecule is connected to two electrodes. But, due to the metallic screening in the electrodes, the charge and potential perturbation caused by the molecular adsorption extend over a finite region into the metal surface. Beyond this region the charge and potential distribution are the same as that of bulk materials. Therefore, the central region in molecular devices always includes the molecule itself and some electrode atoms adjacent to it, which is called the extended molecule. As long as the extended molecule includes sufficient electrode atoms, its influence on the electrodes can be neglected. In other words, by construction the surfaces of the extended molecule are identical to the two electrodes, charge transfer and atomic relaxation are limited in the deep inside of the extended molecule. Thus, the electronic structure of the electrodes is bulk-like, the Hamiltonian matrix and the density matrix can be obtained from separate calculations for the corresponding periodic structure. Via this partition, the only unknown part in a molecular device is the extended molecule region, as shown in Figure 3. It can be seen that the extended molecule has a finite scale but an open boundary condition, which greatly simplifies the calculation of molecule devices. Another advantage of introducing the notation of the extended molecule is that the charge transfer at the interfaces between the molecule and metal electrodes and the resulting lineup of molecular energy levels with respect to the Fermi level of the electrodes can be determined accurately.

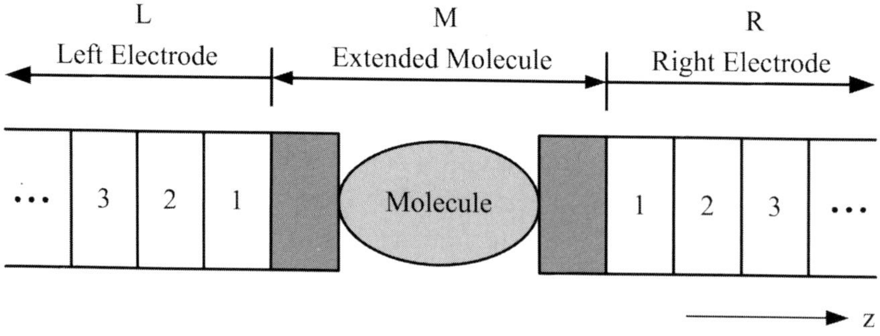

Figure 3. An open system containing a molecule sandwiched between two crystalline metallic electrodes. The extended molecule is composed of the molecule itself and some adjacent electrode atoms. The two electrodes are one dimensional and semi-infinite, extending to $-\infty$ and $+\infty$ along the *z* axis, respectively. The electrodes can be divided into periodic principal layers (PLs) (denoted by numbers 1, 2, 3 . . .).

There are two crucial steps in the self-consistent NEGF+DFT approach: the first one is obtaining the density matrix of the extended molecule from its Hamiltonian matrix under the open boundary condition; the second one is obtaining the Hamiltonian matrix of the extended molecule from its density matrix under the influence of two electrodes. Unlike the conventional DFT in which the density matrix is constructed by diagonalizing the Hamiltonian matrix and filling the occupied states, the density matrix of the extended molecule is calculated from Green's functions in the self-consistent NEGF+DFT approach, which is called the Green's function part. The KS effective potential is then obtained from the density matrix, which is called the DFT part. Iterating these two procedures until self-consistence, we can get the final Hamiltonian matrix and Green's functions used for calculating transport properties. Several different implementations of the NEGF+DFT approach have been developed recently [37-45]. According to the way of treating the extended molecule along the transport direction, these practical implementations can be roughly classified into two categories. In one category, the extended molecule is isolated into a cluster [37-41], and quantum chemistry codes like Gaussian are often employed to do DFT calculations on the extended molecule; in the other category, the extended molecule is put into a supercell and periodic boundary conditions (PBC) are adopted [42-45], DFT calculations of the extended molecule are carried out using solid-state codes like Siesta. The Green's function part in these two kinds of implementations is almost the same, the effects of the two electrodes are incorporated via the concept of the self-energy. But the DFT part has some significant differences. If periodic boundary conditions are adopted, large parts of the electrodes should be included in the extended molecule so that the interaction between the molecule and its images can be screened off by the metallic electrodes in between, which will increase the computational cost. When a bias voltage is applied in the non-equilibrium case, the unphysical electrostatic potential jump between supercells is also a troublesome problem. This needs additional treatments, for example, it has been addressed by having an independent solution of the Poisson equation [42]. In contrast, the implementation of the cluster model for the extended molecule is much more straightforward and simpler, what we need is how to deal with the electrostatic potential induced by the charge (ionic and electronic) distribution in the electrodes. The two electrodes can be treated at the DFT level or by a tight binding approach. Though the tight binding treatment for the electrodes is easily implemented [39-41], there may be artificial scattering at the interfaces between the tight binding part of the two electrodes and the DFT part of the electrodes included in the extended molecule. Hence, we will not discuss the tight binding approach any further. In the work of Taylor and Guo [38], the electrostatic potential is obtained by solving the Poisson equation in real space with bulk boundary conditions. Due to singular points of the electrostatic potential at nuclear positions, generating space grids is a delicate art. Numerically solving the Poisson equation at every iteration step is also a tremendous work. Recently, we have developed a fully self-consistent NEGF+DFT approach [37], which is different from previous implementations. The extended molecule and the two electrodes are treated at the same footing. In the DFT part, we introduce an electrostatic correction for the external potential induced by the electrodes, which is much computationally cheaper than solving the Poisson equation. Another correction for the exchange-correlation potential is also included to eliminate errors caused by the spatial diffuseness of local basis functions, which is called the exchange-correlation correction. Our approach is simple and efficient while still rigorous.

For the convenience of partition, a localized basis set such as a Gaussian-type orbital (GTO) is introduced to linearly expand the KS orbitals, that is, the linear-combination-of-atomic-orbitals (LCAO) ansatz is adopted. This converts the KS equations from a non-linear complicated system of coupled integro-differential equations into an algebraic one, which can be efficiently solved by computer programs. In what follows, we investigate the two procedures of this self-consistent iteration: the Green's function part and the DFT part.

3.2.1. Green's Function Part: Calculating the Density Matrix in an Open System

In the theoretical simulations of an electrode-molecule-electrode junction, the metal electrode can be modeled by either a nanowire with a finite cross section or an infinite surface. For the case of electrodes with a finite cross section, the whole system is quasi-one dimensional. However, special care must be taken when choosing the cross section and orientation of the electrodes in order to avoid quantum confinement effects and the quantum waveguide effect [49]. As a rule of thumb the linear dimension of the cross area should be several times the Fermi wavelength of the material forming the electrodes. To simulate an electrode with an infinite surface, the supercell approach is always adopted in which a finite computational cell consisting of the molecule and two electrodes is repeated periodically in the directions perpendicular to the transport direction. Due to the periodicity of the system in the transverse direction, all quantities including the Green's functions, the Hamiltonian matrix, the density of states (DOS) and the transmission function are defined on the 2-dimensional Brillouin zone of the plane perpendicular to the transport direction. In other words, these quantities are all k-dependent. The transmission per supercell should be evaluated as:

$$T(E) = \frac{1}{\Omega_{BZ}} \int_{1BZ} T(\vec{k}; E) d\vec{k} \ , \ (21)$$

where Ω_{BZ} is the area of the first Brillouin zone (1BZ). Clearly quantum confinement effects can be eliminated in this case, but one should be careful in order to eliminate the interference between the molecule and its images. Therefore, a rather large supercell must be employed. In practice, the integration in Eq. (21) is often reduced to evaluate the transmission function only at the Γ point of the first Brillouin zone. However, if the supercell is not large enough, an insufficient k-point sampling will introduce strong but unphysical features in the transmission function which can be traced to the presence of van Hove singularities in the electrode [50].

It should be noted that the following definitions and equations in this section are all k-dependent if the supercell approach is adopted for the electrodes, though explicit reference to the wavevector k is omitted to simplify the notations. The most acceptable and efficient way of calculating the density matrix from a Hamiltonian matrix is the matrix Green's function (MGF) method [41]. The retarded Green's function matrix of a system with a Hamiltonian matrix H is defined as:

$$G^r(E) = (E^+ S - H)^{-1} \ , \ (22)$$

where $E^+ = E + i\eta$ and η is a positive infinitesimal. Partition the Hamiltonian matrix H of the molecular junction as:

$$H = \begin{pmatrix} H_L & H_{LM} & 0 \\ H_{ML} & H_M & H_{MR} \\ 0 & H_{RM} & H_R \end{pmatrix} . \ (23)$$

Note that the Hamiltonian matrix elements between two electrodes are zeros due to the localization of the basis functions. The similar partition is also applied to the overlap matrix S. In this way, we can deduce the retarded Green's function matrix for the extended molecule as [47]:

$$G_M^r(E) = (E^+ S_M - H_M - \Sigma_L^r(E) - \Sigma_R^r(E))^{-1} \ , \ (24)$$

where

$$\begin{aligned}\Sigma_L^r(E) &= (E^+S_{ML} - H_{ML})g_L^r(E^+S_{LM} - H_{LM}) \\ \Sigma_R^r(E) &= (E^+S_{MR} - H_{MR})g_R^r(E^+S_{RM} - H_{RM})\end{aligned}, \quad (25)$$

and

$$\begin{aligned}g_L^r &= (E^+S_L - H_L)^{-1} \\ g_R^r &= (E^+S_R - H_R)^{-1}\end{aligned}. \quad (26)$$

In the above formulas, g_L^r (g_R^r) is the Green's function of the isolated left (right) electrode. Though the electrode Green's function is the inversion of a semi-infinite matrix for the periodic electrode, only the block near the surface called the surface Green's function makes contribution to the self-energy matrix due to the use of localized basis functions, which can be calculated efficiently using the renormalization or recursion methods [51]. Σ_L^r and Σ_R^r are the non-Hermitian self-energy matrices that incorporate the effect of the two semi-infinite electrodes, and they have the same dimension as the Hamiltonian matrix H_M of the extended molecule. The Hermitian part of $\Sigma_L^r(E)$ and $\Sigma_R^r(E)$ describes the energy shift of molecular orbitals and their anti-Hermitian components,

$$\begin{aligned}\Gamma^L(E) &= i(\Sigma_L^r(E) - \Sigma_L^a(E)) \\ \Gamma^R(E) &= i(\Sigma_R^r(E) - \Sigma_R^a(E))\end{aligned}, \quad (27)$$

represent the broadening of molecular orbitals due to the coupling of the extended molecule to the electrodes.

The spectral function of the extended molecule can be expressed as:

$$A(E) = i[G_M^r(E) - G_M^a(E)] = -2\,\mathrm{Im}\,G_M^r(E)\,. \quad (28)$$

At equilibrium, all states are filled according to the Fermi distribution function and the Fermi level E_F of the whole system is determined by the electrode. Therefore, the density matrix of the extended molecule at equilibrium is given by:

$$\rho = \int_{-\infty}^{+\infty} \frac{dE}{2\pi} A(E) f(E - E_F) = -\frac{1}{\pi}\,\mathrm{Im}\int_{-\infty}^{+\infty} \frac{dE}{2\pi} G_M^r(E) f(E - E_F)\,. \quad (29)$$

The analytic feature of the retarded Green's function enables the above integration to be done accurately with a rough integration mesh along a contour C_{eq} in the complex plane instead of the path along the real axis. The contour C_{eq} is usually composed of an arc C and a line L, which is shown in Figure 4a. The lower bound of the contour C_{eq} should be below the lowest occupied states of the system considered, and the upper bound is usually set to be several kT above the Fermi level due to the exponential decay of f(E-E_F), according to the residue theorem:

$$\rho = -\frac{1}{\pi}\,\mathrm{Im}\Big(\int_{C_{eq}} G_M(z) f(z - E_F)\,dz - 2\pi ikT \sum_{z_n} G_M(z_n)\Big)\,. \quad (30)$$

Here $G_M(z)$ defined in the upper half complex plane is obtained from the retarded Green's function $G_M^r(E)$ by analytic continuation, and $z_n = E_F + i(2n+1)\pi kT$ are the poles of the Fermi distribution function f(z-E_F) in the complex plane. To do the contour integration numerically, the Newton-Cotes formulas are a useful and straightforward family of numerical integration techniques [52], in which function values are evaluated at a set of equally spaced points, multiplied by certain aptly chosen weighting coefficients. However, this method usually cannot obtain the accuracy as high as the formula order, unless the integrand is very smooth. Instead of the Newton-Cotes formulas, the Gaussian quadrature is usually adopted in practice, which gives the freedom to choose not only the weighting coefficients but also the location of the abscissas at which the function is to be evaluated [52]. As the result of the increasing number of degrees of freedom, the Gaussian quadrature can achieve the order, essentially twice than that of the Newton-Cotes formulas with the same number of function evaluations. Figure 4b gives the typical points for Gaussian quadrature on the contour, along with the poles of the Fermi function. It should be pointed out that the location of these points in integration is fixed if the order of the formula is determined, independent of the system calculated. This makes the integration accuracy unsteady for different systems, since the spectral function sensitively depends on the system studied. Another problem for the ordinary Gaussian quadrature is that the points of function evaluation for different order formulas are completely different, which makes obtaining of the quadrature error estimation ineffective by comparing the results with increasing order. Motivated from this, the Gauss-Kronrod formulas [52], developed firstly by Kronrod in 1964, give the way to increase the formula order by adding new nodes on the previous ones and re-selecting the weights of all the nodes. Based on this method, the adaptive Gaussian quadrature has been extended, in which the points of function evaluations in integration are chosen automatically in order to achieve the given quadrature error. Figure 4c gives the points in the contour integration of a model system consisting of a single 4,4' bipyridine molecule connecting to two semi-infinite monatomic gold chains, with the absolute error tolerance of 1e-4. As we can see, the points approaching the real axis near the Fermi level are much denser than the others, indicating that the total spectral function changes drastically at that position.

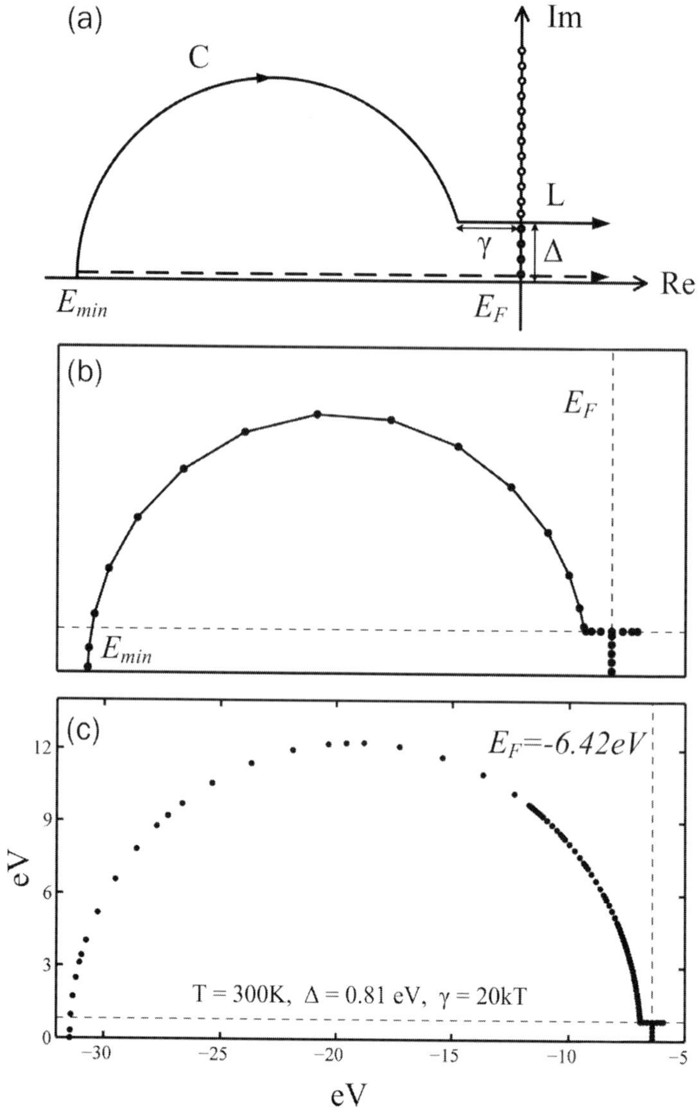

Figure 4. Schematic of the (a) contour of the equilibrium integration composed of an arc C and a line L, along with the poles of the Fermi distribution function $f(z\text{-}E_F)$in the complex plane (black dots), (b) typical points for the Gaussian quadrature on the contour, and (c) points on the contour for the equilibrium integration of a gold-4,4' bipyridine-gold system, using the adaptive Gaussian quadrature based on the Gauss-Kronrod formulas with an absolute error tolerance of 1e-4.

When a bias voltage V is applied, the device is driven out of equilibrium. The charge distribution in the extended molecule will be different from that at equilibrium. The electrodes are always assumed to be in local equilibrium but not in equilibrium with each other, ignoring the self-consistent response to the current [53]. The local Fermi level $\mu_{L(R)}$ of the left (right) electrode is taken to be $\mu_{L(R)} = E_F \pm eV/2$. The only effect of the bias voltage on the electrodes is that of a rigid shift of their Hamiltonian matrices. Thus, the Hamiltonian matrix for the whole device out of equilibrium takes the form:

$$H = \begin{pmatrix} H_L + S_L eV/2 & H_{LM} + S_{LM} eV/2 & 0 \\ H_{ML} + S_{ML} eV/2 & H_M & H_{MR} - S_{MR} eV/2 \\ 0 & H_{RM} - S_{RM} eV/2 & H_R - S_R eV/2 \end{pmatrix}. \quad (31)$$

Note that the coupling matrices between the electrodes and the extended molecule are also not modified by the bias voltage, since by construction the charge density in the surface planes of the extended molecule matches that of the electrodes exactly. Hence, the self-energy matrices are also the same as those at equilibrium except a rigid shift of energy:

$$\begin{array}{l} \Sigma_L^r = \Sigma_L^r(E - eV/2) \\ \Sigma_R^r = \Sigma_R^r(E + eV/2) \end{array}. \quad (32)$$

Since a uniform Fermi level does not exist in this case, we cannot obtain the density matrix from the energy integration of the spectral function any more. Instead, we attempt to get the density matrix from the lesser Green's function:

$$\begin{aligned} \rho &= \frac{1}{2\pi i}\int_{-\infty}^{+\infty} G_M^<(E)dE = \frac{1}{2\pi}\int_{-\infty}^{+\infty}[G_M^r(E)\Gamma^L G_M^a(E)f(E-\mu_L) + G_M^r(E)\Gamma^R G_M^a(E)f(E-\mu_R)]dE \\ &= \frac{1}{2\pi}\int_{-\infty}^{+\infty}[A_L(E)f(E-\mu_L) + A_R(E)f(E-\mu_R)]dE \end{aligned}. \quad (33)$$

Here, the left (right) spectral function $A_{L(R)}(E) = G_M^r(E)\Gamma^{L(R)}G_M^a(E)$ are introduced for convenience [54], the retarded Green's function $G_M^r(E)$ of the extended molecule is given again by Eq. (24) where now we replace $\Sigma_{L(R)}^r(E)$ with $\Sigma_{L(R)}^r$ given in Eq. (32), and the broadening functions are $\Gamma^{L(R)} = \Gamma^{L(R)}(E \mp eV/2)$. The computational cost of the integration in Eq. (33) is quite expensive, because the lesser Green's function is always singular on the real axis. And, what is more important, the density contributed by bound states in the extended molecule region is not included in Eq. (33), which is an indispensable part of the total density matrix [42,55]. Therefore, in order to evaluate the integral in Eq. (33) more efficiently, the density matrix is usually split into two parts: the equilibrium part and the non-equilibrium part:

$$\rho = \rho^{eq} + \rho^{ne}$$
$$\rho^{eq} = \frac{1}{2\pi}\int_{-\infty}^{+\infty}[A_L(E)+A_R(E)]f(E-\mu_R)dE \quad . (34)$$
$$\rho^{ne} = \frac{1}{2\pi}\int_{-\infty}^{+\infty}A_L(E)[f(E-\mu_L)-f(E-\mu_R)]dE$$

For the equilibrium part, instead of $A_L(E)+A_R(E)$, the total spectral function $A(E)$ is often used in practical implementations [41-44], which includes the contribution of both the bound states and the scattering states. Just as we calculate the density matrix of the extended molecule at equilibrium, the equilibrium part of the density matrix can also be calculated using the contour integration:

$$\rho^{eq} = -\frac{1}{\pi}\mathrm{Im}(\int_{C_{eq}} G(z)f(z-\mu_R)dz - 2\pi ikT\sum_{z_R} G(z_R)) \, . (35)$$

It should be noted that the poles of the Fermi function in the complex plane are $z_R = \mu_R + i(2n+1)\pi kT$. Due to the exponential decay of the Fermi function $f(E-\mu_R)$, it can be considered approximately that all the bound states below the right Fermi energy have been included in the total density matrix via the equilibrium integration, but the bound states within the bias window are still ignored. Such bound states can be correctly described only if their occupation information is given additionally, which however is extremely difficult to obtain. Fortunately, the situations with bound states within the bias window are rare and seldom encountered in practice, especially for low bias voltages.

The integration for the non-equilibrium part of the density matrix can only be evaluated near the real axis, and a very fine integration mesh is needed due to the singularities of the left/right spectral function. Even so, this integration may still be inaccurate. In order to estimate the error of this integration, an additional contour C_{ne} can be introduced to evaluate the integration (as shown in Figure 5):

$$\Omega = \frac{1}{2\pi}\int_{-\infty}^{+\infty}A(E)(f(E-\mu_L)-f(E-\mu_R))dE$$
$$= -\frac{1}{\pi}\mathrm{Im}\left(\int_{C_{ne}} G(z)[f(z-\mu_L)-f(z-\mu_R)]dz - 2\pi ikT\left[\sum_{z_L} G(z_L) - \sum_{z_R} G(z_R)\right]\right) . (35)$$

where $z_{L/R} = \mu_{L/R} + i(2n+1)\pi kT$ are the poles of $f(E-\mu_{L/R})$. Similar with the equilibrium integration, this contour integration can also be done accurately. Ignoring the existence of bound states, the total error $\Delta\Omega$ of the non-equilibrium integration can be approximately picked by subtracting the left spectral integral Ω^L and the right spectral integral Ω^R from Ω:

$$\Delta\Omega = \Omega - (\Omega^L + \Omega^R)\text{, where } \Omega^{L,R} = \frac{1}{2\pi}\int_{-\infty}^{+\infty} A_{L,R}(E)[f(E-\mu_L) - f(E-\mu_R)]dE\,. \quad (36)$$

The total error $\Delta\Omega$ includes the errors of both the left and right spectral integral, so the remaining task is how to extract the error of the left spectral integral. In general, this extraction can be formally written as the product of the corresponding elements of the total error matrix and an aptly chosen weight matrix $[\Delta\Omega_L]_{ij} = W_{ij}\Delta\Omega_{ij}$, and the weight matrix can be determined according to a certain physical assumption. For instance, supposing the standard deviation of the left/right integration error is proportional to its own spectral integral, i.e., $\sigma(\Delta\Omega_{L/R}) \propto \Omega^2_{L/R}$, the weight matrix can be defined as $W_{ij} = (\Omega^L_{ij})^2 / ((\Omega^L_{ij})^2 + (\Omega^R_{ij})^2)$ in order to minimize its variance [42]. If the left/right error is assumed to be proportional to the mean of the spectral integral at each atomic site, one can define the weight coefficient for the a-th atom $w_a = Tr[\Omega_L]_a / (Tr[\Omega_L]_a + Tr[\Omega_R]_a)$ and the weight matrix can be defined as its geometric mean $W_{ab} = \sqrt{w_a w_b}$. However, the answer to what is the best weight matrix is not known in general.

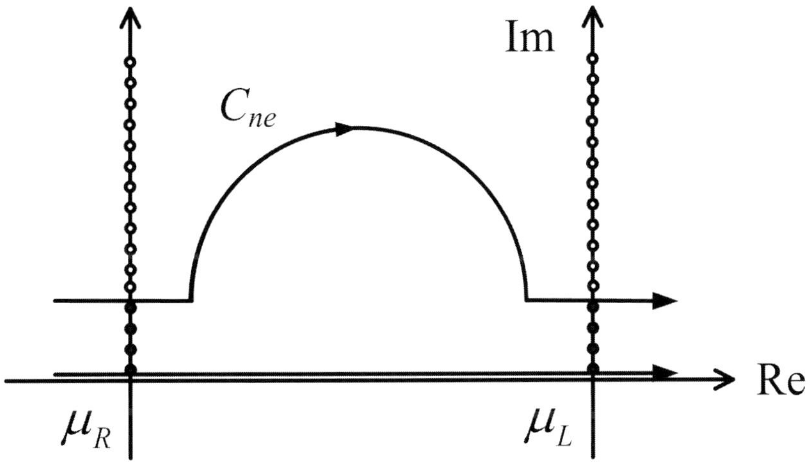

Figure 5. The non-equilibrium integration path and an additional contour C_{ne} used for the estimation of integration errors.

3.2.2. DFT Part: Calculating the KS Hamiltonian Matrix from the Density Matrix

Getting the Hamiltonian matrix of the extended molecule from its density matrix is a more delicate procedure. We consider the electrodes with a finite cross section at first [37]. Embedded in an infinite system along the transport direction, the Hamiltonian matrix of the extended molecule is also relevant to the electrodes.

According to the Kohn-Sham equation, the KS Hamiltonian matrix of the extended molecule can be written as [32]:

$$H_M = T + V_{Ne} + V_H + V_{XC} \, . \quad (37)$$

Given the localized basis set $\{\phi_i\}$ for the extended molecule, elements of the kinetic energy matrix T are

$$T_{ij} = \int \phi_i(\vec{r})(-\frac{1}{2}\nabla^2)\phi_j(\vec{r})d\vec{r} \, . \quad (38)$$

Obviously, the kinetic energy matrix is only relevant to the selected basis set, there is no need to consider the electrodes.

The nuclear attraction matrix V_{Ne} is a sum of nuclear attraction integrals (NAIs):

$$(V_{Ne})_{ij} = \sum_n \int \phi_i(\vec{r}) v_n(\vec{r}) \phi_j(\vec{r}) d\vec{r} \, , \quad (39)$$

where $v_n(\vec{r})$ is the potential of the nth nucleus or the nth ion (a nucleus plus core electrons). We use NAI to denote the integral $\int \phi_i(\vec{r}) v_n(\vec{r}) \phi_j(\vec{r}) d\vec{r}$. The Hartree matrix (electron repulsion matrix) V_H is a sum of electron repulsion integrals (ERIs) multiplied with density matrix elements:

$$(V_H)_{ij} = \sum_{\alpha\beta} (ij,\beta\alpha)\rho_{\alpha\beta} \, , \quad (40)$$

where the ERI $(ij,\beta\alpha)$ is defined as:

$$(ij,\beta\alpha) = \iint \phi_i(\vec{r})\phi_j(\vec{r}) \frac{1}{|\vec{r}-\vec{r}'|} \phi_\beta(\vec{r}')\phi_\alpha(\vec{r}') d\vec{r} d\vec{r}' \, , \quad (41)$$

It should be noted that the summations in Eqs. (39) and (40) should be over all the open system, including the two electrodes.

V_{Ne} and V_H represent potentials caused by the electrostatic interaction. The total electrostatic potential in the extended molecule region is composed of two parts: the potential caused by the extended molecule itself and that induced by the electrodes. We call the latter the external electrostatic potential. In Eqs. (39) and (40), the external electrostatic potentials correspond to the summation of the NAIs and ERIs related to electrodes. Due to their electrostatic nature, NAIs have a 1/d form, where d denotes the distance between the center of a basis function belonging to the extended molecule and a nucleus in the electrode. So do ERIs, where d denotes the distance between the center of a basis function belonging to the extended molecule and that belonging to the electrode. Summing these NAIs or ERIs individually will lead to a harmonic series that is divergent, that is, V_{Ne} and V_H are both infinite and we cannot approximate them by simply truncating the summation range. However, we can consider the nuclear attraction potential and the electron repulsion potential simultaneously. Since every principal layer in the electrode is charge neutral, the total external electrostatic potential induced by one principal layer decays rapidly. It can be proved rigorously that the external electrostatic potential is fully determined by only several principal layers near the surface [37]. Therefore, in order to get $V_{Ne}+V_H$, we need to sum the NAIs and ERIs only over the extended molecule plus finite principal layers near the surface of electrodes. Under this truncation approximation, an accurate $V_{Ne}+V_H$ with a controllable error can be obtained, in the sense that the accuracy of $V_{Ne}+V_H$ can be systematically improved to check for convergence towards the exact result.

NAIs, ERIs and density matrix elements related to the electrodes are constant in the iteration procedure, and they are determined from separate DFT calculations for the bulk systems corresponding to the bulk of electrodes. Hence we can complete the summation relevant to electrodes in Eqs. (39) and (40) before the iteration. We call this part the electrostatic correction matrix. At every step of the iteration, we add the electrostatic correction to the $V_{Ne}+V_H$ obtained from the isolated extended molecule. With the electrostatic correction, we can include the influence of the external electrostatic potential due to electrodes on the extended molecule completely.

The exchange-correlation matrix V_{XC} in Eq. (37) is written as:

$$(V_{XC})_{ij} = \int \phi_i(\vec{r}) v_{xc}(\vec{r}) \phi_j(\vec{r}) d\vec{r} \,, \quad (42)$$

where the exchange-correlation potential $v_{xc}(r)$ is determined by the electron density $\rho(r)$. Within local density approximation (LDA) or generalized gradient approximation (GGA), $v_{xc}(r)$ in the extended molecule region is only relevant to the $\rho(r)$ inside the same region. The electron density takes the following form in the LCAO scheme

$$\rho(\vec{r}) = \sum_{\alpha\beta} \phi_\alpha(\vec{r}) \rho_{\alpha\beta} \phi_\beta(\vec{r}) \quad (43)$$

Due to the spatial diffuseness of localized basis functions, limiting the calculation in the basis functions for the extended molecule is not exact. There are two subtleties that should be handled carefully. First, $v_{xc}(r)$ outside the extended molecule must be considered in Eq. (42), i.e., we must know ρ(r) in a larger region. Second, because the summation in Eq. (43) is over all the system, we should include some parts of the electrodes when we calculate the electron density in the region of the extended molecule. Fortunately, the basis functions all decay rapidly away from their centers, so this work can be limited in a finite range near the surface of both electrodes. Compared with the V_{XC} obtained merely from the basis set of the extended molecule, this modification is called the exchange-correlation correction. Different from the electrostatic correction that has a clear physical origin, the exchange-correlation correction is only mathematically caused by the spatial diffuseness of basis functions. Due to the nonlinear relationship between $v_{xc}(r)$ and ρ(r), this correction is also relevant to the density matrix of the extended molecule, so it must be calculated at every iteration step.

With both the electrostatic correction and the exchange-correlation correction, the KS Hamiltonian matrix of the extended molecule can be calculated from its density matrix correctly. Plus the tractable Green's function procedure from a KS Hamiltonian matrix to a density matrix, the total self-consistent iteration provides an accurate approach for dealing with open systems, which is different from that treating the extended molecule as an isolated cluster or as a supercell in a periodic structure. In our approach, except for the approximations inherent in the DFT, only two main approximations are left: the screening approximation and the truncation approximation in the electrostatic and exchange-correlation correction. By increasing the scale of computing, both of them can be limited at an arbitrary small level.

The above treatments on the electrodes with a finite cross section can be easily extended to the case of the electrodes with an infinite surface. To describe such systems, the basis set can be selected as either the set of localized atomic functions $\{\varphi_{i\vec{R}}\}$ for each cell $\vec{R}$ or the 'crystalline orbitals' $\{\varphi_{i\vec{k}}\}$ which have the proper translational symmetry. Here, both $\vec{R}$ and $\vec{k}$ are the index along the transverse directions. These two forms of the basis set have the relationship:

$$\varphi_{i\vec{k}}(\vec{r}) = \frac{1}{\sqrt{N}}\sum_{\vec{R}}\varphi_{i\vec{R}}(\vec{r})\exp(2\pi i\vec{k}\cdot\vec{R}) = \frac{1}{\sqrt{N}}\sum_{\vec{R}}\varphi_i(\vec{r}-\vec{R})\exp(2\pi i\vec{k}\cdot\vec{R}), \quad (44)$$

where N is the cell number, i is the index of the basis functions in one cell, $\vec{R}$ indicates each cell in the transverse directions. Due to the orthogonality of $\{\varphi_{i\vec{k}}\}$ at different $\vec{k}$ points, both the density matrix and the Hamiltonian matrix in the Green's function part should be written under the basis set $\{\varphi_{i\vec{k}}\}$: $A_{ij}^{\vec{k}} = \langle \varphi_{i\vec{k}} | \hat{A} | \varphi_{j\vec{k}} \rangle$, where $\hat{A} = \hat{H}$ or $\hat{A} = \hat{\rho}$. However, this basis set is not convenient for DFT calculations. In practice, most DFT packages do not calculate the Hamiltonian matrix $H^{\vec{k}}$ at a given $\vec{k}$ point directly from the k-space density matrix $\rho^{\vec{k}}$, but prefer to do this in the real space, that is, from $\rho^{\vec{R}}$ to $H^{\vec{R}}$ under the basis set $\{\varphi_{i\vec{R}}\}$ where $A_{ij}^{\vec{R}} = \langle \varphi_{i0} | \hat{A} | \varphi_{j\vec{R}} \rangle$. According to Eq. (44), it is easy to switch between these two spaces via basis transformation:

$$\begin{aligned} A_{ij}^{\vec{k}} &= \sum_{\vec{R}} A_{ij}^{\vec{R}} \exp(2\pi i\vec{k}\vec{R}) \\ A_{ij}^{\vec{R}} &= \frac{1}{\Omega_{BZ}} \int_{BZ} A_{ij}^{\vec{k}} \exp(-2\pi i\vec{k}\vec{R}) d\vec{k} \end{aligned} . \quad (45)$$

In the real space, the KS Hamiltonian matrix of the extended molecule can be written as $H_M^R = T^R + V_{Ne}^R + V_H^R + V_{XC}^R$. Just as the case of the electrodes of a finite cross section, the electronic kinetic energy matrix T^R is related only with the selected basis functions of the extended molecule. However, the evaluation of the electron-nucleus attraction potential V_{Ne}^R, the Hartree potential V_H^R and the exchange-correlation potential V_{XC}^R should be extended to include the contributions of supercells in the transverse directions. Fortunately, the sum of supercells in these evaluations can all be truncated at a finite size. In this case, the electron density $\rho(\vec{r})$ are obtained from the real space density matrix $\rho^{\vec{R}}$:

$$\rho(\vec{r}) = \sum_{\alpha\beta} \sum_{\vec{R}\vec{R}'} \varphi_{\alpha\vec{R}}(\vec{r}) \rho_{\alpha\beta}^{\vec{R}'} \varphi_{\beta\vec{R}+\vec{R}'}(\vec{r}) . \quad (46)$$

Since the basis functions $\{\varphi_{i\vec{R}}\}$ are localized, the product of $\varphi_{\alpha\vec{R}}(\vec{r})$ and $\varphi_{\beta\vec{R}+\vec{R}'}(\vec{r})$ will decay to zero very quickly following the increase of either $\vec{R}$ or $\vec{R}'$. Thus, the sum over $\vec{R}$ and $\vec{R}'$ to evaluate the electron density can always be truncated at a proper cell number.

3.2.3. Achieving Self-Consistency

Now it is time to combine the above two parts together to give a full picture of the self-consistent procedure. First, a trial density matrix ρ^0 of the extended molecule is given to the DFT part to calculate the electron density $\rho(\vec{r})$ and to construct the KS Hamiltonian matrix H_M. Then a new density matrix ρ^1 is calculated from the Green's function part. This procedure is repeated until the input density matrix ρ^j agrees with the output density matrix ρ^{j+1} within a preset tolerance. After the self-consistency has been achieved, the current-voltage characteristics of the molecular junction can be calculated via Eq. (20).

In order to speed up the self-consistent procedure to converge, acceleration methods for self-consistent convergence are generally needed. The quasi-Newton method is efficient for the solution of such nonlinear equations in the form of **F**(**x**)=0 with a finite, possibly large dimension N [52]. The approximate **F**(**x**) at each iteration can be written in the first order of the Taylor series:

$$\mathbf{F}(\mathbf{x}) \cong \mathbf{F}(\mathbf{x}^{(m)}) + J^{(m)}(\mathbf{x} - \mathbf{x}^{(m)}) \text{ , (47)}$$

where $\mathbf{J}^{(m)} = \partial\mathbf{F}/\partial\mathbf{x}^{(m)}$ is the Jacobian matrix. Hence **F**(**x**)=0 implies

$$\mathbf{x}^{(m+1)} = \mathbf{x}^{(m)} - [J^{(m)}]^{-1}\mathbf{F}(\mathbf{x}^{(m)}) \text{ . (48)}$$

As it is too expensive to use a numerical difference to approximate the true Jacobian $J^{(m)}$, many different methods to approximate $J^{(m)}$ have been developed. The best-performing algorithm in practice results from Broyden's formula [56], which requires that one generate an approximation $J^{(1)}$ to the initial Jacobian $J^{(1)}$ and for $m>1$ get $J^{(m)}$ following the next two conditions:

$$\delta\mathbf{F}^{(m)} - J^{(m)}\delta\mathbf{x}^{(m)} = 0 \text{ , (49)}$$

$$\left\| J^{(m)} - J^{(m-1)} \right\|^2 \text{ be minimized, (50)}$$

where $\delta\mathbf{F}^{(m)} = \mathbf{F}^{(m)} - \mathbf{F}^{(m-1)}$ and $\delta\mathbf{x}^{(m)} = \mathbf{x}^{(m)} - \mathbf{x}^{(m-1)}$. Using Lagrange multipliers, the upgrading procedure of $J^{(m)}$ is

$$J^{(m)} = J^{(m-1)} + \frac{(\delta\mathbf{F}^{(m)} - J^{(m-1)}\delta\mathbf{x}^{(m)})\left[\delta\mathbf{x}^{(m)}\right]^T}{\left[\delta\mathbf{x}^{(m)}\right]^T \delta\mathbf{x}^{(m)}} \text{ . (51)}$$

And a simple initial guess of the Jacobian would be a diagonal constant matrix $J^{(1)} = -(1/\alpha)\hat{I}$.

In practical calculations, the dimension N is often very large, and the size of the Jacobian $J^{(m)}$ is N^2. Hence the storage of $J^{(m)}$ is impossible. Another trick must be played on this problem. The following Srivastava's modification formula is often adopted [57],

$$\left[J^{(1)}\right]^{-1} = -\alpha\hat{I} \text{ , (52)}$$

$$\left[J^{(m)}\right]^{-1} = -\alpha\hat{I} + \sum_{i=2}^{m} u^{(i)} v^{T(i)} \text{ . (53)}$$

And Eq. (48) can be expressed as

$$\mathbf{x}^{(m+1)} = \mathbf{x}^{(m)} + \alpha\mathbf{F}^{(m)} - \sum_{i=2}^{m} c_{mi} u^{(i)} \text{ . (54)}$$

In Eqs. (53) and (54) we have

$$u^{(i)} = \alpha\delta\mathbf{F}^{(i)} + \delta\mathbf{x}^{(i)} + \sum_{j=2}^{i-1} a_{ij}u^{(j)} ,$$

$$a_{ij} = v^{T(j)}\delta\mathbf{F}^{(i)} ,$$

$$c_{mi} = v^{T(i)}\mathbf{F}^{(m)} ,$$

$$v^{(i)} = \frac{\delta\mathbf{F}^{(i)}}{\delta\mathbf{F}^{T(i)}\delta\mathbf{F}^{(i)}} .$$

Using the present method, convergence can be easily reached with the storage of only a few vectors of length N. It must be mentioned that the Newton method is a local method, which means it works only when the initial trial density matrix is 'sufficiently good'. Usually, we can obtain a 'sufficiently good' initial guess through a separate DFT calculation of an isolated extended molecule with several layers of electrode atoms on both sides.

At zero temperature, the Fermi function will reduce to a step function, which is much simpler and implemented more easily. Hence, in some implementations of the NEGF+DFT approach [38,41], the temperature is often assumed to be T=0K. However, this may cause the self-consistent procedure not easy to achieve convergence, especially when sharp peaks exist in the DOS near the Fermi level. In the iterative process, the change of the Hamiltonian matrix will lead to variations in the DOS and make the sharp peaks in the DOS fluctuate about the Fermi level randomly. A small change in the Hamiltonian matrix would lead to a large change in the density matrix, so that it is very hard to reach convergence at zero temperature. This problem can be easily solved by introducing finite temperatures, thus these sharp peaks near the Fermi level can be smoothened by the Fermi distribution function.

3.3. Application and Challenge of the NEGF+DFT Approach

The NEGF+DFT approach has been extensively used to investigate the electron transport properties of various molecular electronic devices including metal monatomic chains [39,44,58], small organic molecules [59-84], and carbon nanotubes [85-87]. For a [100]-oriented gold quantum point contact [44], Rocha and coworkers calculated the zero-bias transmission function and found that the transmission function is a rather smooth function with a value of unity for a broad energy range around the Fermi level, indicating that the transport at the Fermi level is dominated by a single low-scattering s channel. Their results are in good agreement with other theoretical calculations and experimental data [88-92]. This is one of the most successful applications of the NEGF+DFT approach. For the junction conductance of small organic molecules sandwiched between two metal electrodes, the performance of the NEGF+DFT approach is not as good as that for metal nanowires. Although qualitatively correct results may still be found in many cases, a systematic quantitative analysis of the transmission function is not possible. The theoretical results calculated with the NEGF+DFT approach tend to be much larger than the experimentally measured values for the zero-bias conductance of organic molecules coupled to gold electrodes via the Au-S or Au-N bonds, often by orders of magnitude, even if various atomic structures of the molecule-electrode contact have been considered.

One possible reason for the failure of the NEGF+DFT approach, is that the two electrodes are always assumed to be in local equilibrium and that the self-energies of the electrodes are also obtained from an equilibrium calculation. In other words the response of the electrodes to the passing current is completely ignored, as pointed out by Mera, Bokes and Godby [53]. Another possible reason is that the exact, non-equilibrium, many-body self-energy operator is approximated by the static DFT exchange-correlation potential. Although this approximation is reasonably good for nonresonant transport where the molecular level broadening is strong and individual molecular levels strongly overlap, just as the case of gold quantum point contact, it is not justified at weak coupling where the wave functions show pronounced localized features and individual resonances are fully developed [93].It is well known that most approximate LDA and GGA exchange-correlation functionals currently in use always underestimate the HOMO–LUMO gap of molecules and also give the too small ionization potentials when calculated from the eigenvalue of the HOMO, which might incorrectly determine the lineup of the molecular levels to the metal Fermi level and lead to an inaccurate junction conductance [94]. There are three possible alternatives to improve the performance of the NEGF+DFT approach: the hybrid functional, the GW method and time-dependent (TD) DFT.

Hybrid functions, which includes a partial contribution of exact HF exchange into the common LDA or GGA exchange-correlation potential, can yield excellent results for molecules. Although conventional hybrid functions like B3LYP are not suited for metals because exact exchange interactions in three-dimensional metals are doomed to fail [33], a recent developed HSE hybrid functional that employs a screened Coulomb potential for the exchange interaction has been demonstrated to have the potential to be applicable to both molecules and metal electrodes [95-99]. The GW method is based on a Green's function approach for calculating quasiparticle excitation energies and wave functions [100-102]. In this method, an approximation solution of the Dyson equation is obtained by using a first-order expression for the self-energy operator in terms of the Green's function G and the screened Coulomb interaction W. Unlike DFT in which the KS gap is not a physically meaningful quantity [33], the GW method can produce a true quasiparticle excitation spectrum and allow for the treatment of filled and empty states on the same footing, which is an important advantage for understanding charge transport across molecule-metal interfaces. However, compared with DFT calculations with hybrid functionals, the GW method is too time-consuming. Excitation energies of interacting systems can also be obtained via TDDFT [103-105], in which the time-dependent density of an interacting system moving in an external, time-dependent local potential is calculated via a fictitious system of non-interacting electrons moving in a local, effective time-dependent potential. Although TDDFT with the adiabatic local density approximation of the time-dependent exchange-correlation potential is a valid and simple method when studying isolated systems like clusters and molecules, simple kernels that work also for infinite systems are still missing [106]. Therefore, even though there have been several attempts beyond the conventional static DFT + NEGF method [107-111], it is still a long-term goal to develop robust modeling methods that can correctly predict the current-voltage characteristics of molecular electronic devices.

4. Conclusion

Many experimental advances in molecular electronic devices have been achieved in recent years, especially a molecular transistor has been demonstrated using the electrochemical gating approach. However, the detailed atomic structure of the molecule-electrode contacts cannot be measured or controlled in current experiments, which greatly hamper further progress in molecular electronics. Therefore, first-principles theoretical approaches for calculating electron transport through molecular electronic devices are needed to supplement and guide experiments. Currently, the most popular theoretical approach is NEGF+DFT, which employs a set of localized basis functions.

Implementations of the self-consistent NEGF+DFT approach have been discussed in detail. In reference to the Green's function part that calculates the density matrix of the extended molecule from its Hamiltonian matrix, it should be noted that bound states in the bias window are not included if the density matrix is obtained from the integration of the lesser Green's function. In the DFT part that calculates the Hamiltonian matrix from the density matrix, both the electrostatic correction and the exchange-correlation correction are included to account for the effects of the electrodes. Because the exact many-body self-energy operator, due to the electron-electron interaction in the extended molecule region, is approximated by the static DFT exchange-correlation potential in the NEGF+DFT approach, only qualitatively correct results are obtained for the conductance of organic molecules. In order to improve the performance of the NEGF+DFT, three possible alternatives including the hybrid functional, the GW method and TDDFT have been discussed.

References

1. A. Nitzan, M.A. Ratner, Science, 300 (2003) 1384.
2. J.R. Heath and M.A. Ratner, Physics Today, 56(5) (2003) 43.
3. P. Qi, A. Javey, M. Rolandi, Q. Wang, E. Yenilmez, H. Dai, J. Am. Chem. Soc., 126 (2004) 11774.

4. X. Guo, J.P. Small, J.E. Klare, Y. Wang, M.S. Purewal, I.W. Tam, B.H. Hong, R. Caldwell, L. Huang, S. O'Brien, J. Yan, R. Breslow, S.J. Wind, J.Hone, P. Kim, C. Nuckolls, Science 311 (2006) 356.
5. M.A. Reed, C. Zhou, C.J. Muller, T.P. Burgin, J.M. Tour, Science, 278(1997)252
6. B. Xu, N.J. Tao, Science, 301 (2003) 1221.
7. B. Xu, X. Xiao, N.J. Tao, J. Am. Chem. Soc., 125(2003)16164.
8. X. Xiao, B.Xu, N.J. Tao, Nano Lett., 4 (2004) 267.
9. B. Xu, X. Xiao, X. Yang, L. Zang, N. Tao, J. Am. Chem. Soc., 127(2005)2386.
10. J. Ulrich, D. Esrail, W. Pontius, L. Venkataraman, D. Millar, L.H. Doerrer, J. Phys. Chem. B, 110 (2006) 1462.
11. L. Venkataraman, J.E. Klare, I.W. Tam, C. Nuckolls, M.S. Hybertsen, M.L. Steigerwald, Nano Lett., 6 (2006) 458.
12. L. Venkataraman, J.E. Klare, C. Nuckolls, M.S. Hybertsen, M.L. Steigerwald, Nature, 442 (2006) 904.
13. X.D. Cui, A. Primak, X. Zarate, J. Tomfohr, O.F. Sankey, A.L. Moore, T.A. Moore, D. Gust, G. Harris, S.M. Lindsay, Science, 294 (2001) 571.
14. E. Wierzbinski, K. Slowinski, Langmuir, 22 (200) 5205.
15. W. Haiss, R.J. Nichols, H. van Zalinge, S.J. Higgins, D. Bethell, D.J. Schiffrin, Phys. Chem. Chem. Phys., 6 (2004) 4330.
16. T. Dadosh, Y. Gordin, R. Krahne, I. Khivrich, D. Mahalu, V. Frydman, J. Sperling, A. Yacoby, I. Bar-Joseph, Nature, 436 (2005) 677.
17. X. Xiao, L.A. Nagahara, A.M. Rawlett, N. Tao, J. Am. Chem. Soc., 127 (2005) 9235.
18. L. Cai, M.A. Cabassi, H. Yoon, O.M. Cabarcos, C.L. McGuiness, A.K. Flatt, D.L. Allara, J.M. Tour, T.S. Mayer, Nano Letters, 5 (2005) 2365.
19. B. Xu, X. Li, X. Xiao, H. Sakaguchi, N.J. Tao, Nano Letters, 5 (2005) 1491.
20. F. Chen, J. He, C. Nuckolls, T. Roberts, J.E. Klare, S. Lindsay, Nano Lett., 5 (2005) 503.
21. Y. Taur and T. Ning, Fundamentals of VLSI Devices, Cambridge University Press, Cambridge, UK, 1998.
22. E. Tran, M.A. Rampi, G.M. Whitesides, Angew. Chem. Int. Ed., 43 (2004) 3835.
23. X. Li, B. Xu, X. Xiao, X. Yang, L. Zang, N.J. Tao, Faraday Discussions, 131 (2006) 111.
24. B. Kim, J.M. Beebe, Y. Jun, X.-Y. Zhu, C.D. Frisbie, J. Am. Chem. Soc., 128 (2006) 4970.
25. D.R.T. Zahn, G.N. Gavrila, M. Gorgoi, Chem. Phys., 325 (2006) 99.
26. E. Itoh, M. Iwamoto, M. Burghard, S. Roth, Jpn. J. Appl. Phys., 39 (2000) 5146.
27. S. Datta, Quantum transport: atom to transistor, Cambridge University Press, 2005.
28. Y. Meir and N.S. Wingreen, Phys. Rev. Lett., 68 (1992) 2512.
29. K.S. Thygesen, Phys. Rev. B, 73 (2006) 035309.

30. A. Ferretti, A. Calzolari, R. Di Felice, F. Manghi, Phys. Rev. B, 72 (2005) 125114.
31. R.G. Parr and W. Yang, Density-functional Theory of Atoms and Molecules, Oxford University Press, Oxford, 1989.
32. Wolfram Koch, Max C. Holthausen, A Chemist's Guide to Density Functional Theory, 2nd edition, Wiley-VCH, Weinheim, 2001.
33. Richard M. martin, Electronic Structure: Basis Theory and Practical Methods, Cambridge University Press, 2004.
34. P. Hohenberg and W. Kohn, Phys. Rev., 136 (1964) B867.
35. W. Kohn and L.J. Sham, Phys. Rev., 140 (1965) A1133.
36. N.D. Mermin, Phys. Rev., 137 (1965) A1441.
37. J. Zhang, S. Hou, R. Li, Z. Qian, R. Han, Z. Shen, X. Zhao, Z. Xue, Nanotechnology, 16 (2005) 3057.
38. J. Taylor, H. Guo, J. Wang, Phys. Rev. B, 63 (2001) 245407.
39. J.J. Palacios, A.J. Pérez-Jiménez, E. Louis, E. SanFabián, J.A. Vergés, Phys. Rev. B, 66 (2002) 035322.
40. E. Louis, J.A. Vergés, J.J. Palacios, A.J. Pérez-Jiménez, E. SanFabián, Phys. Rev. B, 67 (2003) 155321.
41. Y. Xue, S. Datta, M.A. Ratner, Chem. Phys., 281 (2002) 151.
42. M. Brandbyge, J.-L. Mozos, P. Ordejón, J. Taylor, K. Stokbro, Phys. Rev. B, 65 (2002) 165401.
43. S.-H. Ke, H.U. Baranger, and W. Yang, Phys. Rev. B, 70 (2004) 085410.
44. A. R. Rocha, V. M. García-Suárez, S. Bailey, C. Lambert, J. Ferrer, S. Sanvito, Phys. Rev. B, 73 (2006) 085414.
45. Y.H. Kim, J.Tahir-Kheli, P.A. Shultz, W.A. Goddard III, Phys. Rev. B, 73 (2006) 235419.
46. H. Haug, A.-P, Jauho, Quantum Kinetics in Transport and Optics of Semiconductors, Springer, 1998.
47. S. Datta, Electronic transport in mesoscopic systems, Cambridge University Press, 1995.
48. S. Hou, R. Li, Z. Qian, J. Zhang, Z. Shen, X. Zhao, Z. Xue, J. Phys. Chem. A, 109 (2005) 8356.
49. S.-H. Ke, H.U. Baranger, W. Yang, J. Chem. Phys., 123 (2005) 114701.
50. K.S. Thygesen, K.W. Jacobsen, Phys. Rev. B, 72 (2005) 033401.
51. J. Velev, W. Butler, J. Phys.: Condens. Matter, 16 (2004) R637.
52. W.H. Press, B.P. Flannery, S.A. Teukolsky, and W.T. Vetterling, Numerical Recipes in C: The Art of Scientific Computing, 2nd ed.,Cambridge University Press, Cambridge, England, 1992.
53. H. Mera, P. Bokes, R.W. Godby, Phys. Rev. B, 72 (2005) 085311.
54. S. Datta, Quantum transport: atom to transistor, Cambridge University Press, 2005.

55. R. Li, J. Zhang, S. Hou, Z. Qian, Z. Shen, X. Zhao, Z. Xue, submitted to Phys. Rev. B.
56. C. G. Broyden, Math. Comput. 19 (1965) 577.
57. G. P. Srivastava, J. Phys. A: Math. Gen. 17 (1984) L317-L321
58. P.S. Damle, A.W. Ghosh, S. Datta, Phys. Rev. B, 64 (2001) 201403.
59. Y. Xue, M.A. Ratner, Phys. Rev. B, 68 (2003) 115406.
60. Y. Xue, M.A. Ratner, Phys. Rev. B, 68 (2003) 115407.
61. Y. Xue, M.A. Ratner, Phys. Rev. B, 69 (2004) 085403.
62. Y. Xue, M.A. Ratner, Phys. Rev. B, 70 (2004) 081404.
63. S.-H. Ke, H.U. Baranger, W. Yang, J.Am.Chem.Soc., 126 (2004) 15897.
64. R. Liu, S.-H. Ke, H.U. Baranger, W. Yang, J. Chem. Phys., 122 (2005) 074703.
65. S.-H. Ke, H.U. Baranger, W. Yang, J. Chem. Phys., 122 (2005) 074704.
66. R. Liu, S.-H. Ke, H.U. Baranger, W. Yang, Nano Lett., 5 (2005) 1959.
67. R. Liu, S.-H. Ke, H.U. Baranger, W. Yang, J. Chem. Phys., 124 (2006) 024718.
68. R. Liu, S.-H. Ke, H.U. Baranger, W. Yang, J. Am. Chem. Soc., 128 (2006) 6274.
69. C.-C. Kaun, H. Guo, Nano Lett., 3 (2003) 1521.
70. C.-C. Kaun, B. Larade, H. Guo, Phys. Rev. B,67 (2003) 121411.
71. C.-C. Kaun, H. Guo, P. Grütter, R. Bruce Lennox, Phys. Rev. B, 70 (2004) 195309.
72. Y. Hu, Y. Zhu, H. Gao, H. Guo, Phys. Rev. Lett., 95 (2005) 156803.
73. N. Sergueev, D. Roubtsov, and H. Guo, Phys. Rev. Lett., 95 (2005) 146803.
74. D. Waldron, P. Haney, B. Larade, A. MacDonald, H. Guo, Phys. Rev. Lett., 96 (2006) 166804.
75. J. Taylor, M. Brandbyge, K. Stokbro, Phys. Rev. B, 68 (2003) 121101.
76. W. Lu, V. Meunier, J. Bernholc, Phys. Rev. Lett., 95 (2005) 206805.
77. K.-H. Müller, Phys. Rev. B, 73 (2006) 045403.
78. X. Wu, Q. Li, J. Huang, J. Yang, J.Chem.Phys.,123 (2005) 184712.
79. H. Geng, S.W. Yin, K.Q. Chen, Z.G. Shuai, J. Phys. Chem. B, 109(25) (2005) 12304.
80. X. Yin, H. Liu, J. Zhao, J.Chem.Phys.,125 (2006) 094711.
81. H. Kondo, H. Kino, J. Nara, T. Ozaki, T. Ohno, Phys. Rev. B, 73 (2006) 235323.
82. A. Grigoriev, J. Sköldberg, G. Wendin, Ž. Crljen, Phys. Rev. B, 74 (2006) 045401.
83. M. Paulsson, T. Frederiksen, M. Brandbyge, Nano Lett., 6 (2006) 258.
84. C. Zhang, Y. He, H.-P. Cheng, Y. Xue, M.A. Ratner, X.-G. Zhang, P. Krstic, Phys. Rev. B, 73 (2006) 125445.
85. J.J. Palacios, A.J. Pérez-Jiménez, E. Louis, E. SanFabián, J.A. Vergés, Phys. Rev. Lett., 90 (2003) 106801.
86. S.-H. Ke, W. Yang, H.U. Baranger, J. Chem. Phys., 124 (2005) 181102.
87. K. Odbadrakh, P. Pomorski, C. Roland, Rev. B, 73 (2006) 233402.
88. U. Landman, W.D. Luedtke, B.E. Salisbury, R.L. Whetten, Phys. Rev. Lett. 77 (1996) 1362.

89. L.G.C. Rego, A.R. Rocha, V. Rodrigues, D. Ugarte, Phys. Rev. B, 67 (2003) 045412.
90. V. Rodrigues, T. Fuhrer, D. Ugarte, Phys. Rev. Lett. 85 (2000) 4124.
91. H. Ohnishi, Y. Kondo, and K. Takayanagi, Nature, 395 (1998) 780.
92. N. Agraït, A. Levy Yeyati, J. M. van Ruitenbeek, Phys. Rep., 377 (2003) 81.
93. F. Evers, F. Weigend, M. Koentopp, Phys. Rev. B, 69 (2004) 235411.
94. C. Toher, A. Filippetti, S. Sanvito, K. Burke, Phys. Rev. Lett., 95 (2005) 146402.
95. J. Heyd, G. E. Scuseria, M. Ernzerhof, J. Chem. Phys., 118 (2003) 8207.
96. J. Heyd and G. E. Scuseria, J. Chem. Phys., 120 (2004) 7274.
97. J. Heyd, J.E. Peralta, G.E. Scuseria, R.L. Martin, J. Chem. Phys., 123 (2005) 174101.
98. O.A. Vydrov, J. Heyd, A.V. Krukau, G.E. Scuseria, J. Chem. Phys., 125 (2006) 074106.
99. J. Paier, M. Marsman, K. Hummer, G. Kresse, I.C. Gerber and J.G. Ángyán, J. Chem. Phys., 124 (2006) 154709.
100. L. Hedin, Phys. Rev., 139 (1965) A796.
101. M. S. Hybertsen and S. G. Louie, Phys. Rev. B, 34 (1986) 5390.
102. F. Aryasetiawan, O. Gunnarsson, Rep. Prog. Phys., 61 (1998) 237.
103. E. Runge and E. K. U. Gross, Phys. Rev. Lett., 52 (1984) 997.
104. M. Petersilka, U. J. Gossmann, and E. K. U. Gross, Phys. Rev. Lett. 76 (1996) 1212.
105. G. Onida, L. Reining, A. Rubio, Rev. Mod. Phys., 74 (2002) 601.
106. O. Pulci, M. Marsili, E. Luppi, C. Hogan, V. Garbuio, F. Sottile, R. Magri, R. Del Sole, Phys. Stat. Sol. B, 242(13) (2005) 2737.
107. S. Kurth, G. Stefanucci, C.-O. Almbladh, A. Rubio, E.K. Gross, Phys. Rev. B, 72 (2005) 035308.
108. A. Ferretti, A. Calzolari, R. Di Felice, F. Manghi, M.J. Caldas, M. Buongiorno Nardelli, E. Molinari, Phys. Rev. Lett., 94 (2005) 116802.
109. N. Sai, M. Zwolak, G. Vignale, M. Di Ventra, Phys. Rev. Lett., 94 (2005) 186810.
110. M. Koentopp, K. Burke, F. Evers, Phys. Rev. B, 73 (2006) 121403(R).
111. X. Qian, J. Li, X. Lin, S. Yip, Phys. Rev. B, 73 (2006) 035408.

CHAPTER 7

STRUCTURE, PROPERTIES, AND OPPORTUNITIES OF BLOCK COPOLYMER/PARTICLE NANOCOMPOSITES

Lindsay Bombalski[1,2], Jessica Listak[1], Michael R. Bockstaller[1,*]

[1]Department of Materials Science and Engineering, Carnegie Mellon University, 5000 Forbes Ave., Pittsburgh, PA 15213; [2]Department of Chemistry, Carnegie Mellon University, 4400 Fifth Ave. Pittsburgh, PA 15123;
**Email: Bockstaller@cmu.edu*

When different materials are combined to form a heterogeneous structure, the properties of the resulting composite material depend on the properties of the constituent materials, the length scale as well as the chemical and morphological details of the dispersion. Nanocomposites, *i.e.,* heterogeneous materials in which at least one characteristic length scale of the filler material is in the nanometer range, have attracted particular attention and currently represent one of the fastest growing areas in materials science. Inorganic nanoparticle additives are blended with polymers in order to enhance particular optical characteristics, mechanic stability, wear resistance, barrier properties or flame resistance. In many instances the performance of the nanocomposite is intimately related to the location and orientation of the dispersed particle additive. The use of structure-guiding host materials such as block copolymers affords opportunities for controlling the spatial and orientational distribution of filler particles and facilitates more sophisticated property design than classical single-phase host materials. This article reviews recent research in the area of block copolymer/nanoparticle (BCP/NP) composites with particular focus on the relevant parameters that control the structure formation in block copolymer/nanoparticle blends and the structure-property relations of the resulting composite materials. The morphological characteristics of BCP/NP blends are discussed with respect to equilibrium and non-equilibrium conditions and compared to experimental observations in related areas of materials science in order to provide a framework for interpretation. It will be shown that by control of the particle's location within a microstructured host material new properties can emerge that are not inherent in the properties of the constituent materials (and thus are out of the realm of uniform composite materials) but rather are the result of particle-particle interactions that occur due to the morphology of the dispersion. These emergent properties hold the promise to design novel 'chimera' composite materials, *i.e.*, materials that combine multiple properties that in uniform distributions would exclude each other.

1. Introduction

Current polymer composite technology relies on the dispersion of micro- and nanosized filler particles into uniform polymer matrices in order to tailor the materials' optical, thermo-mechanical, or transport properties [1-3]. Nanoscale filler materials have attracted particular attention as additives because of the unique properties such as UV absorption (*e.g.*, ZnO nanocrystals), luminescence (semiconductor quantum dots), supra-paramagnetism (magnetic nanocrystals) or extraordinary mechanical strength (*e.g.*, carbon nanotubes) that result as a consequence of the spatial confinement and the particular bonding situation in nanomaterials [4]. Additional motivation to use nanoscale particle additives derives from the reduced scattering strength of particles that are smaller than the wavelength of light thus facilitating property enhancements without sacrificing *e.g.*, optical clarity. Three major characteristics that are intimately related to the nanoscale dimension of the filler additives are considered to be critical for the performance of nanoparticle/polymer composites:

(1) The particular size-dependent material properties of the filler that result from the spatial confinement of the material to the fundamental length scales of the respective physical properties. For more information on 'solid-state physics on the nanoscale', that one could consider the core of what is commonly known as 'nanoscience', the reader is referred to the recent literature (see *e.g.*, [Ref. 5]).

(2) The morphological features related to the nano-size of the filler particles such as the large interfacial area between the constituents (~ 10^3 m^2/cm^3), the ultra-low percolation threshold (~ 1 vol%) and the small

distances between filler particles that often are in the range of the polymer's radius of gyration. The dependence of the interfacial area on the particle filler size is depicted in Figure 1 that demonstrates the increase in interfacial area by several orders of magnitude as the size of the filler particles is decreased from micro- to nanometer dimensions.

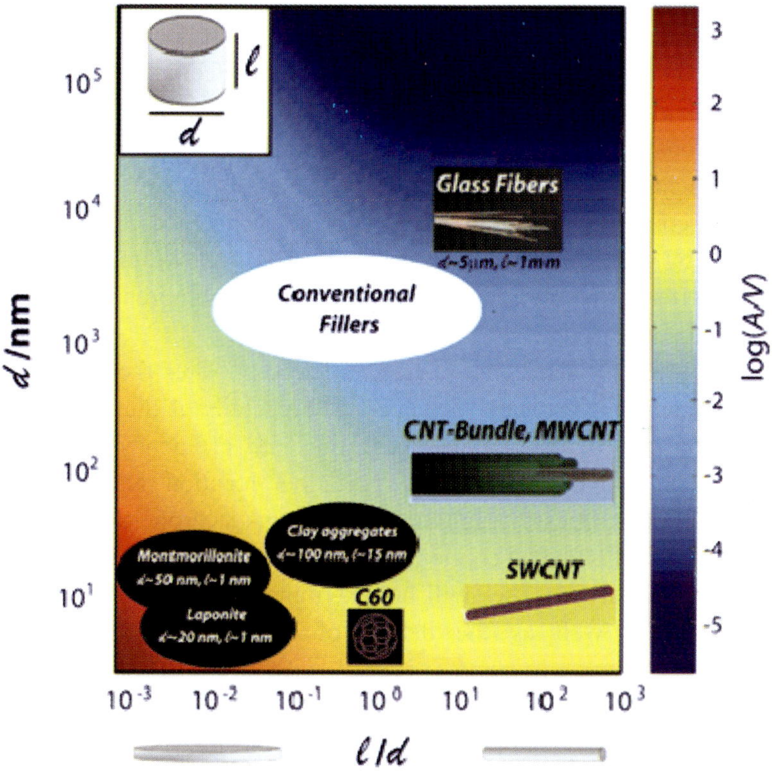

Figure 1. Calculated interfacial area per volume of particles $A/V \sim 1/d + 1/l$ (in nm^{-1}) assuming a right circular cylindrical particle shape for different particle diameter d and aspect ratio l/d. Exfoliated clay (laponite or montmorillonite) particles, spherical nanoparticles, and single wall carbon nanotubes generate up to four orders of magnitude more interfacial area as compared to conventional filler materials like glass fibers of equal volume filling fraction (SWCNT: single-walled carbon nanotube, MWCNT: multi-walled carbon nanotube). (Adopted with permission from [Ref. 6]; copyright 2005, Wiley-VCH).

The consequences of the increased interfacial area represent a major departure from classical micron-sized filler composites since the majority of the matrix polymer is located in the narrow interfacial regions that extend to about a length corresponding to the polymers' radius of gyration away from the particle surface. Because the properties of polymers that are confined within these interfacial regions differ from those of their bulk counterparts (the details depend on the nature of interaction between polymer and interface), the property-prediction is often not accessible to classical two-component effective medium theory (in particular with respect to mechanical properties). The interrelationship between the size, shape and surface chemistry of particle inclusions on their miscibility in homopolymer matrices and the resulting composite properties are subject of intense research as is indicated by some 2000 publications in the topic area within the past five years. The reader is referred to Reference [7] for a comprehensive review of the field.

(3) As for classical micron-sized filler composites the performance of nanocomposites critically depends on the details of the filler particle distribution. However, besides increasing the efficiency of property enhancement, the control of the particle's location and orientation provides opportunities to capitalize on synergistic properties that result from the interaction between the filler particles. Ultimately the properties of polymer nanocomposites will be determined by the properties of the constituents as well as the details of the particle's location and orientation.

The importance of morphological control to fully exploit the opportunities of nanocomposites has fueled research in using structure-guiding host materials such as block copolymers (BCP) as scaffolds to control the spatial location, orientation and mean inter-particle distance of particles within the polymer host. BCP are comprised of distinct polymer chains, covalently joined together at their endpoints to form a single chain. As a consequence of their covalent connection, BCP cannot macrophase separate in the melt like blends of distinct homopolymers but rather segregate on local scale to form periodic microdomain morphologies with symmetries that depend on the interplay between the architecture of the copolymer as well as the size and interactions between the respective blocks. The set of possible microdomain morphologies is particularly well understood for the case of amorphous diblock copolymers where the formation of lamellar (pm), cylindrical (p6mm), spherical ($Im\bar{3}m$) and 3D interconnected microstructures ($Ia\bar{3}d$) is observed [8,9]. These microdomain structures can act as scaffolds for the sequestration of nanoscopic inclusions of appropriate geometry and chemical affinity and – for appropriately matched host polymer and particle symmetry – can impart structural information on the particle distribution. Figure 2 provides an overview of the various microdomain morphologies of amorphous diblock copolymers as well as the particle geometries that are of ‘compatible’ symmetry with the respective polymer template.

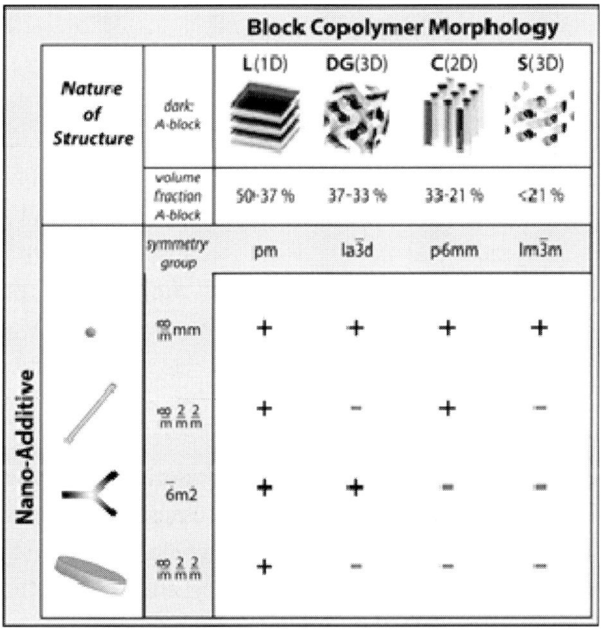

Figure 2. Schematic BCP phase diagram depicting the various microdomain morphologies for amorphous diblock copolymers. The table lists the commensurate symmetry combinations of particle point group and BCP space group representations, respectively. The compatibility of a particle’s point group with the minority domain of the respective BCPs space group is indicated by (+). L = lamellar, DG = double gyroid, C = cylinder, and S = spherical. (Adopted with permission from [Ref. 6]; copyright 2005, Wiley-VCH).

Research in the area of BCP/nanoparticle composites has been pursued in a variety of directions. The first experiments to demonstrate the versatility of BCP to control the formation and organization of inorganic nanocrystals were the in-situ synthesis of gold or silver nanocrystals in lamellar, cylindrical and spherical domains of phosphine-functionalized BCP after reduction of metal salts that are coordinated to the monomeric units [10,11]. Later experiments modified this scheme by using amphiphilic BCP typically derived from poly(ethyleneoxide), poly(acrylic acid) or poly(vinyl pyridine) that can be selectively infiltrated with precursor compounds by immersion of the microstructure within the respective metal salt solutions. Metal nanocrystals then form selectively within the hydrophilic BCP domains after reduction of the precursor salts [12-15]. Similarly, a variety of metal nanocrystals were grown in the precursor-loaded hydrophilic core of amphiphilic BCP micelles in organic solvents (so called 'nanoreactor scheme') [16]. Even though the detailed mechanism is still disputed most of the literature agrees on a reaction-diffusion controlled nucleation and growth process that is initiated by the formation of small clusters that subsequently act as nucleation sites and grow by depleting the precursor from the surrounding polymer matrix [17,18]. Because the particle growth depends on the details of the nucleation and growth process, the stabilization of the particle surface by the surrounding matrix, the homogeneity of the precursor distribution, as well as the change of transport properties during the course of the reaction the *in-situ* approach often results in broad particle distributions. Moreover, because the size of the growing nuclei is much smaller than the length scale of the respective BCP domain, the flux of reactants is isotropic around the growing nuclei and thus results in mostly spherical particle shapes. The limited possibilities to control the surface chemistry and architecture (such as core-shell structures) of the embedded nanocrystals are further drawbacks of the *in-situ* approach. Therefore, while providing an elegant approach to control the spatial formation of inorganic nanocrystals (in particular on surfaces), the *in-situ* approach has not found application in the design of bulk BCP based nanocomposite materials that capitalize on the specific optical and mechanical properties of particle inclusions that in turn are intimately related to the particle's size, shape and architecture [19].

The concurrent organization of *ex-situ* synthesized nanoparticles and block copolymers provides an alternative route towards mesoscopically ordered BCP nanocomposites. The strategy is to separately synthesize and process nanoparticles of desired size, shape and architecture and to tailor the particle's surface chemistry such that preferential chemical affinity will drive the particles into the target BCP domain during the self-organization process of BCP and nanoparticle additives. Preferential chemical affinity is typically achieved by means of surfactants that favorably interact with the respective target domain, *e.g.,* by grafting polymers to the particle surface that are chemically equal to the respective target BCP domain. Current interest in block copolymer/nanoparticles (BCP/NP) blends is driven by two major motivations: from a fundamental perspective, BCP/NP mixtures represent fascinating model systems to study the structure formation in soft/hard heterogeneous materials in which subtle changes of entropic or enthalpic interactions can result in different morphologies. From an application-oriented point of view, BCP/NP composite materials have been shown to exhibit intriguing optical and mechanical properties that arise from the control of nanoparticle location and, in the case of anisotropic particle additives, orientation within the copolymer matrix.

This article reviews recent research efforts in the field of BCP/NP composite materials starting with the theoretical models that have been developed to predict the structure formation in BCP/NP blends. Subsequently, experimental results on the formation of equilibrium and non-equilibrium microstructures in BCP/NP blends will be presented with focus on 0D (*i.e.,* point-like) particle inclusions and the properties of the respective composite materials. The article will conclude with a discussion of some of the opportunities and challenges that derive from the presented results with particular emphasis on 'emergent physical properties', *i.e.,* physical properties of the composite material that are not intrinsic in the constituent materials but rather arise from the detailed microstructure of the nanocomposite. Throughout, the focus will be on bulk BCP/NP composite materials leaving out areas such as continuous organic-inorganic phase microstructures that are obtained *e.g.,* by precursor infiltration, polymerization and subsequent calcination of a BCP, surface patterning techniques based on BCP thin films or micelle-based particle synthesis, for which the reader is referred to the literature [20-25]. For more information on the challenges and opportunities associated with the inclusion of anisotropic particle filler materials within BCP the reader is referred to Reference [6].

2. Structure Formation in BCP Hybrid Materials – Theory and Simulation

Recent advancements in the development of efficient numerical algorithms have resulted in significant insights into self-organization processes in BCP composite materials. However, before presenting major advances in the field, it seems appropriate to point out earlier studies on morphologically related systems such as BCP/solvent and BCP/homopolymer mixtures that bear analogies to BCP/NP blends and that are of significance to current research. The effect of low-molecular inclusions on the structure of block copolymer materials has been pioneered by Hashimoto and coworkers who studied the segregation of organic solvents in concentrated BCP solutions and speculated about an inhomogeneous distribution of solvents that have neutral affinity to the respective polymer domains [26,27]. Later simulations using self-consistent field theory (SCFT) supported these observations [28]. Recently, Spontak *et al.* demonstrated that the SCFT approach provides good agreement to experimental findings on the morphology of BCP/inclusion mixtures even for inclusions that differ by orders of magnitude in size (ranging from low-molecular solvents to nanosized particles) [29]. Shull *et al.* applied a SCFT approach to predict the distribution of homopolymers in BCP/homopolymers blends and suggested increasingly inhomogeneous segregation of the homopolymers with increasing molecular weight, in close analogy to experimental observations [30,31]. Balazs and coworkers were the first to study the structure formation in BCP/NP blends by combining SCFT and density functional theory (DFT) to account for the polymer and particle component, respectively [32-34]. In these studies, the effect of surface-grafted chains (*i.e.,* the particle's corona) was replaced by an effective particle-polymer interaction parameter characterizing the chemical affinity of the particle to the polymer domain. Assuming chemical neutralization of the particles to one of the blocks of a lamellar BCP the authors predicted interfacial segregation of the particles in the limit of small particle sizes ($d/L \ll 1$, with d the particle diameter and L the BCP lamellar spacing) and segregation of the particle inclusions to the center of the compatible domain for larger particle sizes ($d/L > 0.3$). The predicted size-dependence of the particle segregation was rationalized as a consequence of competing entropic contributions arising from the decrease in chain conformational entropy upon distributing particles within the matrix polymer and the increased configurational entropy resulting from the particle dispersion. Buxton *et al.* and Lee *et al.* extended the approach to capture the influence of attractive enthalpic interactions (*i.e.,* $\chi_{PA} < 0$, where χ_{PA} denotes the Flory interaction parameter between the particle 'P' and the compatible domain 'A') and suggested that the equilibrium microdomain morphology of BCP/NP blends depends on a subtle interplay between enthalpic and entropic contributions, indicating a vast parameter space to control and design the morphology of BCP/NP composites [35,36]. Figure 3 depicts the predicted phase diagrams of a BCP/NP blend as a function of polymer-particle interaction parameter, copolymer composition as well as particle size for a given particle concentration. Recent results by Schultz *et al.* using a discontinuous molecular dynamics technique are in qualitative agreement with the SCFT/DFT approach [37]. Liu *et al.* studied the effect of particle size disparity on the structure formation in BCP/NP blends using a dissipative particle dynamics approach indicating that depending on their size, particles will autonomously segregate to different locations within the microstructure (in qualitative agreement to experimental observations, see section 3) [38].

Using a SCFT model, Fredrickson and coworkers investigated the implications of the grafting density and molecular weight of surface-grafted polymer chains on the miscibility of particles within BCP hosts [39]. In this study the surface-grafted polymer chains were found to result in strong distortion of the polymer microstructure resulting in a high miscibility of particles with low grafting density and molecular weight of the respective surface-bound polymers (in analogy to

BCP/homopolymer blends). Even though the study was limited to unrealistic small numbers of the polymer grafting density (assuming only 4 or 6 surface bonded chains), these results provide valuable design criteria for the particle additives. Very recently, Fredrickson and coworkers presented a new simulation algorithm (HPF = Hybrid Particle Field method) that holds the promise to be more versatile that the SCFT/DFT approach and demonstrated excellent agreement with experimental data [40]. Figure 4 depicts a comparison between the predicted and experimentally observed microstructure formation in blends of cylindrical poly(styrene-b-2-vinyl pyridine) and poly(styrene)-coated gold nanoparticles.

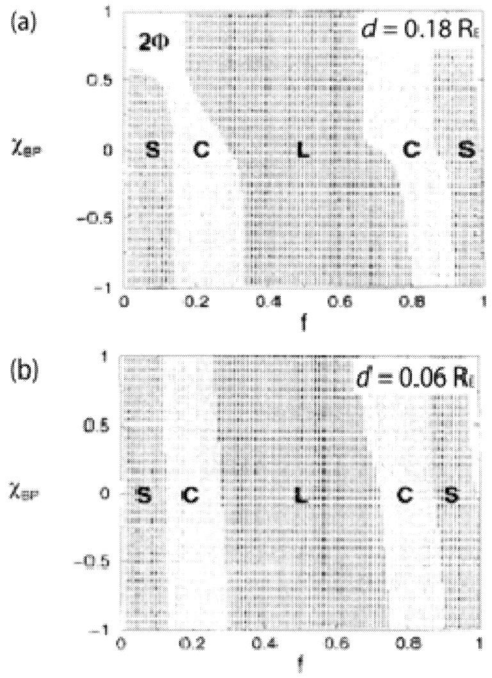

Figure 3. Calculated phase diagram of a BCP/NP blend as a function of polymer-particle interaction parameter and block composition assuming a fixed particle filling fraction of $\phi = 0.15$. Figures correspond to a particle size corresponding to $d = 0.18\ R_E$ (panel a) and $d = 0.06\ R_E$ (panel b), respectively, with R_E denoting the polymer end-to-end distance. Abbreviations: L = lamellar, C = cylindrical, S = spherical. (Adopted with permission from [Ref. 35]; copyright 2002, American Chemical Society).

Analogies with phenomenological related systems such as the penetration of particles into surface-grafted polymer brushes provide additional opportunities for novel insights into the structure formation of BCP/NP composites. For example, Kim *et al.* studied the penetration of particles within end-grafted polymer brushes based on the Alexander-De Gennes (AG) model and suggested various particle distributions depending on the particles size, shape and chemical affinity [41,42]. In a recent study the authors applied a hydrodynamic argument to overcome the limitations of the mean-field AG-model to access the implications of large particle additives on the surrounding polymers conformation and free energy [43]. While this approach did not specifically address the morphology of BCP/NP composite materials it provides a different viewpoint and likely provides ideas and conclusions that will be relevant to BCP/NP systems.

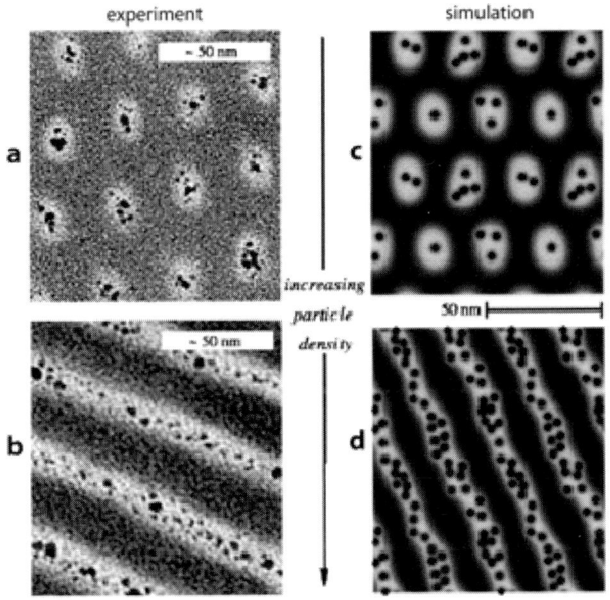

Figure 4. Comparison between the predicted and experimentally observed cylindrical-to-lamellar order-order transition in blends of cylindrical poly(styrene-b-2-vinyl pyridine) and poly(styrene)-coated gold nanoparticles. (Adopted with permission from [Ref. 40]; copyright 2006, American Physical Society).

The theoretical understanding of the various parameters in BCP/NP systems that determine the structure formation process remains a challenge and more experimental and theoretical work is needed to develop a foundation based on which chemists will be enabled to make efficient decisions about the optimum characteristics of a system to achieve particular target morphologies.

3. Structure Formation of BCP Hybrid Materials - Experiments

3.1. Equilibrium BCP / Particle Composite Morphologies

In the context of BCP/NP composites the notion of '0D-particle inclusions' refers to particle additives with characteristic length scales much smaller than the BCP microstructure such as semiconductor quantum dots and metal nanocrystals. BCP/particle structures close to equilibrium are typically obtained by evaporation of the solvent from a BCP/particle solution and subsequent vacuum annealing of the composite film at temperatures above the glass transition temperature of the BCP but below the order-disorder transition temperature. Research in BCP/0D-NP composites is motivated by the intriguing opportunities for engineering novel material properties that result from combining the particular optical or magnetic characteristics of nano-sized matter with the mechanical properties and processability of a microstructured polymer template material. For example, BCP/0D-NP composites have been proposed as material platform for polymer-based photonic crystals or high-efficiency nonlinear optical materials [44,45].

The first to mention the preparation of BCP/0D-NP composites were Hamdoun *et al.* who reported in a series of papers the selective segregation of poly(styrene) (PS) functionalized ferrite nanoparticles within the PS domains of a symmetric poly(styrene-b-butyl methacrylate) copolymer of molecular weight M = 80 kg/mol [46]. Based on neutron scattering analysis Pasyuk *et al.* later deduced that particle inclusions with diameter $d > 6$ nm segregate to the center of the PS domain whereas particles with $d < 4$ nm were expected to segregate to the interface between the adjacent domains [47]. Due to the small difference in particle size (that is additionally blurred by the inherent size disparity of the nanocrystals) as well as the small BCP domain spacing ($L/2 \sim 16$ nm) it is not clear if indeed distinct morphologies were observed, however, the authors clearly stimulated much of the later research. Motivated by the idea to increase the refractive index difference between adjacent BCP domains by selective sequestration of metal nanocrystals to one of the polymer domains, Bockstaller *et al.* demonstrated in 2001 the preferential sequestration of PS-coated gold nanocrystals within high molecular weight lamellar poly(styrene-b-ethylene propylene) (PS-PEP) (M = 800 kg/mol) [45]. PS-coated gold nanocrystals (d = 3.5 nm) with a molecular weight of the grafted PS of M_{PS} = 900 g/mol were found to preferentially disperse within the PS domain of the BCP and to randomly distribute within the domain. Interestingly, nanocrystals of equal size, composition and volume filling fraction, when intermixed with lower molecular weight PS-PEP (M = 80 kg/mol) were found to sequester to the center of the PS domain, in agreement with theoretical arguments that entropic effects related to the relative characteristic length scales of particles and polymers determine the location of the inclusions within the microstructure. Transmission electron micrographs of both morphological types along with schematic illustrations of the microstructure are depicted in Figure 5. Note that the implications of the material length scales on the final composite morphology are symmetric with respect to particle and block copolymer size. Figurc 6 dcpicts the various microstructures that are observed for blends of high molecular weight PS-PEP (M = 800 kg/mol) with aliphatic-coated gold and silica particles of similar surface chemistry but distinct size. Aliphatic-coated gold nanocrystals (d = 3 nm) are found to segregate to the interface between the domains, however, increasing the particle's size results in center-alignment of the nanoparticles, suggesting an analogous effect of increasing the particle's diameter than reduction of the BCP domain spacing. This is in support of numerical simulations that suggest the particle-to-polymer domain size ratio d/L as the structure determining quantity. As the particle size continues to increase, particle aggregation and macrophase separation of the filler particles is observed as can be seen in Figure 6c for a particle-to-polymer domain size ration of d/L = 0.15. The different morphological characteristics of the respective particle additives can be combined to form hierarchically ordered multicomponent nanocomposite materials [48]. Figure 7 depicts the microstructure of a ternary blend of PS-PEP and gold as well as silica particle additives. Both particle constituents retain the morphological characteristics that are observed in the respective binary blends (see Figure 6) – gold nanocrystals segregate to the IMDS and silica nanocrystals to the center of the PEP domain. Analogous effects of particle size on the composite morphology have been observed in blends of PS-PEP with aliphatic-coated gold ($d = 3$ nm) and cobalt nanocrystals ($d = 20$ nm) as illustrated in the STEM micrographs depicted in Figure 8. The exclusive dependence of the morphology on the characteristic length scales of the filler particles (for similar surface chemistry) demonstrates the versatility of the self-organization approach to engineer multicomponent composite materials in which the various nanoparticle filler particles assume specific locations within the microstructure depending on their size and surface chemistry.

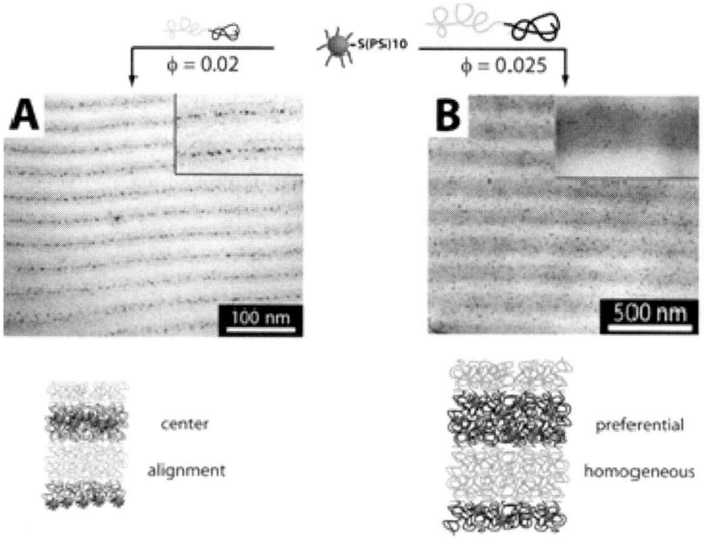

Figure 5. Bright-field transmission electron micrograph and schematics of the respective microstructure demonstrating the dependence of the composite morphology on the block copolymer molecular weight. Panel A: center morphology (PS-PEP, 80 kg/mol intermixed with AuSPS, $\phi \sim 2\%$ w/v). Panel B: homogeneous preferential morphology (PS-PEP, 800 kg/mol intermixed with AuSPS, $\phi \sim 2\%$ w/v).

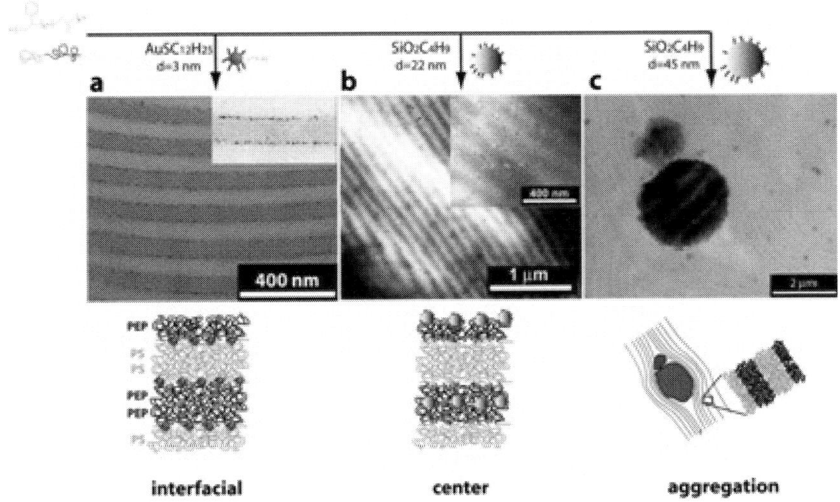

Figure 6. Bright-field transmission electron micrograph and schematics of the respective microstructure demonstrating the dependence of the composite morphology on the particle size. Panel a: interfacial segregated morphology (PS-PEP, 800 kg/mol intermixed with Au-$SC_{12}H_{25}$, d = 3 nm, $\phi \sim$ 2% w/v). Panel b: center morphology (PS-PEP, 800 kg/mol intermixed with SiO_2-C_4H_7, d = 22 nm, $\phi \sim$ 2% w/v) and Panel c: particle aggregation (PS-PEP, 800 kg/mol intermixed with SiO_2-C_4H_7, d = 45 nm, $\phi \sim$ 2% w/v)

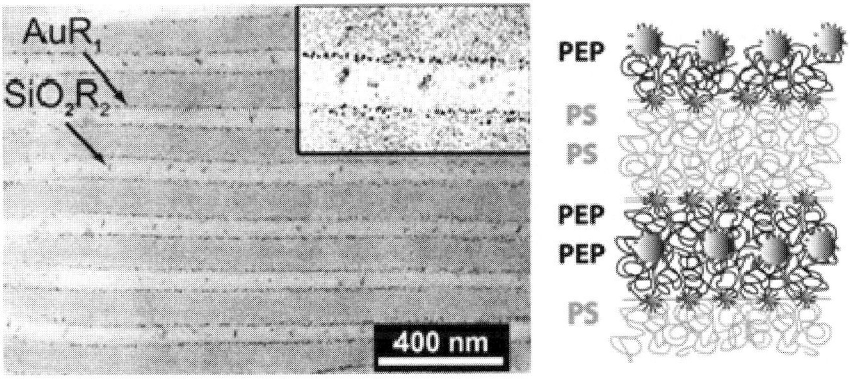

Figure 7. Bright-field transmission electron micrograph and schematics of the hierarchical microstructure of a ternary blend of PS-PEP intermixed with Au-$SC_{12}H_{25}$ (d = 3 nm, $\phi \sim$ 2% w/v) and SiO_2-C_4H_7 (d = 22 nm, $\phi \sim$ 2% w/v). Gold particles segregate to the IMDS, silica particles to the center of the PEP domain (bright). (Adopted with permission from [Ref. 48]; copyright 2003, American Chemical Society).

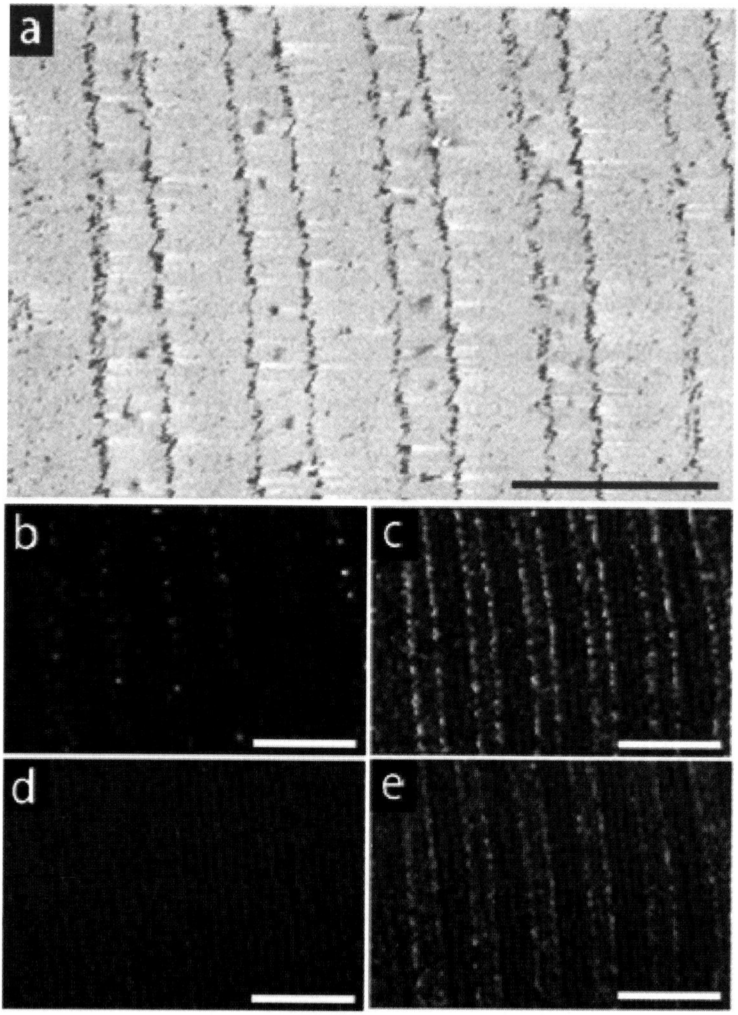

Figure 8. STEM analysis of the PS-PEP/AuSPS/CoTOPO ternary composite microstructure revealing morphology and elemental distribution. Panel a: bright field micrograph revealing the lamellar morphology as well as interfacial and center location of embedded nanoparticles (scanning direction is along picture diagonal). Panel b: X-ray map of cobalt distribution (K-edge signal) confirming center distribution. Panel c: X-ray map of gold distribution confirming interfacial location of Au nanoparticles (M-edge signal). Panel d: X-ray map of silicium confirming low impurity level (K-edge signal). Panel e: X-ray map of sulphur, confirming Au distribution in Panel c (K-edge signal). Scale bar is 250 nm.

The possibility to control of the localization of nanocrystals within BCP microstructures by tailoring the composition of the surface-bound surfactants was recently demonstrated by Kramer and coworkers. Gold nanocrystals of equal size were functionalized with low molecular weight poly(styrene), poly(2-vinylpyridine) (P2VP) as well as an equal mixture of both and subsequently blended with a lamellar PS-P2VP to form microstructured composites in which the nanocrystals were preferentially located at the center of the PS, P2VP or along the inter-material dividing surface (IMDS) separating both domains [49,50]. The latter observation can be rationalized by the gain of interfacial free energy upon sequestration of the mixed-surfactant-coated nanoparticles to the interface, acting similar to surfactant molecules in stabilizing a phase-separated blend. For polymer blends Pieranski demonstrated that particles will segregate to the interface if $|\sigma_{A/NP} - \sigma_{B/NP}| < \sigma_{A/B}$ where $\sigma_{X/Y}$ denotes the interfacial energy between component X and Y [51]. Kramer and coworkers also demonstrated a pronounced effect of the surfactant grafting density on the morphology of particle/BCP composites. Gold particles with various grafting densities of low molecular weight PS were blended with PS-P2VP BCP [52]. For high grafting densities, the sequestration of the particles to the center of the PS-domain was observed whereas decreasing grafting density resulted in segregation to the BCP interface. Figure 9 depicts electron micrographs of the respective particle distributions. Similar results were obtained by experiments on blending PS-functionalized fullerenes (C_{60}) with lamellar poly(styrene sulfonic acid–b–styrene) copolymer that demonstrated center-alignment of particle additives with increasing molecular weight and grafting density, suggesting that a dense packing of a high-molecular weight surface-bound polymer acts to increase the effective particle size similar to an increase of the particle core diameter [53]. This can be rationalized by drawing analogy to the well-studied class of topologically related star polymers that consist of a number f of branches joined together at one junction. Using scaling arguments, Daoud and Cotton as well as Birshtein and Zhulina, demonstrated that star polymers are comprised of three regions: an inner core region extending to $r_{\text{core}} \sim af^{1/2}$, with a being the monomer size, that is characterized by about complete space filling of polymer segments as well as outer regions in which –

depending on the solvent quality – the conformation of the arm polymers is either ideal or swollen [54,55]. Assuming that the grafted polymers are of sufficient length and the number of arms exceeds a threshold value such that $2r_{core} > d$ then increasing the grafting density by some factor x will effectively increase the diameter of the impenetrable core-region by a factor $\sqrt{x}$. Figure 10 illustrates the analogy.

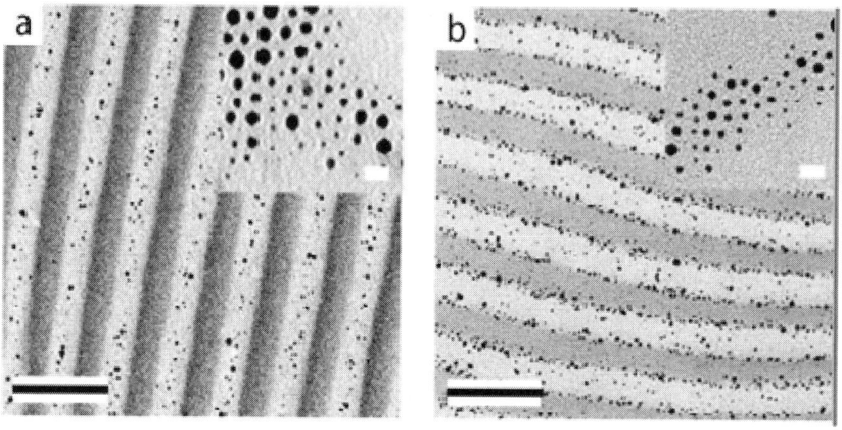

Figure 9. Effect of PS grafting densities on particle location in PS-PVP block copolymer template. Panel a: grafting density is 1.7 chains/nm^2. Panel b: grafting density is 1.2 chains/nm^2. Particles in (a) are dispersed only in the PS domain while those in (b) are located mainly at the PS-b-P2VP interface. Scale bar is 100 nm in the main image and 10 nm in the inset. (Adopted with permission from [Ref. 50]; copyright 2006, American Chemical Society).

When blending polymer-coated particles with a BCP matrix, the addition of particles acts to effectively increase the volume filling fraction of the respective surface-bound polymer component and results in an order-order transition when the concentration of the particle filler exceeds a threshold value (that depends on the BCP composition). For example, Figure 11 shows a lamellar-to-cylinder transition in a blend of PS-P2VP/AuSPS that was reported by Kim *et al.* [56]. Similarly, a cylinder-to-sphere transition was reported by Yeh *et al.* in a blend of poly(styrene-b-ethylene oxide) copolymer with hydroxyl-mercapto functionalized CdS nanocrystals [57,58]. Recently, the same group of authors also demonstrated a cylinder-to-lamellar transition upon blending of mercapto-carboxylic acid-functionalized CdS nanoparticles with the minority P4VP domain of cylindrical PS-P4VP copolymer [59]. This latter approach also introduces the concept of capitalizing on directed interactions between particle-surfactants and the target polymer domain (here: by hydrogen bonding) in order to increase the gain in free enthalpy associated with the particle dispersion and thus to enhance the efficiency and selectivity of the sequestration process.

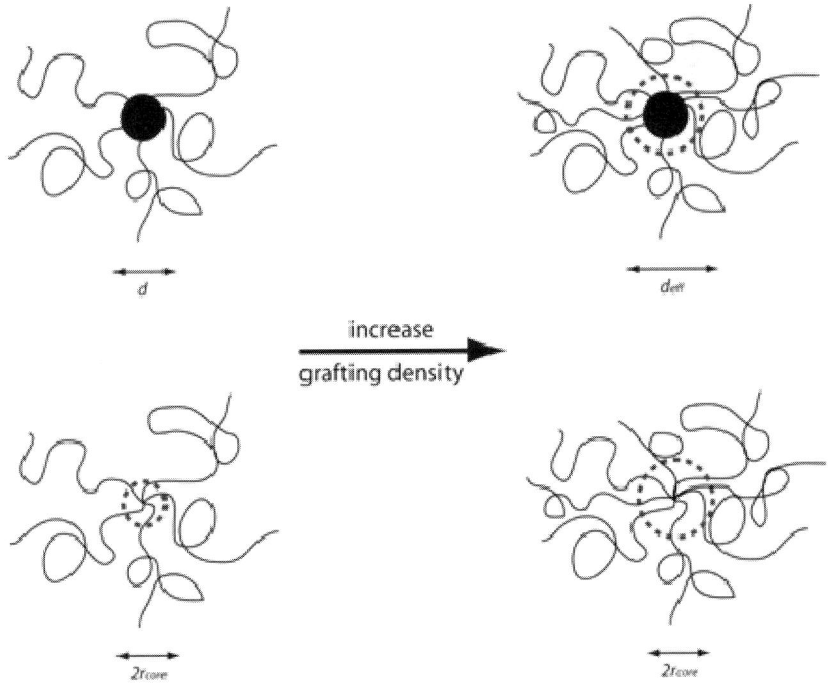

Figure 10. Illustration of analogy between star-polymer architecture and polymer-grafted nanoparticles with high grafting density. Increasing the grafting density of polymer-coated nanoparticles by factor of x increases the effective core diameter by factor $x^{1/2}$.

While the idea of using strong directed interactions to guide the self-organization process is intriguing, the available experimental data is not conclusive and experimental findings on the PS-P4VP/CdS system indicate a rather pronounced aggregation of particles within the P4VP domain. It is not clear whether the aggregation is to be attributed to distortion effects of the particles on the respective polymer domain due to a particle-to-polymer domain size ratio of $d/L \sim 0.4$ (similar to the observed size dependence of the blend morphology in mixtures of PS-PEP/SiO_2, see Figure 6), or to a subtle interplay between particle architecture and particle-particle/particle-polymer interactions. More research will be necessary to establish this approach.

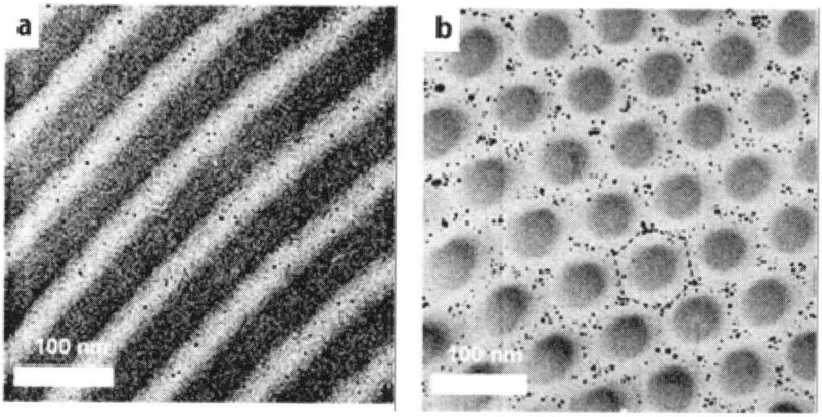

Figure 11. Bright-field electron micrograph of thin films of PS-P2VP with added PS-coated gold nanocrystals. Nanocrystals preferentially sequester within the PS domains (bright) and induce an order-order transition from lamellar to cylindrical morphology at particle filling fractions exceeding a threshold filling factor (here: $f = 40\%$ w/v). Panel a: lamellar morphology. Panel b: cylindrical morphology. Larger gold nanocrystals are found to sequester along the center regions of the PS domains (indicated by dotted line). (Adopted with permission from [Ref. 6]; copyright 2005, Wiley-VCH).

Future progress in the field of BCP/particle composites can greatly benefit by drawing analogy to phenomenological related material systems. The observation of larger particles segregating to the center of the respective polymer domain is reminiscent of the results of earlier studies on the polymer chain distribution in BCP/homopolymer (BCP/hP) blends. For a system comprising blends of PS-PI/hPS and hPI Winey *et al.* demonstrated that with increasing molecular weight, homopolymer additives will tend to concentrate along the center of the respective polymer domain – similar to the segregation characteristics of larger particle inclusions in lamellar BCP (as seen in Figure 6) [60]. By analyzing the curvature of the IMDS, Thomas and coworkers deduced the segregation of high molecular weight hPS additives in a cylinder-domain forming P2VP-I-S triblock copolymer to the center regions of the styrene domain, in analogy to the observed segregation of larger gold nanocrystals to the center regions of the respective polymer domains as shown in Figure 12 [61]. The swelling of BCP with hP additives has been rationalized by Hashimoto and coworkers using a wet-brush/dry-brush model similar to the Alexander - De Gennes brush [42,62-64]. The key parameter in this reasoning is the ratio of molecular weights between the respective BCP-component and the homopolymer additive.

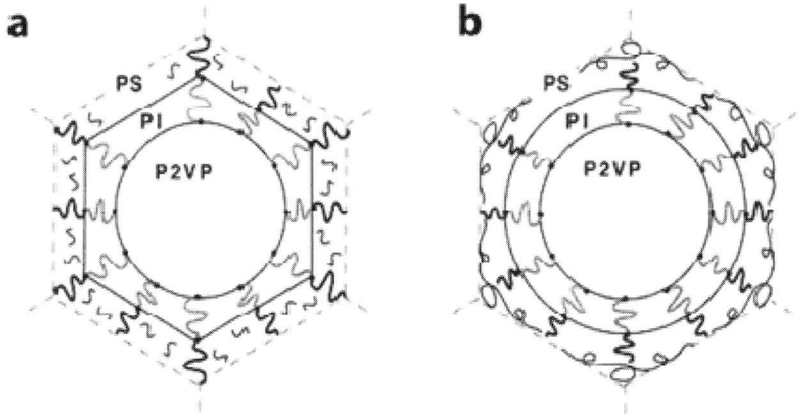

Figure 12. Schematic representations of the block copolymer a) P2VP-PI-PS/hS(4000 g/mol) and b) P2VP-PI-PS/hS(50000 g/mol) chain conformations in a single microdomain observed in parts a and b, respectively. In each schematic, the boundary of the Wigner-Seitz cell is indicated as dashed gray lines. The PI and PS blocks are shown by thick gray and dark curves, respectively. hPS is shown by thin dark lines. Note that the polymer chain trajectories are artificially constrained to lie within one Wigner-Seitz cell. (Adopted with permission from [Ref. 61]; copyright 1998, American Chemical Society).

Three types of segregation are distinguished that act to minimize the free enthalpy penalty that is associated with the non-uniform perturbation of equilibrium chain conformations of the BCP upon accommodation of the additive: small molecular weights ($M_{hS}/M_{BCP\text{-}S} \ll 1$) are predicted to uniformly swell the BCP domain 'S', intermediate molecular weight additives ($M_{hS}/M_{BCP\text{-}S}$ ~ 1) are predicted to segregate to the center of the S-BCP domain whereas for large homopolymer additives ($M_{hS}/M_{BCP\text{-}S}$ >> 1) macrophase separation is predicted. The situation is depicted in Figure 13. Note that the analogy between low molecular, polymer and particle additives is supported by recent numerical simulation that suggest the inclusions excluded volume to be a key parameter in the segregation process, independent of the particular geometry of the sequestered component [29]. It will be interesting for future studies to further explore the similarities between BCP-particle and BCP-homopolymer blends, since a large amount of data has been collected for the latter and the ease of imaging of high-electron density nanoparticles could facilitate the testing of rather speculative conclusions from earlier experiments.

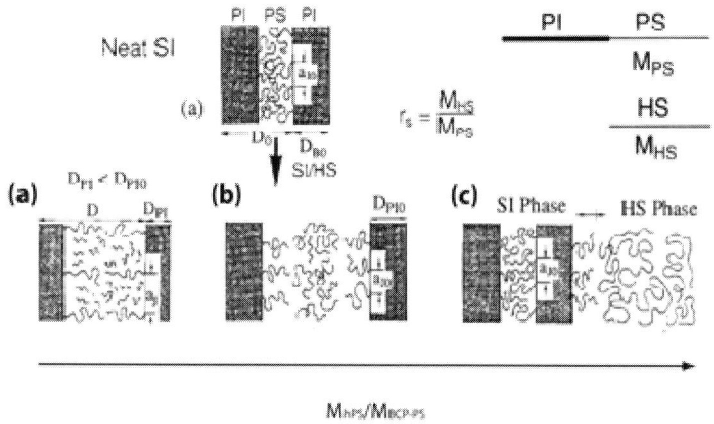

Figure 13. Schematic of the proposed wet-brush/dry-brush transition for BCP/homopolymer blends (here: PS-PI). Panel a: homogeneous distribution of hPS within PS domain (wet-brush) for $M_{hPS}/M_{BCP\text{-}PS}$ <<1. Panel b: center distribution of hPS within PS domain (dry-brush) for $M_{hPS}/M_{BCP\text{-}PS}$ ~ 1. Panel c: incompatibility of hPS in PS-PI domain for $M_{hPS}/M_{BCP\text{-}PS}$ >> 1. (Adopted with permission form [Ref. 63]; copyright 1990, American Chemical Society).

A further analogy that could provide insight into the nature of size-driven particle aggregation in BCP matrices (compare Figures 6c) is provided by membrane-mediated attractive interactions between cell membrane-embedded protein inclusions [65,66]. In these systems linear continuum elastic theory has been applied to demonstrate that the membrane-induced interaction between inclusions varies non-monotonically as a function of the inclusion spacing. The location of the energy minimum depends on the spontaneous curvature and the membrane perturbation decay length, where the latter is set by the membrane moduli and the membrane perturbation energy has been found to increase with the inclusion radius [67]. Figure 14 illustrates the analogy between BCP/particle and membrane/protein scenarios. We note that for platelet-type inclusions such as clay, Fredrickson and coworkers indeed could demonstrate that the elastic interactions from a particle inserted into a lamellar medium propagate over distances of a few micrometers, *i.e.,* spanning several lamellar domains [68].

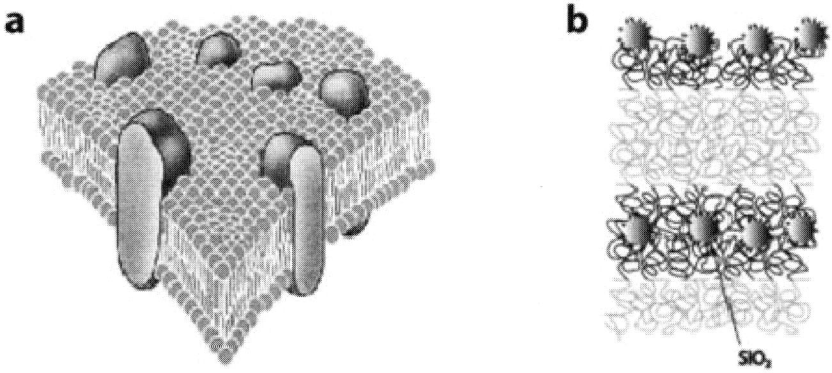

Figure 14. Architectural similarity between a) cell membrane and b) lamellar BCP/NP microstructure. Membrane proteins (gray) embedded in phospholipids bilayer (red/yellow) perturb the planar structure of the membrane similar to particle inclusions in BCP lamellae and can induce attractive interactions.

3.2. Nonequilibrium BCP / Particle Composite Morphologies

Whereas the equilibrium approach facilitates the investigation of the thermodynamic relevant material parameters that determine the structure formation process in BCP/particle blends, the long-time annealing conditions that are necessary to equilibrate the systems will often be inadequate for large-scale processing. It is thus of interest to elucidate the implications of processing conditions – such as solvent selectivity, evaporation kinetics or the application of external fields – on the resulting microstructures. However, despite its relevance, research is just beginning to explore the area of nonequilibrium morphologies in BCP/particle blends. In the following, the major results of few available studies will be presented that hopefully will stimulate further efforts in this direction.

Various morphologies of Fe_2O_3 nanoparticle additives in PS-PI matrices dependent on the solvent selectivity and evaporation kinetics were recently reported by Char and coworkers [69,70]. In neutral solvents (toluene) the formation of body-centered cubic lattice-like structures of particle aggregates was observed whereas in isoprene selective solvents (n-hexane) the sequestration of particles within the isoprene-cylinder domains of the BCP was observed. A schematic of the segregation process is depicted in Figure 15. It is not clear whether the origin of the various polymer/particle morphologies is in the different compatibility or evaporation kinetics of the respective solvents but the results strikingly demonstrate the possibilities to control the microstructure formation process by means of processing conditions.

Since the ultimate performance of BCP/NP composite materials on macroscopic length scales is intimately related to the materials' defect and grain boundary structure, understanding of the implications of particle additives on the structure evolution process will be of critical importance in order to develop strategies towards mesoscopically ordered nanocomposite materials that capitalize on the particular arrangement of the nanoscale inclusions or the polymer matrix. The occurrence of defects and grain boundaries is inherent in the self-organization of BCP (in the absence of external fields) and has been attributed to the nucleation and growth of ordered domains during the structure formation process, the superposition of stress-fields around disclinations and mechanically induced kinking or buckling of domains, *e.g.*, induced by inhomogeneous solvent evaporation during the late processing stages [71,72]. Pioneering studies by Listak *et al.* demonstrated the stabilization of high-angle tilt and twist grain boundary structures in blends of lamellar PS-PEP and PS-coated gold nanocrystals [73]. In these studies particle inclusions were found to preferentially swell free energy 'hot-spots' that occur along the grain boundary due to the local perturbation of the polymer equilibrium microstructure, thus permitting the chains to relax towards their equilibrium conformation.

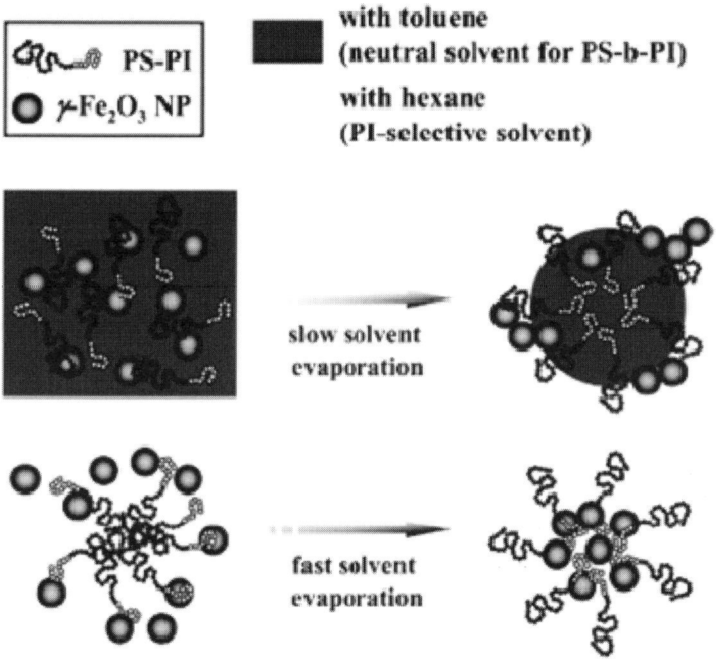

Figure 15. Illustration of solvent effect on the structure formation in blends of PS-PI and Fe_2O_3. (Adopted with permission from [Ref. 69]; copyright 2006, American Chemical Society).

The stabilization is evidenced by an increased frequency of energetically unfavorable asymmetric tilt grain boundaries (so called T-junction GB) that have been shown to be particularly unfavorable in neat BCP [74]. Figure 16 depicts an electron micrograph of a T-junction GB that reveals the aggregation of nanocrystals along the high energy T-regions. The resulting particle structures can be visualized as 3D string-like aggregates that form along the grain boundary surface as illustrated in Figure 17. As in the case of the analogies between the morphologies of BCP/NP and BCP/hP blends, the stabilization of grain boundaries by nanoparticle additives is reminiscent to results of earlier studies by Gido and coworkers of the stabilization of grain boundaries in BCP/hP blends who reported the addition of homopolymer to result in increased frequencies of high-energy grain boundary structures [75].

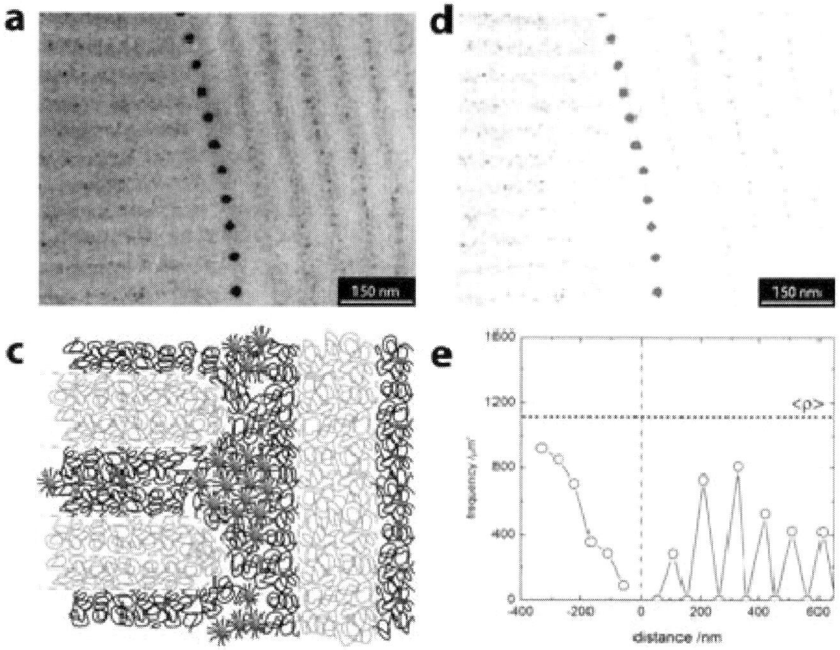

Figure 16. Electron micrograph (a) of the unstained PS-PEP/AuSPS depicting a magnified cross section of an individual T-junction grain boundary. PS constitutes local majority domains, PEP constitutes semicylindrical end-cap regions (local minority domain). Large dark dots along the grain boundary are the particle aggregates that form due to accumulation that is driven by the associated stress relief in the high-energy T-junction regions along the grain boundary (see text). Small dark dots are individual particles. Schematic (b) of microstructure. Contrast enhanced representation (d) of the micrograph shown in (a) depicting particle and aggregate positions only. Particle frequency (e) as a function of the distance and direction from the grain boundary (shown in panel a) calculated by particle counting across rectangular area elements of width Δr = 50 nm parallel to the boundary. The line joining the particle center positions along the grain boundary defines "zero" particle distance. $<\rho> = (1/N)\ \Sigma_i\ (dn_i/da_i)$ denotes the equilibrium particle density with dn_i denoting the number of particles in area section da_i and N the total number of area sections.

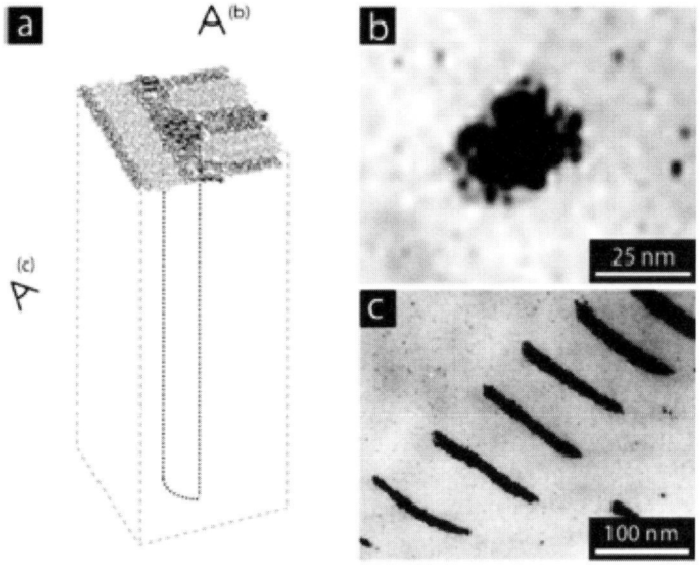

Figure 17. Illustration (a) of the 3D geometry of a T-junction grain boundary. Particles form string-like aggregates along the grain boundary surface. Cross-sectional view (b) of a particle aggregate sectioned normal to the grain boundary surface. Cross-sectional view (c) of particle aggregates sectioned in off-normal direction to the boundary. (Adopted with permission from [Ref. 73]; copyright 2006, American Chemical Society).

The implications and opportunities resulting from the stabilization of non-equilibrium morphologies by particle additives are an open field for research. For example, the ease of visualization of the high-electron density cores could render particle additives suitable tracer-inclusions for the general investigation of the effects of polymer additives such as plasticizers or anti-oxidizing agents on the deformation of thermoplastic elastomers (taking advantage of the universal segregation characteristics of inclusions discussed in Reference [29]). Note that the segregation of particle inclusions to defect and grain boundary regions within a BCP microstructure resembles the 'pinning' of grain boundaries in polycrystalline heterogeneous metal alloys, *i.e.,* the interception of grain growth by particulate segregation to the grain boundaries or to the attraction of solute atoms to dislocation core regions in metal alloys to form a Cottrell atmosphere around the defect zone [76-78]. Since both of these phenomena have been associated with strengthening mechanisms in metal alloys it will be interesting to explore the implications of grain boundary mediated particle aggregation on the mechanical properties of BCP thermoplastic elastomers for which the formation of kink grain boundary structures has been identified as an early failure state [79].

Better understanding of the structure-property relations in BCP/NP composites will greatly benefit from the ability to achieve large area alignment of the composite microstructure. Various techniques have been developed to achieve macroscopic alignment in BCP materials by coupling mechanical flow fields to the materials either during the self-organization process or post-organization, *e.g.,* by extrusion, reciprocating shear, press-molding or roll-casting [80-83]. Despite the obvious opportunities for improving material performance as well as for the preparation of model structures to better understand the implications of particle additives on the properties of BCP composites, the application of flow-field processing techniques to enhance the alignment of BCP/NP blends is a nascent field. Ha *et al.* investigated the microstructure formation of roll-cast lamellar PS-PI-PS/clay-PS composites and reported a flipping transition that was attributed to the structure-guiding effect of the large-aspect ratio particle inclusions that couple to the flow field while the BCP is still in the homogeneous state [84]. Whereas anisotropic particle inclusions impart structure guiding information on the BCP and thus result in more complex ordering phenomena, spherical point-like nanoparticle inclusions are not expected to couple with the flow field. Research will have to show if flow-field processing presents a viable route towards macroscopically ordered BCP/NP hybrid materials.

Related to the observations by Ha *et al.* is the concept of imparting the structural information contained in anisotropic filler particles to control the organization of a BCP matrix material. Even though anisotropic filler particles are out of the scope of the present review article, this concept will be briefly introduced since it is related to the phenomenon of particles stabilizing grain boundary regions mentioned above. The fundamental idea is that the matching of symmetries of particle inclusion and BCP matrix will direct the microphase separation process. Sevink *et al.* demonstrated theoretically that large sheet-like inclusions will accelerate the organization of symmetric BCP [85]. Silvia *et al.* explored the role of clay filler particles on the microstructure formation of PS-PI copolymers and reported a faster kinetics of the order formation in the case of spherical BCP morphologies [86]. The authors reasoned an increased heterogeneous nucleation rate to be the origin of their observations. More recently, Laicer *et al.* demonstrated the possibility to control the nucleation and growth of cylindrical BCP microstructures by particle-surface mediated templating effects. Using micron-sized rod-like filler particles the authors demonstrated the heterogeneous nucleation of a cylindrical PS-PI copolymer and suggested the orientation of the microdomain structure to follow the particle major axis [87]. Future research will have to demonstrate if similar effects can be realized for nanoscale particle inclusions.

As a final note about nonequilibrium morphologies of BCP/NP blends it should be mentioned that the differentiation of both, equilibrium and nonequilibrium states, implies the ability to equilibrate BCP/NP blends – a process that requires annealing of the system for times long enough to assume the minimum free energy configuration. Here, 'long-enough' relates to time scales much greater than the characteristic relaxation time of the material, which in this case will be determined by the diffusion times for polymer and particle constituents. Unfortunately, at present there is insufficient knowledge about the mobility of polymer-coated particles in polymer matrices such as to facilitate accurate *a priori* predictions. Experiments by Cole *et al.* indicate that the diffusion mechanism critically depends on the size of polymer and particle components as well as the mutual interaction and diffusion times can be dramatically different from predictions using the classical Stokes-Einstein relation [88]. In order to provide a quantitative understanding of equilibrium and nonequilibrium phenomena in BCP/NP systems it will thus be essential that future research establishes the dynamic properties of composite polymer materials.

4. Structure-Property Relations and Applications of BCP/NP Hybrid Materials

From an application-oriented point-of-view much of the interest in BCP/NP composite materials is because of the intriguing opportunities that arise from the combination of the particular physical properties of nanoscale materials with the processability and autonomous microdomain formation of BCP. As the length scales of materials shrink below the characteristic length scales of physical processes such as the electron mean free path (for metals), the exciton radius (for semiconductors) or the critical magnetic domain size (for magnetic materials) novel properties arise that can be imparted on a polymer matrix. Figure 18 illustrates the typical range of characteristic length scales for a selection of material properties [89].

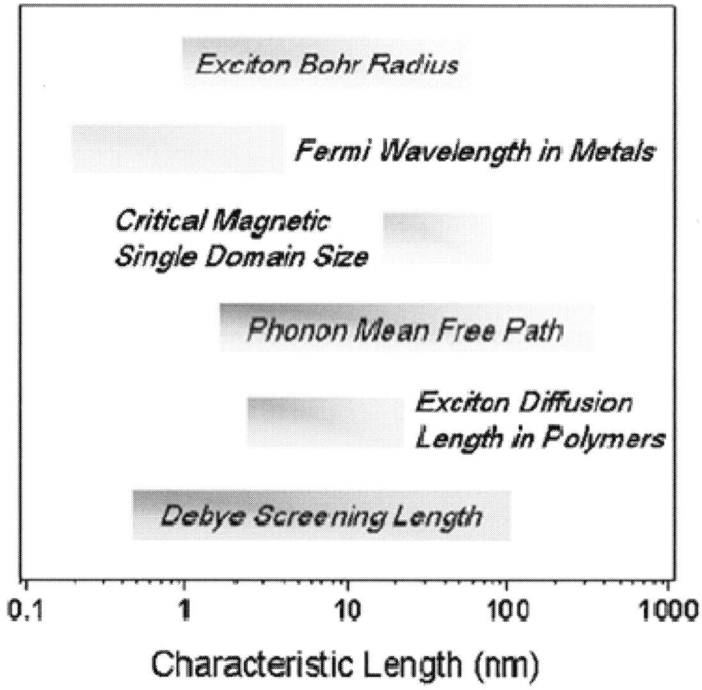

Figure 18. Representation of the characteristic length scales associated with relevant physical properties.

When blending nanoparticle additives with BCP, additional opportunities for the control of the composite properties arise because of the possibility to selectively locate the nano-inclusions within the domain microstructure. In this chapter, recent research on physical properties of BCP/NP composites will be reviewed with particular focus on optical properties of BCP/metal-NP composites that have received by far the most attention. Note that the neglect of anisotropic particle/BCP composites is not meant to be an assertion of lesser relevance but rather follows the spirit of this review. In fact, the opportunities to increase the efficiency of enhancing the tensile modulus or barrier properties by preferentially orienting anisotropic filler particles by means of structure-guiding host materials is a major stimulus for research on BCP composite materials. For a more complete treatise of anisotropic filler/BCP composites the reader is referred to Reference [6].

4.1. Properties Capitalizing on Effective Properties of Randomized NP Inclusions

The optical properties of BCP/nanocrystal composites were investigated experimentally by Bockstaller *et al.* in the context of BCP-based photonic crystal materials [45,90]. The key idea in these experiments was to increase the refractive index contrast between adjacent polymer domains of a BCP by selective sequestration of high-refractive index nanocrystals into one of the polymer domains (preferably the domain with the higher refractive index) and thereby to increase the materials efficiency to reject light. When particles much smaller than the wavelength of light are randomly dispersed in polymer matrices, the optical properties of the resulting composite materials can be approximated by means of weighted volume averages of the properties of its constituents. The theoretical framework behind the averaging process is provided by effective medium theories such as the Maxwell-Garnett or Bruggemann theory or, more recent numerical schemes that extend the application of effective medium concepts to anisotropic particle inclusions [91-94]. Figure 19 illustrates the idea behind the effective medium approach (also homogenization approach) and provides an estimate of the accessible real and imaginary part of the refractive index in a PS/Au-NP composite material with a metal filling fraction up to 20%. While providing a proof of concept, the experiments of Bockstaller *et al.* also point to subtle considerations that need to be taken into account when designing BCP/NP composites for optical applications.

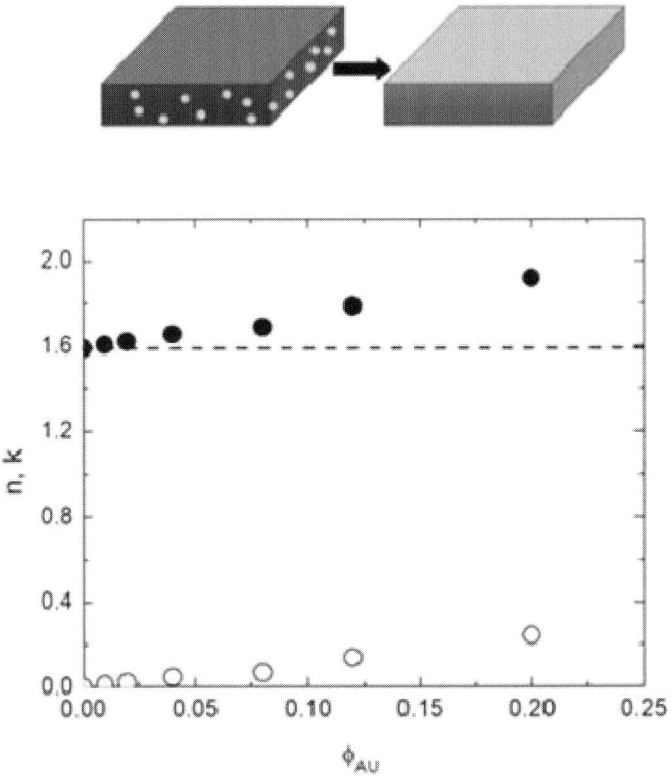

Figure 19. Schematic presentation of the homogenization approach to estimate optical properties of polymer/particle composites by representing heterogeneous material as homogeneous effective medium. Diagram shows the calculated real *n* (●) and imaginary *k* (○) part of the effective refractive index of a blend of PS and gold particles following [Ref. 91] (see text).

Based on the discussion of BCP/NP composite morphologies in section 3, random particle dispersion can only be expected in the limit $d \ll L$, *i.e.,* typically for particle sizes equal or less than about 4 nm. In this size range, properties of materials become a sensitive function of the size and shape of a particle inclusion. For example, in nanoscale metal particles (and similar for semiconductors), the spatial confinement of the free electrons results in additional damping mechanisms that dramatically affect the dielectric properties, as illustrated in Figure 20 [95-97]. The size effect on the dielectric properties of nanoscale materials has dramatic implications on the attainable optical properties of the composite materials. Figure 21 depicts the calculated reflectance spectra of a lamellar stack corresponding to a PS-PEP copolymer with gold-NP filled PS domains assuming equal volume filling fraction (ϕ = 0.1) but a particle size of 3 nm and 30 nm (about equal to the bulk electron mean free path), respectively. The maximum attainable reflectivity is found to decrease by 50% in case of the smaller filler particle additives as compared to the larger particle homologue. Note that the relevance of particle size effects on the properties of polymer composites is not particular for BCP-based nanocomposites but also holds for alternative synthetic routes such as sequential deposition [98].

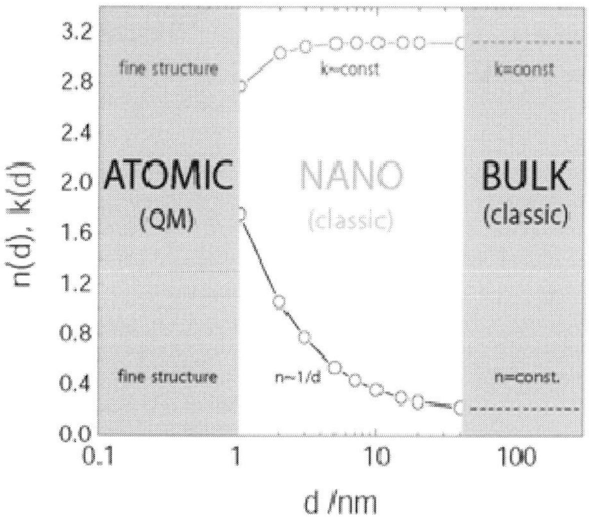

Figure 20. Real n and imaginary k part of the refractive index of gold at a wavelength λ = 630 nm as a function of particle size. Pronounced size dependence is observed when particle sizes are in the range of the electron mean free path, *i.e.*, the characteristic length scale determining optical properties of free electron metals. For particle sizes below the electron mean free path (about 30 nm) additional damping processes result in increasing real part of the refractive index.

Thompson *et al.* studied the effect of spherical nanoparticle additives on the mechanical properties of BCP/NP composites using a combined SCF/DFT/finite element approach and conclude that the addition of particle additives acts to reduce the tensile modulus of the composite [99]. This result is not unexpected, since in the case of point-like particle inclusions stress cannot be transferred from the polymer matrix to the filler particle and thus particles merely dilutes the matrix. Mechanical reinforcement of polymers can be achieved by anisotropic particle filler particles such as clay or carbon nanotubes and the reader is referred to Reference [6] for a more detailed discussion of this topic.

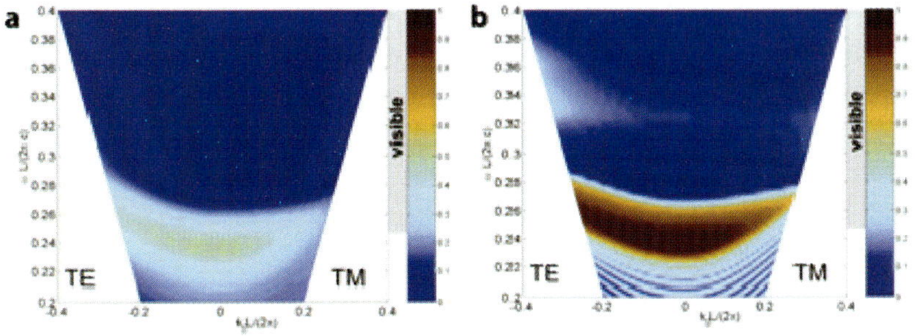

Figure 21. Calculated frequency dependence of reflectivity for TE and TM polarization for a PS-PEP stack with a gold crystal filling fraction of $\phi = 0.2$ in the PS domain. The simulation in panel (a) assumes the corrected dielectric function of the nanocrystals for a particle diameter $d_{core} = 3.2$ nm; that in panel (b) assumes a particle diameter $d_{core} = 42$ nm. The calculation refers to a periodicity of 30 double layers, $L_{PS} = 100$ nm, $n_{PS} = 1.59$, $L_{PEP} = 80$ nm, and $n_{PEP} = 1.49$; the ambient medium is air; $L = L_{PS} + L_{PEP}$ is the lamellar thickness, ω denotes the radial frequency, c is the vacuum speed of light, and $k_{||}$ is the parallel component of the wavevector with respect to the layer orientation. Note the increase in reflectivity for larger particle sizes by factor of two (see text).

4.2. Properties Capitalizing on Cooperative Phenomena of Discrete Particle Arrangements

If metal particles segregate to narrow planar regions within a BCP microstructure (such as in case of the segregation to the IMDS shown in Figure 6a) then closely spaced particle arrangements are formed in which dipolar interactions between the filler particles can give rise to novel collective excitation modes that dramatically impact the composites' optical properties. In general, the interaction between the dipole moments that are induced in adjacent particles (*e.g.*, by irradiation of light) will result in a coupling of the plasmon resonances and a splitting with respect to the incident field direction. The absorbance characteristics of an individual gold particle pair as a function of the particle center-to-center distance are depicted in Figure 22. Although the implications of dipolar interactions on the optical properties of metal nanocrystals are well established, the potential for the development of intriguing technological applications has only recently been recognized. For example, Malynch *et al.* reported a broad-band absorption in 2D arrays of silver nanoparticles that could be attributed to dipolar interactions between adjacent particles [100].

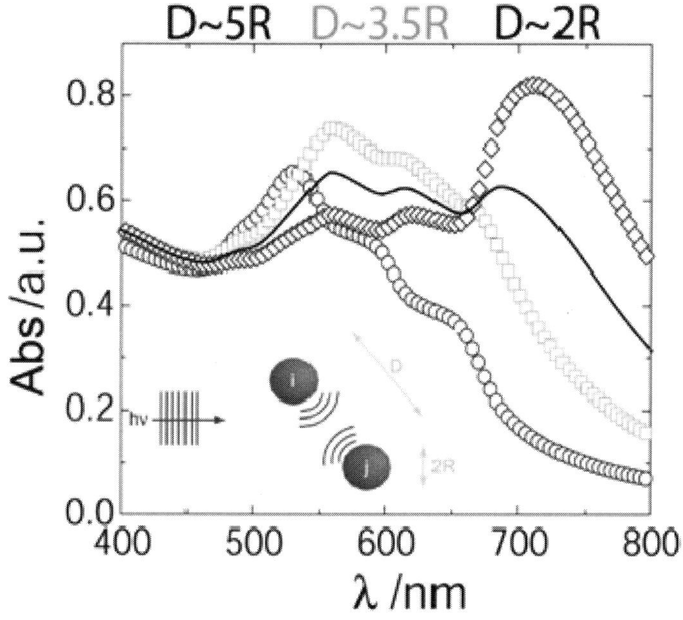

Figure 22. Absorption spectra calculated for a pair of gold nanocrystals ($d = 2R = 3$ nm) for three particle center-to-center distances D. At small distances dipolar interactions result in emergence of longitudinal plasmon absorption (see text). The continuous line corresponds to the average value of the different configurations.

Recently, Bockstaller demonstrated the sensible dependence of the optical properties of BCP/NP composites on the materials microstructure [101]. Figure 23 presents the extinction spectra of PS-PEP/gold nanocrystal ($d \sim 3$ nm) thin films with equal amount gold filler ($\phi \sim 1\%$) but different, *i.e.,* selective layer homogeneous (see Figure 5b) and interfacial segregated (see Figure 6a) as well as those of the neat nanoparticles in solution. Despite the same overall particle filling fraction a pronounced broadening and increase of absorbance characteristics is observed for the interfacial segregated sample that can be related to the small average particle distances within the densely populated interfacial areas. Note that the possibility to segregate particle inclusions to narrow regions within the BCP microstructure (such as the IMDS or lamellar, bicontinuous and cylindrical domain-center regions) facilitates the exploitation of collective excitation modes already at very dilute concentrations.

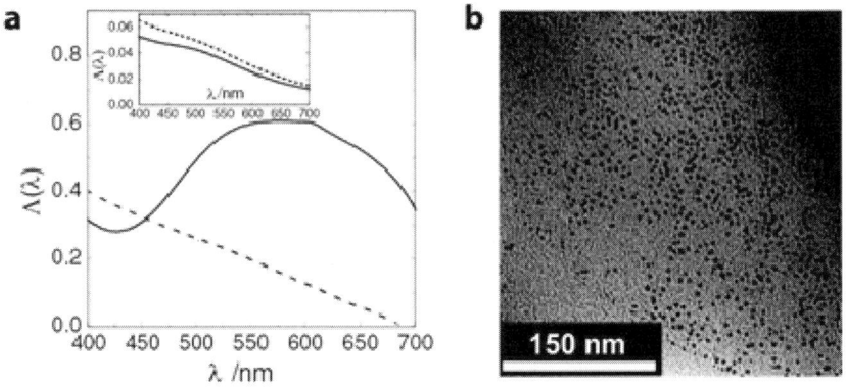

Figure 23. Panel a: absorptance spectra of block copolymer/nanocrystal composite materials with interfacial segregated morphology (dotted line, see Figure 6a) as well as selective-layer uniform morphology (continuous line, see Figure 5b). Inset: absorptance spectra of dilute nanocrystal solutions (concentration = 0.001% (w/v) in toluene) representing $AuSC_{12}H_{25}$ (dotted line) and AuSPS (continuous line), respectively. Panel b: electron micrograph depicting the dense particle arrangement along the IMDS for the interfacial segregated morphology.

New opportunities thus arise for the design of 'chimera composite materials', *i.e.,* materials that combine both, the optical or magnetic properties of densely filled materials with the mechanical properties of the neat polymer matrix. The possibility of a selective location of particle additives within BCP domain microstructures also holds opportunities for the design of polymer-based tunable magnetic materials that derive unique soft/hard magnetic characteristics from the particular structure and properties of the mesoscale polymer template. For example, the preferential sequestration of hard magnetic nanocrystals (such as cobalt or CoFe alloys) within the rubbery spherical domains of a cubic ($Im\bar{3}m$) glassy-rubbery BCP microstructure could facilitate a new platform for high resistive, tunable hard magnetic materials at temperatures below the glass transition temperature (T_g) of the rubbery domain in which minor temperature elevation above the T_g of the minority domain induces a transition to soft magnetic characteristics due to the increased particle mobility within the spherical microdomains. The idea is illustrated in Figure 24.

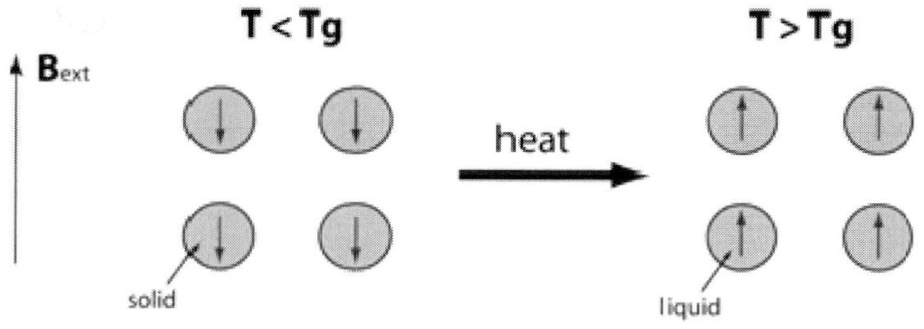

Figure 24. BCP-based tunable soft/hard magnetic materials. Magnetic hard nanocrystals (indicated by arrows) are sequestered within the rubbery spherical domains of an amorphous rubbery-glassy BCP. Below T_g of the rubbery domains the particle positions are fixed and magnetic moments cannot rotate along with external magnetic field B_{ext} (magnetic hard state). Upon heating above T_g the local environment o the particles becomes mobile and particle magnetic dipole moments can adjust by physical rotation within the spherical domains (magnetic soft state).

5. Conclusion

The incorporation of nanoparticles within a BCP matrix facilitates an unprecedented level of control over both, the spatial location and orientation of the filler particles within the matrix polymer and provides new opportunities for scientific insights on the structure formation in soft heterogeneous materials and the structure-property relations in composite materials with structural order on a hierarchy of length scales as well as a vast range of potential technological applications that are only beginning to be explored. The implications of particle additives on the properties of the composite material depend on the particle size and shape as well as the particular morphology of the particle distribution. New collective properties can emerge by interactions between the inclusions within the microstructure that give rise to performance characteristics that could not be achieved in classical homogenized composite materials. The equilibrium morphology of BCP/NP composites is determined by the interplay between the characteristic length scales of the constituents, the particle-polymer interactions as well as the size and density of the grafted ligands. Kinetic control of the microstructure formation by optimizing processing conditions such as solvent selectivity or evaporation rate provides an alternative route to control the composite morphology that holds the promise of higher efficiency and access to novel composite architectures.

In order to fully exploit the potential of BCP/NP materials future research will need to accomplish better understanding of the mechanism of microstructure formation. The chemistry of BCP could be varied to include different chain architectures (such as multiblock or multi-arm) and morphologies (such as crystalline or liquid crystalline) that impart additional structural information on particle additives. The addition of low molecular compounds such as homopolymers to BCP/NP composites could reduce the accumulation of stresses within the BCP microstructure upon particle addition and facilitate higher particle loadings. Attractive interactions between ligands and polymer matrices could potentially be used to increase the range of allowed particle sizes or filling fractions and external fields (electrical, magnetic, mechanical or thermal) to achieve macroscopic order are only beginning to be explored. Finally the range of particle inclusions could be extended to comprise organic and biological functionalities that could provide a platform for novel diagnostic or sensory materials.

Acknowledgements

The authors acknowledge financial support by the Berkman Faculty Development Fund as well as US Steel (US-X Fellowship Program). M.B. is thankful to the many inspirational discussions with Ned Thomas' group at MIT as well as their help in performing STEM measurements.

References

1. J. D. Current, in Handbook of Composites – Fabrication of Composites (Eds: A. Kelly, T. S. Mileiko), Vol. 4, Elsevier, New York 1983, 501-564.
2. J. C. Williams, in Designing, Processing and Properties of Advanced Engineering Materials, (Eds: S.-G. Kang, T. Kobayashi), Vols. 449-552, Trans. Tech. Publications Inc., Enfield, NH,7-12, (2004).
3. D. Hull and T. W. Clyne An Introduction to Composite Materials, 2nd ed., Cambridge University Press, Cambridge, UK, (1996).
4. R. Vaia and H. D. Wagner, Mater. Today, 7, 32 (2004).
5. C. P. Poole and F. J. Owens, Introduction to Nanotechnology, Wiley, Hoboken, NJ, (2003).
6. M. R. Bockstaller, R. A. Mickiewicz and E. L. Thomas, Adv. Mater. 17, 1331, (2005).
7. ACS Symposium Series 804 – Polymer Nanocomposites, (Eds: R. Krishnamoorti, R. Vaia), Oxford University Press, 2002.
8. F. S. Bates and G. H. Fredrickson, Annu. Rev. Phys. Chem. 41, 525, (1990).
9. F. S. Bates, Science 251, 898, (1991).
10. Y. N. Chan, R. R. Schrock and R. E. Cohen, J. Am. Chem. Soc. 114, 7295, (1992).
11. Y. N. Chan, R. R. Schrock and R. E. Cohen, Chem. Mater. 4, 24, (1992).
12. C. C. Cummins, R. R. Schrock and R. E. Cohen, Chem. Mater. 4, 27, (1992).
13. B. H. Sohn, and B. H. Seo, Chem. Mater. 13, 1752, (2001).
14. R. Saito, S. Okumara and K. Ishizu, Polymer 33, 1099, (1992).
15. S. Joly, R. Kane, L. Radzilowski, T. Wang, A. Wu, R. E. Cohen, E. L. Thomas and M. F. Rubner, Langmuir 16, 1354, (2000).
16. J. P. Spatz, A. Roescher, S. Sheiko, G. Krausch, M.Moeller, Adv. Mater. 7, 731, (1995).
17. R. S. Kaue, R. E. Cohen and R. Silbey, Langmuir, 15, 39, (1999).
18. S. King, K, Hyunh and R. Tannenbaum, J. Phys. Chem. B 107, 12097, (2003).
19. M. A. El-Sayed, Acc. Chem. Res. 34, 257, (2001).
20. I. W. Hamley, Angew. Chem. 42, 1692, (2003).
21. M. Lazzari and M. A. Lopez-Quintela, Adv. Mater. 15, 1583, (2003).

22. P. Du, M. Q. Li, K. Douki, X. F. Li, C. R. W. Garcia, A. Jain, D. M. Smilgies, L. J. Fetters, S. M. Gruner, U. Wiesner and C. K. Ober, Adv. Mater. 16, 953, (2004).
23. A. Jain and U. Wiesner, Macromolecules 37, 5665, (2004).
24. J. P. Spatz, A. Roescher and M. Moeller, Adv. Mater. 8, 337, (1996).
25. W. A. Lopez and H. M. Jaeger, Nature, 414, 735, (2001).
26. T. Hashimoto, M. Shibayama and H. Kwai, Macromolecules 16, 1093, (1983).
27. M. Shibayama, T. Hashimoto and H. Hasegawa, H. Kwai, Macromolecules 16, 1427, (1983).
28. J. R. Naugthon and M. W. Matsen, Macromolecules 35, 5688, (2002).
29. R. J. Spontak, R. Shankar, M. K. Bowman, A. S. Krishnan, M. W. Hamersky, J. Sameth, M. R. Bockstaller and K. Rasmussen, Nano Letters 6, 2115, (2006).
30. K. R. Shull and K. I. Winey, Macromolecules 25, 2637, (1992).
31. K. I. Winey, L. J. Fetters and E. L. Thomas, Macromolecules 25, 2645, (1992).
32. R. B. Thompson, V. V. Ginzburg, M. W. Matsen and A. C. Balazs, Science 292, 2469, (2001).
33. R. B. Thompson, V. V. Ginzburg, M. W. Matsen and A. C. Balazs, Macromolecules 35, 1060, (2002).
34. J. Huh, V. V. Ginzburg and A. C. Balazs, Macromolecules 33, 8085, (2000).
35. J. Y. Lee, R. B. Thompson, D. Jasnow and A. C. Balazs, Macromolecules 35, 4855, (2002).
36. G. A. Buxton, J. Y. Lee and A. C. Balazs, Macromolecules 36, 9631, (2003).
37. A. J. Schultz, C. K. Hall and J. Genzer, Macromolecules 38, 3007, (2005).
38. D. Liu and C. Zhong, Macromol. Rapid Commun. 27, 458, (2006).
39. E. Reister and G. H. Fredrickson, J. Chem. Phys. 123, 214903, (2005).
40. S. W. Sides, B. J. Kim, E. J. Kramer and G. H. Fredrickson, Phys. Rev. Lett. 96, 250601, (2006).
41. J. U. Kim and B. O'Shaughnessy, Phys. Rev. Lett. 89, 238301, (2002).
42. S. Alexander, J. Phys. France 38, 938, (1977).
43. J. U. Kim and B. O'Shaughnessy, Macromolecules 39, 413, (2006).
44. C. Park, J. Yoon and E. L. Thomas, Polymer 44, 6724, (2003).
45. M. Bockstaller, R. Kolb and E. L. Thomas, Adv. Mater. 13, 1783, (2001).
46. B. Hamdoun, D. Ausserre, S, Joly, Y. Gallot, V. Cabuil and C. Clinard, J. Phys. II 6, 493, (1996).
47. V. Lauter-Pasyuk, H. J. Lauter, D. Ausserre, Y. Gallot, V. Cabuil, E. I. Kornilov and B. Hamdoun, Phys. B (Amsterdam, Neth.), 241, 1092, (1997).
48. M. R. Bockstaller, Y. Lapetnikov, S. Margel and E. L. Thomas, J. Am. Chem. Soc. 125, 5276, (2003).
49. J. J. Chiu, B. J. Kim, E. J. Kramer, D. J. Pine, Journal of the American Chemical Society 127, 5036, (2005).
50. B. J. Kim, J. Bang, C. J. Hawker, and E. J. Kramer, Macromolecules 39, 4108, (2006).
51. P. Pieranski, Phys. Rev. Lett. 45, 569, (1980).

52. B. J. Kim, J. J. Chiu, G. R. Yi, D. J. Pine and E. J. Kramer, Adv. Mater 17, 2681, (2004).
53. B. Schmaltz, M. Brinkmann and C. Matis, Macromolecules 37, 9056, (2004).
54. M. Daoud and J. P. Cotton, J. Phys. (Paris) 43, 531, (1982).
55. T. M. Birshtein and E. B. Zhulina, Polymer 25, 1453, (1984).
56. B. J. Kim, J. J. Chiu, G. R. Yi, D. J. Pine and E. J. Kramer, unpublished.
57. S. W. Yeh, K. H. Wei, Y. S. Sun, U. S. Jeng and K. S. Liang, Macromolecules 36, 7903, (2003).
58. U. S. Jeng, Y. S. Sun, H. Y. Lee, C. H. Hsu, K. S. Liang, S. W. Yeh, K. H. Wei, Macromolecules 37, 4617, (2004).
59. S. W. Yeh, K. H. Wei, Y. S. Sun, U. S. Jeng and K. S. Liang, Macromolecules 38, 6559, (2005).
60. K. Winey, L. Fetters and E. L. Thomas, Macromolecules 24, 6182, (1991).
61. R. L. Lescanec, L. J. Fetters and E. L. Thomas, Macromolecules 31, 1680, (1998).
62. H. Tanaka, H. Hasegawa and T. Hashimoto, Macromolecules 24, 240, (1991).
63. T. Hashimoto, H. Tanaka and H. Hasegawa, Macromolecules 23, 4378, (1990).
64. S. Koizumi, H. Hasegawa and T. Hashimoto, Makromol. Chem. Macromol. Symp. 62, 75, (1992).
65. H. Aranda-Espinoza, A. Berman, N. Dan, P. A. Pincus and S. Safran, Biophys. J. 71, 648, (1996).
66. M. Goulian. R. Bruinsma and P. A. Pincus, Europhys. Lett. 22, 145, (1993).
67. M. S. Turner and P. Sens, Biophys. J. 76, 564, (1999).
68. J. Groenewald and G. H. Fredrickson, Eur. Phys. J. E 5, 171, (2001).
69. M. J. Park, K. Char, J. Park and T. Hyeon, Langmuir 22, 1375, (2006).
70. K. Char, M. J. Park, Abstr. Am. Chem. Soc. 231, 65, (2006).
71. S. P. Gido and E. L. Thomas, Macromolecules 27, 6137, (1994).
72. G. H. Fredrickson and K. Binder, J. Chem. Phys. 91, 7265, (1989).
73. J. Listak and M. R. Bockstaller, Macromolecules 39, 5820, (2006).
74. D. Duque, K. Katsov and M. Schick, J. Phys. Chem. 117, 10315, (2002).
75. E. Burgaz and S. P. Gido, Macromolecules 33, 8739, (2000).
76. A. H. Cottrell, in Report of the Strength of Solids 30-36, The Physical Society, London, (1948).
77. V. Y. Novikov, Scripta Mater. 42, 439, (2000).
78. J. Ando, Y, Shibata, Y. Okjiama, K, Kanagawa, M. Furshoe and M. Tomloka, Nature 414, 893, (2001).
79. Y. Cohen, R. J. Abalak, B. J. Dair, M. S. Capel and E. L. Thomas, Macromolecules 33, 6502, (2000).
80. B. Walker, Handbook of Thermoplastic Elastomers, Van Nostrand Reinhold, New York, (1979).
81. A. Skoulios, J. Polym. Sci. Polym. Symp. 58, 369, (1977).
82. E. Pedemonte, G. Dondero, F. de Candia and G. Romano, Polymer 17, 72, (1976).

83. R. J. Abalak and E. L. Thomas, J. Polym. Sci. Part B: Polym. Phys. 31, 37, (1993).
84. Y. H. Ha, Y. Kwon, T. Breiner, E. P. Chan, T. Tzianetopoulou, R. E. Cohen, M. C. Boyce and E. L. Thomas, Macromolecules 38, 5170, (2005).
85. G. J. A. Sevink, A. V. Zvelindovsky, B. A. C. van Vlimmeren, N. M. Maurits and J. Fraaije, J. Chem. Phys. 110, 2250, (1999).
86. A. S. Silva, C. A. Mitchell, M. F. Tse, H. C. Wang and R. Krishnamoorti, J. Chem. Phys. 115, 7166, (2001).
87. C. S. T. Laicer, T. O. Chastek, T. P. Lodge and T. A. Taton, Macromolecules 38, 9749, (2005).
88. D. H. Cole, K. R. Shull, L. E. Rehn and P. Baldo, Phys. Rev. Lett. 78, 5006, (1997).
89. M. Law, J. Goldberger and P. Yang, Annu. Rev. Mater. Res. 34, 83, (2004).
90. M. R. Bockstaller and E. L. Thomas, Phys. Chem. B 107, 10017, (2003).
91. J. C. Maxwell Garnett, Phil. Trans. Roy. Soc. (London), A 203, 385, (1904).
92. D. A. G. Bruggemann, Ann. Phys. 5, 277, (1909).
93. Note that effective medium concepts can be derived for many other physical properties of composite materials. The book by G. W. Milton, The Theory of Composites, Cambridge University Press, Cambridge, (2002) provides a good overview.
94. M. Maldovan, M. R. Bockstaller, C. Carter and E. L. Thomas, Appl. Phys. B 76, 877, (2003).
95. The optical properties of dielectrics are typically less affected by confinement effects up to very small cluster sizes unless changes in the crystallization behavior are observed.
96. U. Kreibig and M. Vollmer, Optical Properties of Metal Nanocrystals, Springer, New York, (1996).
97. U. Kreibig, Z. Physik 234, 307, (1970).
98. A. J. Nolte, M. F. Rubner and R. E. Cohen, Langmuir 20, 3304, (2004).
99. R. B. Thompson, R. O. Rasmussen and T. Lookman, Nano Lett. 4, 2455, (2004).
100. S. Malynych and G. Chumanov, J. Am. Chem. Soc. 125, 2896, (2003).
101. M. R. Bockstaller and E. L. Thomas, Phys. Rev. Lett. 93, 166106, (2004).

CHAPTER 8

ELECTRO-OXIDATION AND LOCAL PROBE OXIDATION OF NANO-PATTERNED ORGANIC MONOLAYERS

Daan Wouters and Ulrich S. Schubert*

*Laboratory of Macromolecular Chemistry and Nanoscience, Eindhoven University of Technology, P.O. Box 513, 5600 MB, Eindhoven (The Netherlands); *Center for NanoScience, Ludwig-Maximilians-Universität München, Amalienstraße 54, D-80799, München (Germany), Fax: (+31) 40-247-4186, E-mail: u.s.schubert@tue.nl*

Thin self-assembled monolayers are commonly used in nanotechnology applications either as resists or as building blocks in multilayer systems as well as surface energy modifiers and adhesion promoters. The present review discusses the formation and patterning of thiolate and alkylsilane monolayers on flat supports. Emphasis is on post-patterning methods using energetic beams (UV, X-ray, e-beam and ion beams) as well as local probe-base patterning methods. A special focus is directed to the local degradation and chemical conversion of self-assembled monolayers by electro-oxidation processes. The formed surface patterns can be used as templates in subsequent surface modifications reactions. These include wet etching steps but also the assembly of additional monolayers, nanoparticles, (bio)polymers and other materials.

1. Introduction

With the introduction of the Langmuir-Blodgett technique in the 1930s a method to deposit mono- and multilayers of surfactants onto surfaces was introduced [1,2]. By this technique ordered 2D arrays of molecules on a liquid-air interface are transferred to a solid support. More stable, more robust organic molecules may form spontaneously into densely packed ordered arrays onto suitable surfaces. These monolayers (ML) assemble from suitable organic precursors on a variety of surfaces that generally consist of metals, or metal oxides although monolayers may also be observed on other substrates such as graphite or polymers [3]. A number of reviews is available that describe the important mechanisms of the formation of monolayers on surfaces which in general rely on a positive interaction between the headgroup of the molecule with the surface (Figure 1) [4-7].

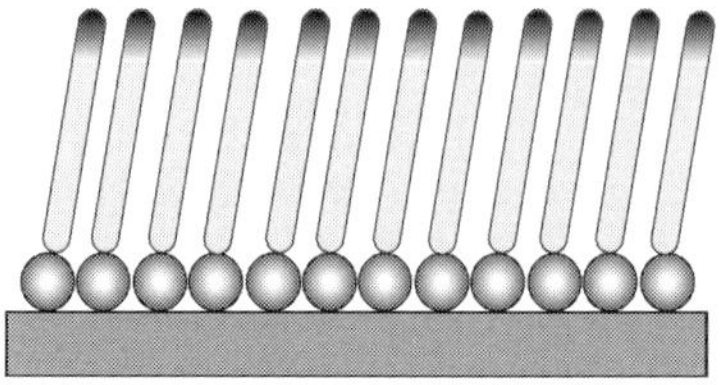

Figure 1. The formation of monolayers is possible on substrates for which a positive interaction exists between the surface active head-group of the molecule and the surface. Densely packed monolayers may be formed for molecules with positive intermolecular interactions and without bulky head groups.

These interactions may be non-covalent interactions as such ionic interactions as, for example, the interaction between carboxylate anions and metaloxide cations in the formation of (mixed) monolayers of fatty acids on metal oxide surfaces. Interactions can also be of more covalent nature as for monolayers of thiolates on gold surfaces or for the coupling of alkenes onto hydrogen terminated silicon surfaces. A dense packing may be obtained in case a positive interaction between the molecules themselves exists within these monolayers. Therefore the heat of association must exceed the net loss of entropy for these molecules when confined in a 2D array. Interactions between the molecules may be based on Van-der-Waals interactions and/or solvophobic effects [8]. Particular suitable for these purposes are long *n*-alkane chains as the free energy gain per $-CH_2$ unit is approximately 3.5 kJ/mol [9]. These systems are also preferable for steric reasons as the diameter of these chains fits reasonable well with the interatomic distance of the anchoring points on the substrate. The presence of bulky endgroups prevents high surface coverage and disturbs the dense packing within these monolayers (see below).

Although self-assembled monolayers are typically only a few nanometers thick they may dramatically change the properties of the substrate. Properties such as surface energy and related properties as wettability and adhesion, but also electrical and optical properties as well as resistivity to environmental influences and reactivity to (bio)chemical reagent may be altered. Applications of monolayers can be found in microelectronics and photolithography processes where MLs may be used as thin resist layers or in column chromatography to reduce a-specific column interactions. The ability of MLs to introduce specific chemical functionality at the substrate-gas interface also allows the substrates to be used in fields such as biosensors [10], as adhesion promoters in coating science, as templates in nanostructure sciences, as stabilizers of nanoparticles [11] or as substrates in fundamental studies into important self-assembly processes in nature.

For many of these applications the ability to pattern and to control the chemical endgroup functionality of monolayers on surfaces is of essential importance. Control in the endgroups of the monolayers can be obtained by selecting suitable starting materials. However, the variety here is limited due to limitations in 1) reactivity of the head-group towards the surface active group and 2) stability or bulkiness of the head group preventing the formation of stable monolayers. An alternative strategy to incorporate these bulky hcadgroups into a monolayer and/or to reduce their concentration is to prepare mixed monolayers. Another strategy to form monolayers with reactive endgroups is relying on chemical endgroup conversion after the monolayer formation of a suitable precursor. Patterning of the monolayers may be obtained already in the monolayer preparation stage by, for example, microcontact printing techniques or by the post-modification of the endgroup by selecting for example suitable masks and suitable chemical reaction mechanisms. The present review deals with the formation of such patterned functional monolayers. By starting with relatively simple and well-controlled monolayers of suitable precursors we will describe the available electro oxidative and reductive methods to pattern such monolayers and to introduce chemical endgroup functionality. The manuscript starts by describing some of the common monolayer preparation and characterization techniques followed by an overview of the various methods to produce electrochemical patterns on these monolayers with lateral dimensions from the micrometer to the nanometer range. Applications of the described electrochemical oxidation processes range from the use of the monolayer as an ultra thin resist to the formation of chemical active surface templates and the subsequent adsorption of a broad range of chemical functional groups.

2. Monolayer Formation

As mentioned before MLs may be formed on a large variety of substrates but undoubtedly the most widely used and most studied example of monolayer formation is the self assembly of thiolates on bare gold substrates. This has to do with the relatively ease of monolayer formation and with the stability and wide availability of a large number of thiol- and disulfide containing organic molecules. The field was started in 1983 by Nuzzo and Allara who reported the adsorption of bifunctional organic disulfides on gold surfaces [12]. The first report on the self-assembly of organic molecules, however, is much older and dates back to 1946 when Zisman *et al.* reported [13] the formation of oleophobic films of eicosyl alcohol, *n*-octadecyl amine, and *n*-nonadecanoic acid onto glass and clean metal (platinum, iron, chromium, etc.) surfaces from dilute hexadecane solutions. This work was followed later by Sagiv with the introduction of alkylchlorosilane monolayers on silicon dioxide in 1980 [14-17]. The more covalent nature of the thiol-gold bond and the silane bonds in the latter two cases results in the formation of more stable monolayers as compared to those of fatty acids and other surfactants on polar substrates such as aluminum oxides. For this reason these two kinds of monolayers are generally preferred for electro-oxidation and patterning applications. Therefore the present review will deal mostly with alkylsilane and thiolate monolayers. Detailed descriptions on the preparation and characteristics of these SAMs can be found in several review articles [5-7,18]. Only a short description will follow below.

2.1. Thiolate Monolayers

The most frequently used example of monolayer formation is the self-assembly of thiolate compounds on gold substrates. Despite the fact that SAMs of thiolates on gold are the most widely studied monolayers, the formation of monolayers is not limited to gold or thio compounds. Thio compounds for example, are also know to form monolayers on silver, palladium, platinum and mercury [19 and references therein]. A variety or detailed reviews that deal with the aspects of formation and characterization [5,18-22] as well as the kinetics of monolayer formation [23] is available.

The formation of monolayers of organic thiolates on gold is facilitated by the high affinity of sulfur to gold. For this rcason a large variety of thiols, disulfides, xantanes, thiocarbamates and thioethers show spontaneous adsorption to gold from either gas phase or from solution.

Although the exact mechanism is still under debate a possible interpretation of the adsorption of thiols to gold involves the oxidative cleavage of the sulphur-hydrogen bond which is accompanied by the formation of gaseous dihydrogen on clean bare gold substrates and by the formation of other reaction products such as water on gold surfaces that contain surface impurities. For dialkyl disulfides the adsorption is characterized by a simple oxidative cleavage of the disulfide bond.

$$\text{R-SH} + \text{Au} \longrightarrow \text{R-S-Au}^{+} + 1/2\, \text{H}_2$$

From the bond energies of the involved materials (i.e., R-SH, H-H and RS-Au; 364, 435, and 167 kJ/mol respectively) the overall reaction energy can be estimated to -21 kJ/mol for thiols and -100 kJ/mol for disulfides. These estimates have been confirmed by electrochemical data [6,24,25]. In fact, adsorption of *n*-alkanethiols and corresponding *n*-alkyl disulfides results in the formation of identical monolayers whereas dialkyl sulfides form less stable monolayers [26,27].

Thiol monolayers may be prepared from the gas phase as well as from solution. The first method results in less densely packed monolayers on the substrate and is limited due to the typical low vapor pressure of most thiol compounds containing long alkyl chains. Therefore thiolate monolayers are usually prepared by immersion of a clean substrate for ~24 h in dilute (~1 mmol/L) ethanolic solutions. Other solvents like acetonitrile, tetrahydrofuran and dimethylforamide may be used as well. The quality of the substrate is important in more than one way. First it should be clean and free of oil and other thiolate residues to enable defect free adsorption of the monolayers and secondly, if characterization by scanning probe methods is to be applied the substrates should be flat. Since the monolayers have thicknesses in the order of only 2 nm and changes within these monolayers are typically even smaller, the surface roughness should be minimized. Flat gold(111) substrates may be obtained from thermally [28] or flame-annealing [29] thin films but terraces are small (< 1 μm) and their relative orientation may vary. Larger flat gold surfaces may be obtained by the stripping method in which gold is evaporated onto a freshly cleaved mica substrate followed by the removal of the mica [30].

2.2. *Alkylsilane Monolayers*

The first report of well-defined silane monolayers on silicon dioxide was published in 1980 by Sagiv [15], who reported the formation and structure of oleophobic monolayers of OTS from organic solvents onto glass substrates. Like monolayers of thiolates on gold substrates monolayers of alkyl silanes on SiO_2 are much more robust than the earlier known LB-films, which in both cases results from a direct chemical coupling of a surface active headgroup to the substrate. Unlike the charge-transfer-like nature of the thiolate-gold bond (R-SH^- Au^+), alkylsilanes ($RSiX_3$, where X = Cl, OMe, OEt) form covalent bonds with hydroxyl groups on the surface. These silanes react with protic materials, e.g. forming alkoxysilanes with alcohols, aminosilanes with amines and silanols with water. Chlorosilanes, being the most water sensitive, forming hydrochloric acids upon reaction will react fastest, followed by methoxy and ethoxy silanes, whose hydrolysis requires the removal of alcohols. The latter two reactions taking hours rather than seconds. The silanols, in turn, will condense with each other or with hydroxylated surfaces. These competing interactions govern the result of the self-assembly process and need to be controlled in a reliable fashion.

Already at the time OTS monolayers were first reported, the importance of water in the process of the monolayer formation was realized. When the reaction is performed in the presence of water, the chlorosilane undergoes a hydrolysis followed by partial polymerization and condensation on the substrate. In case of di- and tri- functional silanes, polymerization will take place. The water may originate from the solvent, the air or being present on the surface as a thin layer.

The protocols for forming well defined monolayers are very diverse in the available literature, though all of these procedures are somehow aimed at controlling and ensuring reproducibility of the amount of water involved in the process as excess water will lead to polymerization of the chlorosilanes. Reaction times may vary from seconds to hours, depending on the amount of water in the solvent and on the degree of hydroxylation on the substrate; often mixtures of solvents (CCl_4, $CHCl_3$, CH_2Cl_2, toluene, benzene, hexanes, and *bis*-cyclohexane) are employed. In experimental conditions using a dry solution of OTS in CCl_4 and dehydrated silica supports no reaction was observed [31].

In the first reports it was hypothesized that all individual OTS chains bind to one surface hydroxyl group, the other two being used in a horizontal cross-linking reaction, further increasing the stability of the monolayer (Figure 3a). Early models, for example by Rye (Figure 3b), suggested the alkyl chains in a monolayer are packed in a crystalline-like hexagonal packing [32]. However, long range order within a monolayer has never been observed, unlike in monolayers of OTS prepared from LB techniques [33].

In a theoretical study by Stevens [34] assuming complete coupling of silanes to surface hydroxyl groups is was demonstrated that due to steric reasons full cross-linking of the OTS chains within a monolayers is not possible. A recent view on the structure of the OTS chains within a monolayer was presented by Kajiyama who concluded that a near perpendicular orientation of the OTS chains in the monolayers is facilitates by extensive lateral cross-linking within the monolayers accompanied by occasional coupling to the substrate [33-38]. This latter view was confirmed later in a study on organosilane multilayers [39].

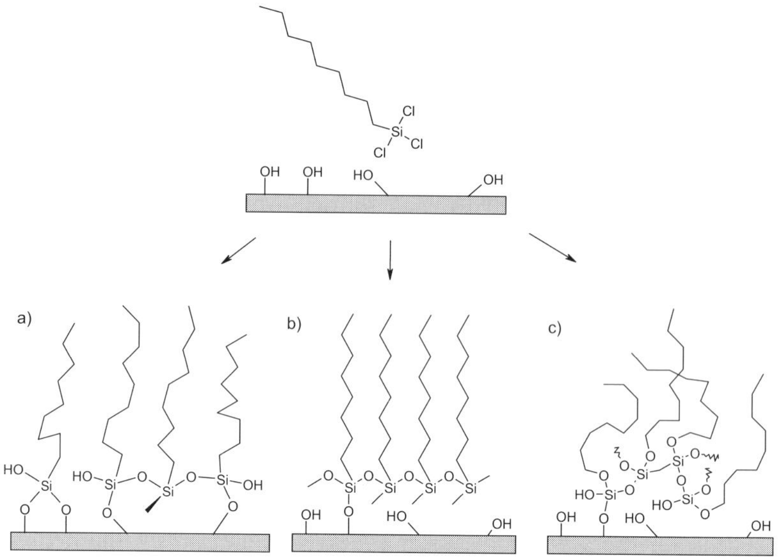

Figure 2. The resulting structure of the silane monolayer depends on the amount of water in the solvent and on the surface. Image shows a) covalently attached chains, b) horizontally polymerized islands, and c) three-dimensional polymerization.

The mechanism of the monolayer formation and the formation of partial monolayers is an important tool in the study into monolayer composition as well. Historically two mechanisms for monolayer formation from solution exist. The first mechanism supports fluid-like monolayer formation in which, in the early stages of monolayer formation, chains are randomly adsorbed onto the surface and these chains initially possess large degrees of freedom [40,41]. Upon prolonged deposition times the chains will orient perpendicular to the surface and the density of the monolayer is increased [42]. The second method suggests an island-type growth in which OTS chains orient in densely packed, dendritically shaped, domains which grow in time [43-49]. These domains may be formed directly at the substrate or may be pre-formed in solution (especially in aged solutions with solvents containing traces of water) to be deposited as a whole on the substrate upon immersion [50]. Areas in between the islands will subsequently be filled in time with more or less amorphous ordered chains.

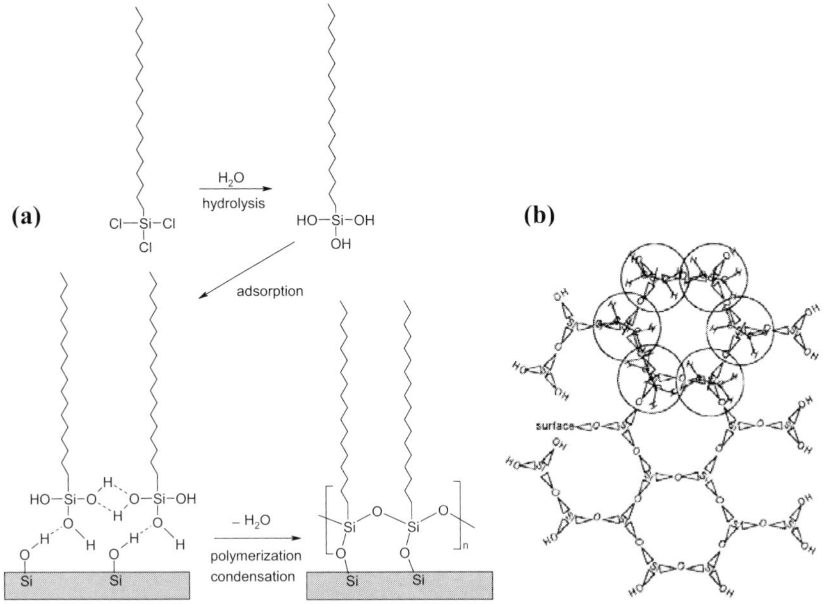

Figure 3. Schematic of (a) the original view of OTS monolayer formation on hydroxylated glass support, and (b) OTS on the hydroxylated surface [Figure 3a is adapted with permission from reference 15 (copyright American Chemical Society, 1980), Figure 3b reprinted with permission from Ref. 32 (copyright American Chemical Society, 1997)]

When the resulted monolayers are imaged by SPM techniques the presence of the islands may go unnoticed since the height difference between these different types of ordering is minimal (~ 0.1 nm) [50]. Resonant mode tapping imaging of seemingly perfect monolayers under water revealed (next to physisorbed air bubbles) small island structures within the monolayers. For many applications the presence of these islands may be not important but there are subtle differences in the properties of the monolayers inside and next to these islands for example in local oxidation processes (Figure 4).

The fact that the formation of these islands within complete OTS monolayers is difficult to observe and that the formation of them depends critically on the temperature [51], the amount of water in the solution as well as on the substrate, makes the comparison between the various procedures from different groups and the resulting quality of the monolayers very difficult.

Next to the assembly of organic monolayers on silicon dioxide it is also possible to assemble organic monolayers directly onto bare silicon supports by direct formation of Si-C or Si-O-C bonds from suitable precursor molecules. The absence of the native oxide layer in between the silicon and the organic monolayers may be advantageous for electronic applications. There are various functionalization routes available but all of them start from a bare hydrogenated silicon support which becomes available after removing the native oxide from the silicon support (i.e., by etching aqueous HF or NH_4F solutions [52,53]). From there on the surface Si-H bonds may be used directly to couple organic molecules having an unsaturated bond on the hydrogen-terminated Si by hydrosilylation reactions [54].

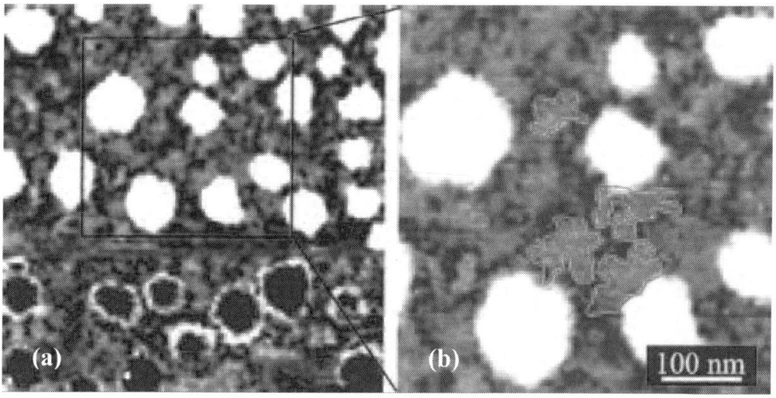

Figure 4. Liquid mode tapping images of an OTS monolayer under water. At (a) low force conditions (top half of the image), air bubbles are visible on the surface which appear with negative contrast at high force imaging conditions (bottom half of the image). At (b) high magnification and with improving contrast, island-type structures can be observed within the monolayer. A few islands are outlined by the blue line. [Image reprinted with permission from Ref. 50 (copyright American Scientific Publishers, 1997)]

This coupling process of vinyl terminated organic molecules proceeds via a free radical mechanism. The radicals in the process may originate from peroxide initiators [54], the hemolytic cleavage of the Si-

H bonds at elevated temperatures [55] or may be promoted by UV irradiation [56,57].

Alternatively, the hydrogen terminated silicon substrate can be activated by halogenation (SiX, X = Cl, Br, I) [58-61]. By activating the surface in this way monolayers of the type Si-C may be formed directly from suitable Grignard of lithiated reagents (RMgX or LiR), whereas monolayers of the Si-O-R type may be formed by reacting the halogenated silicon surface with suitable alcohols [62-64].

3. Monolayer Patterning

The patterning of monolayers can be divided into three different categories. The first, selective assembly, relies on the selective deposition of different materials (Figure 5). This may be either spontaneous phase separation of the components deposited as a mixture or this may be controlled, step-wise addition of the components by various soft lithography and deposition methods. The last method clearly has advantages over spontaneous phase separation in terms of control over desired geometries and positioning (site-selectivity). Site selective deposition may be performed on different length scales through the use of masks and stamps, i.e. negative patterning or in positive mode or by the selective deposition of material by inkjet or other printing technologies [65] on large scales as well as by dip-pen nanolithography [66,67,68,69] on the nanometer scale.

Better control over the positioning may be obtained from so-called soft lithography [70,71] processes like ink(jet) deposition methods [72-76], micro contact printing (μCP) [77-79], and dip-pen nanolithography (DPN), which allows to target samples with macroscopic dimensions (inkjet), centimeter (μCP) and millimeter dimensions (DPN) with resolutions down to several micrometers (inkjet), tens to hundreds of nanometer (μCP) and sub-100 nanometer (DPN).

Spontaneous phase separation is an equilibrium process in which chemically different species on a surface desorb and reabsorb to maximize favorable interactions. These interactions may be Van-der-Waals forces in which a mixed monolayer of, for example, short and long alkylthiolates phase separate [80-83]. Phase separation may also be caused by other interactions like intramolecular hydrogen bonding [80,84-86]. Disadvantage of the spontaneous phase separation method is the lack of site specific control, i.e. the ability to have specific functionality on specific locations of the surface only. Also the ratios of adsorbents on the surface is difficult to predict beforehand since deposition from a 1:1 mixture in solution does not necessary lead to the formation of a ML with the same ratio due to more favorable absorption of one of the compounds in the solution [87,88].

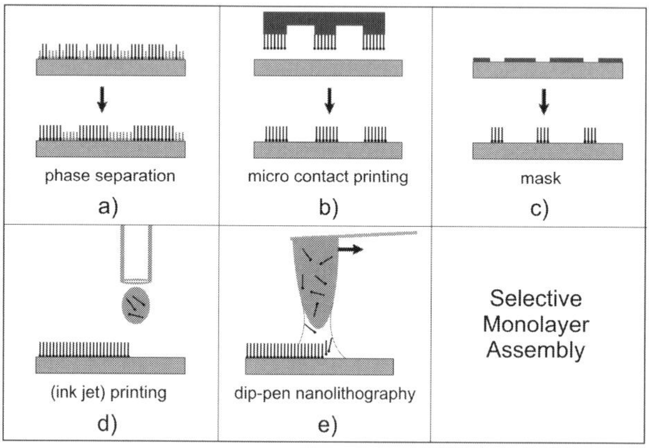

Figure 5. Schematic representation of different forms of additive and selective monolayer assembly, showing a) spontaneous phase separation, b) micro contact printing, c) monolayers deposition on a pre-patterned surface or through a mask, d) deposition by ink jet printing, and e) dip-pen nanolithography.

The second category involves the partial removal of complete monolayers. This removal may be physical by damages induced by applying forces to the monolayer or may involve degradation of the monolayer and desorption of its reaction products.

Degradation of the monolayer (Figure 6) may be induced in a positive mode by means of energetic beams (UV, X-ray, e-beam, ion beam, Figure 6a) or in negative mode (Figure 6b) by applying masks. Many of these processes involve oxidative processes or electro-induced cleavage of bonds (i.e. electro-induced desorption of thiolate monolayers (Figure 6c). An example of this mode is the application of monolayers as ultra thin resists in the patterning of silicon substrates (Figure 6d). By selectively removing parts of the monolayer etching procedures may be applied to etch the underlying substrate whereas the remaining monolayer protects the rest of the substrate. Alternatively, the parts of the monolayer that are opened up upon partial ML degradation may be filled with a second monolayer providing different or complementary chemical functionality. These processes will be described later in more detail. Another widely applied form of patterning monolayers by force is by means of the so called nano-shaving method (Figure 6e) [89,90,91]. This method is mostly applied to thiolate monolayers on gold and involves the movement of an AFM tip over the surface with high contact force in the presence of a solution of another thiol-compound. As the moving tip removes material from the surface the gap is filled in with molecules from the solution, like for example a thiolate with an initiating group for the controlled polymerization of poly(*N*-isopropylacrylamide) (pNIPAAM) brushes [92]. The nano-shaving method has also been reported as a method to pattern lipid bilayers [93].

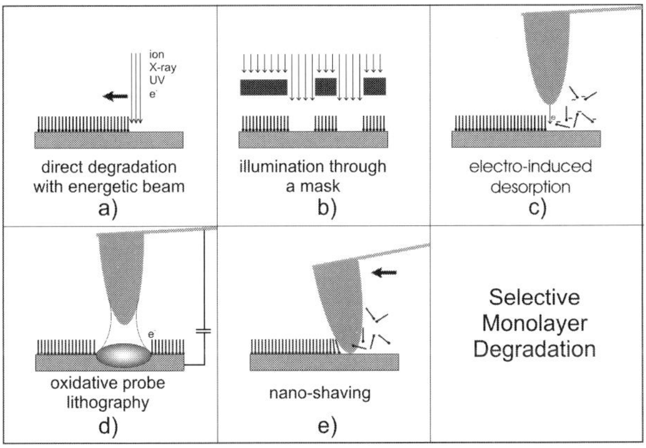

Figure 6. Schematic representation of different forms of selective monolayer degradation, showing a) degradation by direct writing with energetic beams, b) degradation by illumination through a mask, c) electro-induced desorption of monolayers, d) degradation by electro-oxidation, and e) degradation by force.

The last class of monolayer patterning involves the formation of patterns via controlled end-group modifications. Like in the scheme for the monolayer degradation (Figure 6), changes may be induced by energetic beams, for example by targeting a photo cleavable group in the monolayer. Changes in the monolayer may also be induced by chemical endgroup modification (i.e., substitution reactions) or by local oxidation processes. Most electro chemical oxidation processes use water and its degradation products as reagents. A general reaction scheme may be presented with the following reactions taking place at the anode (1 and 2) and cathode (3) [94]. Like in the case of monolayer degradation electrons may be supplied locally to the system by means of e-beam lithography or by the tip of a scanning probe microscope.

$$M + xH_2O \rightarrow MO_x + 2xH^+ + 2xe^- \quad (1)$$

$$2H_2O \rightarrow O_2 + 4H^+ + 4e^- \quad (2)$$

$$2H_2O + 2e^- \rightarrow H_2 + 2OH^- \quad (3)$$

3.1. Monolayer Patterning by Means of Energetic Beams

Energetic beams are commonly used as methods in nanofabrication technology. Energetic beams of electrons, photons and ions are routinely used to locally remove thin films of polymer resists but they may also be used to locally pattern SAMs. The small thickness of SAMs in comparison to polymer films is a distinct advantage in this lithography process. A number of reviews on monolayer patterning by UV, e-beam, and X-ray irradiation is available [95,96].

Patterns on a complete monolayer can be made on a surface either by direct writing with the beam or by illumination through a mask. Alternatively, when obtaining high resolution is not crucial, patterned SAMs may also be obtained by micro contact printing. Like in conventional lithography the exposed areas of the surfaces can be further processed by etching methods whereas the remaining SAM is used as a resist to prevent unwanted surface modifications [97-103]. For example octadecanethiol monolayers may be used as a resist for common gold etchants [104,105] whereas silanes may be used on silica supports to protect the underlying SiO_2 from etching [106-108].

Work in the group of Calvert in the 1990s has demonstrated that deep-UV irradiation (KrF, 248 nm; or ArF, 193 nm) may be used to pattern monolayers of phenyltrichlorosilane (PTCS), benzyltrichlorosilane (BCTS) and amino terminated silanes like 3-aminopropyltrimethoxylsilane (APTS) and aminoethylaminomethyl phenethyltrimethoxysilane (PEDA) [109-111]. Using UV spectroscopy it was found that irradiation with 193 nm UV required a dose of 400 mJ/cm^2 for complete monolayer degradation whereas irradiation with 248 nm wavelength UV required over 50 J/cm^2. The energy of the used photons (620 kJ/mol) is sufficient to break Si-C bonds (368 kJ/mol) within the SAM, cleaving of the entire monolayer exposing the bare substrate which was terminated with hydroxyl groups. A second, fluorinated monolayer could be assembled specifically onto the bare area and, by illumination trough a mask, surface energy patterns were prepared that showed specific adhesion sites consisting of the remaining amine terminated ML for cell adhesion [109,112] or the coupling of DNA chains [113]. Later the process was also demonstrated for a broad range of other silanes containing aromatic functionalities (phenyl-, benzyl-, phenylethyl-, biphenyl-, naphtyl-, antracenyl chlorosilanes and derivatives) [114].

For monolayers of 4-chloromethyphenyl trichlorosilane it was found that at low doses (<50 mJ/cm^2) HCl may be eliminated from the film. IR spectroscopy demonstrated that after exposure aldehydes are present at the SAM. Using dinitrophenylhydrazine (an aldehyde sensitive reagent) the aldehydes on the surface were converted to amines which, in turn, served as anchoring points for Pd catalyst that were used in a nickel plating process (Figure 7) [115]. Metallized nickel patterns could also be obtained in negative mode by degrading a pyridine terminated monolayer by UV exposure [116]. For these monolayers also the use of ion-beam patterning was described [117].

Selective metallization was also reported by UV patterning of chloro-, bromo- and thiol terminated monolayers as templates for CVD depositing of copper films [118]. Similar results were described for silane monolayers adsorbed on polyimide substrates [119].

Besides with UV exposure, soft X-rays were also used to create patterns on an iodo-terminated monolayer [120]. Electrons generated from the surface were suggested to be involved in the cleavage process [121]. Irradiation by soft X-rays leads to the selective cleavage of carbon-halide bonds with the carbon-bromine bond being found the most labile [122].

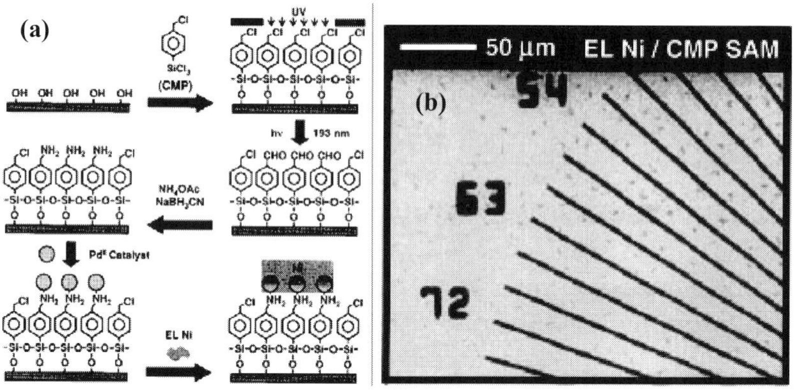

Figure 7. Schematic representation of (a) the deep UV patterning of a chloro-terminated monolayer and subsequent conversion to amine end groups, and (b) the resulting nickel film after adsorption of Pd catalyst and electroless plating of nickel to a patterned monolayer. [Image reprinted with permission from Ref. 115 (copyright American Chemical Society, 1999)]

Exposure of OTS to 50 keV electrons locally degrades OTS monolayers [123]. Dipping in HF to remove SiO_2 followed by etching in KOH gives positive tone. Low energy electron exposure (1 keV) of APTS and PEDA was reported by Craighead *et al.* XPS and friction force atomic force microscopy were used to study the dose needed for ML degradation [124]. The formed patterns could be used as a negative tone template for the adsorption of palladium nanoparticles. OTS monolayers on silicon could also be locally removed by laser irradiation (514 nm, 300 mW) [125,126]. Upon irradiation the substrate is heated inducing local melting.

Later experiments [127] demonstrated that earlier reported monolayers have poor coverage. By exposing a monolayer to a fluorinated chlorosilane, backfilling of up to 44% was reported form XPS studies. By using silanes with long alkyl chains, 11-

phenylundecyltrichlorosilane, a coverage of 94% was reached. Also here UV irradiation cleaves off the phenyl group and part of the alkyl chains. Remaining surface functionalities include aldehydes (see also [112]) but also insoluble decomposition products.

An elegant method of photo patterning using a mixture of orthogonal photosensitive silanes was reported by the group of Jonas [128]. For this purpose functional silanes (-CHO, NH_2 and OH) have been protected with nitroveratyl and 3,5-dimethoxybenzoyn groups. The first being cleavable at wavelengths below 300 nm and the latter, less reactive group, at longer wavelengths up to 420 nm. By using sequential illumination of a mixed monolayer, of protected amines and carboxylic acids terminated monolayer a template was created for the selective adsorption of carboxylated poly(butylmethacrylate) particles (Figure 8) [128].

Patterns of orthogonal binding sites have been made from mixed monolayers containing photo-cleavable groups that can be cleaved at different wavelengths. E-beam irradiation of an imine containing monolayer [129,130] does not immediately cleave off the whole monolayer but the conversion to an amine bond makes the irradiated areas prone to hydrolysis. The amine groups formed after irradiation and hydrolysis can also be used to create surface energy templates for diblock copolymer phase separation experiments by reacting the amines on the surfaces with tribromoacetaldehyde (Figure 9) [131].

X-rays and e-beam irradiation was also found to lead to cross-linking of 1,1':4',1''-terphenyl]-4,4''-dimethanethiol monolayers on gold [132] and hydroxybiphenyl monolayers on hydrogen terminated silicon. Upon irradiation intramolecular cross-links form between the phenyl rings within the ML that stabilize the SAM and increase its resistance against etchants [133]. The latter one was used as a negative tone resist for wet etching of silicon structures with 20 nm resolution.

Monolayers containing terminal nitro groups can be reduced to amines by e-beam irradiation [134-137]. The presence of amines was demonstrated by chemical means via the coupling of aziridine and of succinimidyl functionalized biotine and subsequent coupling with fluorescent (Cy3) labeled straptavidine.

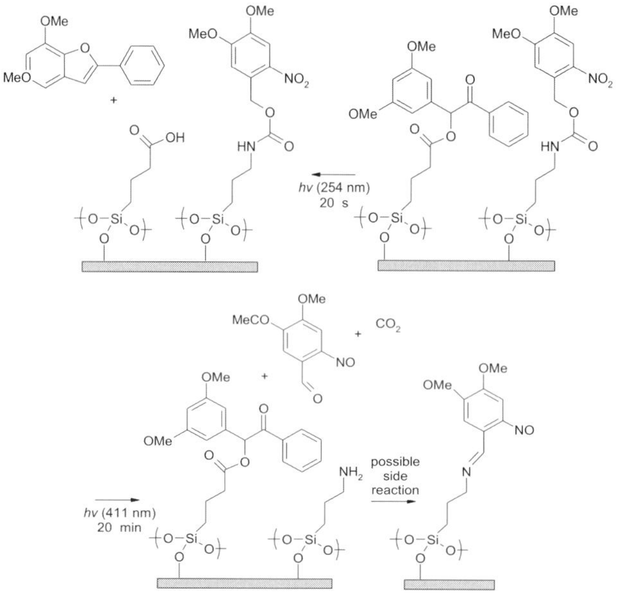

Figure 8. Chemical structures of the photo protected silane monolayers. Different protecting groups may be cleaved at different wavelengths. [Image reprinted with permission from Ref. 128 (copyright Wiley VCH, 2005)]

The process was demonstrated for a SAM on gold [138] and on 3-(4-nitrophenoxy)-propyltrimethoxysilane (NPPTMS) SAMs on silicon which were patterned and used as templates for the adsorption of gold nanoparticles (Figure 10). Templates were made both by direct writing as well as by illumination (5 keV, 290 $\mu C/cm^2$) through a mask [139]. Characterization of the nitro-containing monolayers by XPS revealed that exposure of the ML to X-rays also causes endgroup conversion [139]. This may be used as well to created patterned surface templates for the specific adsorption of oligonucleotides [140].

Disulfide bonds are known to be prone to oxidation/cleavage. This concept is used in a publication from Lieberman *et al.* in which a monolayer of phenyl-(3-trimethoxysilylpropyl)-disulfide on silica is patterned by either local probe anodization or by e-beam lithography. Upon exposure by e-beam the SAM decreases approximately 0.3 nm in height as observed by tapping mode AFM (TM-AFM). An observed reaction between the exposed lines with *N*-(1-pyrene)maleimide indicates that the reaction product is a thiol bond. Performing local probe oxidation experiments on the same monolayers results in monolayer degradation and in the formation of silicon oxide as observed by AFM [141].

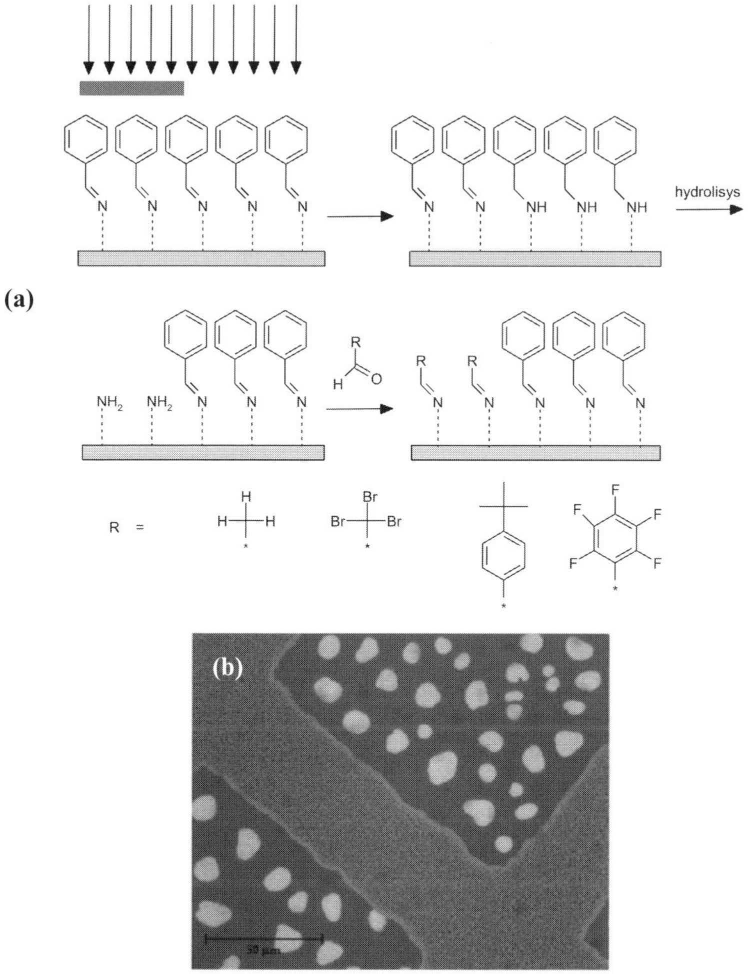

Figure 9. Schematic representation of (a) the irradiation, hydrolysis, and formation of a surface energy template. An optical microscopy image of (b) polymer phase separation on a surface energy patterned surface using a TEM-grid as a mask is also shown. [Image reprinted with permission from Ref. 131 (copyright American Chemical Society, 2006)]

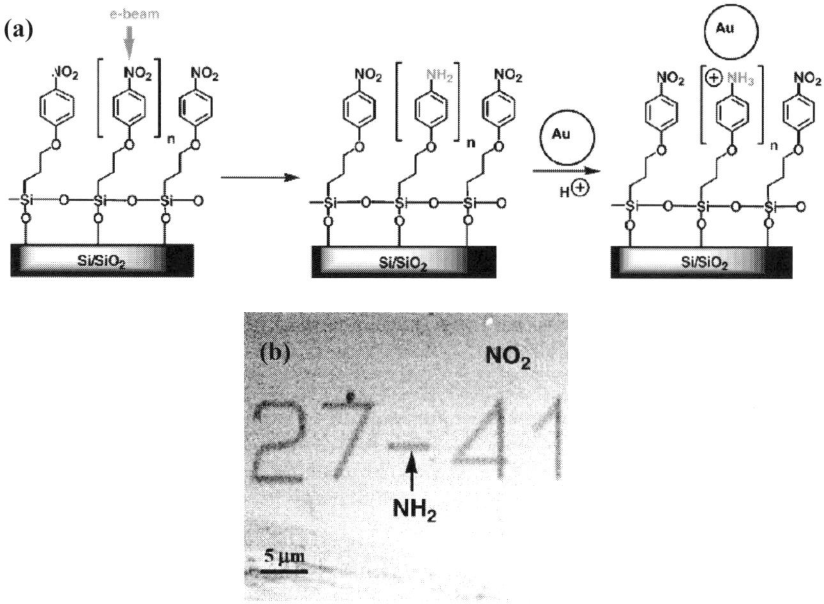

Figure 10. Schematic representation of (a) the reduction of NPPTMS SAMs on silicon by e-beam irradiation. An SEM image of (b) a patterned SAM written by e-beam lithography is also shown. [Image reprinted with permission from Ref. 139 (copyright American Chemical Society, 2004)]

Some comments, however, to the experiments described above can be made from the view-point of monolayer formation. Most of the demonstrated examples use monolayers that contain either short chains or were prepared from methoxy- or ethoxy-silanes. Although the demonstrated patterning processes certainly can be used to form intricate surface patterns and surface modifications, the exact mechanisms of patterning by beam irradiation remains unclear since the formed monolayers are poorly ordered. In addition, characterization by XPS is tricky in many cases and has to be done extremely carefully because it was found that exposure of the mentioned SAMs to X-rays also results in ML damage [95,121,122,142]. For example, in case of aromatic halide compounds, cleavage by X-ray irradiation was found to be X-ray intensity independent process which is selective for the halogens and is caused by secondary electrons originating from the surface [122].

3.2. Monolayer Patterning by Means of Local Probes

As already suggested briefly in the previous section, local probe lithography can also be used to locally pattern monolayers. This can be done in a constructive fashion using dip-pen nanolithography to locally deposit molecules on surfaces but scanning probe microscopy techniques can also be used to expose the monolayers locally to: 1) heat by heatable AFM tips [143-145], 2) photons by scanning near field optical microscopy [146], 3) force (i.e. nanoshaving) and by 4) electrons by means of STM or AFM using a conductive AFM tip. All of these methods have been used to create surface patterns. Several reviews on the topic are available but progress in the local probe oxidation-based patterning of monolayers will be described in more detail [106,147-149].

3.2.1. Local Probe Oxidation

Already in 1985, only 3 years after the invention of STM [150] Güntherodt *et al.* reported the possibility to create structures in $Pd_{81}Si_{19}$ samples by applying voltage pulses during STM imaging [151]. It was realized that STM and AFM can not only be used for imaging of surfaces but when a bias voltage is applied to the scanning tip, surfaces may also be patterned. In electro-chemical patterning experiments, the STM can be used in the two distinct emission modes. In the first mode, electrons from the tip (negative bias on the tip) are used to locally oxidize the material below the tip; a kind of low energy (tunneling) electron beam induces the local oxidation of materials. In the second method (the field emitting mode), the area below the tip is oxidized by the electric field below the tip, material between the tip (e.g. oxygen or water) decomposes and their reaction products oxidize the substrate.

An example of the latter approach is the direct oxidation of hydrogen terminated Si(100) substrates. Examples of patterning on organic self-assembled monolayers were reported by U. Heinzmann [152,153]. In this case hexadecanethiol and *N*-biphenylthiol on Au(111) and octadecyltrichlorosilane on Si(100) were utilized. The pattern was etched into the silicon substrate using two wet etch steps (5% HF, 30% KOH). Earlier experiments by McCord and Pease described also the patterning of PMMA and alkylhalide resists [154,155]. A comprehensive review covering the earlier work of STM induced surface modification was published in 1990 by Shedd and Russell [156].

Monolayers of CMPTS and PEDA, which were also known to be used in low dose e-beam patterning experiments, were also patterned by STM [157]. By removing the native oxide and growing a thinner oxide layer on the bare silicon before assembling the monolayers, threshold voltages of –4 (CMTPS) respectively –8 V (PEDA) were reported. Figure 11a shows a nickel plated substrate on a patterned PEDA monolayer displaying 15 nm features. Patterning was performed at 10 V tip bias, 200 pA current and 0.21 μm/s patterning speed). Electroless nickel plating was performed onto Pd-nanoparticles which were selectively adsorbed onto the unpatterned amine terminated ML. The lines in nickel plated sample were etched by exposure to C_2F_6/O_2 (Figure 11b).

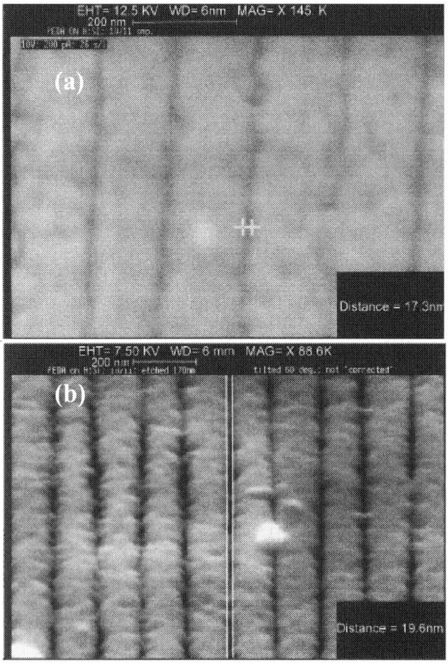

Figure 11. SEM images of (a) the reduction of NPPTMS SAMs on silicon by e-beam irradiation, and (b) a patterned SAM written by e-beam lithography. [Image reprinted with permission from Ref. 157 (copyright American Institute of Physics, 1996)]

Initiated by successful lithography experiments with the STM, several groups transferred the oxidation techniques to AFM. Performing the oxidation using AFM instead of STM has the advantage that a broader range of substrates is available due to less strict restrictions on surface conductivity. Field effect induced oxidation has been applied on a variety of conducting and semi-conducting substrates including silicon [94,158-160], silicon nitride [161-163], zirconium nitride [164], gallium nitride [165], gallium arsenide [166], aluminum [167], titanium [168,169], tantalum [170], chromium [171], molybdenum [172], diamond [173], and other oxidizable materials [174].

Already in 1995, the group of Campbell and Snow reported the creation of one of the first working devices prepared by local probe oxidation, initially using a silicon tip and cantilever in an experiment oxidizing thin (7 nm) titanium films [175]. Using a poorly conducting tip no current could be detected, implying a field-effect induced oxidation of titanium to titanium oxide with adsorbed water. By partly oxidizing titanium wires produced by conventional optical techniques, thin titanium wires (10 nm) and metal oxide junctions were prepared (Figure 11). The I/V characteristics of a 10 nm constricted titanium wire at room temperature and at 4.2 K were compared to those of a pre-modified (large) wire.

The most common nanofabrication process, however, is the local probe oxidation of silicon substrates. In these experiments a bias is applied (usually tip negative) between a conductive AFM tip and the silicon substrate at ambient conditions. In a process called local anodization, the silicon under the tip is oxidized to silicon dioxide. The method was pioneered by Dagata [94,176] and the presence of a water meniscus between tip and substrate is essential for the anodization process (Figure 12). The formed silicon oxide may be used as a negative tone resist in further nanofabrication steps such as wet etching techniques with KOH or as positive tone resist in, for example, electroless plating experiments [177].

Since then a lot of efforts have been reported to understand the mechanism [178-184] of the local anodization process and intricate structures have been made by sequential oxidation and etching steps [185,186]. Recently, the local probe oxidation of a silicon on an insulator was used to fabricate a nano flash device which programming and erasing of the fabricated device was demonstrated [187].

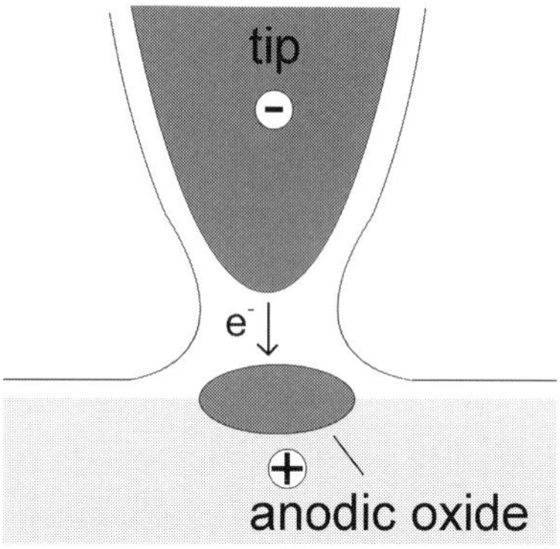

Figure 12. Schematic representation of the local probe anodization of silicon. [Image reprinted with permission from Ref. 174 (copyright American Institute of Physics, 1996)]

In all local anodization processes the formation of a stable water meniscus is essential since it determines the minimum obtainable resolution. Therefore efforts are made in controlling the formation and the size of the meniscus by changing the surface energy of the substrate by the assembly of monolayers. Experiments utilizing resists consisting of (mixed) Langmuir-Blodgett (LB) films [189] and (mixed) self-assembled monolayers (SAMs) [189-193] were reported. In the first case, the application of mixed LB films of hexadecylamine and palmitic acid was reported by Lee *et al.* Different ratios of the LB components in this system were used and it was concluded that the mixed layers led to significantly thinner pattern widths. Although approximately three times thinner lines were observed, no remarks were made on the shape and condition of the used tip. In a publication by Liu *et al.* [193] (Figure 13), a method was reported to create assemblies of gold nanoparticles on patterned SAMs of OTS on SiO_2/Si.

By locally oxidizing the silicon under the organic resist, Liu *et al.* created arrays or silicon oxide dots with diameters of 15 nm and 40 nm spacing. The protruded silicon oxide was used in a subsequent chemical modification step in order to introduce new functionalities. In this step a second silane, aminopropyltrimethoxysilane (APTMS), was reacted with the silicon oxide that was created by the probe oxidation, whereas the unoxidized OTS resist-layer protected the silicon oxide below. The created silicon oxide pattern was transferred and terminal amine groups were introduced to adsorb gold nanoparticles.

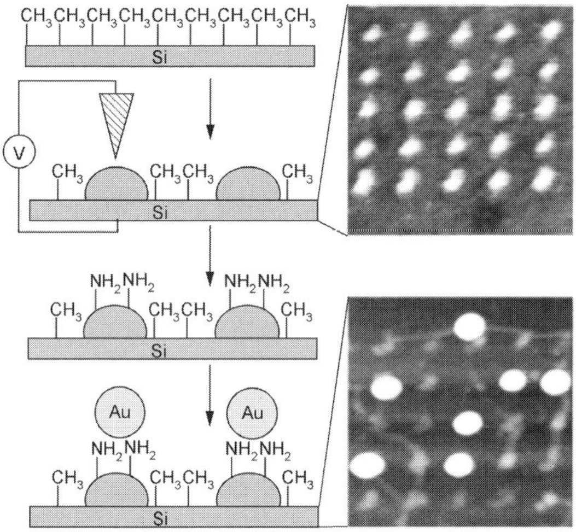

Figure 13. Functionalization scheme introduced by Liu *et al.*. Local probe oxidation of a silicon substrate with a SAM resist layer was used to assemble gold nanoparticles. [Image reprinted with permission from Ref. 193 (copyright American Chemical Society, 2003)]

In more recent literature a large number of publications can be found in which SAMs of alkyl silanes on silicon dioxide [194-200] as well as SAMs assembled on hydrogen terminated silicon [201] are used as resists in the local probe oxidation of the underlying substrate. The remaining ML can be used as a resist in etching steps or as describe above as a blocking layer preventing the assembly of additional monolayers.

3.2.2. Local Probe Electro-Oxidation of SAMs

Beside the use of ML as resists, the monolayers themselves can also be oxidized. Like in the anodization of silica the presence of a water meniscus is essential but in stead of oxidation of the substrate the monolayer itself is oxidized. From the work on using SAMs as resist it may be clear that for successful oxidation of the monolayers the monolayer must be free of defects to prevent short circuits.

The first report on tip induced oxidation of SAMs was published by Sagiv *et al.* in a process called constructive nanolithography. The authors found that tip induced oxidation of organic monolayers introduced surface functionality that could be used in a large number of subsequent functionalization reactions with both organic and inorganic substances yielding a range of nanopatterned functional structures [202-206]. The work principle is summarized in Figure 14. The procedure has first been demonstrated on monolayers of 18-nonadecenyl trichlorosilane on a *p*-type doped silicon wafer (see Figure 14, 1) [203].

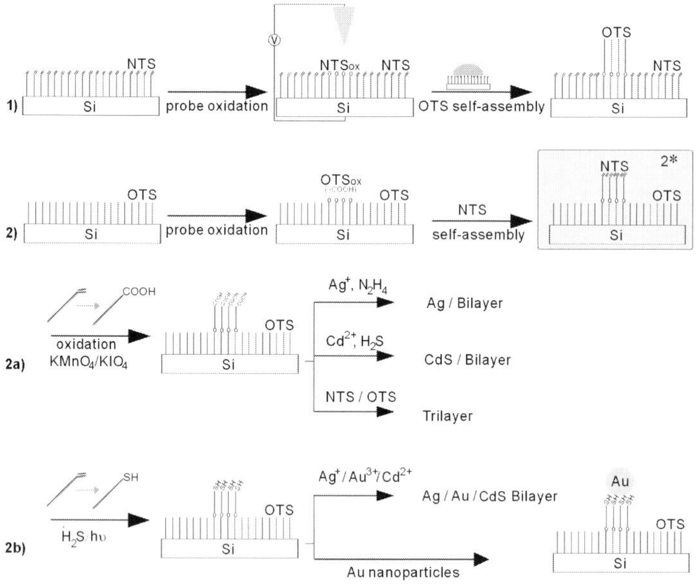

Figure 14. Functionalization of OTS and NTS monolayers by Sagiv *et al.* ML on a silicon wafer (1) is locally oxidized to introduce chemical functionalities to be used as templates for subsequent modification steps, such as with the attachment of a second chlorosilane. The same oxidation procedure was applied to OTS monolayers to introduce the same functionalities. Onto the oxidized template a second ML of NTS was assembled (2*). This ML served again as a basis for further surface modification reactions (2a and 2b). [Image reprinted with permission from Ref. 207 (copyright Wiley VCH, 2004)]

The terminal vinylic end-group could be locally oxidized by a conductive AFM tip with an applied bias voltage of approximately 9 V (tip negative). Oxidation was done at 5 μm/s with a variety of conductive tips (boron doped, diamond coated, tungsten, tungsten carbide, copper- and silver coated silicon nitride and highly doped silicon). The successful selective oxidation of the terminal vinylic end groups to terminal carboxylic acids groups was confirmed afterwards by imaging the area in contact mode with the same probe (without applying a voltage), simultaneously recording height and lateral friction signals. For the unaffected areas the surface remained unchanged, whereas for the oxidized areas the height remained the same but an increase in friction was observed which was attributed to changes in local surface polarity.

Oxidation experiments were verified by partial wet chemical oxidation: the substrate was partially exposed to a 5 mM solution of $KMnO_4$/dicyclohexano-18crown-6 and also to electric oxidation of large surface areas by means of a copper-grid (connected to a 13 V negative bias voltage). Brewster angle infrared spectroscopy indicated the presence of carboxylic acids groups.

In particular interesting is the approach in which a monolayer of 18-nonadecenyl trichlorosilane was attached to locally oxidized OTS (Scheme 1, *). Thus functionalized patterns have been used as templates for the formation of metallic layers by using wet-oxidation and subsequent adsorption and reduction of metal ions (Scheme 1, 2a). The terminal vinyl groups could also be converted to thiols by UV irradiation (Hg lamp, 254 nm) for 10 minutes in a H_2S/Ar (1:1) atmosphere. As a result a stable pattern decorated with thiol end groups, placed in well-defined positions, was created. The formed thiol end groups could be used in subsequent functionalization steps. Application of gold nanoparticles (Au_{55}) to the thiol group resulted in the formation of gold islands connected by lines that were only two nanoparticles wide [202]. The method presented by Sagiv for surface patterning and modification has many possibilities for the production of nanometer-sized devices due to the large number of available chemical and physical modification steps. Possibilities include silane chemistry (silanes with terminal functionality), the coupling of amines and cationic species, thiol chemistry as well as conversion to anhydrides.

In later publications, the method was further explored and it was found that not only 18-nonadecenyl trichlorosilane SAMs could be used but also inert SAMs of octadecyltrichlorosilane could locally be oxidized to carboxylic acid end groups (Scheme 1, 2), opening new avenues for chemical modification [206]. Since then a large number of publications has appeared in which the local probe oxidation of OTS was used to assemble nanoparticles, nanotubes, bilayers and proteins, on surfaces with high accuracy and high resolution [202,206,208-212,215]. A number of these surface modifications will be exemplified later.

Also for the local probe oxidation of OTS SAMs, contact mode height and friction force imaging can be used to follow the local oxidation process. A clear friction signal is observed for the patterned areas and no increase or decrease in height of the ML is expected since the conversion of the terminal methyl group to a carboxylic acids group should lead to negligible changes in height. However, in contrast to this often a small but detectable height contrast can be observed simultaneously. This height contrast reverses sign upon changing of the scanning direction and it was explained by an artifact of the contact mode AFM-imaging. The high friction forces on the patterned areas cause besides lateral deflection, also a small vertical deflection. By laterally exciting the cantilever (in the fast scan direction) while recording the friction images the artifact was averaged out for patters consisting of oxidized ML, whereas on areas where silicon dioxide was formed a consistent height increase was found (Figure 15).

For this purpose a silicon oxide dot was generated inside an electro-oxidized patterned OTS monolayer. The pattern inside the square consists of –COOH terminated ML and the dot in the center is formed by longer oxidation time and consists of protruded silicon dioxide. When imaged with lateral modulation switched off both structures have a positive apparent height. When imaged with lateral modulation switched on the observed height for the –COOH terminated areas vanishes, whereas the oxide dot remains visible, supporting the interpretation that no true change in height of the ML takes place upon local oxidation to -COOH groups [213].

By systematic variation of oxidation conditions (bias voltage and pulse duration) and by carefully studying the forward and backward traces of the resulting height images it was found that for a certain tip a narrow window was found in which methyl groups of the OTS ML are converted to carboxylic acid groups (Figure 16). At higher voltages a clear increase in height is observed indicating the formation of silicon oxide as described earlier. For lower voltages and short pulse durations no change in the ML was detected [213].

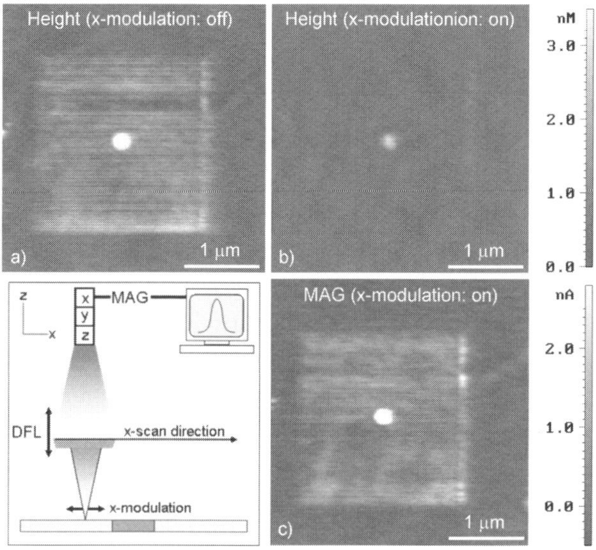

Figure 15. Image of a) the normal contact mode height image of electro-oxidized OTS ML on silicon, b) the same area imaged with lateral modulation switched on, and c) the corresponding friction image. [Image reprinted with permission from Ref. 213 (copyright Wiley VCH, 2005)]

The presence of a narrow regime in which COOH-groups are formed was confirmed by performing force spectroscopic investigations during monolayer oxidation [214]. In these experiments force distance curves for a biased tip approaching on OTS monolayer were recorded up to 50 times on the same spot. Analysis of the snap-in part of the force-distance curve provides information on the onset of COOH formation as well as on the start of the oxidation of the underlying substrate.

Like for the oxidation of NTS SAMs the local oxidation of OTS may be used to create surface patterns that can be used as templates in subsequent surface modification reactions. An increasing number of publications are being presented in which functional nanostructures are made by either direct assembly of material onto the patterns or by the assembly of multilayers and/or endgroup modification reactions. Figure 17 presents an overview of some of the possible surface modification reactions that have been made on locally oxidized OTS monolayers.

Starting from a pattern of carboxylic acids groups surrounded by chemically inert OTS molecules a second monolayer of trimethyl octadecyl ammonium bromide can be assembled, as can be observed in Figure 17, 1 [208]. The terminal –COOH groups may also be used to physisorb positively charged nanoparticles. Examples of assemblies of CeSe/ZnS core/shell (Figure 17, 2) [216] iron (Figure 17, 3) [212] and gold (Figure 17, 3) [208] are presented. Multi component assemblies can be made by sequential oxidation and functionalization steps [208]. Assemblies of various metal particles may be prepared by assembly of pre-formed nanoparticles. The nanoparticles, however, can also be formed directly onto the –COOH terminated patterns. An example is shown in Figure 17, 6 in which an assembly of iron particles was formed by dipping the oxidized template into an iron acetate solution followed by reduction using hydrazine vapor [210]. Magnetic characterization without (Figure 17, 7 bottom) or with (Figure 17, 7 top) external magnetic field demonstrates the super paramagnetic character of the formed particles. Besides the direct assembly of material onto the templates the assembly of a multilayer of (functional) silanes can be used. Figure 17, 8 shows the assembly of a NTS bilayer [205]. The vinyl endgroups of this bilayer can be converted to thiols (H_2S/ UV) [202,206] or to amines (formamide/ BH_3) [215] both of which may serve as anchoring points for the assembly of gold nanoparticles (Figure 17, 9 and 10). Alternatively APTS may be assembled directly onto an oxidized template. Although this will not lead to the formation of a well-defined bilayers the amine groups can be used to assemble carbon nanotubes (Figure 17, 11) [217]. In addition, the bromine terminated monolayer may be assembled onto the surface template. The terminal bromine groups in the monolayer may be used as initiating groups in the controlled polymerization of styrene forming polymer brush-like structures Figure 17, 12) [218].

As demonstrated, the local probe oxidation of OTS and NTS SAMs is a technique which provides a chemically active surface template for a large number of surface modification reactions with high spatial resolution. However, like all other probe patterning techniques, the small areas onto which these modifications are performed makes spectroscopic characterization difficult. In some cases (e.g. iron nanoparticles) magnetic or fluorescence (CeSe nanoparticles) properties may be used to demonstrate the chemical functionalization but for most characterization techniques (IR, XPS, etc) the signal intensity is too low. The patterning of larger areas was demonstrated in a publication by Hoeppener *et al.* [219].

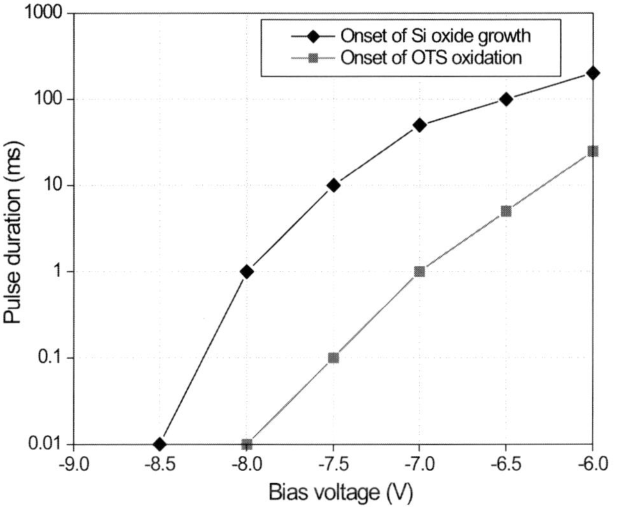

Figure 16. A window of operation was found for the local probe oxidation of OTS monolayer on silicon. Selecting oxidation conditions above the upper line leads to destruction of the OTS monolayer accompanied by the growth of silicon dioxide. Selecting conditions between the blue and the magenta line leads to the conversion of terminal methyl groups of the OTS monolayer to carboxylic acids end groups. At conditions below the magenta line no changes in the OTS monolayer are induced. [Image reprinted with permission from Ref. 213 (copyright Wiley VCH, 2005)]

In this publication the conductive AFM tip is replaced and instead a copper TEM-grid is used as the electrode. By applying a bias pulse between surface and grid the OTS monolayer directly under the grid is oxidized. Contact between the rough TEM-grid bar and the OTS surface is facilitated by condensing a small water layer on the grid before applying the bias voltage. The formation of surface acid groups can be demonstrated by AFM imaging, or, more conveniently, by condensing water vapor onto the patterned surfaces as the water preferentially wets the hydrophilic –COOH terminated parts of the substrate. In an elegant publication from the same group it was demonstrated that the thusly formed patterns can be used themselves as electrodes from the oxidation of another SAM-covered substrate [220].

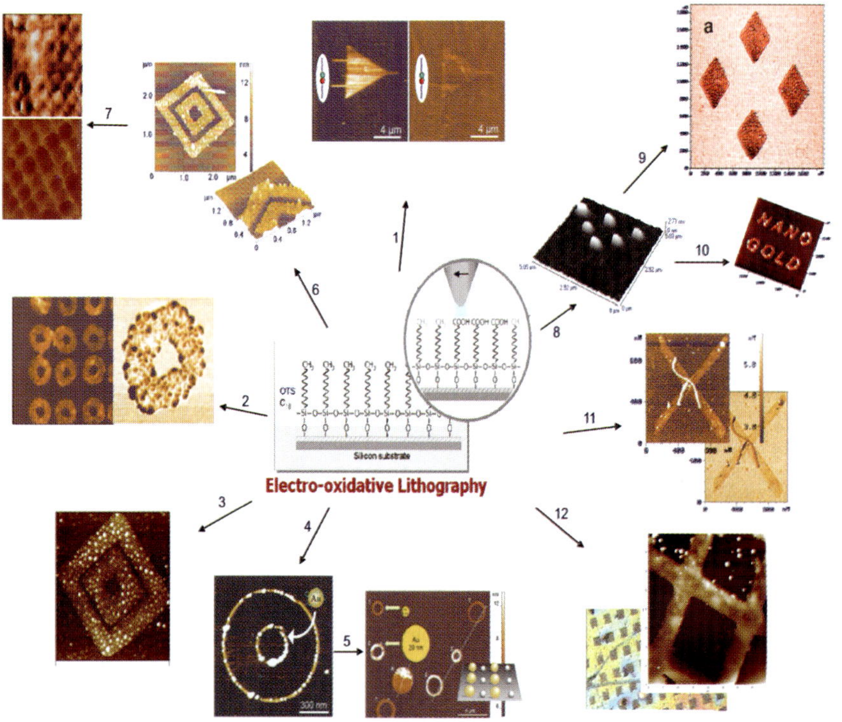

Figure 17. An overview of various surface patterns that are made by local oxidation of OTS SAMs. See text for a more detailed explanation on the formed patterns. [Image adapted with permission from Refs. 206,208, 210,212,215,216,217,218].

The local oxidation of OTS monolayers on silicon supports has been used by Checco *et al.* to study the wetting behavior of ethanol and octane on nanopatterned surfaces [221]. Using the tip-based oxidation (-8 V bias, 9 μm/s), COOH-terminated lines with a width down to 40 nm were created. In an environmental chamber vapors of ethanol and octane were found to condense specifically onto the oxidized lines. The amount of condensing liquid was controlled by a Peltier cooled sample holder. The condensed liquid on the lines was imaged in non-contact mode. Depending on the temperature, different amounts of liquid were found to adsorb onto the stripes. At 15 °C below room temperature a contact angle of 12 degrees was found whereas at 10° C above room temperature the lines were nearly dried-up completely (Figure 19). For octane a similar observation and a contact angle of 8° was observed.

Cai *et al.* in the same group reported the combination of local probe oxidation and dip-pen nanolithography by using a inked-tip during monolayer oxidation, so-called electro-pen nanolithography (EPN) [222]. The authors suggested the use of two distinct inks; quaternary ammonium salts and mercaptopropyltrimethoxysilane (MPTMS). Upon oxidation of the OTS SAM it was claimed that these inks are transferred onto the surface which can be observed by an increase in height in TM-AFM images. For MTPMS the adsorption of Au nanoparticles was demonstrated. However, adsorption was not very specific and, more over, multiple oxidation passes over a single point revealed a linear increase in height up to 2.2 nm after three passes. This was explained by the formation of multi-layers; however, the formation of silicon oxide cannot be excluded.

The group of Fréchet prepared monolayers of special synthesized cleavable protected silanes (Figure 20) [223,224]. Local oxidation of these monolayers is believed to cause cleaving of the protecting group revealing the buried amine [223] or thiol function [224]. These functionalities were used in the subsequent assembly of dendrimers and gold nanoparticles. Since in both cases a bias of 12 V and a monolayer generated from methoxysilanes (poor coverage) were used, because spectroscopic characterization is missing and the height of the formed structures was in both cases under 7 nm, the formation of silicon oxide cannot be excluded.

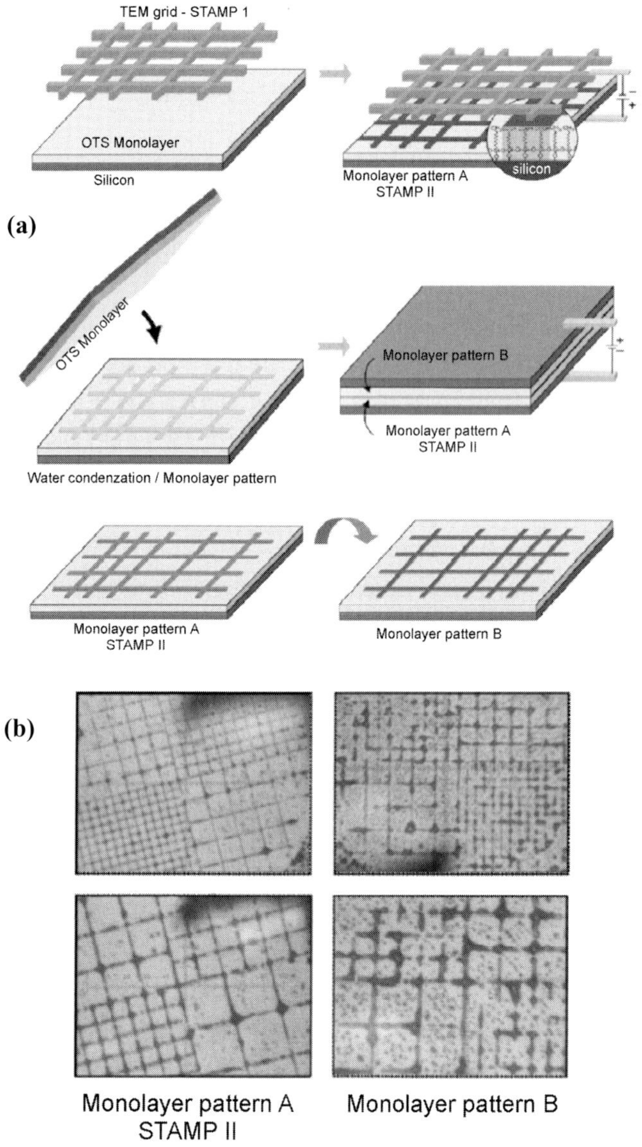

Figure 18. Schematic representation of (a) the oxidation of an OTS monolayer by a TEM grid and the use of the formed pattern as an electrode in a second oxidation process, replicating a mirror image of the original pattern. Optical microscopy pictures of (b) water condensed onto the oxidized patterns. [Image reprinted with permission from Ref. 220 (copyright Wiley VCH, 2006)]

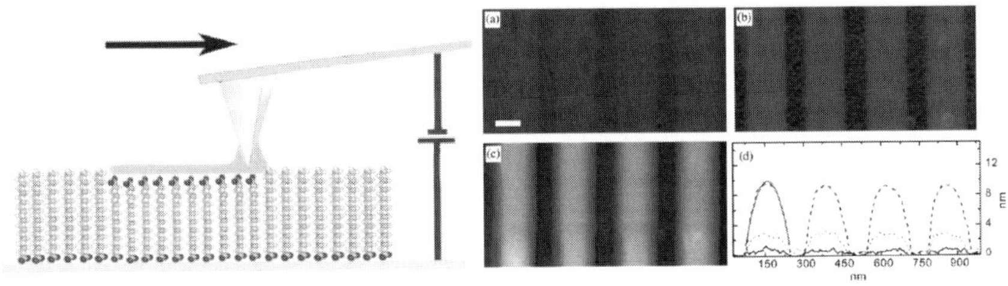

Figure 19. Schematic representation of the oxidation process (left). Non contact mode AFM images (right) of EtOHA condensed onto oxidized lines at (b) room temperature, (a) +10°C, and (c) -15°C. [Image reprinted with permission from Ref. 221 (copyright Elsevier, 2006)]

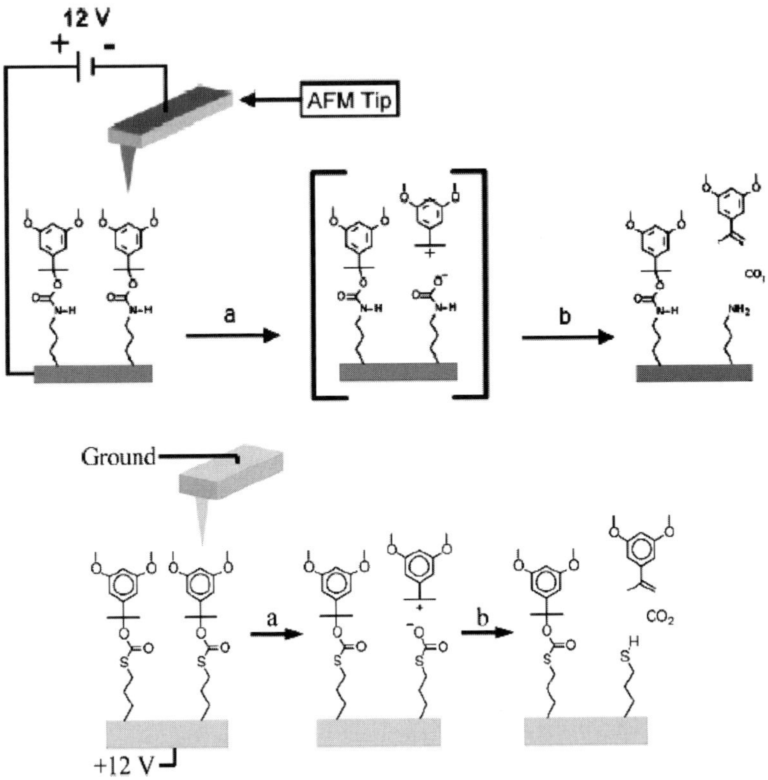

Figure 20. Electro-oxidative cleavage of the protecting group reveals a) amine endgroups or b) thiol endgroups. [Images reprinted with permission from Refs. 223,224 (copyright American chemical Society, 2004-2005)]

3.2.3. Other Examples of Local Probe Electro-Oxidation

The group of Mirkin reported the use of ferrocenyl derivatized thiol inks. Using DPN patterns of two different ferrocene containing inks were placed on a gold substrate. Since both have distinct oxidation and reduction potentials (in order of 250 and 500 mV, respectively) each molecule could be addressed separately using a biased conductive AFM-tip. Onto the patterned ferrocene lines oligonucleotide functionalized nanoparticles (diameter 5 and 13 nm) could be adsorbed specifically [225].

Using a tip coated with a platinum salt the group of Liu was able to deposit platinum lines onto a silicon wafer by electrochemical dip-pen nanolithography [226]. By applying a bias voltage (+4 V to the tip, 10 nm/s) while depositing the ink a Pt-line is formed which is stable to heating in air at temperatures up to 300 °C.

By applying a bias voltage to a tip inside a fluid cell, containing an *n*-octane, etch resistant lines could be written onto the surface of a silicon wafer (Figure 21). Although the exact nature of these lines is still under investigation they do not consist of silicon dioxide as demonstrated by a negative tone transfer by a $NH_4F/H_2O_2/H_2O$ etch treatment [227].

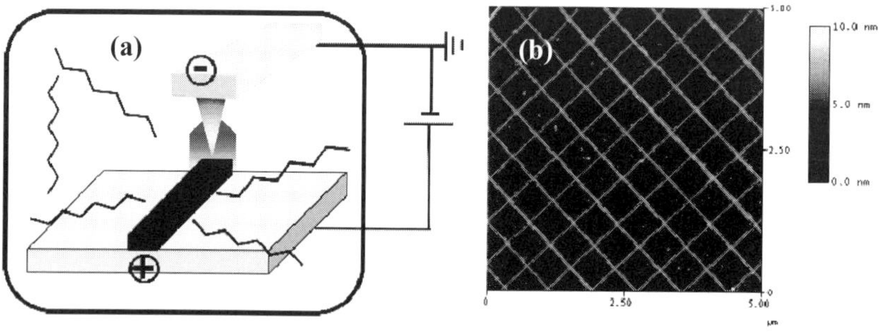

Figure 21. Schematic representation of (a) the formation of etch resistant lines by (b) applying a bias voltage to tip surrounded by *n*-octane. [Image reproduced with permission from Ref. 227 (copyright American Chemical Society, 2005)]

Besides the local oxidation of monolayers and monolayers and (semi)conductors the group of Chrétien reported the local oxidation on single crystal of (tetramethyltetraselenafulvalene)$_2$PF$_6$ salts [228].

In a publication by Martín *et al.* the local probe oxidation of a thin (25 nm) PMMA layer is reported. Upon applying a bias (-30 V to tip, 5 μm/s) 25 nm deep trenches in the polymer film are formed. The remainder of the film was subsequently used as a mask for the deposition of aluminum as well as in wet-etching of the underlying substrate [229].

Local oxidation of polymers was also performed on a polymer film of a polynorbornene with pendant terthiophene groups (Figure 22) [230,231]. Upon applying a potential of 1.4 V by a conductive AFM tip lines can be written onto the surface. The width scales with the writing 120, 170 and 240 nm for speeds of 60, 30 and 15 μm/s, respectively.

Terthiophenes are known to be polymerizable by electrochemistry and therefore the lines have been ascribed to the formation of a conductive polymer.

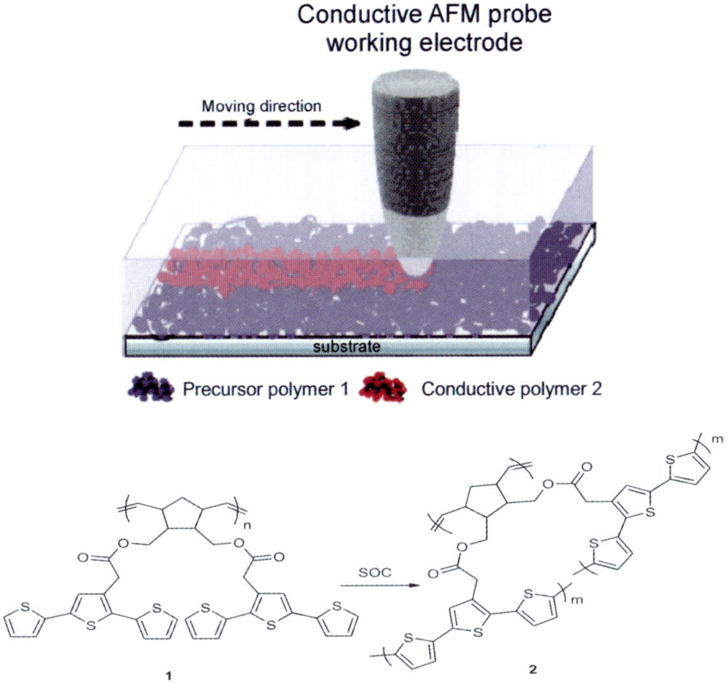

Figure 22. Schematic representation of the local probe electro-oxidation of terpyrylene containing polymer films. [Image reproduced with permission from Ref. 331 (copyright American Chemical Society, 2004)]

4. Summary

The local oxidation of self-assembled monolayers is a powerful and versatile technique for nanopatterning of a broad range of (semi)conductors as well as self assembled monolayers. The oxidation process itself may be performed indirectly by illumination through a mask of directly by using focused beams. The energy needed for the local oxidation may be provided by electron beams or by applying a bias voltage to a local probe. Application of UV and X-rays have also been reported although in the latter case secondary electrons originating from the substrate are expected to play an important role. The reaction products may either completely degrade the monolayer, exposing the underlying substrate or may cause local changes within the monolayer itself. Both mechanisms may be used in subsequent surface modification. In the first the remaining monolayer may serve as a resist for various etching steps or the holes in ML may be filled in with other molecules. If a conversion of the endgroup of the ML occurs, chemical derivatization routines may be used to further convert these groups into reactive species or secondary ML containing chemical functionality may be assembled on top of the patterns.

The local probe electro-oxidation of OTS and NTS monolayers presents an exceptionally useful platform for surface modification due to the versatility of the available functionalization routines. The local oxidation itself requires water and care must be taken not to degrade the monolayer. Good monolayer quality and characterization is important if results from different groups are to be compared. In general, it should be noted that a large number of experimental procedures (solution, vapor, temperature, time, etc.) exist for the preparation of silane monolayers, not all off which per definition lead to the formation of good quality densely packed monolayers. In addition, characterization of patterned monolayers is not straight forward and care should be taken with techniques such as XPS as these may also induce changes in the monolayer. Nevertheless, an increasing number of groups use these techniques to create intricate structures with (chemical) functionality on the nanometer scale.

Acknowledgements

The authors would like to thank NWO (VICI-grant USS), the Dutch Polymer Institute (DPI) and the Fonds der Chemischen Industrie for their financial support, and Dr. S. Hoeppener for helpful discussions.

References

1. K. B. Blodgett, J. Am. Chem. Soc. 55, 495 (1934).
2. I. Langmuir, J. Franklin Inst. 218, 153 (1934).
3. P. Boehme, G. Vedantham, T. Przybycien and G. Belfort, Langmuir 15, 5323 (1999).
4. D. H. McCullough III and S. L. Regen, Chem. Commun. 2787 (2004).
5. J. C. Love, L. A. Estroff, J. K. Kriebel, R. G. Nuzzo and G. M. Whitesides, Chem. Rev. 105, 1103 (2005).
6. A. Ulman, Chem. Rev. 96, 1533 (1996).
7. S. Onclin, B. J. Ravoo and D. N. Reinhoudt, Angew. Chem. Int. Ed. 44, 6282 (2005).
8. M. D. Porter, T. B. Bright, D. L. Allara and C. E. L. Chidsey, J. Am. Chem. Soc. 109, 3559 (1987).
9. D. K. Chattoraj and K. S. Birdi, Adsorption and the Gibbs Surface Excess. Plenum Press, New York (1984).
10. F. Davis and P. J. Higson, Biosens. Bioelectron. 21, 1 (2005).
11. A. C. Templeton, W. P. Wuelfing and W. Royce, Acc. Chem. Res. 33, 27 (2000).
12. D. L. Nuzzo, J. Am. Chem. Soc. 105, 4481 (1983).
13. W. C. Bigelow, D. L. Pickett and W. A. Zisman, Coll. Interf. Sci. 1, 513 (1946).
14. L. Netzer and J. Sagiv, J. Am. Chem. Soc. 105, 674 (1983).
15. J. Sagiv, J. Am. Chem. Soc. 102, 92 (1980).
16. R. Maoz and J. Sagiv, Langmuir 3, 1034 (1987).
17. R. Maoz and J. Sagiv, Langmuir 3, 1045 (1987).
18. R. K. Smith, P. A. Lewis and P. S. Weiss, Prog. Surf. Sci. 75, 1 (2004).
19. J. C. Love, L. A. Estroff, J. K. Kriebel, R. G. Nuzzo and G. M. Whitesides, Chem. Rev. 105, 1103 (2005).
20. X.-M Li, J. Huskens and D. N. Reinhoudt, J. Mater. Chem. 14, 2954 (2004).
21. C. Vericat, M. E. Vela and R. C. Salvarezza, Phys. Chem. Chem. Phys. 7, 3258 (2005).
22. V. Kriegisch and C. Lambert, Springer-Verlag Berlin Heidelberg, Top. Curr. Chem. 258, 257 (2005).
23. D. K. Schwartz, Annu. Rev. Phys. Chem. 52, 107 (2001).
24. L. H. Dubois and R. G. Nuzzo, Ann. Phys. Chem. 43, 437 (1992).
25. J. B. Schlenoff, M. Li and H. Ly, J. Am. Chem. Soc. 117, 12528 (1995).

26. H. Takiguchi, K. Sato, T. Ishida, K. Abe, K. Yasea and K. Tamada, Langmuir, 16, 1703 (2000).
27. H. A. Biebuyck, C. D. Bain and G. M. Whitesides, Langmuir 10, 1825 (1994).
28. M. Wanunu, A. Vaskevich and I. J. Rubinstein, J. Am. Chem. Soc. 126, 5569 (2004).
29. J. L. Vossen, J. L. Physics of Thin Films; G. Haas, M. H. Francombe and R. W. Hoffman (Ed.), Academic Press: New York, (1977).
30. M. Hegner, P. Wagner and G. Semenza, Surf. Sci. 291, 39 (1993).
31. M. L. Hair, C. P. Tripp, Coll. Surf. A 105, 95 (1995).
32. R. R. Rye, Langmuir 13, 2588 (1997).
33. K. Kojio, S. R. Ge, A. Takahara and T. Kajiyama, Langmuir 14, 971 (1998).
34. M. J. Stevens, Langmuir 15, 2773 (1980).
35. K. Kojio, A. Takahara, K. Omote and T. Kajiyama, Langmuir 16, 3932 (2000).
36. C. P. Tripp and M. L. Hair, Langmuir 11, 1215 (1995).
37. P. Silberzan, L. Leger, D. Ausserre and J. J. Benattar, Langmuir 7, 1647 (1991).
38. D. L. Allara, A. N. Parikh and F. Rondelez, Langmuir 11, 2357 (1995).
39. A. Baptiste, A. Gibaud, J. F. Bardeau, K. Wen, R. Maoz, J. Sagiv and B. M. Ocko, Langmuir 18, 3916 (2002).
40. S. R. Wasserman, Y.-T. Tao and G. M. Whitesides, Langmuir 5, 1074 (1989).
41. N. Tillman, A. Ulman, J. S. Schildkraut and T. L. Penner, J. Am. Chem. Soc. 110, 6136 (1988).
42. G. E. Poirier and M. Tarlov, Langmuir 10, 2853 (1994).
43. R. Maoz and J. Sagiv, J. Colloid Interface Sci. 100, 465 (1984).
44. S. R. Cohen, R. Naamana and J. Sagiv, J. Phys. Chem. 90, 3054 (1986).
45. K. Bierbaum, M. Grunze, A. A. Baski, L. F. Chi, W. Schrepp and H. Fuchs, Langmuir 11, 2143 (1995).
46. M. M. Sung, C. Carraro, O. W. Yauw, Y. Kim and R. Maboudian, J. Phys. Chem. B 104, 1556 (2000).
47. A. N. Parikh, D. L. Allara, I. B. Azouz and F. Rondelez, J. Phys. Chem. B 98, 7577 (1994).
48. J. Buseman-Williams and J. C. Berg, Langmuir 20, 2026 (2004).
49. T. Vallant, H. Brunner, U. Mayer, H. Hoffmann, T. Leitner, R. Resch and G. Friedbacher, J. Phys. Chem. B 102, 7190 (1998).
50. D. Wouters, S. Hoeppener, J. P. E. Sturms and U. S. Schubert, J. Scann. Probe. Microsc. 1, 45 (2006).
51. J. B. Brzoska, I. B. Azouz and F. Rondelez, Langmuir 10, 4367 (1994).
52. G. S. Higashi, R. S. Becker, Y. J. Chabal and A. J. Becker, Appl. Phys. Lett. 58, 1656 (1991).
53. G. S. Higashi, Y .J. Chabal, G. W. Trucks and K. Raghavachari, Appl. Phys. Lett. 56, 656 (1990).
54. M. R. Linford and C. E. D. Chidsey, J. Am. Chem. Soc. 115, 12631 (1993).

55. M. M. Sung, J. Kluth, O. W. Yauw and R. Maboudian, Langmuir 13, 6164 (1997).
56. R. L. Cicero, M. R. Linford and C. E. D. Chidsey, Langmuir 16, 5688 (2000).
57. J. Terry, M. R. Linford, C. Wigren, R. Cao, P. Pianetta and C. E. D. Chidsey, Appl. Phys. Lett. 71, 1056 (1997).
58. A. Bansal, X. Li, I. Lauermann and N. S. Lewis, J. Am. Chem. Soc. 118, 7225 (1996).
59. B. J. Eves and G. P. Lopinski, Surf. Sci. 579, L89 (2005).
60. D. Narducci, L. Pedemonte and G. Bracco, Appl. Surf. Sci. 212, 649 (2003).
61. J. M. Lauerhaas and M. J. Sailor, Science 261, 1567 (1993).
62. D. K. Aswal, S. Lenfant, D. Guerin, J. V. Yakhmin and D. Vuilaume, Anal. Chim. Acta 568, 84 (2006).
63. J. H. Song and M. J. Sailor, J. Am. Chem. Soc. 120, 2376 (1998).
64. N. Y. Kim and P. E. Laibinis, J. Am. Chem. Soc. 121, 7162 (1999).
65. B.-J. de Gans, P. Duineveld and U. S. Schubert, Adv. Mater, 16, 203 (2004).
66. M. Jaschke and H.-J. Butt, Langmuir 11, 1061 (1995).
67. D. S. Ginger, H. Zhang and C. A. Mirkin, Angew. Chem. Int. Ed. 43, 30 (2004).
68. F. Stellacci, Adv. Funct. Mater. 16, 15 (2006).
69. S.-W. Chung, D. S. Ginger, M. W. Morales, Z. Zhang, V. Chandrasekhar, M. A. Ratner and C. A. Mirkin, Small 1, 64 (2005).
70. A. Kumar and G. M. Whitesides, Appl. Phys. Lett. 63, 2002 (1993).
71. J. A. Rogers and R. G. Nuzzo, Mater. Today 8, 50 (2005).
72. L. Pardo, W. C. Jr. Wilson and T. Boland, Langmuir 19, 1462 (2003).
73. A. Bietsch, M. Hegner, H. P. Lang and C. Gerber, Langmuir 20, 5119 (2004).
74. A. Y. Sankhe, B. D. Booth, N. J. Wiker and S. M. Kilbey, Langmuir 21, 5332 (2005).
75. B. Michel, A Bernard, A. Bietsch, E. Delamarche, M. Geissler, D. Juncker, H. Kind, J.-P. Renault, H. Rothuizen, H. Schmid, P. Schmidt-Winkel, R. Stutz and H. Wolf, IBM J. Res. Develop. 45, 697 (2001).
76. B. Michel, Industr. Phys. 8, 16 (2002).
77. Y. Xia, Adv. Mater. 16, 1245 (2004).
78. X. Younan and G. M. Whitesides, Ann, Rev. Sci. Mater. 28, 153 (1998).
79. Y. Xia and G. M. Whitesides, Ann. Rev. Mater. Sci. 28, 153 (1998).
80. N. J. Brewer and G. J. Leggett, Langmuir 20, 4109 (2004).
81. T. Ichii, T. Fukuma, K. Kobayashi, H. Yamada and K. Matsushige, Jpn. J. Appl. Phys. 43, 4545 (2004).
82. J. P. Folkers, P. E. Laibinis and G. M. Whitesides, J. Adhes. Sci. Technol. 6, 1397 (1992).
83. J. P. Folkers, P. E. Laibinis and G. M. Whitesides, Langmuir 8, 1330 (1992).
84. R. K. Smith, S. M. Reed, P. A. Lewis, J. D. Monnell, R. S. Clegg, K. F. Kelly, L. A. Bumm, J. E. Hutchison and P. S. Weiss, J. Phys. Chem. B 105, 1119 (2001).

85. B. Luessem, L. Mueller-Meskamp, S. Karthaeuser, R. Waser, M. Homberger and U. Simon, Langmuir 22, 3021 (2006).
86. P. H. Phong, Y. Ooi, D. Hobara, N. Nishi, M. Yamamoto and T. Kakiuchi, Langmuir 21, 10581 (2005).
87. C. D. Bain and G. M. Whitesides, J. Am. Chem. Soc. 110, 6560 (1988).
88. C. D. Bain, J. Evall and G. M. Whitesides, J. Am. Chem. Soc. 111, 7155 (1989).
89. J.-F. Liu, S. Cruchon-Dupeyrat, J. C. Garno, J. Frommer and G.-Y. Liu, Nano Lett. 2, 937 (2002).
90. S. Ryu and G. C. Schatz, J. Am. Chem. Soc. 128, 11563 (2006).
91. J.-F. Liu, J. R. Von Ehr, C. Baur, R. Stallcup, J. Randall and K. Bray, Appl. Phys. Lett. 84, 1359 (2004).
92. M. Kaholek, W.-K. Lee, B. LaMattina, K. C. Caster and S. Zauscher, Nano Lett. 4, 373 (2004).
93. B. L. Jackson and J. T. Groves, J. Am. Chem. Soc. 126, 13878 (2004).
94. H. Sugimura and N. Nakagiri, Jpn. J. Appl. Phys. 34, 3406 (1995).
95. M. Zharnikov and M. Grunze, J. Vac. Sci. Technol. B 20, 1793 (2002).
96. J. A. Preece and P. M. Mendes, Curr. Opn. Interf. Sci. 9, 236 (2004).
97. C. O'Dwyer, G. Gay, B. Viaris de Lesegno, J. Weiner, K. Ludolph, D. Albert and E. Oesterschulze, J. Appl. Phys. 97, 114309/1 (2005).
98. J. Xin, K. Mitsunori, S. Taku and Y. Yasushi, Thin Sol. Films, 464-465, 420 (2004).
99. T. Tanii, T. Hosaka, T. Miyake and I. Ohdomari, Jpn. J. Appl. Phys. 43, 4396 (2004).
100. M. Zharnikov and M. Grunze, J. Vac. Sci. Technol. B: 20, 1793 (2002).
101. Y. Xia, X.-M. Zhao and G. M. Whitesides, Microelectron. Eng. 32, 255 (1996).
102. H. Sugimura, H. Sano, K.-H. Lee and K. Murase, Jpn. J. Appl. Phys. 45, 5456 (2006).
103. A. Kuller, M. A. El-Desawy, V. Stadler, W. Geyer, W. Eck and A. Golzhauser, J. Vac. Sci. Technol. B 22, 1114 (2004).
104. A. Kumar, H. A. Biebuyck, N. L. Abbott and G. M. Whitesides, J. Am. Chem. Soc. 114, 9188 (1992).
105. E. Kim, G. M. Whitesides, M. B. Freiler, M. Levy, J. L. Lin and R. M. Osgood, Nanotechnol. 7, 266 (1996).
106. H. Sugimura, Int. J. Nanotechnology 2, 314 (2005).
107. M. J. Lercel, G. F. Redinbo, F. D. Pardo, M. Rooks, R. C. Tiberio, P. Simpson, H. G. Craighead, C. W. Sheen, A. N. Parikh and D. L. Allara, J. Vac. Sci. Technol. B 12, 3663 (1994).
108. P. M. StJohn and H. G. Craighead, J. Vac. Sci. Technol. B 14, 69 (1996).
109. C. S. Dulcey, J. H. Georger Jr, V. Krauthamer, D. A. Stenger, T. L. Fare and J. M. Calvert, Science 252, 551 (1991).
110. D. A. Stenger, J. H. Georger, C. S. Dulcey, J. J. Hickman, A. S. Rudolph, T. B. Nielsen, S. M. McCort and J. M. Calvert, J. Am. Chem. Phys. 114, 8435 (1992).

111. M.-S. Chen, C. S. Dulcey, L. A. Chrisey and W. J. Dressick, Adv. Mater. 16, 774 (2006).
112. A. C. Friedli, R. D. Roberts, C. S. Dulcey, A. R Hsu, S. W. McElvany and J. M. Calvert, Langmuir 20, 4295 (2004).
113. L. A. Chrisey, O'Ferral, B. J. Spargo, C. S. Dulcey and J. M. Calvert, Nucl. Acid Res. 24, 3040 (1996).
114. C. S. Dulcey, J. H. Georger, M.-S. Chen, S. W. McElvany, E. O'Ferral, V. I. Benezra and J. F. Calvert, Langmuir 12, 1638 (1996).
115. S. L. Brandow, M.-S. Chen, R. Aggarwal, C. S. Dulcey, J. M. Calvert and W. J. Dressick, Langmuir 15, 5429 (1999).
116. W. J. Dressick, C. S. Dulcey, J. H. Georger and J. M. Calvert, Chem. Mater. 6, 148 (1993).
117. E. T. Ada, L. Hanley, S. Etchin, Meingailis, W. J. Dressick, M.-S. Chen and J. M. Calvert, J. Vac. Sci. Technol. 13, 2189 (1995).
118. P. Doppelt and M. Stelze, Microelectron. Eng. 33, 15 (1197).
119. A. Hozumi, S. Asakura, A. Fuwa, N. Shirahata and T. Kameyama, Langmuir, 21, 8234 (2005).
120. D. Suh, J. K. Simons, J. W. Taylor, J. M. Calvert and T. S. Koloski, J. Vac. Sci. Technol. B 11, 2850 (1993).
121. R. L. Graham, C. D. Bain, H. A. Biebuyck, P. E. Laibinis and G. M. Whitesides, J. Phys. Chem. 97, 9456 (1993).
122. J. H. Moon, Y.-H. La, J. Y. Shim, B. J. Hong, K. J. Kim, T.-H. Kang, B. Kim and H. Kang, J. W. Park, Langmuir 16, 2981 (2000).
123. M. J. Lercel, R. C. Tiberio, P. F. Chapman, H. G. Craighead, C. W. Sheen, A. N. Parikh and D. L. Allara, J. Vac. Sci. Technol. B 11, 2823 (1993).
124. C. K. Harnett K. M. Satyalakshmi and H. G. Craighead, Appl. Phys. Lett. 76, 2466 (2000).
125. T. Balgar, S. Franzka and N. Hartmann, Appl. Phys. A 82, 689 (2006).
126. T. Balgar, S. Franzka, E. Hasselbrink and N. Hartmann, Appl. Phys. A 82, 15 (2006).
127. A. C. Friedli, R. D. Roberts, C. S. Dulcey, A. R. Hsu, S. W. McElvany and J. M. Calvert, Langmuir 20, 4295 (2004).
128. A. del Campo, D. Boos, H. W. Spiess and U. Jonas, Angew. Chem. Int. Ed. 44, 4707 (2005).
129. Y. J. Jung, Y.H. La, H. J. Kim, T.-H. Kang, K. Ihm, K.-J. Kim, B. Kim and J. W. Park, Langmuir 19, 4512 (2003).
130. Y. J. Jung, J.-I. Kim, T.-H. Kang, K. Ihm, K.-J. Kim, B. Kim and J. W. Park, J. Coll. Interf. Sci. 282, 241 (2005).
131. C. O. Kim, D. H. Kim, J. S. Kim and J. W. Park Langmuir 22, 4131 (2006).
132. Y. Tai, W. Eck, M. Grunze and M. Zharnikov, Langmuir 20, 7166 (2004).
133. K. Kueller, M. A. El-Desawy, V. Stadler, W. Geyer, W. Eck and A. Goelzhaeuser, J. Vac. Sci. Technol B 22, 114 (2004).

134. Y.-H. La, H. J. Kim, I. S. Maeng, Y. J. Jung and J. W. Park, Langmuir 18, 301 (2002).
135. A. Goelzhaeuser, W. Eck, W. Geyer, V. Stadler, T. Weimann, P. Hinze and M. Grunze, Adv. Mater. 13, 806 (2001).
136. W. Geyer, V. Stadler, W. Eck, A. Goelzhaeuser, M. Grunze, M. Sauer, T. Weimann and P. Hinze, J. Vac. Sci. Technol. B 19, 2732 (2001).
137. A. Biebricher, A. Paul, P. Tinnefeld, A. Goelzhaeser and M. Sauer, J. Biotechnol. 112, 97 (2004).
138. W. Eck, V. Stadler, W. Geyer, M. Zharnikov, A. Goelzhaeuser and M. Grunze, Adv. Mater. 12, 805 (2000).
139. P. A. Mendes, S. Kacke, K. Critchley, J. Plaza, Y. Chen, K. Nikitin, R. E. Palmer J. A. Preece, S. D. Evans and D. Fitzmaurice, Langmuir 20, 3766 (2004).
140. Y.-H La. Y. J. Jung, H. J. Kim, T.-H. Kang, K. Ihm, K.-J. Kim, B. Kim and J. W. Park, Langmuir 19, 4390 (2003).
141. X. Wang, W. Hu, R. Ramasubramaniam, G. H. Bernstein, G. Snider and M. Lieberman, Languir 19, 9758 (2003).
142. K. Heister, M. Zharnikov, M. Grunze, L. S. O. Johansson and A. Ulman, Langmuir 17, 8 (2001).
143. H. J. Mamin and D. Rugar, Appl. Phys. Lett. 61, 1003 (1992).
144. B. W. Chui, T. D. Stone, T. W. Kenny, H. J. Mamin, B. D. Terris and D. Rugar, Appl. Phys. Lett. 69, 2767 (1996).
145. P. Vettiger, M. Despont, U. Drechsler, U. Dürig, W. Häberle, M. I. Lutwyche, H. Rothuizen, R. Stutz, R. Widmer and G. K. Binnig, IBM J. Res. Develop. 44, 323 (2000).
146. S. Sun and G. J. Leggett, Nano Lett. 4, 1381 (2004).
147. A. A. Tseng, A. Notargiacomo and T. P. Chen, J. Vac. Sci. Technol. B 23, 877 (2005).
148. M. Geissler and Y. Xia, Adv. Mater. 16, 1249 (2004).
149. S. Krämer, R. R. Fuierer and C. B. Gorman, Chem. Rev. 103, 4367 (2003).
150. G. Binnig, H. Rohrer, Ch. Gerber and E. Weibel, Phys. Rev. Lett. 49, 57 (1982).
151. M. Ringger, H. R. Hidber, R. Schlögl, P. Oelhafen and H.-J. Güntherodt, Appl. Phys. Lett. 46, 832 (1985).
152. J. Hartwich, M. Sundermann, U. Kleineberg and U. Heinzmann, Appl. Surf. Sci. 144-145, 538 (1999).
153. U. Kleineberg, A. Brechling, M. Sundermann and U. Heinzmann, Adv. Func. Mat. 11, 208 (2001).
154. M. A. McCord and R. F. W. Pease, J. Vac. Sci. Technol. B 5, 430 (1987).
155. M. A. McCord and R. F. W. Pease, J. Vac. Sci. Technol. B 6, 293 (1988).
156. G. M. Shedd and P. Russell, Nanotechnology 1, 67 (1990).
157. F. K. Perkins, E. A. Dobisz, S. L. Brandow. J. M. Calvert, J. E. Kasokowski and C. R. K. Marrian, Appl. Phys. Lett. 68, 550 (1996).

158. J. A. Dagata, J. Schneir, H. H. Harary, C. J. Evans, M. T. Postek and J. Bennet, Appl. Phys. Lett. 56, 2001 (1990).
159. P. M. Campbell, E. S. Snow and P. J. McMarr, Appl. Phys. Lett. 63, 749 (1993)
160. M. Yasutake, Y. Ejiri and T. Hattori, Jpn. J. Appl. Phys. 32, L1021 (1993).
161. S. Gwo, T. Yasuda and S. Yamasaki, J. Vac. Sci. Technol. A 19, 1806 (2001).
162. S. Gwo, J. Phys. Chem. Solids 62, 1673 (2001).
163. Y. Kim, I. Choi, S. K. Kang, K. Choi and J. Yi, Microelectron. Eng. 81, 341 (2005).
164. N. Farkas, J. R. Comer, G. Zhang, E. A. Evans, R. D. Ramsier and J. A. Dagata, J. Vac. Sci. Technol. A 23, 846 (2005).
165. B. W. Maynor, J. Li. C. Lu and J. Liu, J. Am. Chem. Soc. 126, 6409 (2004).
166. Y. Matsuzaki, N. Ota, A. Yamada, A. Sandhu and M. Konagai, J. Cryst. Growth 251, 276 (2003).
167. Z. J. Davis, G. Abadal, O. Hansen, X. Borisé, N. Barniol, F. Pérez-Murano and A. Boisen, Ultramicroscopy 97, 467 (2003).
168. K. Matsumoto, Y. Gotoh, T. Maeda, J. A. Dagata and J. S. Harris, Jpn. J. Appl. Phys. 38, 477 (1999).
169. S. C. Minne, Ph. Flueckiger, H. T. Soh and C. F. Quate, J. Vac. Sci. Technol. B 13, 1380 (1995).
170. H. Sugimura, T. Uchida, N. Kitamura and H. Masuhara, Appl. Phys. Lett. 63, 1288 (1993).
171. H. Sugimura and N. Nakagiri, Jpn. J. Appl. Phys. 34, 3406 (1995).
172. M. Rolandi, C. F. Quate and H. Dai, Adv. Mater. 14, 191 (2002).
173. T. Kondo, M. Yanagisawa, L. Jiang, D. A. Tryk and A. Fujishima, Diamond Related Mater. 11, 1788 (2002).
174. H. Sugimura and N. Nakagiri, J. Vac. Sci. Technol. A 14, 1223 (1996).
175. E. S. Snow and P. M. Campbell, Science 270, 1639 (1995).
176. J. A. Dagata, Science 270, 1625 (1995).
177. H. Sugimura and N. Nakagiri, Thin Sol. Films 281-282, 572 (1196).
178. J. A. Dagata, T. Inoue, J. Itoh and H. Yokoyama, Appl. Phys. Lett. 73, 271 (1998).
179. F. Pérez-Murano, K. Birkelund, K. Morimoto and J. A. Dagata, Appl. Phys. Lett. 75, 199 (1999).
180. F. Pérez-Murano, K. Birkelund, K. Morimoto and J. A. Dagata, Appl. Phys. Lett. 75, 199 (1999).
180. H. Kuramochi, K. Ando, T. Tokizaki and H. Yokoyama, Appl. Phys. Lett. 84, 4005 (2004).
181. J. A. Dagata, F. Perez-Murano, C. Martin, H. Kuramochi and H. Yaokohama, J. Appl. Phys. 96, 2386 (2004).
182. F. Pérez-Murano, C. Martin, N. Barniol, H. Kuramochi, H. Yokoyama and J. A. Dagata, Appl. Phys. Lett. 82, 3086 (2003).

183. J. A. Dagata, F. Perez-Murano, C. Martin, H. Kuramochi and H. Yaokohama, J. Appl. Phys. 96, 2393 (2004).
184. E. S. Snow, G.G. Jernigan and P. M. Campbell, Appl. Phys. Lett. 76, 1782 (2000).
185. Y. Y. Zhang, J. Zhang, G. Lou, X. Zhou, G. Y. Xie, T. Zhu and Z. F. Liu, Nanotechnol. 16, 422 (2005).
186. F. S.-S. Chien, W.-F. Hsieh, S. Gwo, A. E. Vladar and J. A. Dagata, J. Appl. Phys. 91, 10044 (2002).
187. J. T. Sheu, C. C. Chen, K. S. You and S. T. Tsai, J. Vac. Sci. Technol. B 22, 3154 (2004).
188. S. J. Ahn, Y. K. Jang, S. A. Kim, H. Lee and H. Lee, Ultramicroscopy 91, 171 (2002).
189. H. Sugimura, T. Hanji, K. Hayashi and O. Takai, Ultramicroscopy 91, 221 (2002).
190. W. Lee, E. R. Kim and H. Lee, Langmuir 18. 8375 (2002).
191. H. Sugimura, O. Takai and N. Nakagiri, J. Electroanal. Chem. 473, 230 (1999).
192. H. Sugimura, T. Hanji, K. Hayashi and O. Takai, Adv. Mat. 14, 524 (2002).
193. Q. Li, J. Zheng and Z. Liu, Langmuir 19, 166 (2003).
194. M. He, X. Ling, J. Zhang and Z. Liu, J. Phys. Chem. B 109, 10946 (2005).
195. X. Ling, X. Zhu, J. Zhang, T. Zhu, M. Liu, L. Tong and Z. Liu, J. Phys. Chem. B 109, 2657 (2005).
196. M. Yang, Z. Zheng, Y. Liu and B. Zhang, J. Phys. Chem. B 110, 10365 (2006).
197. M. Yang, Z. Zheng, Y. Liu and B. Zhang, Nanotechnology 17, 330 (2006).
198. M. Shin, C. Kwon, S. Y. Kim, H. J. Kim, Y. Roh, B. Hong, J. B. Park and H. Lee, Nano Lett. 6, 1334 (2006).
199. I. Choi, S. K. Kang, J. Lee, Y. Kim and J. Yi, Biomater. 27, 4655 (2006).
200. B. Kim, G. Pyrgiotakis, J. Sauers and W. M. Sigmund, Coll. Surf. A 253, 23 (2005).
201. M. Ara, H. Graaf and H. Tada, Jpn. J. Appl. Phys. 41, 4894 (2002).
202. S. Liu, R. Maoz, G. Schmid and J. Sagiv, Nano Lett. 2, 1055 (2002).
203. R. Maoz, S. R. Cohen and J. Sagiv, Adv. Mat. 11, 55 (1999).
204. R. Maoz, E. Frydman, S. R. Cohen and J. Sagiv, Adv. Mat. 12, 424 (2000).
205. R. Maoz, E. Frydman, S. R. Cohen and J. Sagiv, Adv. Mat. 12, 725 (2000).
206. S. Hoeppener, R. Maoz, S. R. Cohen, L. Chi, H. Fuchs and J. Sagiv, Adv. Mat. 14, 1036 (2002).
207. D. Wouters and U. S. Schubert, Angew. Chem. Int. Ed. 43, 2480 (2004).
208. D. Wouters and U. S. Schubert, Langmuir 19, 9033 (2003).
209. D. Wouters, J. P. E. Sturms and U. S. Schubert, Trans. MRS-J, 29 (2004).
210. S. Hoeppener and U. S. Schubert, Small 1, 628 (2005).
211. D. Wouters and U. S. Schubert, J. Mater. Chem. 15, 2353 (2005).
212. S. Hoeppener, A. S. Susha, A. L. Rogach, J. Feldmann and U. S. Schubert, Current Nanosci. 2, 135 (2006).

213. D. Wouters, R. Willems, S. Hoeppener, C .F. J. Flipse and U. S. Schubert, Adv. Func. Mater. 15, 938 (2005).
214. S. Hoeppener, J. H. K. Van Schaik and U. S. Schubert, Adv. Funct. Mater. 16, 76 (2006).
215. S. Liu, R. Maoz and J. Sagiv, Nano Lett. 4, 845 (2004).
216. Unpublished results. G. Wei, S. Hoeppener, U. S. Schubert (2004)
217. Unpublished results. D. Wouters, U. S. Schubert (2005)
218. C. R. Becer, C. Haensch, S. Hoeppener and U. S. Schubert, Small, in press (2007).
219. S. Hoeppener, R. Maoz and J. Sagiv, Nano Lett. 3, 761 (2003).
220. S. Hoeppener, R. Maoz and J. Sagiv, Adv. Mater. 18, 1286 (2006).
221. A. Checco, Y. Cai, O. Gang and B. M. Ocko, Ultramicroscopy 106, 703 (2006).
222. Y. Cai and B. M. Ocko, J. Am. Chem. Soc. 127, 16287 (2005).
223. Z. M. Fresco, I. Suez, S. A. Backer and J. M. J. Fréchet, J. Am. Chem. Soc. 126, 8374 (2004).
224. Z. M. Fresco and J. M. J. Fréchet, J. Am. Chem. Soc. 127, 8302 (2005).
225. A. Ivanisevic, J.-H. Im, K.-B. Lee, S.-J. Park, L. M. Demers, K. J. Watson and C. A. Mirkin, J. Am. Chem. Soc. 123, 12425 (2001).
226. Y. Li, B. W. Maynor and J. Liu, J. Am. Chem. Soc. 123, 2105 (2001).
227. I. Suez, S. A. Backer and J. M. J. Fréchet, Nanolett. 5, 321 (2005).
228. O. Schneegans, A. Moradpour, H. Houzé, A. Angelova, C. H. de Villeneuve, P. Allongue and P. Chrétien, J. Am. Chem. Soc. 123, 11486 (2001).
229. C. Martín, G. Rius, X. Borrisé and F. Pérez-Murano, Nanotechnol. 16, 1016 (2005).
230. S.-Y.Jang, M. Marquez and G. A. Sotzing, Synth. Metals 152, 345 (2000).
231. S.-Y. Jang, M. Marquez and G. A. Sotzing J. Am. Chem. Soc. 126, 9476 (2004).

CHAPTER 9

RECENT DEVELOPMENT OF ORGANOGELS TOWARDS SMART AND SOFT MATERIALS

Norifumi Fujita, Pritam Mukhopadhyay, Seiji Shinkai*

*Department of Chemistry and Biochemistry, Graduate School of Engineering, Kyushu University, 744 Moto-oka, Nishi-ku, Fukuoka 819-0395, Japan; *Phone: +81-92-802-2818, Fax: (+81) 92-802-2820, E-mail: seijitcm@mbox.nc.kyushu-u.ac.jp*

Gels are generally classified into organogels, hydrogels, polymer gels and aerogels. In the present article, we will focus on the different aspects of organogels with an emphasis on the modulation of gelation properties and functions in the presence of a stimulus. Organogels are viscoelastic, thermoreversible materials consisting of low molecular-weight compounds [1]. They form a continuous three-dimensional entangled network in the solvent, which in turn prevents the liquid from flowing. The self-assembly of these low molecular-weight organogelators (LMWG) into fibrous networks is driven by multiple, weak interactions such as dipole–dipole, van der Waals, π-stacking and hydrogen-bonding interactions. The self-assembly process from a single molecule to fibers and finally to entangled network structures is thus completely thermodynamically reversible. The network structure is constituted of well-ordered arrays of molecules, which can be of several micrometers in length and can have intriguing architectures like tapes, ribbons, rods, fibers, sheets, cylinders, etc. [2]. Depending upon the operative driving forces for molecular aggregation the organogelators can be broadly classified into two groups: a) non-hydrogen-bond based gelators and b) hydrogen-bond based gelators. Cholesterol or steroid derivatives are examples of the former group while aliphatic amide, urea and saccharide containing gelators represent the latter group. Recently, a new designing approach has come into limelight which utilizes π-π stacking as one of the non-covalent forces to form 1-D organogels [3]. When the gelator molecule shows very high gelation ability, i.e., a very small concentration of the compound (generally less than 0.1 wt%) is needed to gel the solvent, the compound is called a 'supergelator'. On the other hand, when a gelator molecule can gel a wide range of solvents it is called a 'versatile gelator'. One of the prominent examples for a versatile gelator is that of a cholesteryl appended pyridine ligand that gels a large spectrum of common organic solvents [4]. With these lines of information in hand, organogels are now donated to the fields of supramolecular chemistry and materials science.

1. General Introduction

In this review article, we will describe the plethora of building blocks that can be used to design organogelators (Section 2). In Section 3, we will focus on the gels that show adaptability and multi-responsiveness to stimuli. Considering the enthralling properties of these multi-responsive gelator molecules and the enormous amount of interest they have continuously received, we have been inspired to term these classes of molecules as 'Second Generation Gels.' This section has been divided into several subsections, which deal with the various stimuli and interactions (host-guest, metal ions, charge-transfer, etc.) that regulate the properties and functions of the gels. In Section 4 we will discuss some interesting properties of miscellaneous gels. Our final section (Section 5) comprises a discussion on gels that are potential candidates for biomedical applications.

2. First Generation Organogels

We have termed those organogelators as *First Generation Organogels*, which are primarily designed to study new gelation properties. In general, this category of gelators has limited higher order functions. In this section, we will look at different organogels that have been reported with the main aim of finding new structural motif for gelation. We have tried to have a detailed description of the functional properties of gelator molecules in Section 3 under the heading of *Second Generation Organogels*.

Although many of the discoveries of new motifs for gelation have been found accidentally, researchers can now quite successfully design molecules with gelation properties, taking a leaf from the vast library of present day organogels. On the other hand, a successful prediction whether a newly designed molecule would gel a particular solvent still remains elusive to us. This might be due to our lack of knowledge and understanding on the physics: that is, how a solvent molecule interacts with the gelator in the gel phase, and what their kinetics, thermodynamics and their subsequent packing modes are in the bulk phase.

2.1. Steroid-Based Gelators

Steroids form an important class of gelator molecules. In 1979, the first steroid gelator **1** (D-3β-hydroxy-17,17-dipropyl-17a-azahomoandrostanyl-17a-oxy) to gel hydrocarbons, was found accidentally by Martin-Borret *et al.*, while they were synthesizing D-homosteroidal nitroxide free radical [5]. Intrigued by this important finding, the steroid unit became an attractive choice for researchers to design efficient gelator molecules. We have demonstrated that a variety of steroidal gelators could be synthesized, in which azobenzene, crown-ethers, etc. could be appended at the C_3 position of the steroidal unit [6]. These gelators **2** – **5** were found to gelate a wide range of common organic solvents and show interesting phenomena like light and metal ion responsiveness.

2.2. Anthracene-Based Gelators

Weiss *et al.* reported anthryl and anthraquinone appended steroid-based gelators **6** and **7** that could gel hydrocarbons, alcohols, amines, etc. [7]. Brotin *et al.* have reported that dioxyanthracene **8** and anthraquinone **9** molecules when derivatized with alkyl groups of proper chain length could gel alcohols, aliphatic amines, alkanes, etc. [8]. It is considered that in these gelators, both the van der Waals interaction among the cholesterol groups and the π-π stacking interaction among the aromatic groups are operative cooperatively.

Chart 1

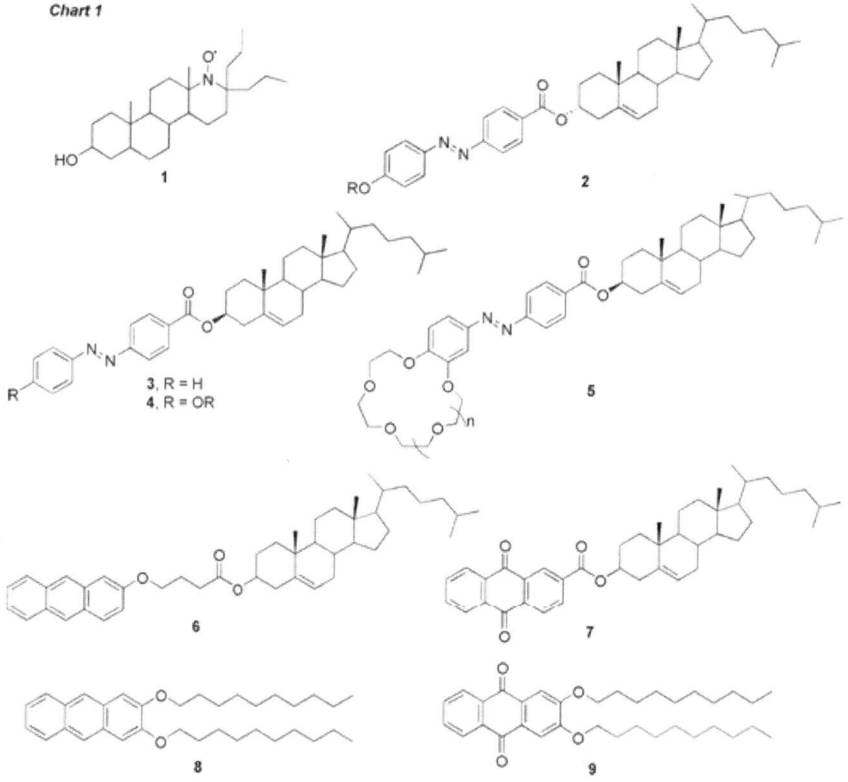

2.3. Amino Acid and Ammonium Carbamate-Based Gelators

Menger *et al.* exploited the oligomeric α–amino acid gelators **10**, which could gel a wide variety of organic liquids [9]. Amino acid type gelators like 2-octadecyl 4-[[(1-naphthylamino)carbonyl]amino]benzoate **11** have been designed by Campbell *et al.* [10]. Hanabusa and co-workers have subsequently reported amino acid-based gelators, for example, N-benzyloxycarbonyl-L-alanine 4-hexadecanoyl-2-nitrophenyl ester **12**, N-benzyloxy-carbonyl-L-valyl-L-valine n-octadecyl amide **13**, both of which were shown to gel common organic solvents [11]. They have also reported a novel class of cyclodipeptides **14** having four hydrogen-bonding sites per molecule, which could gel organic liquids [12]. De Vries reported a new class of depsipeptide molecules cyclo[CH_2CO-L-leucyl-L-leucine] **15**, which could gel diethyl ether, acetonitrile but not alkanes or alcohols [13]. Interestingly Bhattacharya *et al*. have further explored the N-acyl derivatives of short amino acids such as N-lauroyl-L-alanine alanine **16** and have found that these compounds could gelate the oil phase selectively from a two-phase mixture of water and oil [14].

Chart 2

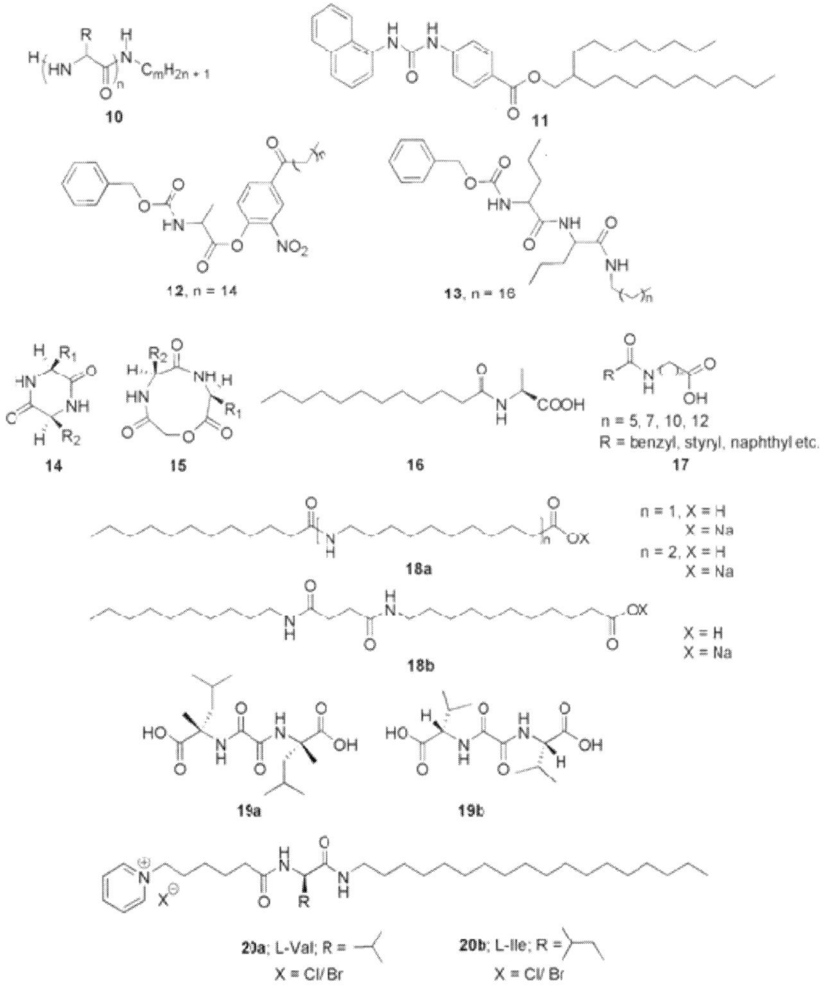

Vögtle and co-workers have shown that gelators based on N-acyl-1,ω–amino acid scaffolds can gelate DMF in the mM domain. Thus, a diverse number of gelator molecules **17** were synthesized by varying the pendant n-acyl group (with R being benzyl, styryl, naphthyl, dendron, adamantly, furyl, etc.), amino acid chain length, and the ionization state of the carboxylic acids [15]. This group has further extended their study by utilizing 1-aminoundecanoic acid as the building block to prepare (oligo) amide gelators **18a** and **18b** capable of gelating organic solvents and water [16]. Small organic gelators based on bis(amino acid) oxalamide units have been reported by Zinic *et al.* These gelators **19a** and **19b** can gelate solvents of low polarity and the thermal properties of the gels could be altered substantially by slight changes in the solvent composition and polarity. These gels remain stable without any apparent gel-to-sol transition, 40-50°C above the boiling point of the gelled solvent [17]. L-valine and L-isoleucine amphiphiles **20a** and **20b**, having a positively charged terminal group, have been shown to be a versatile gelator as it gels organic solvents, oil, pure water, and aqueous solutions containing inorganic acids [18]. These results consistently support the view that amino acid oligomers can serve as powerful structural segments to design organogelators. These examples demonstrate that the amide group can act as a driving force for gel formation utilizing the hydrogen-bonding interaction. We consider, therefore, that this chemistry has a profound relation with the β-sheet structure in proteins and the formation of amyloids related to Alzheimer disease.

Weiss *et al.* have reported novel ammonium carbamate-based reversible organogelators **21b**, based on the spontaneous and isothermal uptake of CO_2 by the corresponding amine and subsequent loss of CO_2 from **21b** upon heating (Scheme 1). The ammonium carbamate-based organogels have been found to have greater thermodynamic stability than the corresponding amines [19]. These results are interesting firstly because of reversible formation of gel-forming amide-like groups and secondly because of the novel approach toward CO_2 fixation.

Scheme 1

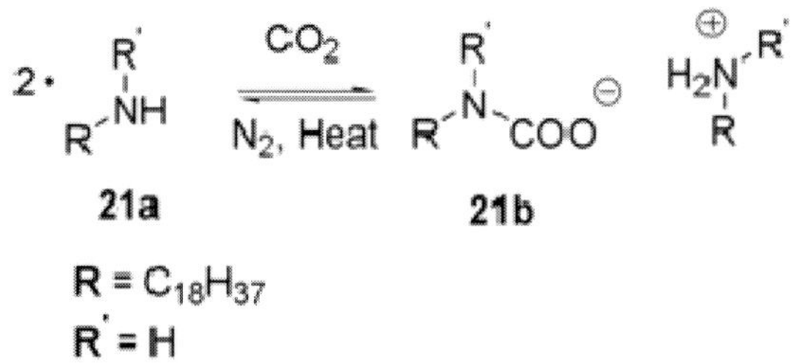

2.4. Sugar-Based Organogels

Sugar-based organogels primarily form the superstructure with the help of H-bonding interactions. Using a variety of monosaccharides as a molecular library, one can design many potential organogelators. We have skipped a discussion on these important molecules and would suggest the readers to refer to the comprehensive review on sugar-based organogels by Shinkai *et al.* [20].

2.5. Chiral Gelators

Hanabusa *et al.* reported the first examples of cyclohexanediamine-based organogels [21]. The motivation for molecular design of this class of gelators seems to be the β-sheet structure in proteins. Subsequently, there have been a few reports on this interesting class of molecules. Lehn *et al.* reported an isomeric, alternating pyridine-pyridazine based heterocyclic compounds **22** which yielded a helical structure with twelve heterocycles per turn. Gel formation was observed in pyridine and dichloromethane [22]. Huc *et al.* reported gemini surfactants **23a-b** based on dimers of cetyltrimethylammonium ions which formed gels in most of the chlorinated solvents. The chirality was introduced by the anionic salts of L-tartarate [23]. It is interesting that the gigantic helical superstructures can be created by non-covalently assembling small low molecular-weight compounds.

2.6. Glycoluril-Based and Macrocycle-Based Gelators

Menger *et al*. reported the first glycoluril-based gelator **24**, which gels benzyl alcohol. Interestingly, the authors observed an unusual transformation from macroscopic fibers to sheet-like structures during the preparation of the xerogel of this gelator molecule [24]. As a new class of gelators, one can utilize the macrocyclic building blocks in designing novel gelator molecules. Macrocycles like calix[n]arenes, crown-ethers, porphyrins, and cyclodextrins are important molecules due to their ability to complex a vast range of guest molecules. These macrocyles, when functionalized with suitable groups, have been found to show interesting gelation properties towards common organic solvents.

Chart 3

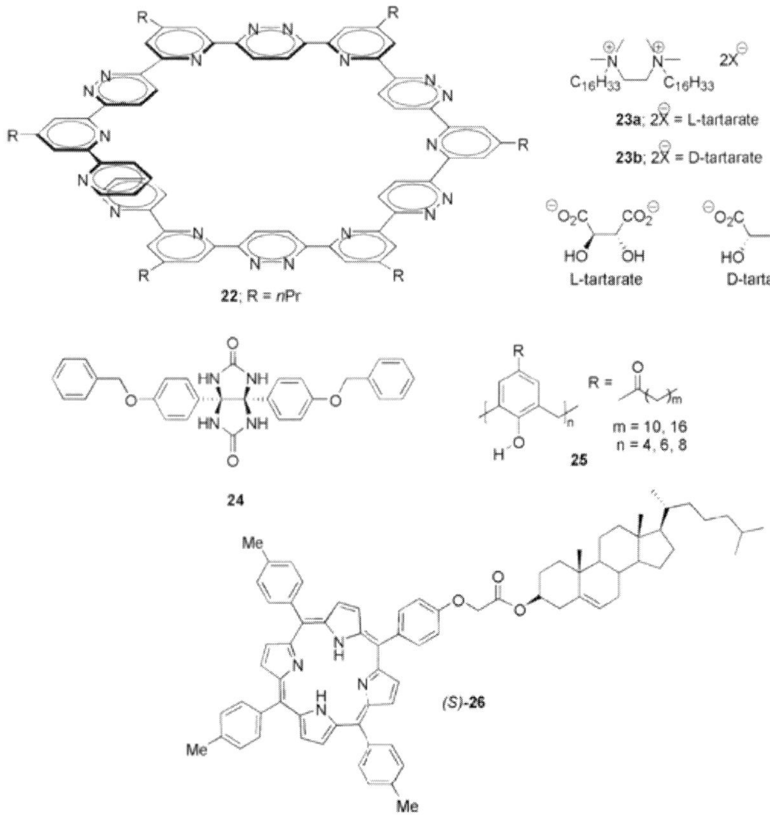

Our group reported the first example of calix[n]arene (n = 4 or 8) based gelators **25**, having alkanoyl chains at the para positions of the phenolic ring (m = 12 or 18) which could gel a wide range of organic solvents [25]. Interestingly, calix[6]arene with a ring size in between that of calix[4] and calix[8]arenes could not gel organic solvents. Taking an advantage of the macrocyclic structure, one can expect that their gelation properties can be changed by the host-guest type interaction: in other words, it is possible to design a new class of stimuli-responsive functional gel system. Similarly, cholesterol-appended porphyrins **26** can also act as gelators of organic solvents [26]. In these organogels, the porphyrin core is located around a central column of cholesterol moieties resembling a spiral staircase. Other important macrocyclic structures like phtalocyanines [27] and cyclodextrins [28] have also found to act as gelators of some specific organic solvents.

2.7. Gelators Based on Complex Building Blocks

Gelator molecules that are formed from complex building blocks requires a set of cooperative interactions like H-bonding, π-stacking, etc. for the self-assembly process and in turn the forms organogels. Ajayaghosh *et al.* reported the first phenylenevinylene-based organogels **27a-b**, which undergo gelation as a consequence of cooperative hydrogen-bonding and π–stacking interactions [29]. In a recent communication, Park *et al.* have demonstrated that simple conjugated organogels **30** can be designed without the incorporation of long alkyl chains, urea/amide moieties, sugar units or steroidal substituents [30]. Both of the above examples have been discussed in Section IV. Finn *et al.* have recently reported that formamidines **28** and **29** can be efficient gelators of alcoholic solvents requiring as little as 0.3% of the gelator molecule [31]. Linear 1H-imidazole amphiphiles **31a-c** have been recently shown to gel various mixtures of organic solvents (1 wt %). Interestingly, it has been found that a hydrophilic antibiotic drug like Norfloxacin could be entrapped into the strands of these self-assembled amphiphiles [32]. The finding suggests that these gels may act as potential drug carriers.

Chart 4

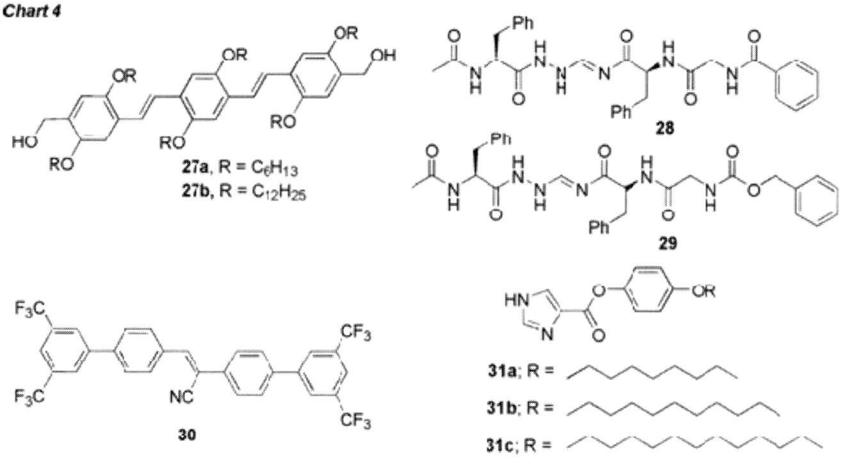

3. Second Generation Organogels

To design effective functional materials like sensors, actuators, molecular devices, etc. molecules have to exhibit responsiveness to stimuli. In this context, it becomes advantageous to use LMWGs because they show reversible and sharp sol-gel transition in the presence of thermal stimuli. We and other groups have shown that by judicious designing, a plethora of stimuli like guest molecules, complementary H-bonding, metal ions, light, oxidizing/reducing agents can be used for modulating the properties and functions of gelator molecules. Gelator molecules possessing such diverse properties and inherent ability to adapt and respond can be called as *Second Generation Organogels*.

In this section, we will look at a variety of such tailored gelator molecules and will try to have an answer for the following questions: 1) what is the basic designing principle required to create gelator molecules with tunable function and properties, 2) what are the various properties and functions of a gel that can be modulated as a response to stimuli, and 3) what is the extent to which these properties can be modulated.

3.1. Host-Guest Interaction

Host-guest interaction has been known for a long time in solution and in crystals. With the advent of sophisticated instruments and a better knowledge of the aggregational properties of gels, researchers have started to utilize this long-known non-covalent tool to modulate the gelation ability, thermodynamic stability and morphology upon guest binding. Thus, one may apply various host molecules such as crown-ethers, aza macrocycles, calixarenes, etc. in designing this class of gelators and various guests and metal ions as stimulating signal species (Figure 1).

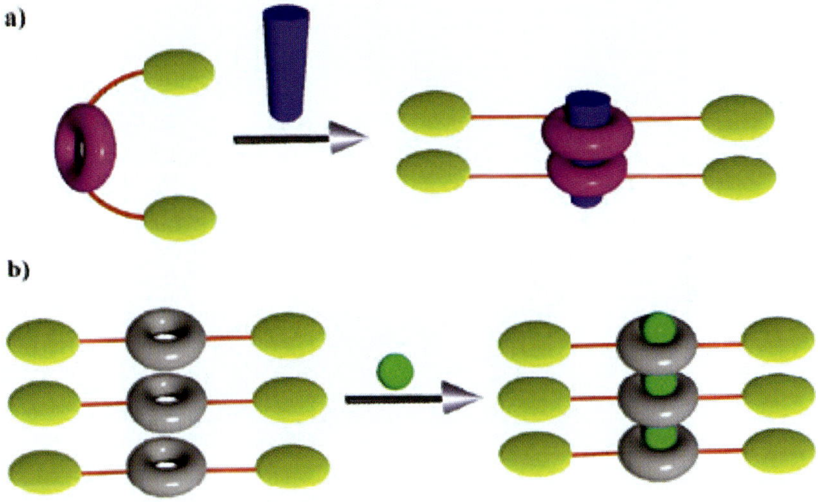

Figure 1. Concept for different host-guest binding strategies (a and b) that have been applied to organogels to modulate its conformation, gel stability, function, and morphology.

We have recently reported that a porphyrin molecule equipped with programmed H-bonding sites can dimerize into a 'capsule' using a circular array of intermolecular H-bonds [33]. The cavity of the 'capsule' has a dimension exactly comparable to the size of [60]fullerene (C_{60})

(Figure 2). Thus the complex $(\mathbf{32})_{2n}\bullet(C_{60})_n$ could gel benzene, toluene, p-xylene and anisole while **32** itself could gel only benzene and toluene. Similarly, the gel-to-sol phase transition temperature, i.e., T_{gel} of **32** in benzene increased sharply from 79°C to 120°C upon addition of C_{60}. The SEM and TEM images of the xerogel of **32** reveal a two-dimensional sheet-like structure unfavorable to gel formation, which is transformed in the presence of C_{60} into a one-dimensional fibrous network structure, which is suitable for gel formation. The additional stability gained by the gelator **32** in the presence of C_{60} is thus solely due to the formation of a very stable 2:1 host-guest $(\mathbf{32})_{2n}\bullet(C_{60})_n$ complex.

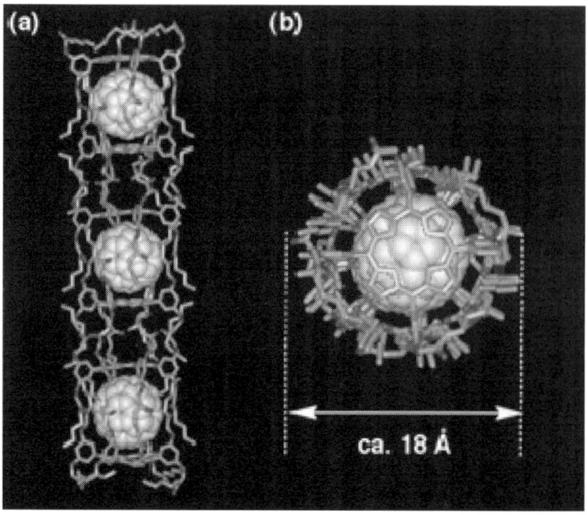

Figure 2. An energy-minimized structure of a porphyrin capsule that encapsulates C_{60} in its one-dimensional matrix. The image shows (a) the side view and (b) the top view.

It has also been shown from our laboratory that the gelation ability of the host Zn(II) porphyrin-appended cholesterol molecules can be efficiently modulated by the addition of C_{60} guest molecules [34]. Thus, at 5°C compounds **33a** (n = 2) and **33c** (n = 4) with even number of spacer units could gelate aromatic hydrocarbons such as benzene, toluene and p-xylene, while they were converted into sols at 20°C. Interestingly, upon the addition of 0.5 equiv of C_{60} the gel structure of **33a** and **33c** could be well maintained even at 20°C. This significant stabilization effect of the gel structure is possible due to the 2:1 sandwich complex formed by two Zn(II) porphyrin planes and one C_{60} molecule. In sharp contrast, **33b** and **33d** with odd number of spacer units (n = 3 and 5, respectively) could not gelate these solvents even in the presence of C_{60}. This novel even-odd relationship is attributed to the difference that the compounds with an even number of spacer units tend to adopt an extended structure while those with an odd number tend to adopt a bent structure. It can be suggested that the extended structure, in which the Zn(II) porphyrin plane is perpendicular to the cholesterol plane, helps to build up the one-dimensional column necessary for formation of the gel structure.

The well-known chemistry of the threading of alkyl ammonium or bipyridinium ions by large crown ether molecules, developed by Stoddart′s group, can be profitably applied to generate a novel host-guest-based gelator system. For instance, a dibenzo-24-crown-8 derivative bearing two cholesterol moieties **34** is either insoluble or undergoes precipitation from most organic solvents whereas in the presence of a diammonium guest $\mathbf{G_1}$ the pseudorotaxane complex can efficiently gelate these solvents [35]. Furthermore, the T_{gel} values increased monotonously with the addition of the guest alkyl ammonium ions. A 2:1 host:guest ratio was confirmed by various spectroscopic methods. The SEM pictures of the xerogel clearly depicted a conformational change of **34** from a three-dimensional aggregational structure (Figure 3a) to a one-dimensional fibrous structure with 40-250 nm diameters (Figure 3b) triggered by the addition of alkyl ammonium guest molecule $\mathbf{G_1}$.

We have also reported recently a dibenzo-crown-based gelator bearing two citrolloyl moieties **35a** and **35b**, which exhibit better gelation ability and higher thermodynamic stability after binding to bipyridinium-based ions [36]. T_{gel} values of **35a** and **35b** increased sharply by 15°C and 20°C, respectively, in the presence of $\mathbf{G_2}$ at a stoichiometric ratio of 1:2 (guest/host gelator). One can regard, therefore, that these host-guest type interactions change the conformation of the host/guest complexes to stabilize the organogels by the synergic effect of H-bonding, hydrophobic and donor-acceptor interactions.

Chart 5

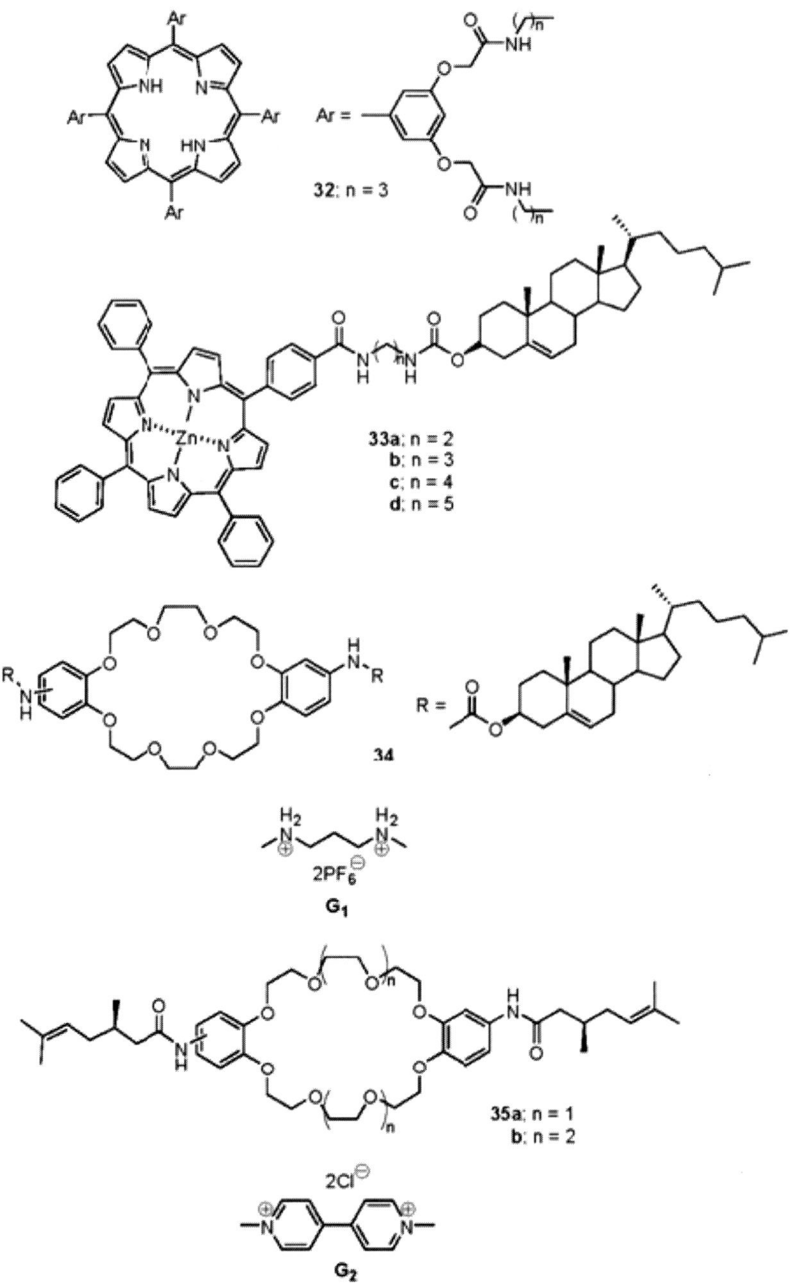
Ar
NH
N
HN
Ar =
NH
32; n = 3
33a; n = 2
b; n = 3
c; n = 4
d; n = 5
Zn
34
R =
H_2
2PF$_6$
G$_1$
35a; n = 1
b; n = 2
2Cl
G$_2$

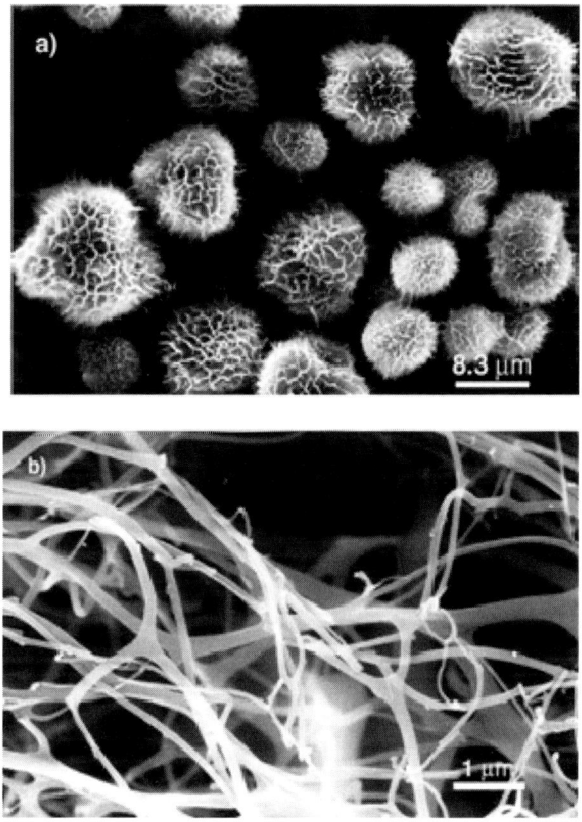

Figure 3. SEM images showing morphological changes induced upon guest addition, showing the morphology of crown-ether **34** a) before the addition of guest, and b) after the addition of guest molecule $\mathbf{G_1}$.

3.2. H-Bonding Interaction

It is well-known that nature has evolved with time a remarkably rich and natural supramolecular chemistry. This stimulates us with tremendous inspiration and motivation to seek for more sophisticated nonbiological analogues that can carry out functions and possess properties not found in nature. Towards such a goal, molecular recognition events occurring in biomolecules like nucleobases or other artificial supramolecular motifs like pyridine-carboxylic acid interactions can be applied to the gel phase to modulate gelation properties and to develop new functions. In this sub section, we will look at some recent examples of complementary H-bonding interactions, which have been used to tune the gelation properties. Figure 4 illustrates the different designing strategies used to design complementary H-bonding-based organogel systems. In the first approach, both the H-bond donor and the acceptor moieties are equipped with gelator molecules like cholesterol units or long-chain alkyl groups. Whereas, in the second approach, either the H-bond donor or the acceptor can have gelator groups integrated with itself. Moreover, a 1:1 or 2:1 H-bonded systems can be designed using these aforementioned principles.

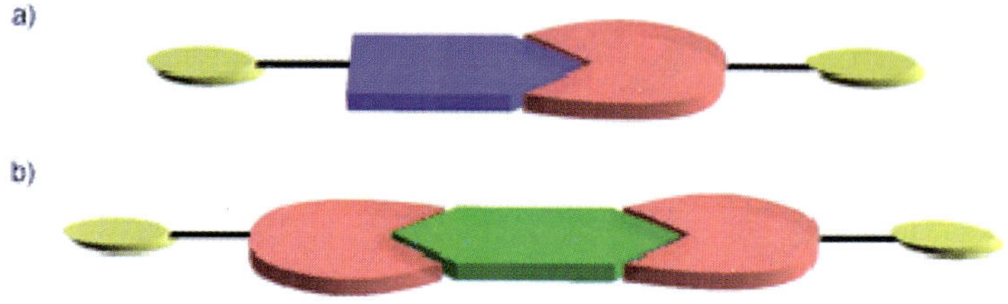

Figure 4. Different designing strategies for H-bonding based gelator systems (a and b) that have been applied to organogels to modulate its gel stability, and morphology (pink disc: H-bond acceptor, blue and green wedge: H-bond donor, yellow disc: gelator, and black line: spacer).

We have recently demonstrated that gel stability and morphology of an uracil appended cholesterol gelator **36** can be controlled in different ways by the addition of complementary **41** and non-complementary nucleobase derivatives **37** – **40** [37]. Thus, by the addition of a complementary compound **41** bearing an adenine and a cholesterol moiety the gel stability increases by 37°C at a 1:1 molar ratio. In contrast, the non-complementary partners can sharply decrease the stability of the gel. This effect is most pronounced in case of the guanine derivative which reduces the T_{gel} of **36** by 92°C. Moreover, morphology of the gel is transformed from fiber-like aggregates to sheet-like structure in the presence of **41**. Kim and co-workers have also reported that thymidine-based organogelator **42** – **46** can show good gelation ability towards non-polar solvents utilizing H-bonding, van der Waals and π-π stacking interactions [38]. On the other hand, Shimizu *et al.* have reported a novel finding based on a hydrogelator that forms DNA-like nanofibers, which have a double-helical arrangement of A-T base pairs [39]. These superstructures are formed as a result of the complementary oligonucleotide-templated self-assembly of thymidine-appended bolaamphiphiles with oligoadenylic acids.

Chart 6

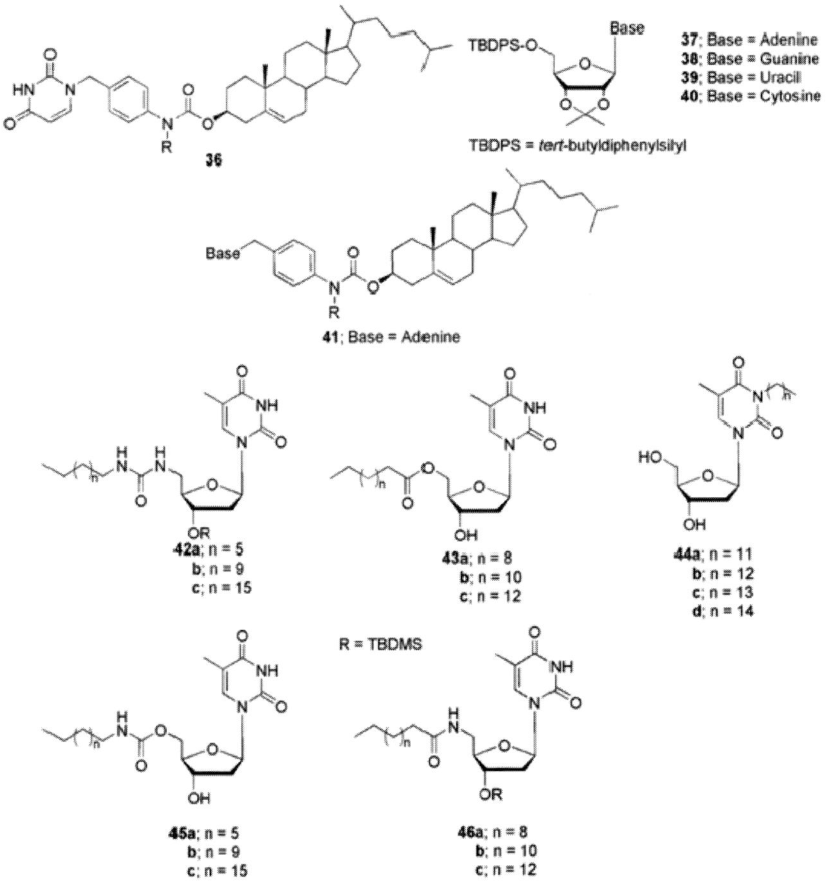

Working on similar lines, we have also shown that it is indeed possible to drastically modulate the aggregation mode and the gelation property of 5-esterified thymidine derivative by the addition of the complementary H-bonding partner, i.e., RNA [40]. The opaque gel formed by **47** changed to a transparent gel on addition of the complementary poly(A)/lipid **48** composite, while the gel remains unchanged on addition of the non-complementary partner, i.e., poly(C)/lipid **49** composite. In addition, electron micrograph images show sharp changes in the morphology of the thymidine gelator molecule upon H-bonding with **48**. The SEM and TEM images show a plate-like crystalline structure for gelator **47** and the mixture of **47** and poly(C)/lipid. The crystalline structure of the gelator molecule **47** quite sharply changes into a 3-D entangled fiber structure in the presence of **48**. Such drastic modulation of the gelation property as well as that of the morphological architecture is believed to arise due to the complementary A-T base pairing.

Chart 7

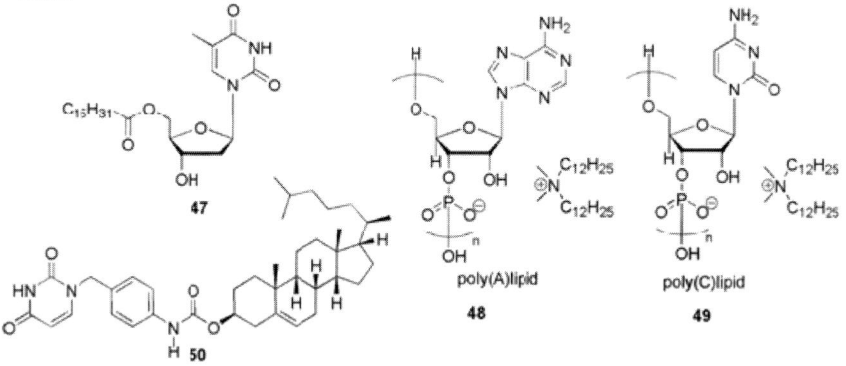

In a recent communication, we have reported that when uracil-appended cholesterol gelator molecule **50** is mixed with poly(A) or poly(C), addition of RNA strongly affects not only the T_{gel} values but also the morphology of the gel fiber structure [41]. The complementary poly(A) addition stabilizes the gel system and also induces the helical structure in the original gel fiber. In contrast, although the mixture of **50**

and poly(C) gives a fibrous network, the helical structure could not be formed due to the non-specific **50**-poly(C) interaction.

Scheme 2

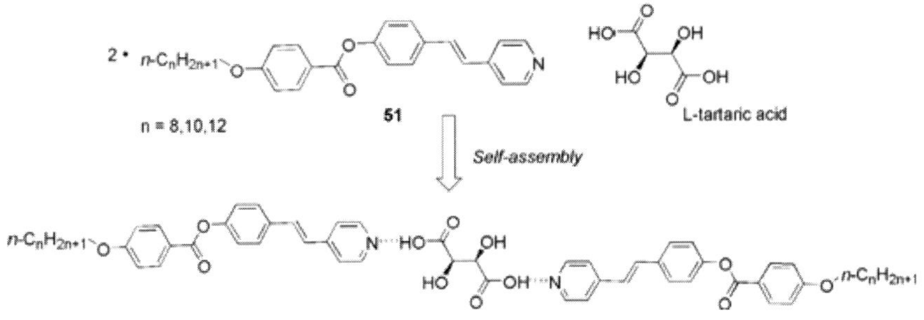

Other synthetic supramolecular complementary interactions like those between H-bond donor carboxylic acid and H-bond acceptor pyridine moieties have attracted considerable interest from crystal engineers and material chemists due to their applications in SHG active NLO materials. Utilizing this motif, Lu *et al.* have recently reported an elegant example of a binary organogel (Scheme 2). They have shown that 4-(4-alkoxybenzoyloxy)-4′-stilbazole **51** can gel most of the chlorinated solvents but only in the presence of L-tartaric acid. Interplay of multiple H-bonds and π-π_interactions drives the gel formation process [42]. A remarkable 1000-fold increase in the fluorescence intensity was observed in the gel phase due to this complementary H-bonding interaction. A synergistic effect of limited molecular motions and formation of J-aggregates are thought to be the reason for this unusual fluorescence enhancement. Porphyrin-based organogelators appended with pyridine and carboxylic acids **52** can show interesting tunable morphological properties depending on the polarity of the solvent chosen for gelation. The aggregation of **52** is governed by three different forces, that is, pyridine-carboxylic acid H-bonding interaction, porphyrin-porphyrin π-π stacking interaction and van der Waals interaction among alkyl chains and the various architectures are created as a result of the relative strength of these three forces. In this example, we have shown that the pyridine-carboxylic acid H-bonding interaction induces a 2-D aggregational process in non-polar solvents like cyclohexane, although the porphyrin moiety itself is known to have a strong tendency to stack in a 1-D fashion [43]. The 2-D aggregation process can be explained by the fact that a combination of π-π stacking and H-bonding interactions is operational. This aggregational property can be nicely tuned depending on the polarity of the solvent. Thus, in methanol the complementary H-bonding interactions of **52** can be switched off, which results in a 1-D aggregate driven by the π-π stacking interaction. This phenomenon was confirmed from the SEM images, which shows a 2-D sheet-like structure in non-polar solvents like cyclohexane and fiber-like structure in methanol.

We have reported in a few examples that cholesterol-based organogels are applicable to chiral molecular recognition [44,45]. The cholesteryl phenylboronic acid **53** and its complexes with monosaccharides (**53**:monosaccharide = 2:1) were prepared [44]. It was shown that many of the isolated complexes form gels in a variety of solvents, some of which display D vs. L chiral discrimination. This forms the first example of chiral discrimination in a saccharide-based gel system. The xerogel fibers prepared from **53** or its D- or L-xylose complexes show the chiral discrimination ability in the re-binding of saccharides and partially retain the memory for the original imprinted saccharide [45]. These novel findings indicate that the sol-gel transition in the organogel is utilizable to the chiral discrimination and even to the molecular imprinting. Feringa *et al.* reported chiral recognition in organogels driven by co-operative interactions between a 1,2-bis(ureido)cyclohexane derivative **54** and a co-aggregating guest with azobenzene appended bis(ureido)cyclohexane framework **55** [46]. The association constants for both of the co-assemblies (R)-**55** with (R)-**54** and (R)-**55** with (S)-**54** was determined, which showed that the dimerization constant K_2 is considerably smaller than the association constant K for the higher aggregates indicative of a strong cooperative association process. It was also found out that the dimerization values $K_{2,RR}$ and $K_{2,RS}$ are of the same magnitude. Interestingly, the K_{RS} value was found to be almost twice as that of K_{RR}, which confirms that chirality does not affect the dimerization process but enantiomeric discrimination can take place in the formation of larger aggregates. One can thus regard that in this system cooperativity results in chiral recognition.

Chart 8

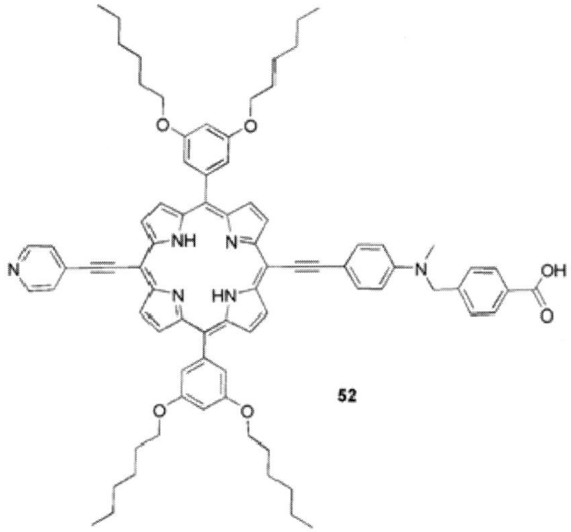

52

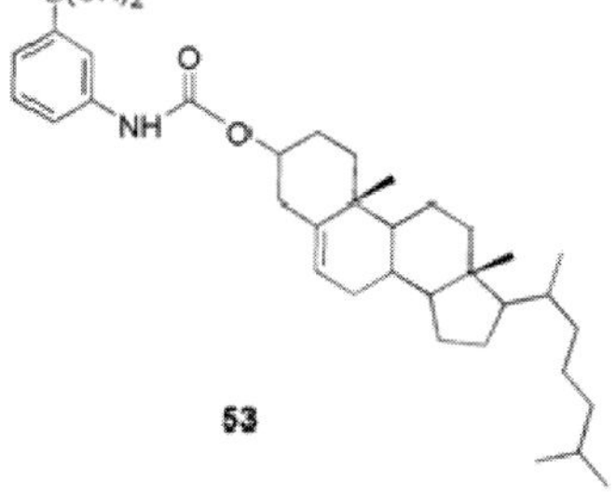

53

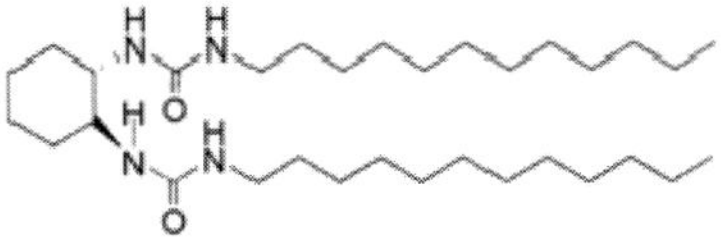

(S)-**54** *S, S*

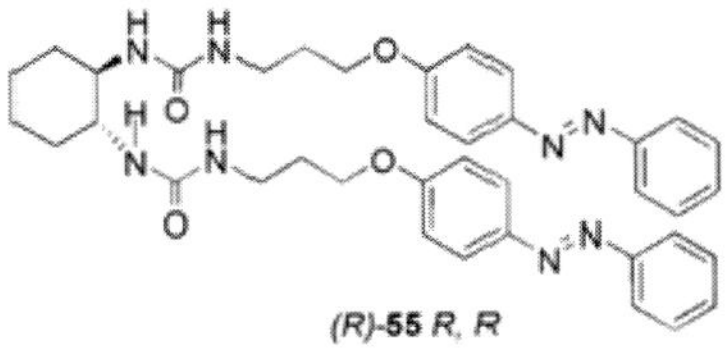

(R)-**55** *R, R*

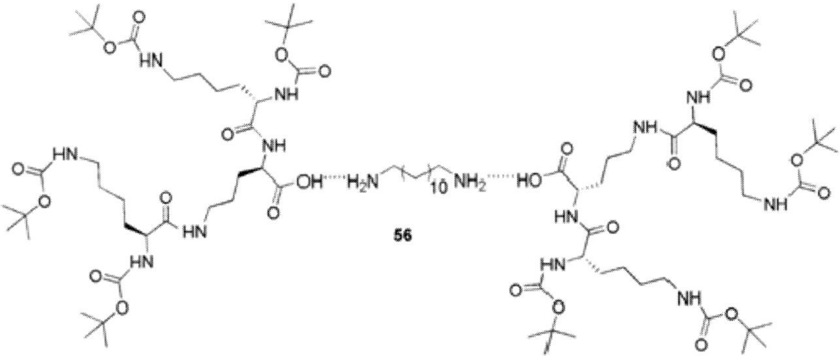

56

A binary dendritic gelator **56** was reported from the group of Smith *et al.*, where they have demonstrated a two-component gelation system that uses the dendritic building block based on L-lysine in combination with an aliphatic diamine [47]. The morphologies of the gels were found to be dependent on the molar ratio of the two components. They also found that the gel properties could be tuned by varying the core aliphatic spacer unit. Moreover, this group has investigated the transcription of the stereochemistry of the individual dendritic building blocks.

3.3. Donor-Acceptor Interaction

The first report on donor-acceptor mediated two-component gelator system came from the group of Maitra and co-workers [48]. They have demonstrated that aromatic-donor-substituted bile acid derivatives **57** and **58** could gel several organic solvents, but only in the presence of an acceptor, trinitrofluorenone **59**. The gels formed from colorless solutions of **57** and **58** and pale yellow **59** are significantly colored as a result of inter-molecular charge transfer (ICT) band. Thus, it is clear that the 1:1 donor-acceptor interaction plays a pivotal role in the gelation process.

In a different approach, we have employed two sugar-based gelator system, one containing an acceptor group **60** and the other a donor group **61** [49]. This novel dual component system was found to gel water, octanol, and diphenyl ether. A sharp color change from colorless to yellow was observed for the dual component gel upon cooling to room temperature. A charge-transfer band at λ_{max} = 420 nm was observed from UV-Vis spectroscopy arising out of the extensive stacking of the donor-acceptor molecules in the gel fibers. In diphenyl ether the dual component gel showed a maximum T_{gel} value at a 1:1 molar ratio and an enhanced stabilization of 30-40°C with respect to the gels of the single components. The TEM images reveal a fibrous network for the single component gels in diphenyl ether, while a novel helical fibrous bundle-like structure was observed for the dual component gels. The difference suggests that when gelator molecules are densely packed, for example, by operation of the charge transfer interaction, the original gelator structure is more strongly reflected by the resultant assembly.

Chart 9

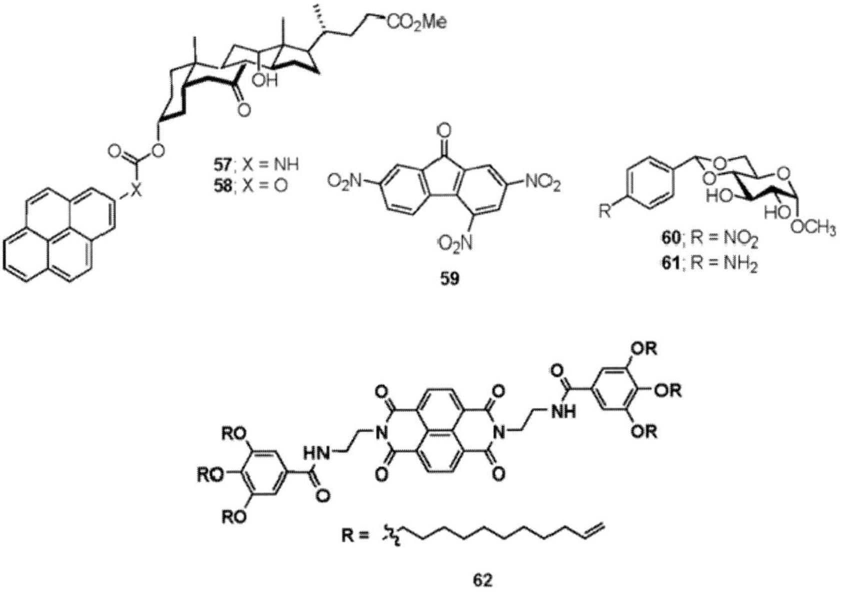

Recently, it has been reported that colorimetric sensing system which can differentiate seven positional isomers using CT complex gel[50]. Naphthalenediimide is known as a strong electron acceptor molecule which can interact with electron donor molecules such as alkoxynaphthalenes. Naphthalenediimide-based gelator (**62**) has been designed and it can successfully gel various common organic solvents ranging from hydrocarbons and aromatic solvents to ethers and alcohols. In the gel state, compound **62** forms one-dimensional fibrous structure in which donor molecules can be bound. Actually, a sharp color change has been observed in the gel state of **62** when dihydroxynaphthalene is mixed with **62** in cyclohexane to prepare the gel. A 1:1 stoichiometry of the host-guest ratio was confirmed by UV-Vis spectral study when various ratios (0~1.2 molar equivalent) of the guest molecule was mixed with the host **62** in the gel phase. Surprisingly, for all seven structural isomers of dihydroxyhaphthalenes, the organogel **62** shows different color. A 1-D matrix of organogel fibers of **62** can act as a one-dimensional matrix for precise molecular recognition phenomenon. Naked-eye differentiation becomes possible as a result of hydrogen-bonding-driven recognition that significantly amplifies the charge transfer interaction in the gel state.

3.4. Metal-Responsive Organogels

Incorporation of metal ions into organogels is highly attractive as metal ions can have a wide range of geometry, oxidation state, and reactivity. Thus, a new set of redox or magnetic properties can be anticipated or there can arise unusual geometries of metal ions due to the restricted environment in the gel phase. Along with this design concept, the existing organogel properties like gelation ability and thermodynamic stability can obviously be tuned when an organogel interacts with a metal ion.

To explore this working hypothesis, we designed cholesteryl-appended benzo crown-ether-based gelator molecules and studied its response towards various metal ions [51]. The T_{gel} of compound **5** increased with the addition of Li^+, Na^+, K^+, Rb^+, and NH_4^+ ions which could form 1:1 complex with the gelator while it decreased in the presence of Cs^+ ions due to the formation of a 2:1 (**5**:Cs^+) sandwich complex. Thus the organogel **5** was found to respond in a specific manner depending on the hole-size selectivity between the crown-ether and the metal ion.

Rowan and co-workers have reported an impressive metallo-supramolecular organo gelator system, which shows response to a variety of stimuli. This group has exploited the ability of the tridentate ligand bis(2,6-bis(1'-methyl-benzimidazolyl)-4-hydroxypyridine) to form polymeric aggregates in the presence of transition and lanthanide metal ions [52]. Self-assembly to the gel-like material happens spontaneously with the addition of La(III)/Eu(III) nitrate followed by the transition metal ions Co(II)/Zn(II) perchlorate salts to a solution of **63** in $CHCl_3/CH_3CN$. The transition metal ions endorse linear chain extension units while the lanthanide metal ions promote the cross-linking. All of the four gel materials **63**•Co/La, **63**•Co/Eu, **63**•Zn/La, and **63**•Zn/Eu show thermo-responsive and mechano-responsive behaviors. The **63**•Zn/Eu shows ligand as well as metal centered luminescence. Interestingly, the Eu(III) centered luminescence intensity undergoes sharp reduction on heating the gel material while the ligand emission remains nearly intact. The **63**•Zn/Eu also shows chemo-responsiveness in the presence of a small amount of formic acid. The formate anion captures Eu(III) resulting in switching off the metal centered emission. Kimizuka *et al.* have reported interesting gelation properties of lipophilic cobalt(II) complexes of 4-alkylated 1,2,4-triazoles. The compounds **64** and **65** form a blue gel-like phase in chloroform due to the tetrahedral coordination of cobalt(II) [53]. The lipophilic triazole complex Co(**64**)$_3$Cl$_2$ was found to be an efficient gelator as it could gel chloroform with a minimum concentration as low as 0.007 wt% while Co(**65**)$_3$Cl$_2$, requires 10 wt % to gelatinize chloroform. The authors surprisingly found that the blue gel-like phase turned into a solution by cooling below 25°C. A pink colored solution was obtained at 0°C, characterizing the formation of octahedral complex. The thermochromic transition was found to be totally reversible. The formation of gel-like networks by heating is an unprecedented finding with respect to the conventional organogels, which form gel-like structures upon cooling. Therefore, a novel gel-system was found, which undergoes a reversible thermochromic transition induced by the geometrical changes around the metal ion ($T_d \Leftrightarrow O_h$).

For the first time, oxidative and reductive stimuli have been effectively applied to a coordination gelator to induce a reversible chromatic and sol-gel phase transition. We have been able to develop a system based on a 2,2'-bipyridine derivative bearing two cholesteryl groups **66**, which complexes Cu(I) and spontaneously gels solvents like benzonitrile, 1-butyronitrile and THF/acetonitrile [54]. 1-butyronitrile solution of Cu(I)•**66**$_2$ complex forms a gel upon cooling, with a subsequent color change from reddish brown to greenish-blue during the phase transition. This unusual color of the Cu(I) complex possibly arises as a result of the structural distortion from the ideal tetrahedral geometry taking place in the confined environment inside the gel fibrils.

Redox responsiveness of this system was tested in the presence of oxidizing ($NOBF_4$) and reducing (ascorbic acid) agents. The reduction of Cu(II)•**66**$_2$ complex by ascorbic acid reverts to the Cu(I) state with the formation of the greenish blue gel, while oxidation of a greenish blue gel, the greenish blue gel in the Cu(I) state by $NOBF_4$ transforms to a sol phase with Cu in the (+II) state (Figure 5). Thus, stability of organogels could be nicely tuned by applying the simple redox chemistry of transition metal ions.

Chart 10

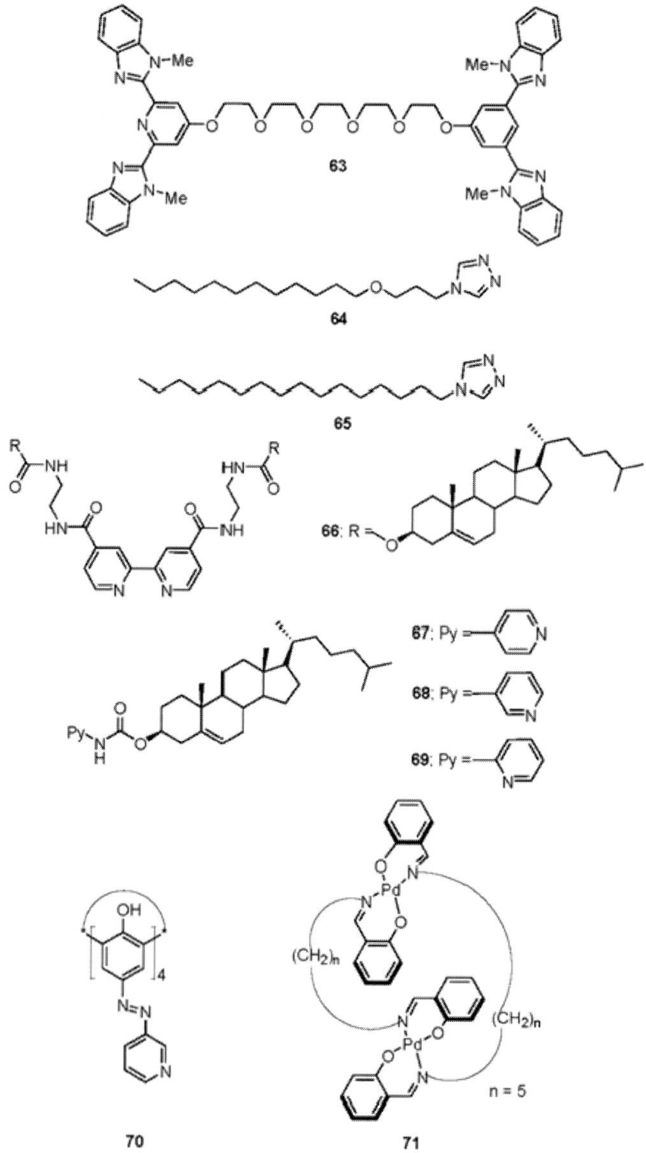

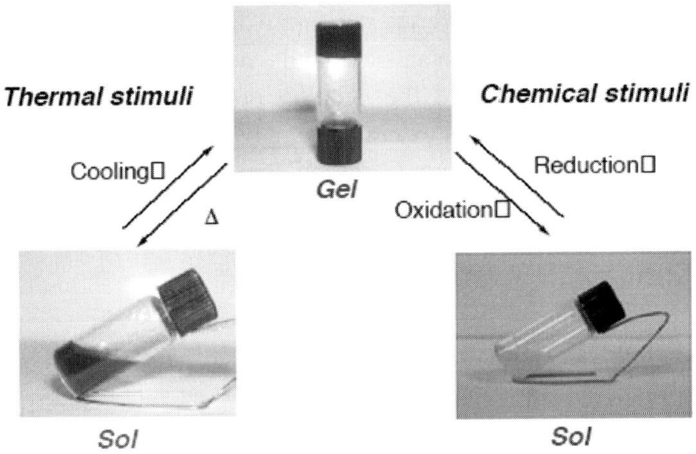

Figure 5. Pictures of response of Cu(II)•$\mathbf{66}_2$ complex in the presence of thermal and chemical stimuli.

A 2-, 3-, or 4-substituted pyridyl group bearing a cholesterol moiety (**67** – **69**) has been shown from our group to form stable organogels in the presence of various organic solvents [55]. All of these compounds were checked for their gelation ability in the presence of Ag(OTf). Interestingly, only the 3-substituted pyridine molecule showed improved gelation ability as a result of the Ag(I)-pyridine interaction. The Ag(I) ion is able to modulate the morphology of this gelator from rod-like clusters to well-defined fibrillar aggregates.

Bing Xu *et al.* have recently synthesized a 3-pyridine azo-calix[4]arene molecule **70** which gels DMSO in the presence of Pd(II) ions and could partition organic molecules from aqueous phase [56]. This organogel was found to be stable in water over a wide range of pH (1-13) and even at elevated temperatures of 100°C. It was also stable in most of the hydrophilic and hydrophobic organic solvents. The exceptional stability of this gel material allows its efficient uptake of non-ionic organic molecules like toluene and chlorobenzene from aqueous phase. The partitioning ability of this gel material was comparable to active carbons, which is a commonly used as an absorbent. Recently, Naota *et al.* have reported a dinuclear Pd(II) complex **71** that can undergo spontaneous gel formation upon irradiation with ultrasound [57]. This forms the first example whereby the authors have demonstrated instantaneous and remote controlled sol-gel processes. The initiation of the gelation process took place only by sonication, whereas other external stimuli did not initiate the gel formation. The sol-gel phase transition upon heating and subsequent cooling could be repeated indefinitely and the transition arises from a simple conformational change of the Pd(II) complex. In this case, the sonication-induced gelation was rationalized by the initiation of aggregational polymerization.

Therefore, we have learnt from the foregoing section that metal-based organogel system can present us with unprecedented and unforeseen results compared to that of the solid as well as the solution states. Incorporation of metal ions in an organogel system gives us a tremendously attractive platform as new electronic, geometric and magnetic properties can be found, due to the restricted and isolated nature of the molecules in the gel state. Moreover, the reversibility of the sol-gel phase transition in an organogel system gives us a further tool to maneuver all of these properties.

3.5. Gels with Novel Optical Properties

A rationally designed triphenylene derivative **72** has been shown from our group to efficiently gelate hydrocarbon solvents [58]. Triphenylene derivatives in general form a helical or staggered π-electron overlap in the liquid crystal phase or crystal systems. In the present system, a unique eclipsed π-electron overlap of the triphenylene moieties has been achieved due to the synergistic play of H-bonding among the amide groups and van der Waals interaction among the long alkyl chains in compound **72**. As a consequence of this extraordinary eclipsed geometry the organogel **72** in cyclohexane exhibit an excimer emission (Figure 6).

In a recent communication, Park *et al.* have demonstrated that simple conjugated organogels can be designed without the incorporation of long alkyl chains, urea/amide, or steroidal substituents. Thus, compound 1-cyano-*trans*-1,2-bis-(3,5-bistrifluoromethyl-biphenyl)ethylene has been shown to gel toluene, chloroform, and 1,2-dichloroethane at room temperature [59]. The origin of this gelation phenomenon can be explained by the strong π-π interactions of the rigid rod-like aromatic segments and the strong secondary bonding forces induced by the properly positioned CF_3 groups. A remarkable 170-fold increase in the fluorescence intensity was observed for the gel state. This group has also reported that benzene-1,3,5-tricarboxamide derivative bearing three 2,5-diphenyl [1,3,4]oxadiazole arms **73** can effectively gelate a wide range of aprotic organic solvents. Here, the nonfluorescent monomer unit exhibits a switching "ON" effect of the fluorescence upon gel formation [60]. A face-to-face intermolecular H-bonding in molecule **73** facilitates the formation of a supramolecular aggregate, which in turn induces a strong fluorescence emission. This aggregation induced fluorescence emission takes place as a result of the significant singlet-triplet splitting reducing the rate of intersystem crossing (ISC). These examples consistently demonstrate that the gel formation as well as the sol-gel phase transition is particularly useful to modulate optical properties of chromophoric functional groups.

In an elegant example, it has been shown that sol to gel phase transition can successfully modulate the twisted intramolecular charge transfer (TICT) state of p-dimethylaminobenzoate (p-DMAB)-appended cholesterol molecule **74** [61]. Conventionally, p-DMAB molecule in the solution state shows a dual fluorescence emission around 350 and 500 nm arising due to the coplanar and orthogonal conformation of the dimethylamino group and the benzoate plane, respectively. Interestingly, only the shorter wavelength fluorescence emission could be observed in the gel phase of **74**, while the sol phase exhibits the normal TICT based emission properties. This fluorescence emission property could be tuned several times in a reversible fashion by heating (sol phase) and cooling (gel phase) recycling processes.

A phosphorescent organogel has been recently reported by Aida *et al.* that can reversibly switch the RGB luminescence color upon sol-gel transition [62]. A trinuclear Au(I) pyrazolate complex appended with long alkyl chains **75** forms an organogel in hexane utilizing a Au(I)-Au(I) metalophilic interaction. This self-assembly process gives rise to a red-luminescence band at 640 nm while doping it with 0.01 equiv of Ag^+

results in a blue luminescence band at 458 nm keeping the gel stability intact. The original red-luminescence could be completely recovered when treated with cetyltrimethylammonium chloride. Gel-to-sol phase transition of the nondoped system results in a loss of the red-luminescence while the blue-luminescence of the Ag^{+}-doped system turns green, due to the disruption of the metallophilic interaction. Original luminescence was recovered in both cases during cooling to the gel phase.

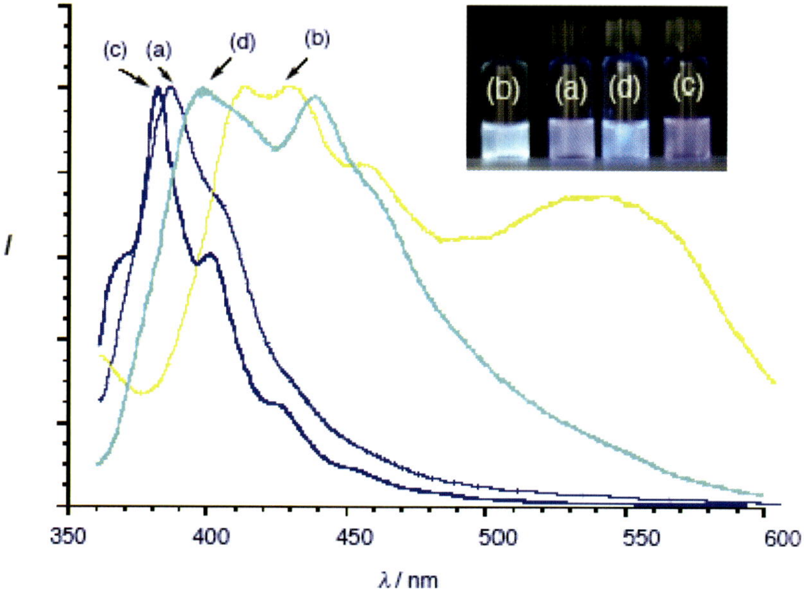

Figure 6. Fluorescence spectra of **72**, showing (a) [**72**] = 5 x 10^{-5} mol dm^{-3}, in chloroform solution, (b) [**71**] = 5 x 10^{-5} mol dm^{-3}, in cyclohexane gel, (c) [Reference non-gelator compound] = 5 x 10^{-5} mol dm^{-3}, in chloroform solution, and (d) [Reference non-gelator compound] = 5 x 10^{-5} mol dm^{-3}, in cyclohexane as a partial gel. Inset photograph demonstrates the samples excited by a UV lamp at λ_{ex} = 365 nm.

We have recently found that 3,4,5-tris(n-dodecyloxy)benzoylamide substituents-appended 8-quinolinol platinum(II) ligand **76** can efficiently gelate various organic solvents [63]. This gelator exhibits interesting thermo and solvatochromism of visible color as well as a color change in the phosphorescence emission triggered by the sol-gel phase transition.

Due to the novel environment in the gel phase of this molecule, the deactivation of the excited triplet states by collision with dioxygen molecules could be efficiently inhibited. This result can be rationalized due to the fact that the gelator phase is well segregated from the solvent molecules containing dioxygen acting as a quencher of the triplet states.

Chart 11

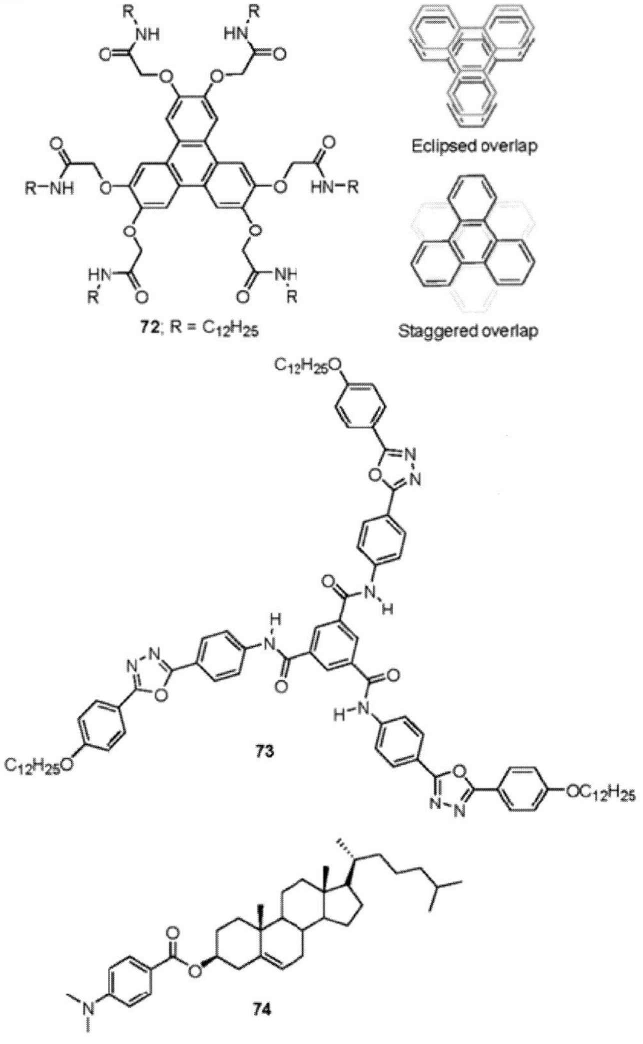

Chart 12

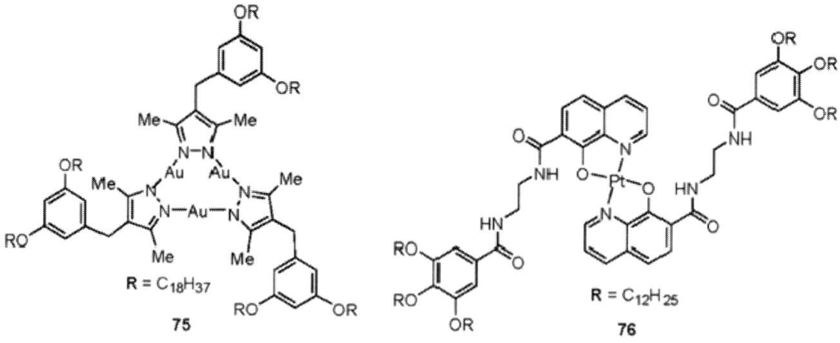

3.6. Photo-Responsive Organogels

We had demonstrated earlier that azobenzene molecules (**2** – **5**) appended with cholesteryl moieties can show interesting photo-responsive behaviors [64]. Recently, we have shown that a combination of anthracene carboxylate and alkyl ammonium ions of specific chain lengths leads to formation of a cyclohexane gel. The binary gelator **77** shows a very interesting phenomenon of photoresponsiveness [65].

Photoirradiation of the gel **77** by a Hg-lamp ($\lambda > 300$ nm) results in a decrease of the UV-Vis bands assignable to the monomeric anthracene and transformation into a sol phase in a very short span of time (Scheme 3). This change is rationalized in terms of the reduced π-π stacking effect and the resultant disordering of the molecular packing. The gel phase could not be regenerated directly as precipitation occurs due to thermal isomerization. The gel form could only be generated back by heating the sol phase at bp of cyclohexane and subsequent cooling. The photoinduced morphological change was monitored by dark-field optical microscopy (Figure 7). At the initial gel phase (at 0 s) fibrillar bundles could be recognized. With increasing photoirradiation time they gradually disappear, and such a structure cannot be detected in the screen after 120 s when the gel is changed to the sol. This forms a unique example where the phase changes of a gelator system was tuned by a light mode and a heat mode.

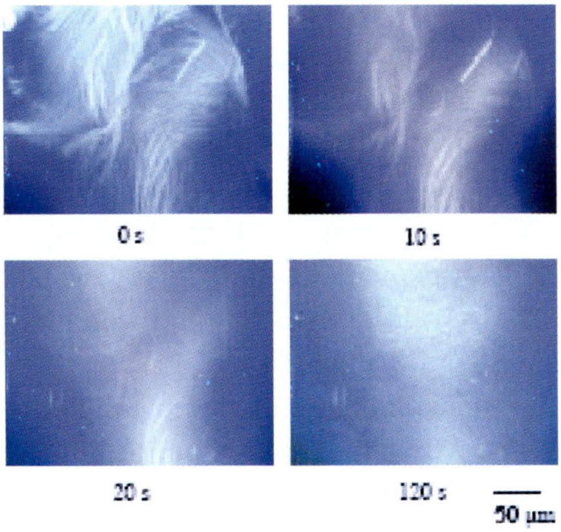

Figure 7. Photographs of CLSM images to study the morphological changes at various time scales.

Feringa and van Esch *et al.* have shown in a wonderful piece of work that a reversible, photoresponsive, self-assembling molecular system can be successfully designed in which the molecular and supramolecular chirality communicate [66]. A dithienylethene photochromic unit functionalized with (R)-1-phenylethylamine-derived amides was synthesized, which exists as two antiparallel, interconvertible open forms **78** with P- and M-helicity, which cyclize in a reversible manner upon irradiation with ultraviolet (UV) light to two diastereomers of the closed product **79**. The light induced switching between **78** and **79** was followed by changes in electronic properties and conformational flexibility of the molecules. The amide groups incorporated in **78** induce gel formation in organic solvents at RT.

Interestingly, the chirality present in **78** was expressed in a supramolecular aggregated system. The open switch **78** in solution was found to be CD silent, whereas upon gel formation a strong CD absorption was observed due to the locking of the M- or P-helical conformation of the open form **78** in the gel state. It was found that the photoactive supramolecular system comprised of two different aggregation states α and β which could include either the open **78** or the closed **79** form, leading to a total of four different states. The aggregation and switching processes by which these four states can be addressed is summarized in Scheme 4. A solution of open form **78** gives a stable gel (α) **78** (P-helicity) upon cooling. Photocyclization gives a metastable gel (α) **79** (PSS) (P-helicity) [PSS: Photostationary State] with high diastereoselectivity (96%), which is fully reversible. On heating the gel (α) **79** (PSS) leads to a solution of **79** which gives a stable gel (β) **79** (PSS) (M-helicity). Irradiation of gel (β) **79** (PSS) with visible light results in metastable gel (β) **78**, which in turn can be reconverted to the stable gel (β) **79** (PSS) by UV irradiation. Lastly, a heating and cooling cycle results in the transformation of the metastable gel (β) **78** to the original stable gel (α) **78** via the solution of **78**. This outstanding ability of this system to control chirality at different hierarchical levels in a synthetic system would attract huge potential from advanced technology such as molecular memory systems and smart materials.

Scheme 3

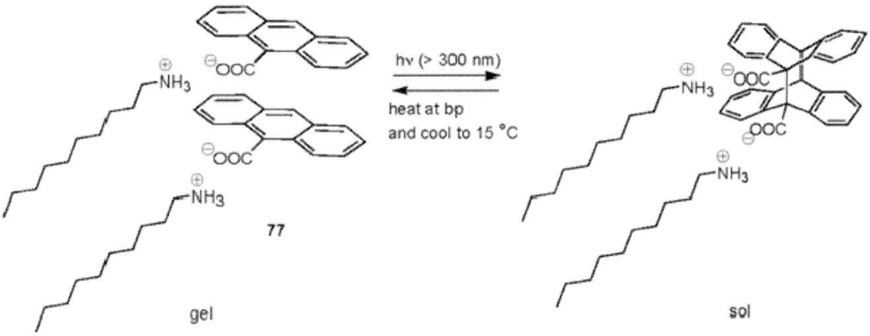

Chart 13

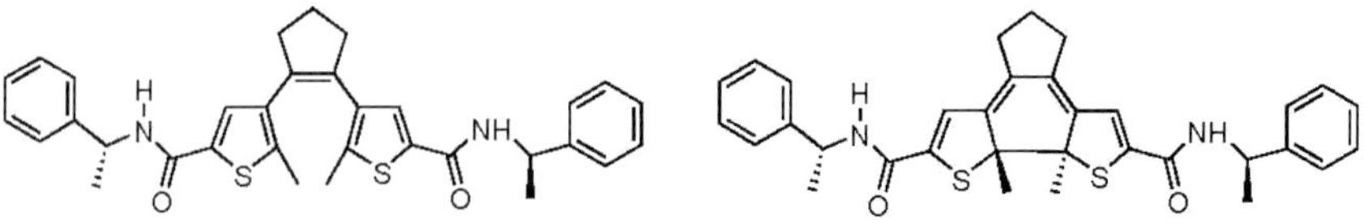

78 (Open form) **79** (Closed form)

Photo-induced color generation and color erasing is one of the sought after goals for materials chemists as it provides an important platform for color storage devices. It is known that alkylammonium polyoxomolybdates exhibit significant photochromism but due to its low solubility and uncontrollable reversibility its practical application in material devices poses a serious challenge. We have shown that alkylammonium polyoxomolybdate, $[NH_2(CH_2CH_2CH_2NH_3)_2]_2 \cdot [Mo_8O_{26}(MoO_4)]$ can be successfully incorporated in a trans-(1R,2R)-1,2-bis(undecylcarbonylamino)cyclohexane based organogel system and subsequently a controllable photo-induced color generation and color erasing organogel system can be developed [67]. The organogel sample as well as the reference solution (as a control experiment) results in a blue species when irradiated with a UV light. A comparison of the reaction rate for this color generation process in the gel state and solution showed that the blue color is generated twice as fast in the gel system. Furthermore, the decoloration process is 26 times slower compared to that of the solution state. Interestingly, this decoloration process was found to be controllable by sol-gel phase transition and also the photochromism in the gel state could be repeated infinite number of times. We thus believe that utilizing this concept novel materials can be designed which would show better operational efficiency in the gel system compared to that of the conventional solution or solid state.

Scheme 4

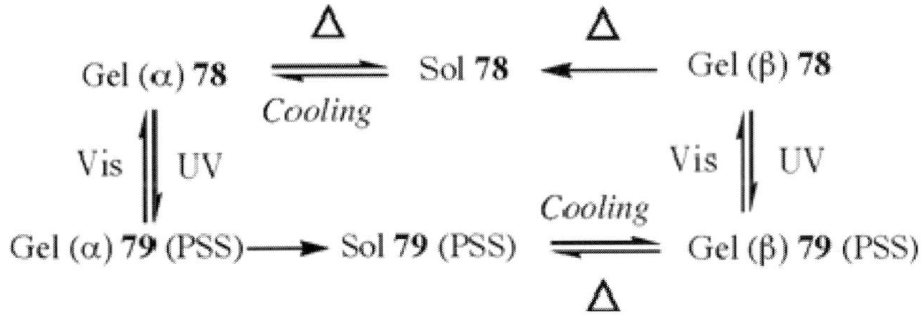

3.7. Redox Active Organogels

A redox stimulus is indispensable for construction of electromechanical soft materials like artificial muscles, electrorheological fluids, etc. In this context, we have demonstrated that 2,2'-bipyridine-Cu(I) complex bearing two cholesteryl groups **66** can undergo sol-gel transition by treatment with oxidizing/reducing agents (for details see Section 4.4). This idea can be extended to redox-active organic molecules when they are appropriately integrated in the gelator molecules. For example, we have recently reported that a suitably designed sexithiophene derivative bearing two cholesteryl groups at the α–position **80** can respond to a redox active stimulus [68]. Thus, when the gelator is treated with an oxidizing agent like $FeCl_3$, the red gel turns into a brown colored solution. Treatment of this solution with a reducing agent like ascorbic acid reversibly converts the solution to the red gel at ambient temperature. This is a rare example of a heating-free sol-gel phase transition that can be induced by a redox stimulus.

Chart 14

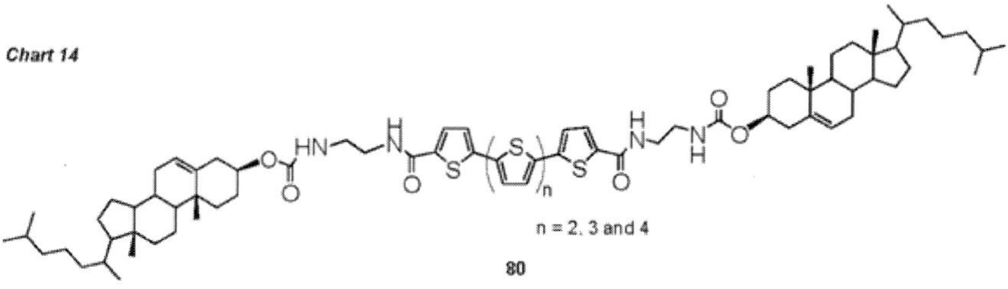

80

Although there have been a large number of reports on reversible redox changes of supramolecular species in solution state, there remains a big opportunity to design new redox-active molecules that would operate in the gel phase.

3.8. Light Harvesting Organogel Systems

There is wide spread interest among researchers to develop artificial photosynthetic systems. Molecular arrays, dendrimers, monolayers have been reported as light-harvesting systems to mimic natural photosynthetic system [69-71]. It has been anticipated that the gel phase would provide a novel environment to arrange the chromophores in order to achieve an efficient, spatially-oriented transfer of the absorbed energy to the actual reaction centers. In a pioneering work, Kimizuka *et al.* have reported that cationic L-glutamate derivatives can form supramolecular light-harvesting hydrogels when bound to fluorescent molecules [72].

Ajayaghosh *et al.* have demonstrated that π-stacking and H-bonding interactions in oligo(phenylenevinylene)s (OPVs) can induce supramolecular assembly and gel formation. The donor (OPV derivatives) **81a,b** and acceptor (Rhodamine B) chromophores **84** are forced to order closely in the gel state compared to that in the solution state [73]. This in turn facilitates efficient energy transfer and light-harvesting from the donor to the acceptor by the fluorescence-resonance energy transfer (FRET) mechanism. In contrast, energy-transfer in case of **82a,b** and **83a,b** was not predominant as these compounds could not form gels. Our group has designed a novel visible light-harvesting system utilizing the one-dimensional aggregational tendency of erylene chromophores appended with cholesterol-based gelators [74].

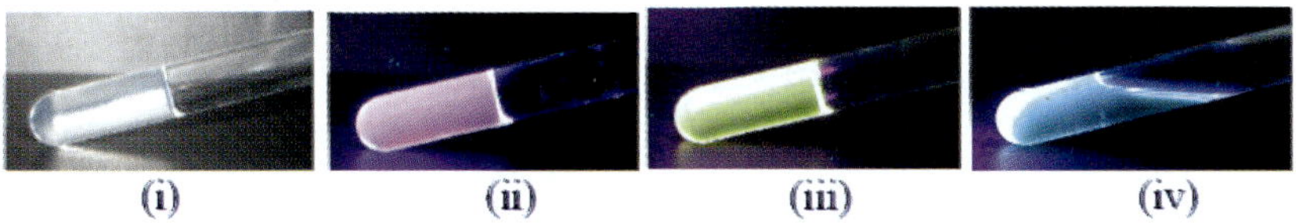

Figure 8. Photographs of (i) **86** + 1-propanol gel, (ii) **86** + 1-propanol gel under UV-light (365 nm), (iii) **86** + 1-propanol gel in the presence of TFA, and (iv) after heating the gel (iii) at 90°C.

A variety of perylene based molecules **85a-d** were synthesized that could absorb a wide range of light energy as a result of subtle modification of the substituents. In a quaternary system composed of **85a-d**, one-dimensional aggregation of the four perylene groups resulted in an energy gradient, as a consequence of which the excited energy of **85a** was transferred to **85b** and **85c** and finally to **85d**. We found that **85d** acts as the energy sink because of its low-emissive nature. Thus, we could harvest light consisting of a broad range of wavelengths to the energy sink, a phenomenon that is significantly assisted by the 1-D stacking nature of the perylene moieties in the gel phase. We have also demonstrated that a 1,10-phenanthroline-appended cholesterol-based gelator **86** can have dual functions; it can act as a proton-sensor as well as a light harvesting system [75]. It was found that a red-shifted emission occurs in the gelator system due to the protonation of the phenanthroline moiety by trifluoroacetic acid. Moreover, we found that efficient energy transfer from neutral **86*** to protonated **86**·H$^+$ takes place resulting in strong fluorescence intensity due to **86**·H$^+$. Figure 8 shows the dramatic color changes at different conditions: The colorless **86** + 1-propanol gel shows a purple color under UV light irradiation. Addition of TFA changes its fluorescence color to greenish-yellow and heating this gel converts it into the sol phase which has a light-blue color.

Chart 15

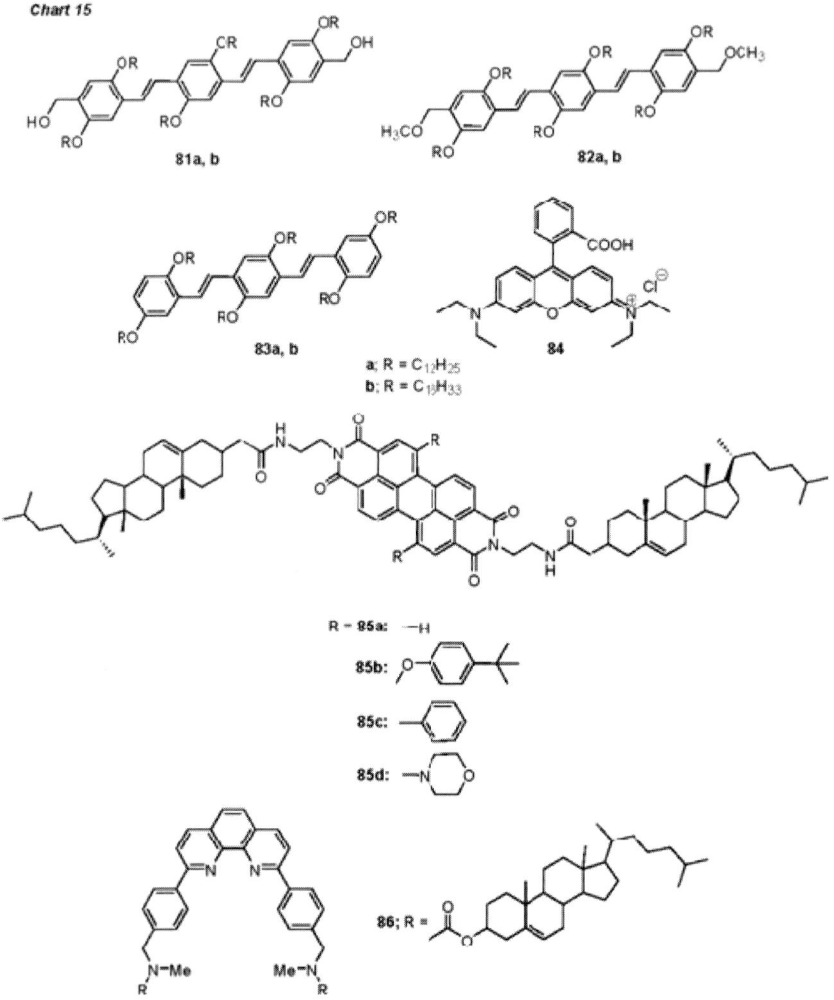

4. Miscellaneous Organogels

In this section, we will look at gelator molecules that show vastly modulated properties upon polymerization [76-78]. Thus, new properties evolve post-polymerization in these gelator molecules. We have recently shown that a porphyrin derivative **87**, containing diacetylene units can be used as a photo-polymerization template. Using this methodology, highly elongated fibers of several μm in lengths constituted of unimolecularly-stacked array of porphyrins could be obtained by in situ polymerization.

Chart 16

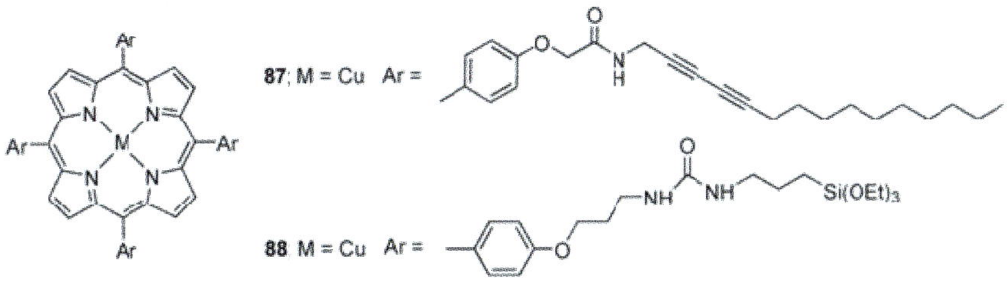

This is one of the first examples of template-directed polymerization of diacetylene molecules in the gel phase which gives rise to conjugated molecular wires. AFM observations (Figure 9) clearly reveal that each fiber is about 3 nm thick consistent with the molecular width of this compound [79].

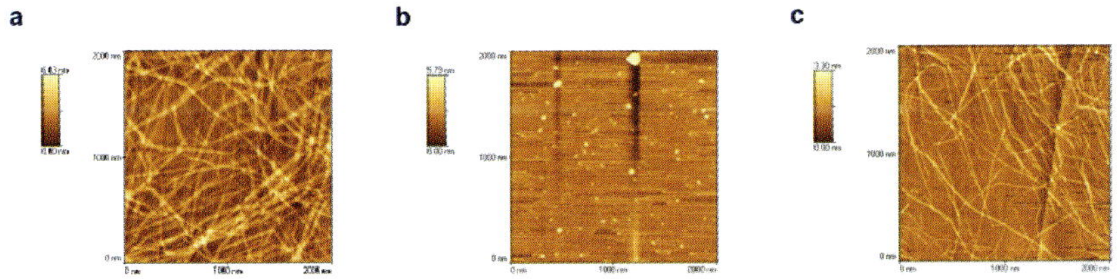

Figure 9. AFM images of a) the Decalin gel of **87**, b) the Decalin gel of **87** without UV irradiation after chloroform rinsing, and c) the Decalin gel of **87** after UV irradiation and chloroform rinsing. All of the above processes were carried out on HOPG with the following conditions: [**87**] = 1.0 g dm^{-3}, 500 W high-pressure mercury lamp, and 25°C.

Recently, we observed a very interesting phenomenon of elasticity upon polymerization of a suitably functionalized 1-D porphyrin molecule

88 [80]. In this example, we carried out a sol-gel polycondensation of the peripheral trimethoxysilyl groups, which subsequently gains exceptionally high thermal stability as well as mechanical strength. The gelatinous elastic material thus obtained after polycondensation showed a dramatic increase in the storage modulus by 14 times compared to that of the original organogel. These findings suggest that post-polymerization is useful not only to immobilize the non-covalently assembled superstructures but also to improve their mechanical properties.

5. Biomedical Applications

One of the major applications of gelator molecules are expected to be in the field of medicine and its efficient delivery in human body. Recently, 1H-imidazole-based amphiphiles have been found to entrap antibiotic drug like Norfloxacin into its strands, therefore, making it attractive as a potential drug carrier (please see Section 2.7). A large number of polymeric and hydro gelators have been reported which find direct applications to the field of biomedicinal chemistry. Unfortunately, the polymeric gels and hydrogelators are out of the scope of this chapter. Nevertheless, a vancomycin-pyrene hydrogelator needs to be mentioned as this is the first example of an antibiotic gelator. The authors have predicted that this gelator can be a potential candidate for developing bio-materials that would serve as an antiseptic matrix and provide new means of drug delivery [81]. Furthermore, Hamachi *et al.* have shown in an elegant example that a hydrogelator based on a glycosylated amino acid building block, a semi-wet (aqueous cavities created in the gel matrix provides a semi-wet reaction medium for substrates) peptide/protein gel array can be constructed in which the peptides or proteins are entrapped in the active form [82]. The authors have shown that several drawbacks associated with the conventional chips can be overcome utilizing their new approach. Their protein/peptide chip is expected to be applicable to protein/enzyme analyses based on high-throughput activity. Hence, there is a vast opportunity for designing and studies of new gelator molecules that would have potential applications in pharmaceutical research.

6. Conclusions and Future Outlook

In summary, we have seen through this chapter that a phenomenal number of gelator molecules can be designed applying various novel strategies. Some of these molecules have the additional properties of responsiveness towards stimuli. This adds tremendous utility to the gelation chemistry as a whole, since these molecules can be further exploited in the field of sensors, new materials, and so on. The field of organogel chemistry has grown in the past few years with a very rapid pace. Due to their well-defined structure, coexistence of highly ordered fibers with a liquid phase, large interfacial area, and ability to entrap solutes within the network pores, organogels are highly promising candidates for sensors, drug delivery, catalysis, membranes, and separation technology. It is extremely likely that with time and ever increasing interests of researchers around the globe, organogels will find a wide and new range of applications, not only in nanotechnology but also in biotechnology, which might be out of our present day understanding and imagination.

References

1. For comprehensive reviews on organogels, please see: (a) P. Terech and R. G. Weiss, Chem. Rev. 97, 3133 (1997). (b) J. Esch, F. Schoonbeek, M. Loos, H. Kooijman, E. M. Veen, E. M.; R. M. Kellogg and B. L. Feringa, In Supramolecular Science: Where It Is and Where It Is Going; R. Ungaro and E. Dalcanale, Eds.; Kluwer: Dordrecht, The Netherlands, 1999; 233. (c) E. Caretti, L. Dei and R. G. Weiss, Soft Matter 1, 17 (2005).
2. (a) S. Shinkai and K. Murata, J. Mater. Chem. 8, 485 (1998). (b) R. E. Melendez, A. J. Carr, B. R. Linton and A. D. Hamilton, Struct. Bonding 31 (2000).
3. A. Ajayaghosh and S. J. George, J. Am. Chem. Soc. 123, 5148 (2001).
4. S. –i. Kawano, N. Fujita, K. J. C. van Bommel and S. Shinkai, Chem. Lett. 32, 12 (2003).
5. O. Martin-Borret, R. Ramasseul and R. Rassat, Bull. Soc. Chim. Fr. 7-8, II-401 (1979).
6. K. Murata, M. Aoki, T. Susuki, T. Harada, H. Kawabata, T. Komori, F. Ohseto, K. Ueda, and S. Shinkai, J. Am. Chem. Soc. 116, 6664 (1994).
7. Y. –C. Lin and R. G. Weiss, Macromolecules 20, 414 (1987).
8. T. Brotin, R. Utermohlen, F. Fages, H. Bouas-Laurent, and J. P. Desvergne, J. Chem. Soc., Chem. Commun. 416 (1991).

9. F. M. Menger and K. S. Venkatasubban, J. Org. Chem. 43, 3413 (1978).
10. J. Campbell, M. Kuzma and M. Labes, Mol. Cryst. Liq. Cryst. 95, 45 (1983).
11. (a) K. Hanabusa, K. Okui, K. Karaki, T. Koyama and H. Shirai, J. Chem. Soc., Chem. Commun. 1371, (1992). (b) K. Hanabusa, J. Tange, Y. Taguchi T. Koyama, and H. Shirai, J. Chem. Soc., Chem. Commun. 390 (1993).
12. K. Hanabusa, Y. Matsumoto, T. Miki, T. Koyama, and H. Shirai, J. Chem. Soc., Chem. Commun. 1401 (1994).
13. E. J. de Vries and R. M. Kellogg, J. Chem. Soc., Chem. Commun. 238, (1993).
14. S. Bhattacharya and Y. K. Ghosh, Chem. Commun. 185 (2001).
15. G. M. Gundert, L. Klein, M. Fischer, F. Vögtle, K. Heuze, J. –L. Pozzo, M. Vallier and F. Fages, Angew. Chem. Int. Ed. 40, 3164 (2001).
16. A. D. Aleo, J-L Pozzo, F. Fages , M. Schmutz, G. M.-Gundert, F. VÖgtle, V.Caplar and M. Zinic, Chem. Commun. 190 (2004).
17. J. Makarevic, M. Jokic, L. Frkanec, D. Katalenic, and M. Zinic, Chem. Commun. 2238 (2002).
18. M. Suzuki, S. Owa, M. Kimura, A. Kurose, H. Shirai, and K. Hanabusa, Tetrahedron Lett. 46, 303 (2005).
19. M. George and R. G. Weiss, J. Am. Chem. Soc. 123, 10393 (2001).
20. For a comprehensive review for sugar-based gelators, see O. Gronwald and S. Shinkai, Chem. Eur. J. 7, 4328 (2001).
21. K. Hanabusa, M. Yamada, M. Kimura and H. Shirai, Angew. Chem. Int. Ed. 35, 1949 (1996).
22. L. A. Cuccia, J. –M. Lehn, J. –C. Homo and M. Schmutz, Angew. Chem. Int. Ed. 39, 233 (2000).
23. R. Oda, I. Huc and S. J. Candau, Angew. Chem. Int. Ed. 37, 2689 (1998).
24. M. Kolbel and F. M. Menger, Chem. Commun. 275 (2001).
25. M. Aoki, K. Murata and S. Shinkai, Chem. Lett. 1715 (1991).
26. H. J. Tian, K. Inoue, K. Yoza, T. Ishi-i and S. Shinkai, Chem. Lett. 871 (1998).
27. (a) C. F. van Nostrum, S. Picken, A. –J. Schouten and R. J. M. Nolte, J. Am. Chem. Soc. 117, 9957 (1995). (b) H. Engelkamp, S. Middelbeek and R. J. M. Nolte, Science 284, 785, (1999).
28. C. De Rango, P. Charpin, J. Navaza, N. Keller, I. Nicolis, F. Villain and A. W. J. Coleman, J. Am. Chem. Soc. 114, 5475 (1992).
29. Please see [Ref. 5].
30. B. –K. An, D. –S. Lee, J. –S. Lee, Y. –S. Park, H. –S. Song and S. Y. Park, J. Am. Chem. Soc. 126, 10232 (2004).
31. D. D. Diaz and M. G. Finn, Chem. Commun. 2514 (2004).
32. S. H. Seo and J. Y. Chang, Chem. Mater. 17, 3249 (2005).
33. M. Shirakawa, N. Fujita and S. Shinkai, J. Am. Chem. Soc. 125, 9902 (2003).
34. T. Ishi-i, R. Iguchi, E. Snip, M. Ikeda and S. Shinkai, Langmuir 17, 5825 (2001).
35. S. –i. Kawano, N. Fujita and S. Shinkai, Chem. Commun. 1352 (2003).

36. J. H. Jung, S. J. Lee, J. A. Rim, H. Lee, T. –S. Bae, S. Lee and S. Shinkai, Chem. Mater. 17, 459 (2005).
37. E. Snip, S. Shinkai and D. N. Reinhoudt, Tetrahedron. Lett. 42, 2153 (2001).
38. Y. J. Yun, S. M. Park and B. H. Kim, Chem. Commun. 254 (2003).
39. R. Iwaura, K. Yoshida, M. Matsuda, M. Ohnishi-Kameyama, M. Yoshida and T. Shimizu, Angew. Chem., Int. Ed. 42, 1009 (2003).
40. K. Sugiyasu, M. Numata, N. Fujita, S. M. Park, Y. J. Yun, B. H. Kim and S. Shinkai, Chem.Commun. 1996 (2004).
41. M. Numata and S. Shinkai, Chem. Lett. 32, 308 (2003).
42. C. Bao, R. Lu, M. Jin, P. Xue, C. Tan, G. Liu and Y. Zhao, Org. Biomol. Chem. 3, 2508 (2005).
43. Tanaka, M. Shirakawa, K. Kaneko, M. Takeuchi and S. Shinkai, Langmuir 21, 2163 (2005).
44. T. D. James, K. Murata, T. Harada, K. Ueda and S. Shinkai, Chem. Lett. 273 (1994).
45. K. Inoue, Y. Ono, Y. Kanekiyo, T. Ishi-i, K. Yoshihara and S. Shinkai, Tetrahedron Lett. 39, 2981 (1998).
46. M. D. Loos, J. van. Esch, R. M. Kellogg and B. L. Feringa, Angew. Chem. Int. Ed. 40, 613 (2001).
47. K. S. Yartridge, D. K. Smith, G. M. Dykes and Y. T. Mc Grail, Chem. Commun. 319 (2001).
48. U. Maitra, P. V. Kumar, N. Chandra, L. J. D'Souza, M. D. Prasanna and A. R. Raju, Chem. Commun. 595 (1999).
49. A. Friggeri, O. Gronwald, K. J. C. van Bommel, S. Shinkai and D. N. Reinhoudt, J. Am. Chem. Soc. 124, 10754 (2002).
50. P. Mukhopadhyay, Y. Iwashita, M. Shirakawa, S.-i. Kawano, N. Fujita and S. Shinkai, Angew. Chem. Int. Ed. 45, 1592 (2006).
51. K. Murata, M. Aoki, T. Nishi, A. Ikeda and S. Shinkai, J. Chem. Soc. Chem. Commun. 1715 (1991).
52. J. B. Beck and S. J. Rowan, J. Am. Chem. Soc. 125, 13922 (2003).
53. K. Kuroiwa, T. Shibata, A. Takada, N. Nemoto and N. Kimizuka, J. Am. Chem. Soc. 126, 2016 (2004).
54. S.–i Kawano, N. Fujita and S. Shinkai, J. Am. Chem. Soc. 126, 8592 (2004).
55. Please see [Ref. 4].
56. B. Xing, M. –F. Choi and B. Xu, Chem.Commun. 362 (2002).
57. T. Naota and H. Koori, J. Am. Chem. Soc. 127, 9324 (2005).
58. M. Ikeda, M. Takeuchi and S. Shinkai, Chem.Commun. 1354 (2003).
59. Please see [Ref. 31].
60. S. Y. Ryu, S. Kim, J. Seo, Y. –W.; Kim, O. –H. Kwon, D. –J. Jang and S. Y. Park, Chem.Commun. 70 (2004).
61. Y. Iwashita, K. Sugiyasu, M. Ikeda, N. Fujita and S. Shinkai, Chem. Lett. 33, 1124 (2004).

62. A. Kishimura, T. Yamashita and T. Aida, J. Am. Chem. Soc. 127, 179 (2005).
63. M. Shirakawa, N. Fujita, T. Tani, K. Kaneko and S. Shinkai, Chem. Commun. 4149 (2005).
64. Please see [Ref. 7].
65. M. Ayabe, T. Kishida, N. Fujita, K. Sada and S. Shinkai, Org. Biomol. Chem. 1, 2744 (2003).
66. J. J. D. de. Jong, L. N. Lucas, R. M. Kellogg, J. H. van Esch and B. L. Feringa Science 304, 278 (2004).
67. T. Yi, K. Sada, K. Sugiyasu, T. Hatano and S. Shinkai, Chem. Commun., 344 (2003).
68. S. –i Kawano, N. Fujita and S. Shinkai, Chem. Eur. J. 11, 4735 (2005).
69. H. S. Cho, H. Rhee, J. K. Song, C. K. Min, M. Takase, N. Aratani, S. Cho, A. Osuka, T. Joo and D. Kim, J. Am. Chem. Soc. 125, 5849 (2003).
70. V. Vicinelli, P. Ceroni, M. Maestri, V. Balzani, M. Gorka and F. Vogtle, J. Am. Chem. Soc. 124, 6461 (2002).
71. N. Kimizuka and T. Kunitake, J. Am. Chem. Soc. 111, 3758 (1989).
72. T. Nakashima and N. Kimizuka, Adv. Mater. 14, 1113 (2002).
73. A. Ajayaghosh, S. J. George and V. K. Praveen, Angew. Chem. Int. Ed. 42, 332 (2003).
74. K. Sugiyasu, N. Fujita and S. Shinkai, Angew. Chem. Int. Ed. 43, 1229 (2004).
75. K. Sugiyasu, N. Fujita, M. Takeuchi, S. Yamada and S. Shinkai, Org. Biomol. Chem. 1, 895 (2003).
76. The gel formation followed by polymerization of gelators has been reported by several groups: M. Matsuda, T. Hanada, K. Yase, T. Shimizu, Macromolecules 31, 9403 (1998).
77. K. Inoue, Y. Ono, K. Kanekiyo, S. Kiyonaka, I. Hamachi and S. Shinkai, Chem. Lett. 225 (1999).
78. M. Barboiu, S. Cerneaux, A. van der Lee and G. Vaughan, J. Am. Chem. Soc. 126, 3545 (2004).
79. M. Shirakawa, N. Fujita and S. Shinkai, J. Am. Chem. Soc. 127, 4164 (2005).
80. T. Kishida, N. Fujita, K. Sada and Seiji Shinkai, J. Am. Chem. Soc. 127, 7298 (2005).
81. B. Xing, C. –W. Yu, K. –H. Chow, P. –L. Ho, D. Fu and B. Xu, J. Am. Chem. Soc. 124, 14846 (2002).
82. S. Kiyonaka, K. Sada, I. Yoshimura, S. Shinkai, N. Kato and I. Hamachi, Nature Mater. 3, 58 (2004).

CHAPTER 10

BIOSENSORS BASED ON GOLD NANOPARTICLE LABELING

Robert Möller[#] and Wolfgang Fritzsche*

*Institute for Physical High Technology, Photonic Chip Systems Department, Albert-Einstein-Str. 9, 07745 Jena, Germany; [#]Current address: Friedrich-Schiller-University, Institute for Physical Chemistry, Helmholtzweg 4, 07743 Jena, Germany; *E-Mail: fritzsche@ipht-jena.de*

The fast, reliable and cost efficient detection of biomolecules and their interaction is one of the major challenges for the coming years. Even so the knowledge about biomolecules, their interaction, and their importance for diagnostics is growing fast; the use of biomolecules as markers in modern diagnostics is still limited. Gold nanoparticles are an interesting alternative for the standard fluorescence labelling of biomolecules used today. These labels enable a variety of detection schemes that might make the analysis of biomolecules faster, more cost efficient and even more sensitive than the systems used today. This paper reviews different detection schemes that have been described in recent years for the detection of biomolecules using gold nanoparticles as labels. It will focus especially on chip-based detection methods, but other methods are discussed as well.

1. Introduction

The unique properties of gold nanoparticles, especially regarding their optical, electric and catalytic behaviors, make them excellent labels for the detection of biomolecular interactions [1-6]. The ever growing knowledge about the interaction of biomolecules and their diagnostic importance creates a need for new biomolecular detection schemes that are faster, more reliable and cheaper than the ones we are using today. Based on the unique properties of gold nanoparticles a number of different detection schemes have been developed in the last years. Some of these detection schemes may proof to be useful in point-of-care diagnostics.

One of the rising diagnostic tools is the DNA- or gene-chip, which allows researchers to conduct hundreds of different DNA sequence tests on a single chip [7,8]. This parallelization of experiments makes the DNA-chip not only interesting for research and development regarding medical diagnostics applications, but also for applications in food safety, pathology, detection of biological weapons, and other fields. The main labeling technique for DNA-chips today is the labeling with fluorescent dyes. Fluorescence labeling of DNA-chips has the high sensitivity and selectivity that is necessary for the detection even of small amounts of target molecules in a mixture of similar molecules. Additionally, fluorescent labeling also allows the multiplexed detection of different single-stranded DNA molecules using different fluorescent dyes. So it is possible to hybridize DNA from different sources to the same chip. Given the great potential that DNA-chips have one might wonder why they are not more common in today diagnostic applications. One of the main limitations of DNA-chips is probably the fluorescent labeling and its readout. There are some disadvantages that hamper the development of point of care diagnostic tools that are based on fluorescence detection. Fluorescent labels are expensive, they tend to bleach, and the equipment needed for the readout is expensive and hard to miniaturize as needed for point of care tools. As result, DNA-chips with fluorescent labeling have been quickly adopted in research laboratories, while other applications that require cost efficient, rapid and easy-to-handle measurement equipment have not yet taken advantage of the DNA-chip technology.

Gold nanoparticles can be used as labels for biomolecules instead of fluorescent dyes. Because of their unique properties a number of different detection schemes are based on gold nanoparticle labeling. These detection schemes exploit the unique physical properties like the large extinction and scattering coefficients, surface electronics, their weight, and efficient Brownian motion in solution. More importantly the gold nanoparticle labeling might solve some of the above described problems using fluorescent labeling, like cost efficiency, ease of use, and may even increase the sensitivity and specificity of biomolecular tests.

This review documents the development of gold nanoparticles as labels in biosensor applications mainly for the detection of DNA, but other applications will also be mentioned. It will focus on the different detections schemes that have been described in recent years to detect the binding of biomolecules using gold nanoparticles as labels in homogeneous as well as heterogeneous (here especially DNA-chip) detection formats. The review will mainly discuss gold nanoparticles even though other metal nanoparticles (Ag, Pt, Pd and others) have also been used as labels for biomolecules and will only be mentioned where appropriate.

2. General Features of Gold Nanoparticles: Synthesis and Bioconjugation

For the chemical synthesis of gold nanoparticles or gold colloids gold (III) salts are commonly used. The synthesis can be done in aqueous [9,10] as well as in organic solvents [3,11,12]. A number of different approaches and reducing agents [13] have been described. The most common method to produce gold nanoparticles is probably the citrate reduction of $HAuCl_4$ [14]. But other reducing agents like substituted ammonias, formaldehyde, organic acids, and others are described as well [14]. Today, gold nanoparticle solutions are available from a number of commercial distributors like Ted-Pella, British Biocell, Nanoprobes, Sigma-Aldrich, and others. To stabilize the nanoparticles they can be covered with a layer of surfactant molecules. These surfactant molecules can greatly influence the modification of gold nanoparticles with biomolecules. So, in order to have complete control over the composition of the surfactant layer and the stability of the nanoparticles one has also to do custom particle synthesis.

Gold nanoparticles synthesized by the citrate reduction of gold ions as well as commercially available particles are not very stable. Concentrations of 100 mM NaCl and higher normally cause these particles to aggregate. This aggregation of particles can be easily detected because the surface plasmon-based color of the solution changes from red to blue and after a while a black precipitate can be observed. This particle aggregation is not reversible. To stabilize the particles they can be modified with surfactant molecules like phosphines, or carboxyles [15]. These surfactant molecules add negative charges to the particle surface. The negatively charged particles strongly repel each other electrostatically, which leads to an increased stabilization of the nanoparticle solution.

To nonspecifically modify proteins or DNA with gold nanoparticles it is possible to use either negatively or positively charged gold nanoparticles. The positively charged gold nanoparticles are commercially available as Genogold™ (British Biocell) and will bind to the negatively charged phosphate backbone of the DNA. Negatively charged gold colloids will bind to positive charges in proteins. These particles are also commercially available as Protogold™ (British Biocell). The problem with labeling biomolecules using just charged particles is that there is no sequence specificity at all and that the labeling depends just on the charge distribution on the biomolecules. So these gold nanoparticle solutions are normally used to stain all biomolecules (DNA or protein) in electrophoresis gels or blot membranes.

To use gold nanoparticles as specific labels for biomolecules it is necessary to modify the particles with specific biomolecules. The simplest method would be the unspecific adsorption of a specific capture biomolecules to the nanoparticles surface [16]. These adsorption-based methods do not lead to a stable covalent linkage of the biomolecule and the gold nanoparticle as it would be desirable for a biosensing application. The modification of biomolecules with biotin enables their binding to surfaces modified with steptavidin or avidin, and can also be used for the modification of gold nanoparticles. DNA–nanoparticle conjugates have been formed by attaching biotinylated oligonucleotides to streptavidin modified particles [17,18]. These conjugates were then used to successfully label a DNA-chip. This approach can also be applied for detecting biotinylated proteins or other biomolecules.

In order to use gold nanoparticles as labels in DNA detection applications the use of thiol-modified oligonucleotides for gold nanoparticle modification is a common method. This approach takes advantage of the high attraction of gold surfaces and thiol groups. Alivisatos *et al.* and Mirkin *et al.* both have described an easy method to modify gold nanoparticles with a specific thiolated oligonucleotide

[19,20]. Under the right conditions the incubation of the thiolated oligonucleotides with colloidal gold leads to functionalized gold nanoparticles, which are stable for years, if stored under adequate conditions (exposure to light, high temperatures, oxidants and microbes needs to be minimized). The DNA–gold nanoparticle conjugates also show a higher tolerance to increased salt concentration in the solution and can easily be distinguished from unmodified particles using this particular feature. To improve the stability of the thiol bond different ligands have been designed, using multiple thiol groups to bind one oligonucleotide [4,21]. These anchor molecules will be necessary if one plans an experiment using temperatures above 60 °C or the use of other thiol reagents. It has been described that those conditions can lead to a dissociation of DNA–gold nanoparticle conjugates [21,22]. Park *et al.* use incubation with mercaptohexanol to control the orientation of the oligonucleotides bound to a nanoparticle and could improve the binding of DNA – gold nanoparticle conjugates using this procedure [23].

Other metal nanoparticles (Ag, Cu, Pd, and Pt) can also be modified with DNA using thiolated oligonucleotides. However, due to weaker bond energies for silver and the other metals, the conjugates are not as stable as gold nanoparticle – DNA conjugates. To create a stable DNA modification of those nanoparticles the core shell approach might be a solution. Thereby, a thin layer of gold can be grown around the particles, to be used for the coupling of thiolated oligonucleotides, taking advantage of the higher bond energies of gold and sulphur. If the gold layer is thin enough (~ 1nm) the particle properties remain determined by the core [24,25].

A true covalent coupling between the biomolecules and the gold nanoparticles can be realized using commercially available Nanogold™ (Nanoprobes Inc.). These small (1.4 nm) gold nanoparticles bear a functional group for defined coupling of biomolecules to the gold nanoparticles. The maleimide or N-hydroxysuccinimde group can specifically react with thiol or amino groups [26]. This makes the controlled formation of 1:1 nanoparticle:biomolecule conjugates possible that later can be used as labels in a biosensor application.

3. Detection of Gold Nanoparticle – DNA Conjugates

Since Mirkin and Alivisatos first described the modification of gold nanoparticles with thiolated oligonucleotides a vast number of different detection schemes have been described using gold nanoparticles as label to detect biomolecular interactions. All the described detection methods exploit the unique properties of gold nanoparticles for detection. The detection schemes using gold nanoparticles as labels can be divided into three main groups: optical detection, taking advantage of the unique optical properties of gold nanoparticles, micromechanical detection, using cantilevers or microbalances, and electrical detection, generating a direct electrical signal through the nanoparticle binding.

3.1. Optical Detection

Colloid gold solutions display intense colors and have quite different properties than bulk gold. The properties of gold nanoparticle solutions are mainly influenced by their large surface-to-volume ratio. The bright colors displayed by gold and many other metal nanoparticle solutions are one of the most interesting properties of metal colloids. A collective resonance of their conductive electrons (surface plasmon resonance) is responsible for the intense colors [2,27]. The optical properties of metal colloids are mainly influenced by the particle size, shape, and composition [28] (Figure 1). The interesting optical properties make a number of detection schemes possible using absorbance, scattering, and other optical effects for the detection. Besides the method of detection, gold nanoparticle-based tests for biomolecular interactions can further be divided in homogeneous and heterogeneous assays.

3.1.1. Homogeneous Detection

The color of a colloid solution is also influenced by the interparticle distance. A gold colloid solution with 15 nm sized particles is red, whereas the color changes from red to blue if the particles aggregate. This effect has been used in a homogeneous assay for DNA detection by reducing the interparticle spacing through the binding of DNA. Mirkin and co-workers used two batches of gold nanoparticles, each modified with a different oligonucleotide. If those two solutions were mixed and single-stranded DNA was added that was complementary to both oligonucleotides immobilized on the gold nanoparticles, the particles were connected into a network and the interparticle distance was decreased to less than the average particle's diameter, resulting in a color shift from red to purple [29,30]. Other than the color shift observed when nanoparticles aggregate in high salt buffer, this color change is reversible through heating the solution above the melting temperature (T_m) of the formed DNA duplex. This leads to a release of the gold nanoparticles out of the network and to an increased interparticle spacing (Figure 2a-d).

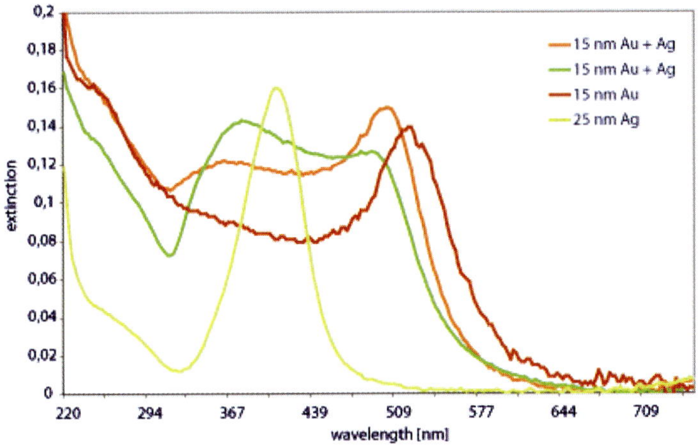

Figure 1. Graph of an absorption measurement of nanoparticle solutions with different nanoparticle sizes and compositions. The characteristics of the core-shell particles (gold core with silver shell) differ with the dimension of the shell. While gold nanoparticles with thin silver shell (orange line) have a similar absorption spectrum like pure gold nanoparticles (red line), a thicker silver shell (green line) leads to a clear shift in the absorption spectrum and similar characteristics like the pure silver nanoparticles (yellow line).

Interestingly, the melting profiles of the duplex DNA in the gold nanoparticle-DNA conjugates revealed extra-ordinarily sharp melting transitions, when compared to conventional fluorophore-labelled DNA [29-32]. This phenomenon is intriguing not only from a scientific standpoint, but also from a technological one. It allows one to very accurately differentiate single nucleotide polymorphisms (SNP) or point mutations from complementary targets on the basis of color and temperature [5,30,33].

As basis for the sharp melting transitions Mirkin *et al.* postulated that the dense loading of oligonucleotides on the surface of the nanoparticles and their ability to bind DNA in a highly cooperative manner was responsible for this effect [31]. A permanent record of those tests could be achieved if droplets of the gradually heated hybridized gold nanoparticles were transferred to a reverse-phase silica plate. This also allowed an easy visualization of the hybridization state at any given temperature. Moreover, the drying on the solid support enhanced the color differentiation due to an increased aggregation of pre-organized nanoparticles [29]. Subsequent the group could show that one could detect target DNA with single mismatches regardless of position using a "tail to tail" alignment of the DNA-gold nanoparticle probes rather then a "head to tail" arrangement [30]. A comparison with conventionally fluorophore-labeled targets showed a three times higher selectivity for the gold nanoparticle-labeled probes [34].

The same effect of the spectral shift of a colloidal solution due to biomolecular recognition events was used in a similar assay to discriminate a single-base mismatch. Upon adding single-stranded DNA the above described network formed. In a second step a ligase was added. If a perfect duplex was formed between the target DNA and the oligonucleotides immobilized on the gold nanoparticles the ligase would covalently link the particles together. In case of a mismatch or other imperfection the ligase would not link the particles together. So the differentiation of perfect match vs. mismatch could be achieved by heating the solution above melting temperature (T_m) of the formed DNA duplex (Figure 2e-f). The network of not covalently linked particles would dissociate and the color shifts back to red, while the covalently linked particles could not be separated by heating and no color shift back to red was observed [35].

Furthermore, the nonreversible aggregation of nanoparticles was applied in another approach to discriminate single mismatches in a DNA duplex. Thereby, Sato *et al.* used also thiol oligonucleotides to modify gold nanoparticles. When the target DNA was added it formed a duplex with the immobilized oligonucleotides. If a perfect duplex was formed the repulsive forces between the gold nanoparticles were reduced, causing the particles to aggregate in high salt buffer, which can be detected by a subsequent color shift (Figure 2g-h). In case of the formation of an imperfect duplex the reduction of the repulsive forces was lower and the gold nanoparticle more stable. So no aggregation occurred and no color shift could be detected [36,37].

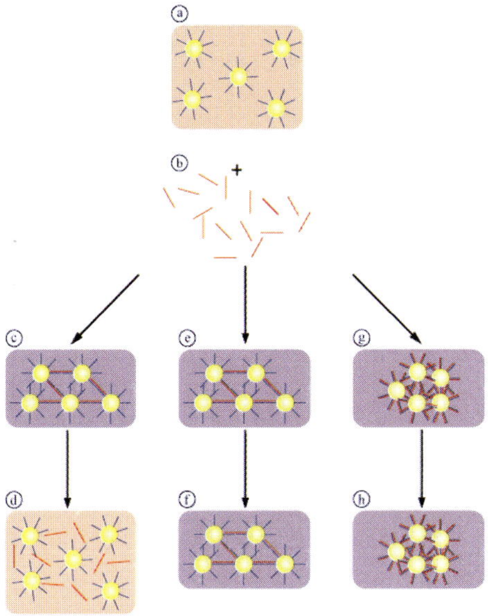

Figure 2. Homogeneous detection of DNA using gold nanoparticle solutions. When complementary target DNA (b) is added to gold nanoparticles modified with capture DNA (a) the formation of the DNA duplex leads to a formation of a network and reducing the interparticle spacing causing a detectable color shift (c). The particles can be released from the network by increasing the temperature above the melting temperature(T_m) of the formed DNA duplex, releasing the particles from the network (d). For a better mismatch detection a ligase can be added to the formed network (e) leading to a covalent coupling of the nanoparticles. No color shift can be observed when the temperature is increased (f). The reduction of the interparticle spacing can also be achieved by reducing the repulsive forces between the particles upon binding the complementary target DNA (g). This particle aggregation is also not reversible by increasing temperatures (h).

To use the color shift of aggregating gold nanoparticles the covalent modification of gold nanoparticles with thiol-modified oligonucleotides is not even required. Short oligomers of a specific sequence can be nonspecifically absorbed to the gold nanoparticles and stabilize the particles against aggregation in high salt buffers. If the complementary single-stranded target DNA is added the short oligonucleotides dissociate from the nanoparticles and hybridize with the complementary target DNA. This destabilizes the gold nanoparticles and they aggregate causing a color shift [38,39]. This approach can also be used for a fluorescent based detection. For this the short oligonucleotides were modified with a fluorescent dye. The fluorescence is quenched by the gold surface as long as the dye remains in close proximity to the gold surface. Due to the dissociation and binding to the complementary target DNA the quenching is stopped and a fluorescence signal can be detected [40].

3.1.2. Heterogeneous Detection

3.1.2.1. Spectral Shift

In an effort to achieve highly multiplexed detection systems, there has been a tendency to develop detection schemes with surface-based readouts. In those chip-based schemes a variety of different detection methods has been used.

The same effect of a spectral shift upon aggregation or binding of gold nanoparticles, which has been used in the homogeneous assays, can also be utilized in a chip-based setup. Hütter and Pileni used chip substrates with nanoscale metal islands. These metal islands absorb a particular spectrum of light. This spectrum will change if gold or other metal nanoparticles are bound to those islands due to a biomolecular recognition [41]. With this method as little as 400 nanoparticles / μm^2 could be detected.

3.1.2.2. Optical Absorbance

To fully take advantage of a microarray application the quantification of the signal is necessary. That means for nanoparticles as labels that the number of bound nanoparticles has to be quantified. Individual bound nanoparticles on a chip surface can be studied using scanning force or scanning electron microscopy [42]. However, these techniques are not very useful for analyzing an entire micro array, because analysis is time-consuming and the equipment is very expensive. Reichert *et al.* demonstrated that by measuring the optical absorbance of gold nanoparticles bound to a transparent surface by a molecular recognition, one could measure the amount of target DNA [43]. For this test chips with microstructured capture DNA-spots were used. The spots with capture DNA were recognized by target DNA labeled with gold nanoparticles and the gold nanoparticles were specifically bound to the chip surface with a low background in nonfunctionalized areas. The structures in the micrometer range could be detected depending on the spot size. While larger structures (60 μm squares) were visible with the naked eye after gold nanoparticle labeling, the smaller structures (4 μm squares) were visualized using a standard light microscope. The spots could be detected either in transmission or reflection. Digital imaging allowed for absorbance measurements from the DNA spots with readout times of a few milliseconds [44].

This method works only for DNA concentrations down to the nanomolar range. However, in practical applications higher sensitivities are needed. But target DNA concentrations in the femto- or picomolar range resulted in a low gold nanoparticle coverage on the surface that could not be detected by simple imaging techniques. But the signal of surface-bound gold nanoparticles can be greatly enhanced by using the gold nanoparticles as "seeds" for a silver deposition. In this autocatalytic reaction a silver shell is formed around the gold nanoparticle, causing an increase in the particle's diameter and a strong enhancement of the absorbance. This process has been used for decades in immunocytochemical application to increase the signal from individual gold nanoparticles for light- and electron microscopy [45]. The process of silver deposition on the bound nanoparticles is quite similar to the one used in monochrome photography. A silver (I) salt and a reducing agent like hydroquinone are normally the main components of such a silver enhancement solution. The growth of the bound gold nanoparticles due to the deposition of silver enables the detection of lower concentrations of target DNA. It even makes the detection of individually bound gold particles possible that are to small to be detected by light microscopy without an enhancement step [46]. The detection of target DNA concentration in the femtomolar range could be achieved taking advantage of the silver enhancement of surface bound gold nanoparticles [34]. Very similar results were achieved when instead of gold nanoparticles modified with thiol oligonucleotides, streptavidin gold nanoparticles and biotinylated target DNA were used [18] (Figure 3).

Based on this detection scheme Li and co-workers realized an integrated device including both the microarray and its detection as it is preferred for a point-of-care application. They combined the DNA-chip with an array of photodiodes. With these integrated photodiodes (one under each DNA spot) they were able to measure the binding and the growth of gold nanoparticles. Using standard optical illumination they could detect the change in light intensity measured with the integrated photodiodes [47]. The ease of this approach makes the detection of optical absorbance an interesting readout scheme for simple DNA chip application.

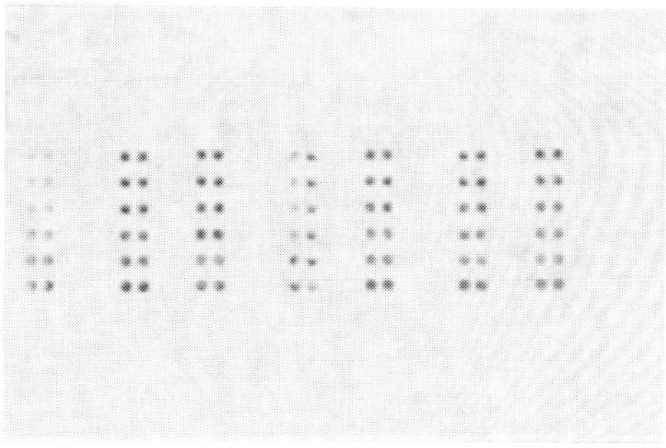

Figure 3. Picture of a DNA-Chip labeled with gold nanoparticles and subsequent silver deposition on the bound particles. The individual spots can be clearly distinguished and easily be detected by the naked eye or a conventional flatbed scanner.

3.1.3. Optical Scattering

Metal nanoparticles scatter light very efficiently; this ability can also be used for the detection of biomolecular recognition. The flux of scattered light from a single 80 nm particle has been estimated to be equivalent to the light flux from 6 x 10^6 colocalized individual fluorescein molecules [1,2,48,49]. Therefore, the scattered light from gold nanoparticle-labeled biomolecules can be imaged and even quantified extremely sensitive. Even the detection of a single surface-bound nanoparticle is possible via scattered light using an oblique illumination source. Like the color of a nanoparticle solution the spectrum of scattered light is determined as well by the particles size, shape, and composition.

A normal microscope slide can be used as internal reflecting wave guide in order to efficiently image surface-bound gold nanoparticles. When using this approach it is very interesting that only gold nanoparticles that are directly bound to the surface of the wave guide scatter light and produce a signal. Unbound particles are not within the evanescent wave depth of the wave guide and are consequently not illuminated and not imaged. This feature make the detection of gold nanoparticle binding to the wave guide's surface possible while the wave guide is incubated with gold nanoparticle-labeled biomolecules. So the system allows a real-time detection of binding events as well as studying the hybridization and denaturation of DNA. Using selenium nanoparticle conjugates and biotinylated target DNA, Stimpson and co-workers were able to show that a DNA detection with this setup was possible [50]. They also compared the system with the best fluorescence-based system available at the time and showed comparable results.

Because the spectrum of scattered light depends on the particles sizes and composition the creation of a multilabel detection system is possible (Figure 4). This is a main advantage over the use of absorbance as detection method. A first indication how a multilabel system could work was given by Taton *et al.* They used two gold nanoparticle sizes (50 and 100 nm) and modified them with specific thiolated oligonucleotides. Both nanoparticle solutions were then incubated on a wave guide that was used as DNA-chip substrate. The DNA gold nanoparticle conjugates would then bind to their complementary sequences, which were immobilized on the wave guide's surface (Figure 5). Upon illumination the 50 nm surface-bound nanoparticles glowed green, while the 100 nm surface-bound particles glowed orange [51]. It was also demonstrated that the dissociation of the gold nanoparticles could be studied if the system was heated above melting temperature.

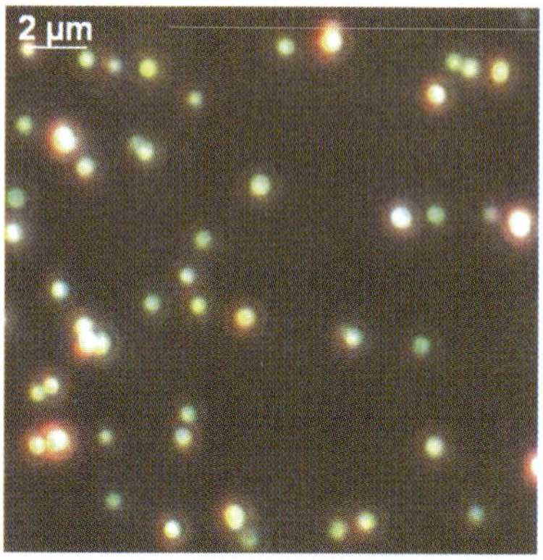

Figure 4. Picture of surface bound nanoparticles scattering light depending on their size. The nanoparticles were immobilized on a microscope slide, which was used as internal wave guide. By the color of the scattered light two sizes of gold nanoparticles can be distinguished. While gold nanoparticles with 60 nm glow green bigger particles glow yellow to red.

The change of interparticle spacing can also be detected using light scattering. Storhoff and co-workers used the above described approach of a homogenous assay, in which networks of gold nanoparticles were formed by the formation of DNA duplexes. These formed aggregates were then spotted on a microscope slide, which later served as a wave guide. When the slide was illuminated with white light the formed aggregates glowed yellow to orange, due to the plasmon band red shift. Control spots with particle solutions that were not aggregated, scattered green light because of the larger interparticle spacing [52]. This approach uses the microscope slide only for the detection but not as DNA-chip substrate, because there are no capture molecules bound on the surface of the wave guide. The biomolecular recognition is performed in a homogeneous assay, and the solutions are only transferred to a solid substrate for analysis.

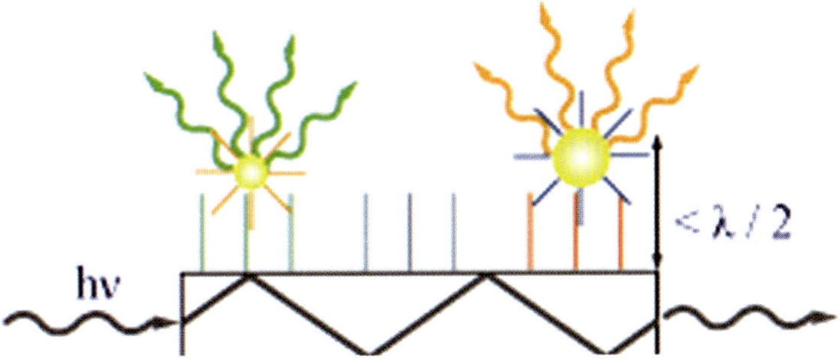

Figure 5. Scheme using gold nanoparticles as labels and optical scattering for the detection of binding events on a DNA-chip. Only the surface bound nanoparticles are within the evanescent field of the wave guide. The spectrum of the scattered light depends on the size, shape, and composition of the nanoparticles.

The possibility to create a multilabel system using metal nanoparticles of different size and compositions makes light scattering an interesting option for more complex DNA-arrays. So this detecting scheme can actually be used for gene expression or single polymorphism analysis. A comparison with fluorescence-based detection schemes showed a 60-fold increase in sensitivity when using light scattering on surface-bound nanoparticles [53]. When utilized in expression analysis the nanoparticle labeling and detection of the scattered light proved to be very effective. With metal nanoparticles as labels more genes were quantified in comparison to standard fluorescence-labeled DNA chips at all used target concentrations. For this comparison transcripts from human lung cancer and leukemia RNA samples were investigated on a DNA chip [54].

3.1.4. Raman Scattering

In addition, gold and other metal nanoparticles can enhance the scattering signal from absorbed Raman-active molecules [55,56].

Especially gold and silver nanoparticles have been described in being very effective at enhancing the Raman signal, and enhancement factors of up to 10^{14} have been described using metal nanoparticles and absorbed Raman-active molecules [57]. The effect of surface enhanced Raman spectroscopy (SERS) can also be used in biological applications. Because of the vast number of Raman active molecules (each with its specific spectrum) the number of possible labels is huge. So SERS could enable the user to create a true multilabel system with a great number of labels used simultaneously.

A first indication how a SERS based detection scheme for DNA could look like was given by Cao *et al.*[58]. They used 13 nm gold nanoparticles and modified them with special thiol-modified oligonucleotides. Fluorescent dyes were used as Raman active labels and additionally integrated in the thiol oligonucleotides and each dye was chosen for a specific sequence. These special DNA-gold nanoparticle conjugates were then hybridized on a DNA chip. After the conjugates had bound to their complementary capture sequences on the chip no specific Raman signal was detected. As reason for this effect the author assumed that the nanoparticles were not closely enough packed on the surface to give an electromagnetic field enhancement. In order to reduce the distance between the individual bound nanoparticles, a silver deposition was used. The bound particles were enlarged by this procedure and an increase in the Raman signal was detected (Figure 6a). This test was performed with six different DNA targets and six different Raman-labeled gold nanoparticles, giving a first indication of the multiplexing capabilities of this system. The chip-based Raman detection has the same sensitivity and specificity as other chip-based DNA detection systems, but adds multiplexing capabilities that can not be achieved with other conventional labeling techniques. This approach has also been described for the detection of proteins [59]. So the Raman-based detection of biomolecules is an interesting method for future applications in medical diagnostics and biological imaging [56].

The multiplexing capabilities of a Raman-based detection scheme can also be utilized in a solution-based assay. Instead of using a chip substrate with multiple capture DNA spots for the immobilization of the DNA-gold nanoparticle conjugates, glass beads were used in this approach. Each glass bead was modified with a specific capture DNA sequence. These modified glass bead bind then target DNA and the complementary gold nanoparticles (Figure 6b). After the necessary silver enhancement of the bound gold nanoparticles, each glass bead is analyzed and the Raman signal is detected [60]. Interestingly, the authors reported that they also could differentiate defined mixtures of Raman-active dyes, which further enhances the number of possible labels.

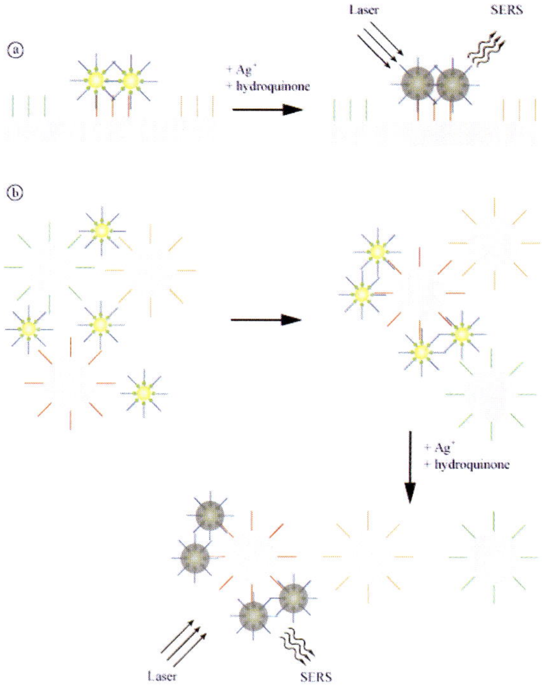

Figure 6. Schematic drawing of the detection of biomolecular recognition using Raman spectroscopy and surface enhanced Raman spectroscopy (SERS). In a chip-based format the gold nanoparticles modified with the specific labeled DNA molecules bind to their complementary strand immobilized on the chip surface (a). The binding can not be detected by Raman spectroscopy. Upon deposition of silver on the bound gold nanoparticles a SERS signal can be detected. The same detection scheme can also be utilized using glass microbeads instead of a chip surface (b). Each microbead is modified with a specific sequence. Through the binding of the complementary DNA specially modified gold nanoparticles are bound to the microbead and after a silver deposition the SERS signal can be detected.

In a different approach not the detection of a Raman signal was used for the detection of a target molecule, but rather the weakening and vanishing of the Raman signal. Therefore, a DNA stem-loop structure was immobilized on gold nanoparticles via a thiol modification. The other end of this stem-loop structure was modified with a Raman-active dye. As long as the stem-loop structure was intact the Raman signal for the dye was detected. When complementary single-stranded target DNA was added the stem-loop structure opens and the dye was removed from the surface of the nanoparticle. Because the dye was no longer in close proximity to the metal surface of the nanoparticle no spectroscopical signal was detected [61].

3.1.5. Surface Plasmon Resonance (SPR) Imaging

The SPR technique is a quite sensitive detection method with a wide variety of chemical and biological applications [62]. In contrast to all other detection schemes described so far the SPR method is well established and a variety of different instruments is commercially available today. The first instrument using the SPR effect for the detection of binding events was introduced to the market in 1990 under the name of BIAcore™. Since then a lot of improvements has been made increasing the ease of use as well as the number of measurement spots. Like the micromechanical applications that will be discussed later a labeling with metal nanoparticles is not necessary to detect binding events with SPR. However, gold nanoparticle labeling increases the sensitivity of this detection.

For the SPR detection planar polarized light is totally reflected from a thin (± 50nm) gold film. The light leads to an evanescent field that penetrates the gold film. The evanescent field excites electromagnetic surface plasmon waves in the gold film, which leads to a loss of energy in the reflected light. If a molecule is bound to the gold surface the propagation of the evanescent electric field is modified. This causes a change in the deflection angle of the light, which is used as indicator for the binding of molecules to the surface [62]. In a typical measurement setup a laser beam is aimed at the back of a gold film under a specific angle. Binding events of molecules on the top of the gold film can then be detected as shift in the deflection angle.

Because SPR is a relatively "old" detection method the instruments are well developed and integrate miniaturized flow systems for efficient sample delivery to the sensor and a control over the reaction conditions. The SPR system also allows real time measurements, enabling the study of binding kinetics and other parameters in biomolecular recognition [63]. This makes the technique an important tool in studying biomolecule–biomolecule and biomolecule–ligand interactions. The real-time detection of binding of a variety of unlabeled target molecules has been achieved and investigated using SPR. It has also been shown that the binding of unlabeled DNA to a complementary surface-bound capture DNA can be measured using a SPR detection device [64,65]. Through the integration of a higher number of measurement spots the readout of DNA chip-like substrates has been made possible. The studies of DNA hybridization however showed that the detection limit was only in the nanomolar range [66]. The change in the deflection angle, which is detected as signal in SPR is directly influenced by the mass of the bound molecules. So an increase of the mass of a target through a specific label would lead to an increased sensitivity of the system (Figure 7). Gold nanoparticles can be used as such labels and with their high specific weight have a great influence on the refractive index. They can be integrated in a "sandwich procedure" and increase the sensitivity [67]. When using gold nanoparticles as labels with their large impact on the refractive index target DNA concentrations down to the picomolar range could be detected [17].

Gold nanoparticles can also be used to enable a label-free detection of biomolecular interactions by reproducible self-assembly of gold nanoparticles from solution on glass substrates. Thus making the production of cheap and fast-to-reproduce SPR sensors possible that can be easily multiplexed for high throughput screening in a biochip format for genomics, proteomics, or drug discovery [68].

Glass surfaces were also be used for a modified SPR monitoring method measuring the shift in transmission surface plasmon resonance (T-SPR). Gold nano islands were structured on the glass surface and functionalized with thiol oligonucleotides. The binding of the capture molecules as well as the hybridization of complementary target DNA could be detected as shift in the T-SPR signal. The integration of gold nanoparticles further enhanced the sensitivity of the system [69].

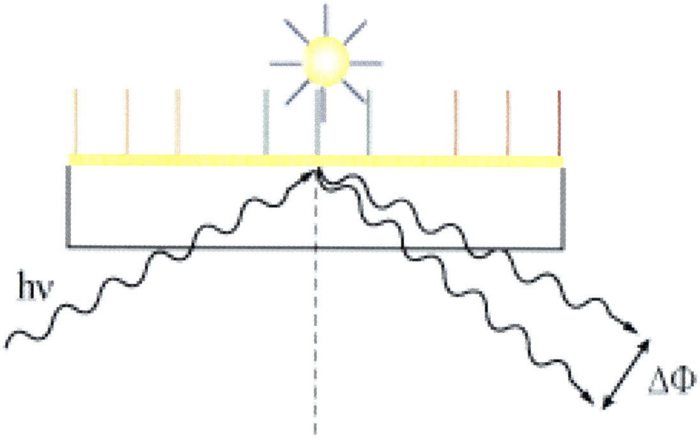

Figure 7. Scheme of the enhanced SPR detection by using gold nanoparticle labels. The binding of molecules to the surface can be detected by SPR as change in the resonant reflection angle Φ. The shift in the resonant reflection angle Φ can be greatly enhanced by using gold nanoparticles as labels, enhancing the sensitivity of SPR imaging.

3.1.6. Photothermal Imaging

The photothermal detection of surface-bound gold nanoparticles was proposed by Tokeshi *et al.* [70,71], who could detect very low concentrations of absorbing molecules in liquid solutions. The photothermal effect can be detected when gold nanoparticles are illuminated with laser light of the appropriate wave length (5 nm gold nanoparticles absorb light at ca. 514 nm). Boyer *et al.* used a green argon-ion laser at 514 nm to heat surface-bound 5 nm gold nanoparticles and a red helium-neon laser at 633 nm for probing [72]. This detection scheme was further developed to measure the binding of gold nanoparticles to a surface through the formation of a duplex DNA. Gold nanoparticle-tagged target DNA was bound to surface-bound single-stranded capture DNA molecules and the binding could be detected using photothermal imaging [73]. With this setup target DNA concentrations as low as picomolar could be detected without any silver deposition on the surface-bound nanoparticles.

3.2. Micromechanical Detection

While all the above described optical methods used the unique optical properties (for the detection of biomolecular recognition), the micromechanical detection exploit the high specific mass of gold nanoparticles for monitoring binding events. Because a weight change is detected as signal for a binding event the micromechanical detection is often also referred to as micro gravimetry. Micromechanical resonators are capable of detecting femtogram masses of material [74].

3.2.1. Quartz-Crystal Microbalances (QCM)

QCMs detect characteristic changes in the frequency of the oscillating sensor surface, which can be correlated to small mass changes. One big advantage of QCMs is that the entire sensor can be immersed in a solution containing target molecules, making QCMs perfect sensors to study dynamics of binding reactions. Okahata and co-workers described the label-free detection of single-stranded target DNA using a QCM with immobilized capture DNA on its sensing surface [75]. Although this was a working demonstration of a label-free detection of DNA hybridization, the sensitivity was only in the nanomolar range and needed to be improved. A 50-fold increase in sensitivity compared to an unlabeled system could be achieved when gold nanoparticles were added as labels. To further enhance the sensitivity a second layer of nanoparticles was added in a dendritic fashion [76,77]. The impact of the gold nanoparticle label could also be increased by a silver deposition on the bound gold particles [78,79]. Instead of the classical hybridization approach Willner and co-workers used the polymerase I activity to couple biotinylated bases complementary to a single-point mutation. The biotin labeled nucleotides were introduced in a primer extension reaction and afterwards 10 nm avidin-gold nanoparticle conjugates were added, bound to the biotin modification, and detected [78]. Further investigations showed that 45 nm gold nanoparticles are optimal labels for a QCM-based detection [80]. It might not be practical to integrate QCMs in a dense micro array format with hundreds of measurement spots. But simple detection with a limited number of measurement sites can be realized. Another problem is that the production of microbalances is not yet compatible with the materials and processes used in standard micro-electronics industry.

3.2.2. Microcantilever

Unlike QCMs it is possible to make arrays of microcantilevers of 1000 or more measurement points. These microcantilever arrays can be mass produced and each cantilever can monitor the binding of a different analyte, when each cantilever is modified with a different capture molecule [81-83]. The binding of molecules cause a bending of the cantilever and can be detected with two methods: first, the frequency change due to the additional mass loading or a change in the force constant is detected (the cantilever is used like a microbalance); second, the bending of the cantilever is detected, which results in a change of the deflection of a laser beam aimed against the surface of the cantilever. Because measurements using just one of the detection principles are very prone to disturbances like temperature changes, vibrations or surface effects caused by the adsorption of molecules, modern detection systems combine the two methods to circumvent those problems [84].

To use microcantilever arrays as highly paralleled biosensors the modification of individual cantilevers with specific capture molecules has to be achieved. This has been demonstrated with single-stranded capture DNA molecules. Upon incubation of different target DNA molecules specific signals were detected [83,85]. Because the detected signal greatly depends on the mass of the bound molecule the sensitivity of the system could be enhanced by using labels with a high specific mass. Su *et al.* could demonstrate a 200-fold increase in sensitivity of a cantilever-based biosensor by using gold nanoparticles as labels and silver deposition on the bound particles [86]. Because microcantilevers like QCMs can be immersed in solution and simultaneously detect a signal, they are suitable for real-time monitoring of binding events.

3.3. Electrical Detection

Biosensors based on an electrical detection scheme are very attractive, because it combines the biomolecular recognition with the advantages of electrical techniques like high sensitivity, low cost, minimal power requirements, simple design, suitability for microfabrication, and portability. When using gold nanoparticles as labels for an electrical detection two main detection schemes can be differentiated: first, an electrochemical detection scheme exploiting the electrochemical activity of gold for the detection of a binding event; and second, resistance (conductivity) or capacity measurement due to the conductivity of the nanoparticles.

3.4. Electrochemical Detection

Since the discovery of the electroactivity of nucleic acids by Palecek there has been an ever growing interest in developing DNA sensors based on electrochemical detection [87]. A wide variety of methods and electrochemical labels has been used for detection and is described in the literature [88,89]. Besides the electroactivity of nucleic acids different redox-active molecules, enzymes, and also metal nanoparticles have been explored as labels for an electrochemical DNA detection.

When using metal nanoparticles as labels for an electrochemical detection one takes advantage of the high number of oxidizable atoms in every single particle (e. g. 1.7×10^5 gold atoms in a spherical gold nanoparticle with a diameter of 18 nm), making the detection scheme very sensitive. Normally different methods of voltammetry are used for the detection of specifically bound gold nanoparticles.

To directly detect bound gold nanoparticles Ozosz *et al.* used differential pulse voltammetry (DPV). They modified a graphite electrode with single-stranded capture DNA and bound gold nanoparticles modified with the complementary DNA strand to the surface of a graphite electrode. At a stripping potential of ~ + 1.2 V the direct oxidation of the bound nanoparticles was detected [90]. With this method the nanoparticles were not dissolved before the detection like in other detection setups. Through the use of other metal nanoparticles with different stripping potentials the creation of a multilabel system can be envisioned. Again an enhancement of the signal was achieved by silver deposition on the bound gold nanoparticles, thereby increasing the number of oxidizable atoms. For this test glassy carbon electrodes were used [91].

For an indirect detection of the bound nanoparticles the highly sensitive analytical method called anodic stripping voltammetry (ASV) is used [92]. Therefore, gold nanoparticles were bound to surface either directly through DNA-DNA interaction or via the biotin-streptavidin system. This step was followed by a treatment with HBr/Br_2 , which dissolved the bound nanoparticles [93,94]. The so released metal ions were then accumulated at a negatively charged working electrode and reduced to the metal again. The accumulation step prior to the actual detection step accounts for the high sensitivity of this detection method. Finally, the metal is reoxidized by increasing the potential of the electrode. This oxidation can be measured as a specific current signal vs. potential. Because the signal is specific for the metal, the construction of a multilabel system using different metal nanoparticles is possible. Quantification can also be achieved using this signal. For an extremely inexpensive setup disposable carbon-based screen-printed electrodes were used to detect different numbers of bound gold nanoparticles [95]. Wang and co-workers applied a mixture of conjugates of DNA-modified gold nanoparticles and DNA-modified magnetic beads to further increase the sensitivity of the ASV detection. The magnetic beads were used to separate the hybridized target DNA from the sample and afterwards the DNA was analyzed by ASV [93]. Additionally, an 80-fold increase in sensitivity could be achieved when gold was deposited on the bound gold nanoparticles prior to dissolving the bound particles. A detection system with multiple labels was constructed when using different inorganic materials as specific labels. Each of the used inorganic materials showed a specific response in the ASV detection, demonstrating the multiplexing capabilities of such a detection system [96].

A combination of DNA-modified magnetic beads and DNA-gold nanoparticle conjugates was used in solid state chronopotentiometric detection. The formed conjugates of magnetic beads and gold nanoparticles were silver-enhanced and collected at an electrode surface using a magnet and finally analyzed [97].

3.5. Resistive or Capacitive Detection

The binding of gold nanoparticles in a gap between two electrodes can be detected either by conductivity or capacity measurement. Conjugates of gold nanoparticles and biomolecules can be utilized to monitor biomolecular recognition events using this detection scheme.

If a resistance measurement is used for the detection the binding of gold nanoparticles in a gap between to electrodes should lead to a drop in the measured resistance [98]. However, it has not been possible to detect the binding of gold nanoparticles in an electrode gap just by a simple DC resistance measurement. There are two possible reasons for that: first, it has not been possible to bind the gold nanoparticles densely enough in the gap by a molecular recognition, so that a conductive layer is not create; second, the layer of biomolecules that functionalize the nanoparticle might insulate the particle from completing the circuit. To bridge a micrometer gap using gold nanoparticles as labels several research groups have used the deposition of conductive material on the bound particles to overcome the above described problems and to create a conductive connection.

This detection principle was first demonstrated by Velev *et al.* [99]. Thereby, latex microspheres functionalized with capture molecules have been trapped in an electrode gap. After immobilization of the capture molecules in the gap an antigen-antibody reaction was used to bind gold nanoparticles in the gap. After a final step of silver deposition on the bound nanoparticles the biomolecular recognition event was detected as drop in the measured resistance.

The immobilization of capture DNA in the electrode gap enables DNA detection. The capture DNA can be directly immobilized on a solid substrate using the same chemistry as in DNA chip production [8]. The binding of gold nanoparticles modified with DNA complementary to the immobilized capture sequence can then be detected using the above described method (Figure 8). Interestingly, it has been found that there is a direct correlation between the measured resistance, the silver deposition, and the concentration of bound gold nanoparticles, which can be directly correlated to the concentration of target DNA in a sample [100]. This makes the quantification of such a resistance measurement suitable for modern molecular diagnostics. The specificity of this detection method was demonstrated by differentiating single nucleotide polymorphisms using this detection scheme [101]. In an array of electrode gaps the researchers modified different electrode gaps with different capture DNA sequences and could show the selectivity of a resistance-based detection system of up to 10^5:1. This system also took advantage of the increased specificity in the DNA recognition when gold nanoparticles are used as labels. To achieve the above mentioned selectivity the researchers only used washing steps at room temperature, eliminating the stringent temperature control during hybridization and washing, normally needed to reach specificity when incubating target DNA on a DNA chip.

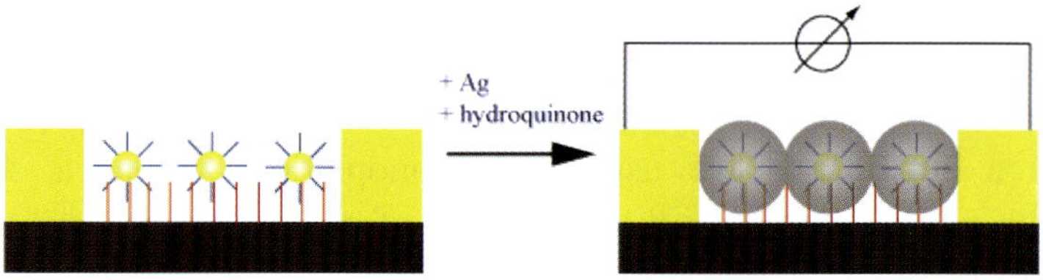

Figure 8. Scheme the electrical detection using gold nanoparticles as labels. The gold nanoparticles are bound into a gap between two electrodes by a specific molecular recognition. Upon silver deposition on the surface bound nanoparticles the gap between the electrodes, causing a drop in the measured resistance.

Because of the ease of detection it is possible to construct an easy-to-use, portable, and robust chip readout device using DNA chips with a detection based on resistance measurement [102]. This DNA Chip Reader combines an embedded PC with an ohmmeter and a multiplexer for a fast and easy readout of the chips. For a measurement the chip is positioned in a special socket, which holds and contacts the chip during the measurement. The results are displayed on an integrated display and can be either stored in the reader or transferred to a desktop PC for further data analysis. In this readout device a microelectrode array with 42 measuring sites can be analyzed. The chip is produced with standard photolithographic procedures and each measurement spot consists of a10 μm wide electrode gap. The addition of a microstructured silicone nitride layer helps to increase the sensitivity of the chip by avoiding crosstalk between the individual measurement spots [103]. A drawback to the system is that no continuous measurement is possible: the reaction has to be stopped and the chip must be dried for each measurement. This bears the risk of under or over enhancing the chip. If the measured signals are to low, the chip is under enhanced; a further step of silver deposition can increase the signal intensity. However, the over enhancement is problematic and the chip can not be recovered. To avoid this problem a constant monitoring of the silver deposition would be needed. Diessel *et al.* described such a system, making the real-time detection of the silver deposition process possible [104].

An AC capacitance measurement enables the direct detection of bound nanoparticles in an electrode gap without an additional step of silver deposition. Using this detection method, however, requires much more sophisticated measurement equipment as well as more elaborate chip architecture. Measurement spots with interdigitated electrodes with gap sizes below one micron are usually used for this detection scheme. The capture DNA is immobilized in the gaps between the interdigitated electrodes and biotinylated target DNA is hybridized. Using the interaction between biotin and streptavidin, streptavidin-modified gold nanoparticles are bound to the chip. The deposition of silver is not needed for detection, but can increase its sensitivity [105].

4. Further Applications of Gold Nanoparticles for Biosensing

As illustrated in the above described applications, there is a wide variety of possible detection schemes using gold nanoparticles as labels. Instead as direct label for the readout of a biomolecular recognition event, gold nanoparticles can also be used to enhance the detection limits of other reactions.

A very interesting phenomenon is that the addition of gold nanoparticles can enhance the sensitivity and specificity of the PCR reaction. This effect has been described independently by two research groups [106,107]. Li *et al.* added commercially available 10 nm gold nanoparticles to the PCR mixture and investigated the influence of the concentration of nanoparticles on the efficiency of the PCR. They were able to show that there is a significant improvement in the PCR reaction by adding gold nanoparticles. However, they showed also that concentration of gold nanoparticles higher than 1nM in the PCR mixture lead to an inhibition of the reaction and lower yields [106]. Another research group achieved similar results when adding 13 nm gold nanoparticles to the PCR mixture. The addition of 0.7 nM gold nanoparticles showed a significant increase in the yield of the PCR reaction. Keeping the yield constant they were further able to reduce the reaction time by increasing the heating and cooling rates. The sensitivity of different tested PCR systems showed an increase of up to 10^4- fold.

The two groups postulated two totally different explanations as reason for this interesting phenomenon. One explanation was that the gold nanoparticles mirrored the activity of the single-strand DNA binding protein (SSB). By binding single-stranded DNA the gold colloid eliminated unspecific products from the reaction [106]. The other explanation was that the surface of gold colloids forms a highly ordered liquid layer, which leads to a higher thermal conductivity. By this the efficiency of the PCR might be enhanced through a faster heat transfer into the DNA by nanoparticles that are close or bound to the DNA [107]. But the actual mechanism remains unclear and needs to be further investigated.

Gold nanoparticles might also help to eliminate the PCR reaction in future application using so-called bio-bar-code particles for a chip-based DNA detection [108-110]. The original bio-bar-code assay used magnetic beads and gold nanoparticles for detection. The magnetic particles bear the capture DNA that binds the target DNA. The target DNA strand is significantly longer than the capture DNA, so that there is a single-stranded DNA to bind DNA-modified gold nanoparticles to the magnetic beads. The gold nanoparticles are modified with two kinds of DNA: one single strand, for binding the target DNA and a double-stranded DNA. After the formation of the complex of gold nanoparticles and magnetic beads, the complexes were separated from the solution by a magnetic field. The double-stranded DNA is then denatured, freeing the bio-bar-code DNA. Because the bio-bar–code DNA was immobilized in great excess on the gold nanoparticles, an enhancement of the number of detectable molecules is achieved. The freed DNA is then used for an analysis on a DNA chip using gold nanoparticle labeling and light scattering for detection. To improve the feasibility of such an assay the number of different oligonucleotides needed to be reduced. So the gold nanoparticles were only modified with one oligonucleotide, which is complementary to the target DNA with one end and its other end is used for the chip-based detection. After the magnetic separation the bio-bar-code DNA is freed by an incubation with dithiothreitol [111]. Because a single gold nanoparticle can bear hundred of oligonucleotides on its surface, very low concentrations of target molecules can be detected. This method is especially interesting for the detection of protein, because proteins can not be amplified by a PCR-like reaction as it is possible for nucleic acids. When, for instance, the bio-bar-code DNA on gold nanoparticles is combined with proteins for the binding of the target molecules this principle can be also used for protein detection [110,112]. By using dye-labeled bio-bar-code oligonucleotides, protein detection has been demonstrated that does not need the final step of analyzing the freed oligonucleotides on a chip. Rather the fluorescence intensity of the solution was measured after the bio-bar-code DNA was freed [113].

The high loading capacities of gold nanoparticles for a reporter molecule can also be exploited for an electrochemical detection scheme. The surface of a gold electrode was modified with capture DNA. Through biomolecular recognition target DNA and DNA-functionalized gold nanoparticle were bound as "sandwich" to the gold electrode. Besides the modification with thiolated oligonucleotides the gold nanoparticles also carried the redox–active reporter molecule $[Co(phen)_3]^{3+/2+}$, which binds electrostatically to DNA. Through the biomolecular recognition the reporter molecule is brought in to close proximity to the electrode surface and was detected either by DPV or cyclic voltammetry [114].

5. Outlook

Gold nanoparticles are very potent labels for the detection of biomolecular interactions. Especially the high number of different detection schemes for the same label and the high sensitivity of detection making them interesting candidates for future diagnostic applications.

In this review, we have described the different detection schemes developed in the last decade for the detection of nanoparticle-labeled biomolecules. The nanoparticles act not always as label that is directly detected. They can also be used for making established detection methods more sensitive (e.g. SPR, SERS, or micromechanical detection). The unique optical properties enable a variety of optical methods for the detection of metal nanoparticle labels. Gold nanoparticles can also be used to develop new robust detection setups, for instance the electrical detection schemes. The ease of detection and the comparable low costs for the equipment needed for detection make gold nanoparticles a promising alternative label for future diagnostic applications. It will also enable the creation of hand held detection devices for a point of care application.

For the future of gold nanoparticles as labels for biomolecules it will be essential to develop stable and easy-to-handle modifications for coupling biomolecules to gold nanoparticles. The continuous search for better coupling groups already resulted in more stable conjugates of DNA and gold nanoparticles that are stable enough to be used in analyzing DNA hybridization and denaturation.

In the future the high number of possible biomolecular tests and the ever growing knowledge about biomolecules, their interactions, and their importance for our well-being will demand an increased number of test systems. Because of the variety of test with different requirement concerning sensitivity, number of measurement spots, costs, and so on there will be a high number of different test systems. Some very simple with only a limited number of test sites and others highly complex with thousands or even millions of test spots like in expression profiling. Some test systems will be of low costs, easy-to-handle and used outside of specialized laboratories, while other applications will remain expensive and bound to laboratory use.

In the last years first tests have been commercialized using gold nanoparticles as labels. Because of their advantages over some of the already established test systems we will see more and more applications where gold nanoparticles are used as labels in a variety of diagnostic applications.

References

1. Yguerabide, J. & Yguerabide, E.E. Light-scattering submicroscopic particles as highly fluorescent analogs and their use as tracer labels in clinical and biological applications. Anal Biochem 262, 157-76 (1998).
2. Yguerabide, J. & Yguerabide, E.E. Light-scattering submicroscopic particles as highly fluorescent analogs and their use as tracer labels in clinical and biological applications. I. Theory. Analytical Biochemistry 262, 137-156 (1998).
3. Pellegrino, T. et al. On the Development of Colloidal Nanoparticles towards Multifunctional Structures and their Possible Use for Biological Applications, Small 1, 48-63 (2005).
4. Fritzsche, W. & Taton, T.A. Metal Nanoparticles as Labels for Heterogeneous, Chip-Based DNA Detection. Nanotechnology 14, R63-R73 (2003).
5. Thaxton, C.S., Georganopoulou, D.G. & Mirkin, C.A. Gold nanoparticle probes for the detection of nucleic acid targets. Clinica Chimica Acta 363, 120-126 (2006).
6. Cheng, M.M. et al. Nanotechnologies for biomolecular detection and medical diagnostics. Curr Opin Chem Biol 10, 11-9 (2006).
7. Schena, M. Microarray analysis, (Wiley-Liss Verlag, Hoboken, New Jersey., 2003).
8. Pirrung, M.C. Die Herstellung von DNA-Chips. Angew. Chem. 114, 1326-1341 (2002).
9. Hayat, M.H. Colloidal Gold: Principles, Methods, and Applications, 680 (Elsevier Science & Technology Books, 1989).
10. Grabar, K.C., Freeman, R.G., Hommer, M.B. & Natan, M.J. Preparation and Characterization of Au Colloid Monolayers. Analytical Chemistry 67, 735-743 (1995).
11. Brust, M., Fink, J., Bethell, D., Schiffrin, D.J. & Kiely, C.J. Synthesis and reactions of functionalized gold nanoparticles. Journal of the Chemical Society, Chemical Communications, 1655-1656 (1995).
12. Brust, M., Walker, M., Bethell, D., Schffrin, D.J. & Whyman, R., J. Chem. Soc., Chem. Commun., 801-802 (1994).

13. Daniel, M.C. & Astruc, D. Gold nanoparticles: assembly, supramolecular chemistry, quantum-size-related properties, and applications toward biology, catalysis, and nanotechnology. Chem Rev 104, 293-346 (2004).
14. Turkevich, J., Stevenson, P.L. & Hiller, J. Discuss. Faraday Soc. 11, 55 (1951).
15. Parak, W.J.G., D.; Pellegrino, T.; Zanchet, D.; Micheel, C.; Williams, S. C.; Boudreau, R.; LeGros, M. A.; Larabell, C. A.; Alivisatos, A. P. Biological applications of colloidal nanocrystals. Nanotechnology 14, R15-R27 (2003).
16. Gearheart, L.A., Ploehn, H.J. & Murphy, C.J. Oligonucleotide adsorption to gold nanoparticles: A surface-enhanced Raman spectroscopy study of intrinsically bent DNA. Journal of Physical Chemistry B 105, 12609-12615 (2001).
17. He, L. et al. Colloidal Au-enhanced surface plasmon resonance for ultrasensitive detection of DNA hybridization. J Am Chem Soc 122, 9071-9077 (2000).
18. Alexandre, I. et al. Colorimetric silver detection of DNA microarrays. Anal Biochem 295, 1-8 (2001).
19. Alivisatos, A.P. et al. Organization of 'nanocrystal molecules' using DNA. Nature 382, 609-611 (1996).
20. Mirkin, C.A., Letsinger, R.L., Mucic, R.C. & Storhoff, J.J. A DNA-based method for rationally assembling nanoparticles into macroscopic materials. Nature 382, 607-9 (1996).
21. Li, Z., Jin, R., Mirkin, C.A. & Letsinger, R.L. Multiple thiol-anchor capped DNA-gold nanoparticle conjugates. Nucleic Acids Res 30, 1558-62 (2002).
22. Letsinger, R.L., Elghanian, R., Viswanadham, G. & Mirkin, C.A. Use of a steroid cyclic disulfide anchor in constructing gold nanoparticle-oligonucleotide conjugates. Bioconjug Chem 11, 289-91 (2000).
23. Park, S., Brown, K.A. & Hamad-Schifferli, K. Changes in Oligonucleotide Conformation on Nanoparticle Surfaces by Modification with Mercaptohexanol. Nano Letters 4, 1925-1929 (2004).
24. Taton, T.A. Nanostructures as tailored biological probes. Trends Biotechnol 20, 277-9 (2002).
25. Cao, Y., Jin, R. & Mirkin, C.A. DNA-modified core-shell Ag/Au nanoparticles. J Am Chem Soc 123, 7961-2 (2001).
26. Hainfeld, J.F., Powell, R.D. & Hacker, G.W. Nanoparticle Molecular Labels in Nanobiotechnology (eds. Niemeyer, C.M. & Mirkin, C.A.) 353-386 (Wiley-VCH, Weinheim, Germany, 2004).
27. Kreibig, U. & Vollmer, M. Optical Properties of Metal Clusters, (Berlin, 1995).
28. Kelly, K.L., Coronado, E., Zhao, L.L. & Schatz, G.C. The optical properties of metal nanoparticles: The influence of size, shape and dielectric environment. J. Phys. Chem. B 107, 668-677 (2003).
29. Elghanian, R., Storhoff, J.J., Mucic, R.C., Letsinger, R.L. & Mirkin, C.A. Selective colorimetric detection of polynucleotides based on the distance-dependent optical properties of gold nanoparticles [see comments]. Science 277, 1078-81 (1997).

30. Storhoff, J.J., Elghanian, R., Mucic, R.C., Mirkin, C.A. & Letsinger, R.L. One-Pot Colorimetric Differentiation of Polynucleotides with Single Base Imperfections Using Gold Nanoparticle Probes. J Am Chem Soc 120, 1959-1964 (1998).
31. Jin, R., Wu, G., Li, Z., Mirkin, C.A. & Schatz, G.C. What controls the melting properties of DNA-linked gold nanoparticle assemblies? J Am Chem Soc 125, 1643-54 (2003).
32. Reynolds, R.A., Mirkin, C.A. & Letsinger, R.L. Homogeneous, Nanoparticle-Based Quantitative Colorimetric Detection of Oligonucleotides. Journal of the American Chemical Society 122, 3795-3796 (2000).
33. Storhoff, J.J. et al. What Controls the Optical Properties of DNA-Linked Gold Nanoparticle Assemblies? Journal of the American Chemical Society 122, 4640-4650 (2000).
34. Taton, T.A., Mirkin, C.A. & Letsinger, R.L. Scanometric DNA array detection with nanoparticle probes. Science 289, 1757-60 (2000).
35. Li, J. et al. A colorimetric method for point mutation detection using high-fidelity DNA ligase. Nucleic Acids Res 33, e168 (2005).
36. Sato, K., Hosokawa, K. & Maeda, M. Non-cross-linking gold nanoparticle aggregation as a detection method for single-base substitutions. Nucleic Acids Res 33, e4 (2005).
37. Sato, K., Hosokawa, K. & Maeda, M. Rapid aggregation of gold nanoparticles induced by non-cross-linking DNA hybridization. J Am Chem Soc 125, 8102-3 (2003).
38. Li, H. & Rothberg, L. Colorimetric detection of DNA sequences based on electrostatic interactions with unmodified gold nanoparticles. Proc Natl Acad Sci U S A 101, 14036-9 (2004).
39. Li, H. & Rothberg, L.J. Label-free colorimetric detection of specific sequences in genomic DNA amplified by the polymerase chain reaction. J Am Chem Soc 126, 10958-61 (2004).
40. Li, H. & Rothberg, L.J. DNA sequence detection using selective fluorescence quenching of tagged oligonucleotide probes by gold nanoparticles. Anal Chem 76, 5414-7 (2004).
41. Hutter, E. & Pileni, M.-P. Detection of DNA Hybridization by Gold Nanoparticle Enhanced Transmission Surface Plasmon Resonance Spectroscopy. Physical Chemistry B 107, 6497-6499 (2003).
42. Csaki, A., Möller, R., Straube, W., Köhler, J.M. & Fritzsche, W. DNA monolayer on gold substrates characterized by nanoparticle labeling and scanning force microscopy. Nucleic Acids Res 29, E81 (2001).
43. Reichert, J., Csaki, A., Köhler, J.M. & Fritzsche, W. Chip-based optical detection of DNA hybridization by means of nanobead labeling. Anal Chem 72, 6025-9 (2000).

44. Köhler, J.M. et al. Selective labeling of oligonucleotide monolayers by metallic nanobeads for fast optical readout of DNA-chips. Sensors and Actuators B-Chemical 76, 166-172 (2001).
45. Hacker, G.W. et al. The use of silver acetate autometallography in the detection of catalytic tissue metals and colloidal gold particles bound to macromolecules. Prog Histochem Cytochem 23, 286-90 (1991).
46. Csaki, A., Kaplanek, P., Möller, R. & Fritzsche, W. The optical detection of individual DNA-conjugated gold nanoparticle labels after metal enhancement. Nanotechnology 14, 1262-1268 (2003).
47. Li, J. et al. A DNA-detection platform with integrated photodiodes on a silicon chip. Sensors and Actuators B 106, 378-382 (2005).
48. Schultz, D.A. Plasmon resonant particles for biological detection. Current Opinion in Biotechnology 14, 13-22 (2003).
49. Schultz, S., Smith, D.R., Mock, J.J. & Schultz, D.A. Single-target molecule detection with nonbleaching multicolor optical immunolabels. Proc Natl Acad Sci U S A 97, 996-1001 (2000).
50. Stimpson, D.I. et al. Real-time detection of DNA hybridization and melting on oligonucleotide arrays by using optical wave guides. Proc Natl Acad Sci U S A 92, 6379-83 (1995).
51. Taton, T.A., Lu, G. & Mirkin, C.A. Two-color labeling of oligonucleotide arrays via size-selective scattering of nanoparticle probes. J Am Chem Soc 123, 5164-5 (2001).
52. Storhoff, J.J., Lucas, A.D., Garimella, V., Bao, Y.P. & Muller, U.R. Homogeneous detection of unamplified genomic DNA sequences based on colorimetric scatter of gold nanoparticle probes. Nat Biotechnol 22, 883-7 (2004).
53. Oldenburg, S.J., Genick, C.C., Clark, K.A. & Schultz, D.A. Base pair mismatch recognition using plasmon resonant particle labels. Anal Biochem 309, 109-116 (2002).
54. Bao, P. et al. High-Sensitivity Detection of DNA Hybridization on Microarrays Using Resonance Light Scattering. Analytical Chemistry 74, 1792-1797 (2002).
55. Freeman, R.G. et al. Self-Assembled Metal Colloid Monolayers: An Approach to SERS Substrates. Science 267, 1629-1632 (1995).
56. Vo-Dinh, T., Yan, F. & Wabuyele, M.B. Surface-enhanced Raman scattering for medical diagnostics and biological imaging. Journal of raman spectroscopy 36, 640-647 (2005).
57. Nie, S. & Emory, S.R. Probing Single Molecules and Single Nanoparticles by Surface-Enhanced Raman Scattering. Science 275, 1102-6 (1997).
58. Cao, Y.C., Jin, R. & Mirkin, C.A. Nanoparticles with Raman spectroscopic fingerprints for DNA and RNA detection. Science 297, 1536-40 (2002).
59. Cao, Y.C., Jin, R., Nam, J.M., Thaxton, C.S. & Mirkin, C.A. Raman dye-labeled nanoparticle probes for proteins. J Am Chem Soc 125, 14676-7 (2003).

60. Jin, R.C., Cao, Y.C., Thaxton, C.S. & Mirkin, C.A. Glass-bead-based parallel detection of DNA using composite Raman labels. Small 2, 375-380 (2006).
61. Wabuyele, M.B. & Vo-Dinh, T. Detection of human immunodeficiency virus type 1 DNA sequence using plasmonics nanoprobes. Anal Chem 77, 7810-5 (2005).
62. Karlsson, R. SPR for molecular interaction analysis: a review of emerging application areas. J Mol Recognit 17, 151-61 (2004).
63. Jonsson, U. et al. Real-time biospecific interaction analysis using surface plasmon resonance and a sensor chip technology. Biotechniques 11, 620-7 (1991).
64. Thiel, A.J., Frutos, A.G., Jordan, C.E., Corn, R.M. & Smith, L.M. In situ surface plasmon resonance imaging detection of DNA hybridization to oligonucleotide arrays on gold surfaces. Anal Chem 69, 4948-4956 (1997).
65. Peterlinz, K.A. & Georgiadis, R.M. Observation of Hybridisation and dehybridisation of Thiol-Tethered DNA Using Two-Color Surface Plasmon Resonance Spectroscopy. Journal of the American Chemical Society 119, 3401-3402 (1997).
66. Nelson, B.P., Grimsrud, T.E., Liles, M.R., Goodman, R.M. & Corn, R.M. Surface plasmon resonance imaging measurements of DNA and RNA hybridization adsorption onto DNA microarrays. Anal Chem 73, 1-7 (2001).
67. Lyon, L.A., Musick, M.D. & Natan, M.J. Colloidal Au-Enhanced Surface Plasmon Resonance Immunosensing. Analytical Chemistry 70, 5177-5183 (1998).
68. Nath, N. & Chilkoti, A. A colorimetric gold nanoparticle sensor to interrogate biomolecular interactions in real time on a surface. Anal Chem 74, 504-9 (2002).
69. Hütter, E. & Pileni, M.-P. Detection of DNA Hybridization by Gold Nanoparticle Enhanced Transmission Surface Plasmon Resonance Spectroscopy. Physical Chemistry B 107, 6497-6499 (2003).
70. Slyadnev, M.N., Tanaka, Y., Tokeshi, M. & Kitamori, T. Photothermal temperature control of a chemical reaction on a microchip using an infrared diode laser. Anal Chem 73, 4037-44 (2001).
71. Tokeshi, M., Uchida, M., Hibara, A., Sawada, T. & Kitamori, T. Determination of subyoctomole amounts of nonfluorescent molecules using a thermal lens microscope: subsingle-molecule determination. Anal Chem 73, 2112-6 (2001).
72. Boyer, D., Tamarat, P., Maali, A., Lounis, B. & Orrit, M. Photothermal Imaging of Nanometer-Sized Metal Particles Among Scatterers. Science 297, 1160-1163 (2002).
73. Blab, G.A. et al. Optical Readout of Gold Nanoparticle-Based DNA Microarrays without Silver Enhancement. Biophysical Journal: Biophysical Letters, L13-L15 (2006).
74. Lavrik, N.V. & Datsko, P.G. Femtogram mass detection using photothermally actuated nanomechanical resonators. Applied Physics Letters 82, 2697-2699 (2003).
75. Okahata, Y. et al. Hybridization of nucleic acids immobilized on a quartz crystal microbalance. J. Am. Chem. Soc. 114, 8299-8300 (1992).

76. Zhou, X.C., O´Shea, S.J. & Li, S.F.Y. Amplified microgravimetric gene sensor using Au nanoparticle modified oligonucleotides. Chem. Commun., 953-954 (2000).
77. Patolsky, F., Ranjit, K.T., Lichtenstein, A. & Willner, I. Dendritic amplification of DNA analysis by oligonucleotide-functionalized Au-nanoparticles. Chem. Commun., 1025-1026 (2000).
78. Weizmann, Y., Patolsky, F. & Willner, I. Amplified detection of DNA and analysis of single-base mismatches by the catalyzed deposition of gold on Au-nanoparticles. Analyst 126, 1502-4 (2001).
79. Willner, I., Patolsky, F., Weizmann, Y. & Willner, B. Amplified detection of single-base mismatches in DNA using microgravimetric quartz-crystal-microbalance transduction. Talanta 56, 847-856 (2002).
80. Tatsuma, T., Watanabe, Y., Oyama, N., Kitakizaki, K. & Haba, M. Multichannel Quartz Crystal Microbalances. Anal. Chem. 71, 3632-3636 (1999).
81. Lutwyche, M.I. et al. High parallel data storage system based on scanning probe arrays. Applied Physic Letters 77, 3299-3301 (2000).
82. Vettiger, P. et al. Ultrahigh density, high-data-rate NEMS-based AFM data storage system. Microelectronic Engineering 46, 11-17 (1999).
83. Fritz, J. et al. Translating Biomolecular Recognition into Nanomechanics. Science 288, 316-318 (2000).
84. Ziegler, C. Cantilever-based biosensors. Anal Bioanal Chem 379, 946-59 (2004).
85. Hansen, K.M. et al. Cantilever-Based Optical Deflection Assay for Discrimination of DNA Single-Nucleotide Mismatches. Anal. Chem. 73, 1567-1571 (2001).
86. Su, M., Li, S. & Dravid, V.P. Microcantilever resonance-based DNA detection with nanoparticle probes. Appl. Phys. Lett. 82, 3562-3564 (2003).
87. Palecek, E. Oscillographic polarography of highly polymerized deoxyribonucleic acid. Nature 188, 656-7 (1960).
88. Drummond, T.G., Hill, M.G. & Barton, J.K. Electrochemical DNA sensors. Nat Biotechnol 21, 1192-9 (2003).
89. Kerman, K., Kobayashi, M. & Tamiya, E. Recent trends in electrochemical DNA biosensor technology. Meas. Sci. Technol. 15, R1 - R11 (2004).
90. Ozsoz, M. et al. Electrochemical genosensor based on colloidal gold nanoparticles for the detection of Factor V Leiden mutation using disposable pencil graphite electrodes. Anal Chem 75, 2181-7 (2003).
91. Cai, H., Wang, Y., He, P. & Fang, Y. Electrochemical detection of DNA hybridization based on silver-enhanced gold nanoparticle label. Analytica Chimica Acta 469, 165-172 (2002).
92. Wang, J. Stripping Analysis: Principles, Instrumentation, and Applications, (VCH, Deerfield Beach, FL, 1985).

93. Wang, J., Xu, D., Kawde, A.N. & Polsky, R. Metal nanoparticle-based electrochemical stripping potentiometric detection of DNA hybridization. Anal Chem 73, 5576-81 (2001).
94. Dequaire, M., Degrand, C. & Limoges, B. An electrochemical metalloimmunoassay based on a colloidal gold label. Anal Chem 72, 5521-8 (2000).
95. Authier, L., Grossiord, C. & Brossier, P. Gold nanoparticle-based quantitative electrochemical detection of amplified human cytomegalovirus DNA using disposable microband electrodes. Anal Chem 73, 4450-6 (2001).
96. Wang, J., Liu, G. & Merkoci, A. Electrochemical coding technology for simultaneous detection of multiple DNA targets. J Am Chem Soc 125, 3214-5 (2003).
97. Wang, J., Xu, D. & Polsky, R. Magnetically-induced solid-state electrochemical detection of DNA hybridization. J Am Chem Soc 124, 4208-9 (2002).
98. Mroczkowski, S.J., Siegesmund, K.A. & Yorde, D.E. Method for electrical detection of a binding reaction. Vol. WO 90/05300 (Midwest Research Technologies US, 1992).
99. Velev, O.D.K., E. W. In Situ Assembly of Colloidal Particles into Miniaturized Biosensors. Langmuir 15, 3963- 3698 (1999).
100. Möller, R., Csaki, A., Köhler, J.M. & Fritzsche, W. Electrical classification of the concentration of bioconjugated metal colloids after surface adsorption and silver enhancement. Langmuir 17, 5426-5430 (2001).
101. Park, S.J., Taton, T.A. & Mirkin, C.A. Array-based electrical detection of DNA with nanoparticle probes. Science 295, 1503-6 (2002).
102. Urban, M., Möller, R. & Fritzsche, W. A paralleled readout system for an electrical DNA-hybridization assay based on a microstructured electrode array. Rev Sci Instr 74, 1077-1081 (2003).
103. Li, J. et al. A High-Density Conduction-Based Micro-DNA Identification Array Fabricated With a CMOS Compatible Process. IEEE Transactions on electron Devices 50, 2165 - 2170 (2003).
104. Diessel, E., Grothe, K., Siebert, H.M., Warner, B.D. & Burmeister, J. Online resistance monitoring during autometallographic enhancement of colloidal Au labels for DNA analysis. Biosens Bioelectron 19, 1229-35 (2004).
105. Moreno-Hagelsieb, L. et al. Sensitive DNA electrical detection based on interdigitated Al/Al2O3 microelectrodes. Sensors and Actuators B-Chemical 98, 269-274 (2004).
106. Li, H. et al. Nanoparticle PCR: Nanogold-Assisted PCR with enhanced Specificity. Angew Chem Int Ed Engl 117, 5230-5233 (2005).
107. Li, M., Lin, Y.C., Wu, C.C. & Liu, H.S. Enhancing the efficiency of a PCR using gold nanoparticles. Nucleic Acids Res 33, e184 (2005).
108. Nam, J.M., Park, S.J. & Mirkin, C.A. Bio-barcodes based on oligonucleotide-modified nanoparticles. J Am Chem Soc 124, 3820-1 (2002).

109. Nam, J.M., Stoeva, S.I. & Mirkin, C.A. Bio-bar-code-based DNA detection with PCR-like sensitivity. J Am Chem Soc 126, 5932-3 (2004).
110. Nam, J.M., Thaxton, C.S. & Mirkin, C.A. Nanoparticle-based bio-bar codes for the ultrasensitive detection of proteins. Science 301, 1884-6 (2003).
111. Thaxton, C.S., Hill, H.D., Georganopoulou, D.G., Stoeva, S.I. & Mirkin, C.A. A bio-bar-code assay based upon dithiothreitol-induced oligonucleotide release. Anal Chem 77, 8174-8 (2005).
112. Georganopoulou, D.G. et al. Nanoparticle-based detection in cerebral spinal fluid of a soluble pathogenic biomarker for Alzheimer's disease. Proc Natl Acad Sci U S A 102, 2273-6 (2005).
113. Oh, B.K., Nam, J.M., Lee, S.W. & Mirkin, C.A. A Fluorophore-Based Bio-Barcode Amplification Assay for Proteins. small 2, 103-108 (2006).
114. Zhang, Z.L., Pang, D.W., Yuan, H., Cai, R.X. & Abruna, H.D. Electrochemical DNA sensing based on gold nanoparticle amplification. Anal Bioanal Chem 381, 833-8 (2005).

CHAPTER 11

QUANTUM DOT APPLICATIONS IN BIOTECHNOLOGY: PROGRESS AND CHALLENGES

Cheng-An J. Lin*[†], Jimmy K. Li[†], Ralph A. Sperling*, Liberato Manna[#], Wolfgang J. Parak*, Walter H. Chang[†]

*[†]Department of Biomedical Engineering, Center for Nano Bioengineering, R&D Center for Membrane Technology, Chung Yuan Christian University, 200, Chung Pei Rd. Chung Li, Taiwan 32023, R.O.C., Email: jclin6924@yahoo.com.tw; *Center for Nanoscience, Ludwig Maximilians University, Amalienstrasse 54, 80799 München, Germany; [#]National Nanotechnology Laboratory (NNL) of CNR-INFM, Universita delgi studi Lecce, Via per Arnesano Km 5, 73100 Lecce, Italy*

Advances in materials chemistry and optical physics have allowed for new nanostructured fluorescent probes for imaging and targeting in cell biology and medical diagnosis. Research on fluorescent semiconductor nanocrystals, also known as quantum dots (Qdots), has shown tremendous advances over the past ten years from pure materials science to biological applications. In the first part of the article we will review current approaches on how to synthesize, water-solubilize, and functionalize Qdots with high quality in a biocompatible way, so that they are "ready to meet biotechnology". Next, we review current challenges of Qdots as cellular probes, as biosensors based on fluorescence resonance energy transfer (FRET), and as *in vivo* probes. In particular, we also review current advances of "greener" quantum dots (GQD) that could overcome the problem of long-term non-toxicity. Finally, we highlight two important directions of Qdots in advanced biotechnology: multifunctional nano-machines created by the assembly of Qdots with other nanostructures for diagnostic or therapeutic purposes and multimodal imaging utilizing Qdots in different types of imaging techniques.

1. Introduction

Probes for labeling are widely used in biomedical research (i.e. cell biology, proteomics, genomics, and clinics) [4-8]. Fluorescent labels are maybe the most common probes for detection. To investigate cellular processes, for example, it is very important to visualize compartments, structures, and even signal transduction within cells by labeling certain involved molecules with fluorescent probes that can be further monitored with an optical fluorescence microscope. Another prominent example is one of DNA sequencing that labels each of the dideoxynucleotide chain-terminators with a separate fluorescent dye, which fluoresces at a different wavelengthT. Nowadays a huge variety of organic dyes is commercially available and there is basically a dye for any desired wavelength of emission. However, several limitations are associated with traditional organic fluorophores. Relatively fast photobleaching [9] (mainly due to irreversible light-induced photo-oxidation) and broad emission spectra limit long-term imaging and multicolor detection. Organic fluorophores are also sensitive to their environment, such as to thermal fluctuation of the solvent [10], and the fluorescence of single fluorophores can reversibly go 'on' and 'off' (the so-called blinking) as a consequence of conformational fluctuations [11].

A decade ago colloidal semiconductor nanocrystals were introduced as new fluorophores that might circumvent some of the above mentioned limitations. These nanometer-sized crystalline particles, also called quantum dots [12] (Qdots) have generated a worldwide interest among scientists of several disciplines. Chemists, physicists and material scientists have developed a multitude of protocols for synthesizing particles of different size, shape, and composition. Researchers involved in life sciences have introduced compatible interfaces between these particles and biological objects, such as proteins or cells. Colloidal Qdots are well dispersed crystalline particles, whose physical dimensions are smaller than the Bohr radius [13,14] of light generated electron-hole pairs (excitons). Upon recombination of the electron-hole pairs fluorescence can be emitted. Due to the spatial restriction of the exciton to the nanoparticle only fluorescence at discrete wavelengths can be emitted. Hereby the wavelength of emission depends of the diameter of the Qdots.

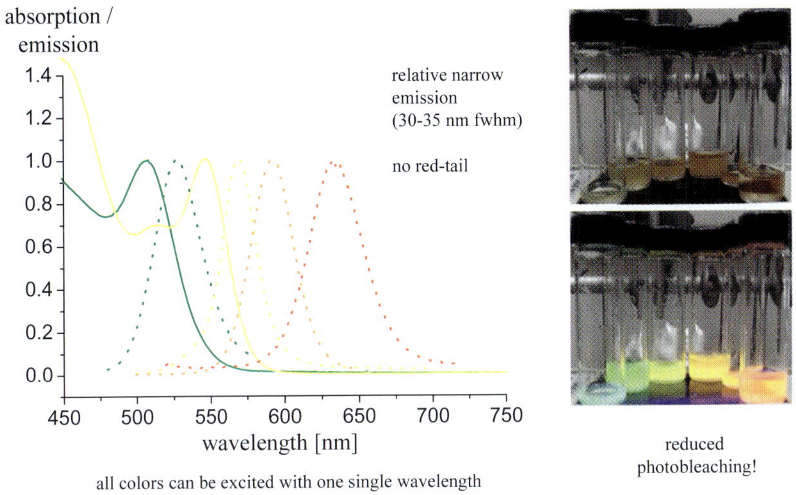

Figure 1. The optical properties of colloidal CdSe nanocrystals strongly depend on their size. The smaller the nanoparticle, the more blue-shifted its fluorescence is. The absorption and emission spectra of four different samples of CdSe nanocrystals with different sizes are shown. The absorption spectra of the nanoparticles are continuous in the UV region and have a peak whose position is shifted to a shorter wavelength as they reduce in size. The fluorescence emission of the nanoparticles is fairly symmetric and narrow (30–35 nm full width at half maximum), the peak position is a few nanometers red-shifted compared to the absorption peak. Because of the continuous absorption spectrum in the UV region, all nanoparticles of different color can be excited at one single wavelength. This figure has been adopted from [Ref. 168].

In this way the emission wavelength can be tuned from ultraviolet to infrared by precisely tuning the size (Figure 1) or by selecting the material of the Qdots. Due to their inorganic nature Qdots posses improved stability compared to organic fluorophores. In particular their reduced photobleaching makes them attractive for time resolved studies.

Currently, the use of Qdots in biology is one of the very fast moving and exciting fields of nanobiotechnology. Various aspects involving Qdots chemistry and early biomedical applications have been already reviewed elsewhere. In 1998, the most important milestones for biological applications of Qdots were made [15,16]. For the first time high quality Qdots could be made water soluble in a stable way and were conjugated to biological molecules. With these Qdots labeled probes first cellular staining experiments have been reported. These fundamental demonstrations have led to promising Qdots based applications in cellular labeling such as the specific labeling of cellular structures and molecules, tracing cell lineage, monitoring physiological events in live cells, measuring cell mobility, deep-tissue imaging, tracking cells *in vivo*, and assay labeling. Therefore, in this chapter we want to summarize the current progress of Qdots in biological and biomedical research. A brief history of Qdot development from physical chemistry to biological applications will also be described, whereby the major barriers that hamper current Qdot technologies will be discussed and the future perspective of the field will be outlined.

2. Quantum Dots: Synthesis and Surface Modification for their Use in Biomedical Research

2.1. Synthesis of Colloidal Semiconductor Nanocrystals

Colloidal nanocrystals are crystalline clusters of a few hundreds up to a few thousands of atoms, which are freely dispersed in a solvent. Colloidal nanocrystals can be prepared from many different materials, including noble metals (e.g. Au, Ag, Cu, Pt) [17], semiconductors (e.g. CdSe, CdTe, ZnS) [18], and insulators (e.g. SiO_2, Fe_2O_3) [19]. Due to their optoelectronic properties, semiconductor nanocrystals are also called quantum dots. Although a large variety of methods allows for growing nanocrystals directly in aqueous solution, which naturally results in hydrophilic and water soluble nanoparticles, for many materials synthesized in organic surfactants is preferred for getting high quality structures [20-23]. The main advantage of growing nanocrystals directly in organic surfactants is that with several of these solvents synthesis at high temperature above the boiling point of water is possible. Obviously annealing of defects in the crystal lattice during the synthesis is more easily achieved at higher temperatures.

The synthesis of monodisperse Qdots, such as CdE (E = S, Se, Te), can be achieved by injecting liquid precursors into hot (300 °C) coordinating organic solvent. One of the first organometallic synthetic procedures of Qdots has been developed by Murray *et al.*[18]. Fundamental research in the 1980s and the early 1990s have allowed for improving particle monodispersity and size tenability [24-28]. These detailed investigations of the reaction conditions have guided the development of the current state-of-the-art Qdot synthesis. The basic principles of Qdots synthesis can be described by referring to a well-studied case, the growth of CdSe nanocrystals as it was used from the beginnings until several years ago. Generally, a mixture of surfactants, trioctylphosphine oxide (TOPO) and a phosphine (tributyl- or trioctylphosphine), is heated up at temperature around 300~350 °C under argon atmosphere. The mixture can be more complex including alkyl amines, phosphonic acids or carboxylic acids. When dimethyl cadmium, a precursor for Cd atoms, is injected into the hot surfactants, free Cd atoms are released by thermal decomposition. In a similar way, free Se atoms can be introduced in a Se precursor solution in which Se is dissolved in phosphine surfactants. Usually the molar ratio of the introduced Cd and Se atoms is 1.4:1.0. Nucleation and growth of CdSe nanocrystals are observed after injection of the precursor solutions and incubation at a specific temperature (200 °C to 350 °C) for controlling the size of the Qdots [18]. Recent advances in the synthesis of CdSe nanocrystals have led to a "greener" method [22], which uses less-hazardous Cd precursors, such as complex compounds of Cd^{2+} ions and alkylphosphonic or carboxylic acids. Such precursor complexes can be directly prepared in the reaction flask by heating cadmium oxide together with alkylphosphonic or carboxylic acid.

The typical criterion of 'quantum size' control involves the dynamics of nucleation and growth of the semiconductor nanocrystals [29]. The Qdot growth can be controlled by considering a number of parameters, such as the surface energy of the Qdots, the concentration of free precursors in solution, and the size of the Qdots. During the growth of CdSe Qdots the thermodynamic equilibrium between Cd and Se atoms assembled on the nanocrystals surface and the corresponding free precursors in solution dominates the size distribution of Qdots. The surface-to-volume ratio of the nanocrystals indicates the overall reactivity in the growth process, as the smaller particles possess a higher surface tension. This means that smaller nanocrystals are thermodynamically less stable than larger ones. At the early state of the growth with high concentrations of free atomic precursors in solution, smaller crystals therefore grow faster than the larger crystals by rapidly incorporating the free precursors because of their high reactivity [30]. The incorporation of atoms from solution into the particles lowers the amount of available precursors and shifts the thermodynamic equilibrium. Over time, the growing condition is starting to exhaust all the free precursors and the fresh monomers are not able to sustain the fast growth of smaller particles. This situation of narrowing the size distribution over time is defined as "focusing regime" [29]. At this point of the reaction the size distribution of the nanoparticles is best. Finally, when the concentration of the free precursors drops further, the small particles start dissolving into free atomic precursors in the solution and then feed the large particles. The overall size distribution now broadens fast over time, a phenomenon known as "Ostwald ripening". To avoid the broadening of the size distribution, Peng *et al.* have demonstrated the refocusing of the size distribution by the injection of additional monomers upon the onset of Ostwald ripening, which shifts the critical size back to a smaller value [31].

For biological applications of CdSe Qdots it is extremely important to passivate them with a layer of e.g. ZnS which can improve the quantum yield up to 50~80%. Such a core/shell structure can partly protect the Qdots against photo-oxidation and reduce cytotoxcity [2,3,20,32-34]. To produce a ZnS capping layer, a precursor mixture of Zn (e.g. dimethylzinc) and S (e.g. hexamethyldisilathiane) is slowly dripped into the reaction flask at low temperature, so that the nucleation of ZnS is prevented and the ZnS precursors instead crystallize on the CdSe surface. The thickness of the ZnS is mediated by the amount of precursor mixture that is injected into the reaction flask. Due to the highly pyrophoric properties of dimethylzinc, zinc salts such as zinc stearate are also alternative precursors for the shell growth. The low lattice parameter mismatch of the ZnS relative to the CdSe core minimizes crystal strain and allows for epitaxial growth. Other core/shell systems such as CdSe/CdS, CdSe/ZnSe, CdTe/CdSe, CdSe/ZnTe, InAs/InP, and InAs/CdSe have also been developed to eliminate core surface defect states, which results in general in a higher quantum yield due to the larger bandgap of the shell materials. Currently shell growth has been improved by the use of air-stable reagents via successive ion layer adsorption and reaction (SILAR). SILAR originally had been developed for the deposition of thin films on solid substrates from solution baths. It has been introduced as a technique for the growth of high-quality core/shell nanocrystals of compound semiconductors first by Peng [35] and co-workers. The size distribution of the core/shell nanocrystals could be maintained even after a shell of five monolayers of CdS had been grown onto the core nanocrystals. The SILAR technique with its precise thickness control without homogeneous nucleation of the shell materials has made it possible to grow high-quality colloidal quantum wells [36-38] or more complicated core/shell1/shell2 nanostructures [39-41]. To appreciate these efforts it has to be pointed out again that the capping layer with a larger band gap to improve the quantum confinement is very important for biological applications of Qdots, as it results in very bright and highly stable (optically as well as chemically) fluorophores [20,32].

2.2. Hydrophilic Modification

Most commonly the synthesis of high quality semiconductor nanocrystals such as CdS, CdSe, CdTe, or CdSe/ZnS [20-23] is carried out at high temperature in organic solvent, which results in hydrophobic surfactant-coated particles. The surfactant layer mediates colloidal stability, whereby its polar head group is attached to the particle surface and the hydrophobic chain protrudes into the organic solvent. Particles with such hydrophobic surface layers are well dispersed in non-polar solvents such as chloroform or toluene but not in aqueous. However, most biological experiments require water-soluble materials and therefore the Qdots' surface has to be rendered hydrophilic before they can be used in biological applications. Since the first introduction of Qdots for biological imaging in 1998 [42,43], the concepts of phase transfer have bridged the communication between Qdots and biomedical research. Many methods have been developed to transfer hydrophobic Qdots into aqueous solution, which include replacing the surfactant layer, or overcoating it with an additional layer. Ultimately the outermost layer needs to bear either electric charge or hydrophilic polymers for mediating solubility in water. Klostranec and Chan [44] have defined the demands for any phase transfer of Qdots in aqueous solution. The ideal coating should A) prevent Qdots from flocculating during long-term storage, B) efficiently convert the solubility of Qdots from organic solvents to aqueous solution, C) maintain the Qdots fluorescence quantum yield, and D) maintain the sub-10 nm size of Qdots. However, the currently available coating strategies do not fulfill all of these requirements. In the following several strategies will be described in more detail.

2.2.1. Ligand Exchange

In practice, current coating strategies involve either a ligand exchange or the adsorption of additional molecules. A ligand exchange consists of replacing the hydrophobic surfactant coating with a hydrophilic one. Conceptually the effective size of the particles does not increase significantly upon ligand exchange. The hydrophilic ligand molecules consist on one end of functional groups that are reactive towards the nanocrystal surface, and hydrophilic groups on the other end, which ensure water solubility. Mercaptohydrocarbonic acids (HS---COOH) such as mercaptoacetic acid [43,45-47], mercaptopropionic acid [48], mercaptoundecanoic acid [49], mercaptobenzoic acid [50], dihydrolipoic acid [51,52], and cysteine [53,54] are the most commonly used ligand molecules. Their thiol (-SH) terminals are reactive to the surface of the semiconductor nanocrystals and their carboxyl (-COOH) terminals are used as hydrophilic head group. The carboxyl groups are deprotonated for pH values above 5 and thus introduce negative charge ($-COO^-$) to the surface of the nanocrystals, so that the particles repel each other. Coulomb repulsion between nanocrystals with surface charge of the same polarity prevents aggregation in water. However, in salt-containing solution the charges are screened and the particles can precipitate, a phenomenon well known in colloid chemistry as "salt-induced aggregation". The thickness of the counter-ion cloud surrounding a charged particle, the so-called diffuse layer, is inversely proportional to the square root of the ionic strength and thus is reduced with increasing ionic strength. The particles flocculate when they approach each other close enough during thermal collisions that the repulsive electrostatic forces are overcome by short range attractive van der Waals forces. The higher the salt concentration is the smaller the repulsive forces become due to screening by counter charges and the easier the particles aggregate. Another example involving ligand exchange is based on synthetic phytochelatin related peptides. Pinaud and his co-worker [55,56] have demonstrated that the peptides could be designed to bind to the nanocrystals via a C-terminal adhesive domain that is composed of multiple repeats of cysteine pairs flanked by hydrophobic 3-cyclohexylalanines. Another terminal with a flexible hydrophilic linker domain then ensures their high colloidal stability. Other ideas to stabilize particles are the use of non-charged ligand molecules such as dithiothreitol [57], organic dendrons [58-60], cyclodextrin [61] and pyridine-functionalized polyethylene glycol (PEG) [62] which cause steric repulsion of the particles and thus prevent particle aggregation. In the case of pyridine-functionalized PEG the pyridine terminal offers the binding site to anchor the PEG to the surface of the semiconductor nanocrystals, whereas the hydrophilic polymer, PEG, stabilizes the nanocrystals in aqueous solution by steric repulsion. Although the ligand exchange from hydrophobic to hydrophilic surfactants is straightforward, there are some disadvantages connected to this method. The most severe one is that the bond of the ligands to the surface of the nanoparticles is not stable enough over extended periods of time. Aldana *et al.* [63] have studied the photochemical instability of CdSe nanocrystals coated with hydrophilic thiols. Photocatalytic oxidation of the thiol ligands on the surface of Qdots ultimately leads to precipitation of the nanocrystals. In other words, the bond between thiol groups and the surface of II-VI Qdots is not strong enough to maintain long term stability and the hydrophilic ligand shell around the nanocrystal is prone to disintegration [63,64]. Recent progress such as reacting various chains of PEG molecules with thioctic acid [65], conjugating recombinant proteins by electrostatic force [51], or further cross-linking with the amino acid (lysine) in the presence of dicyclohexylcarbodiimide to form a stable hydrophilic shells [66] have much reduced the drawbacks from ligand exchange techniques. Therefore, and in particular because of its easiness to apply, ligand exchange is still frequently used to render Qdots hydrophilic, although advanced methods exist.

2.2.2. Surface Silanization

A more laborious but more stable method of converting Qdots into water-soluble particles is surface silanization. The addition of a hydrophilic silica shell around Qdots was first described by the Alivisatos group [42], who also used this method for introducing Qdots as fluorescent biological labels. The first step of surface silanization [64,67-70] is a ligand exchange, which is followed by crosslinking the ligand shell. The generally used molecule for the ligand exchange step is mercaptopropyltrimethoxysilane (MTS). After replacing the original hydrophobic surfactant on the Qdot surface through MPS, whereby the mercapto group (-SH) binds to the Qdot surface, the outer layer with trimethoxysilane terminals can be crosslinked under the formation of siloxane bonds (-Si-O-CH_3 + CH_3-O-Si- + H2O→ -Si-O-Si- +2CH_3OH). In the second step, hydrophilic trimethoxysilane molecules are added to provide water solubility to the silica shell. Hydrophilic trimethoxysilanes are molecules bearing three methoxysilane groups at one and a hydrophilic group at the other end. By cross-linking the methoxy groups through the formation of siloxane bonds the molecules are linked to the silane shell. The hydrophilic groups stabilize the nanoparticles in aqueous phase. This stabilization can be either by electrostatic or steric repulsion. Phosphate or ammonium head groups [69] add sufficient charge so that the particles repel each other in buffer solutions up to a few hundred millimolar ionic strength. The other possibility is to use silane molecules with long hydrophilic polymer chains, such as PEG, which can offer steric repulsion [62,69] at even higher salt concentrations. Also other methods have been developed which instead of surrounding individual Qdots with a silane shell are based on Qdots incorporation into silica spheres. Such hybrid materials could be obtained with reaction protocols in aqueous phase or water-in-oil microemulsions [68,71-75]. However, effectively these materials comprise several aggregated quantum dos inside one silica sphere and therefore are different in nature to individual Qdots. Only surface silanization of individual Qdot results in silica-coated Qdots consisting of single nanocrystals embedded each in a separate silica sphere [64,70]. This seems to be the most straightforward approach in the context of preservation of the original spectral properties. It has to be pointed out that silanized Qdots in aqueous solution have a comparatively low quantum yield compared to the original Qdots in organic solvents. Recently, Bakalova and his co-worker have reduced this problem through a stabilization of single Qdot micelles by a "hydrophobic" silica precursor and an extension of this silica layer to form a silica shell around the micelles using "amphiphilic" and "hydrophilic" silica precursors [76,77]. In this way silica-coated single CdSe/ZnS Qdot micelles were obtained without removing the hydrophobic ligands. This resulted in a comparatively high quantum yield in aqueous solution, a controlled small size, sharp photoluminescence spectra, and absence of aggregation [77]. In summary it can be stated that silica shells give Qdots in aqueous solution improved stability compared to particles that have been transferred to water by ligand exchange. On the other hand surface silanization is significantly more laborious and also often less reproducible. It also leads to thicker shells around the particles.

2.2.3. Amphiphilic Polymer / Surfactant Coating

Recently, a novel strategy which involves coating the surface with amphiphilic polymers has been reported [78-81]. Instead of exchanging the hydrophobic surfactants, a polymer is wrapped around them, whereby the hydrophobic tails of the polymer intercalate the hydrophobic surfactant molecules on the nanocrystal surface and the hydrophilic groups protrude outwards to ensure the water solubility of the particle.

An additional overcoating has the conceptual advantage not to depend on the core material of the nanoparticles, but only on the hydrophobic nature of their original surfactant molecules, which are in most cases alkyl chains. 40% octylamine-modified polyacrylic acid mixed with hydrophobic Qdots in chloroform, for example, is added to hydrophobic Qdots in organic solvent, which can be redispersed in water after evaporation of the solvent [78]. A more complex polymer based on the polyacrylic acid modification has also been demonstrated recently [82]. Another example which involves cross-linked amphiphilic polymer shell is the use of poly(maleic anhydride) derivatives such as poly(maleic anhydride alt-1-tetradecene). After the hydrophobic alkyl chains of the polymer have intercalated the surfactant molecules of Qdots while slowly evaporating the solvent, the anhydride rings are located on the surface of the polymer-coated nanocrystals. The anhydride rings are further cross-linked by amino-terminated crosslinkers by opening the rings and binding to the individual polymer chains. Finally, the surface of the nanocrystals becomes negatively charged after the anhydride rings have opened in aqueous environment under the formation of carboxylic groups. Thus, the particles are stabilized in water by electrostatic repulsion [79]. First results indicate that this shell is thinner and more homogeneous than silica shells, although the wrapping of an additional shell around the inorganic core (surfactant, polymer, counter-ions etc.) makes the overall diameter larger than the corresponding Qdots transferred into water by ligand exchange. In a similar way more complex amphiphilic polymers, for example triblock copolymers, have been used to transfer hydrophobic Qdots into aqueous solution [83]. For example, triblock copolymers which consist of a polybutylacrylate segment (hydrophobic), a polyethylacrylate segment (hydrophobic), a polymethacrylic acid segment (hydrophilic), and a hydrophobic hydrocarbon side chain can encapsulate single TOPO-capped Qdots via a spontaneous self-assembly process and disperse them in aqueous solution [83]. In most cases of polymer coating, the final surface of the particles is rich in carboxy groups. Besides stabilizing the particles via electrostatic repulsion, these groups can be also used as anchor points for further bioconjugation (Section 2.3). In summary we believe that polymer coating will be largely used for phase transfer of Qdots for the transfer from organic to aqueous phase. Although the particles are larger in size than particles solubilized by ligand-exchange, the polymer shell offers enhanced stability in salt containing solution, and is less prone to loss of molecules because of the larger contact area. The experimental procedures are simple and can be applied to particles in large quantity. Future progress can be expected in improved amphiphilic polymers that are tailored towards the respective applications in which the particles are to be used.

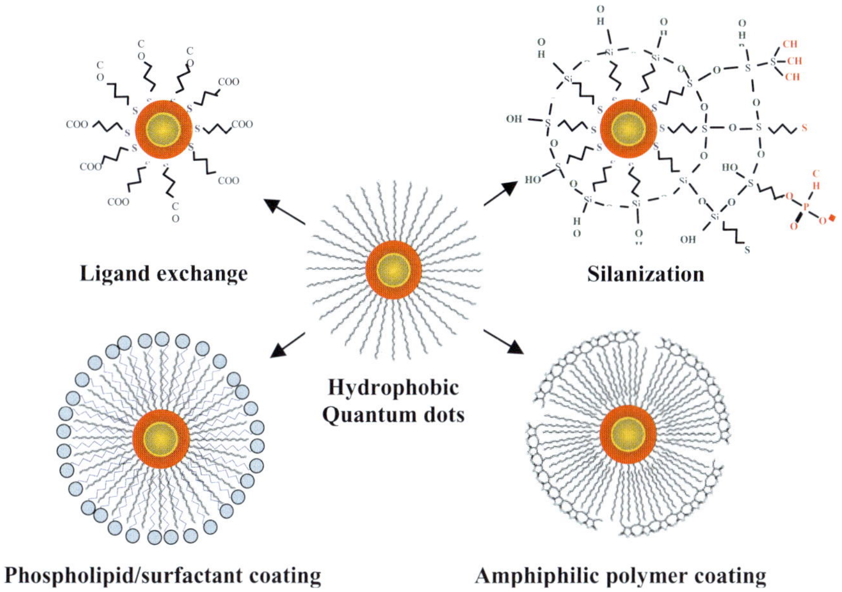

Figure 2. General strategy for the hydrophilic modification of quantum dots: Ligand exchange (Section 2.2.1.), surface silanization (Section 2.2.2.), amphiphilic polymer or surfactant coating (Section 2.2.3.).

Besides the above mentioned amphiphilic polymers also other amphiphilic molecules have been used to stabilize of CdSe/ZnS nanocrystals in aqueous solutions, such as phospholipids [84-86], calixarene [87-89], and cyclodextrin [90,91]. Qdots, for example, could be encapsulated in the hydrophobic core of micelles composed of a mixture of n-poly(ethylene glycol) phosphatidylethanolamine (PEG-PE) and phosphatidylcholine, as first demonstrated by Dubertret and coworkers [84]. Nanocrystals in phospholipid block-copolymer micelles are quite stable in aqueous phase (even in 1 M NaCl), whereby the PEG-block and the two alkyl chains linked to the PE block are mandatory. Surfactants with only a single alkyl chain or only PEG conjugated to the PE block do not seem to form single Qdot micelles.

Conversely, surfactants with two alkyl chains that lack PEG, such as bis(2-ethylhexyl) sulfosuccinate (AOT), allow the water-solubilization of single Qdots, but fail to prevent aggregation at increased salt concentrations [84]. The PEG molecules provide salt stability, biocompatibility and low non-specific adsorption under complex physiological environments [85]. Disadvantages of this technique are the high costs of the coating reagents or the complex procedures for their synthesis [87,92], and the significant increase in overall size of the Qdots after the coating [84].

2.2.4. Conclusions

As a short summary, several reliable methods have been developed to stabilize Qdots in aqueous solution comprising ligand exchange, surface silanization and amphiphilic polymer or surfactant coating (Figure 2). Although the ligand exchange method is reproducible, rapid, and yields Qdots with a regular, well-oriented, thin coating, their colloidal stability is poor [63]. Conversely, the surface-silanized Qdots are quite stable *in vitro* [64], but the coating process is laborious and the coating is sometimes difficult to control. For many applications in which stable particles are required the polymer coating is an attractive possibility, although the effective size of the particles will be bigger than that from a ligand exchange. Also all so far available commercial Qdot products for biolabeling are based on polymer coated particles.

There is no optimum protocol available yet which includes all the advantages of the individual procedures, but state-of-the-art nanocrystals have reached a degree of performance regarding water solubility that is sufficient for biological experiments.

2.3. Bioconjugate Techniques

To fully utilize Qdots as a useful probe in biological applications, it is necessary to conjugate them to biological molecules without disturbing the biological function of the molecule. In Section 2.2, we have introduced some general strategies to create hydrophilic Qdots with abundant functional groups on their surface (such as thiols (-SH), carboxyls (-COOH), or amines (-NH_2)), which can be considered as binding sites to attach biological molecules. Several successful approaches of conjugating Qdots with biological molecules have been used, including thiol chemistry, electrostatic interaction, adsorption, and covalent linkage. Biological molecules containing thiol groups are directly reactive toward the Qdot surface through an ligand exchange process [45,46,48,93-95], in which the surfactant molecules are replaced by the thiol-containing molecules. This is based on the fact that thiols bind with a certain affinity to the surface of semiconductor materials (CdSe, CdS, CdTe, ZnS). If colloidal Qdots are incubated with biological molecules bearing mercapto (-SH) groups, some of the surfactant molecules on the nanocrystal surface will be replaced by these molecules. However, the same problem as already mentioned in Section 2.2.1 occurs: the binding force between CdSe or CdSe/ZnS and thiols is not very strong, and therefore the thiol-mediated attachment of biological molecules is not very stable, as they may eventually detach from the particles. Current strategies to overcome this problem include the introduction of multiple thiol groups as anchor, which has been demonstrated for example with the conjugation of Qdots with peptides that had multiple cysteine residues on one end.[55,56]

Some biomolecules, such as oligonucleotides [96,97] and serum albumins [49], can be directly adsorbed to the hydrophilic shell of Qdots in a nonspecific way, depending on pH, ionic strength, temperature and the surface charge of the molecule. In a more specific way, biomolecules can also be adsorbed onto the nanocrystal surface by electrostatic interaction. Mattoussi and coworkers have pioneered this direction by establishing a method of conjugating proteins to Qdot surfaces through electrostatic attraction. Genetically engineered proteins with a positive zipper on one end can be electrostatically adsorbed to the Qdot surface. The resulting complexes prepared in this way have been shown to be extremely stable in aqueous solution and to have even higher quantum yields compared to non-conjugated particles (which is attributed to the positive charge of the zipper) [52,98-101]. Finally, a more stable linkage is obtained by covalently attaching biological molecules to Qdots by means of cross-linker molecules [42,43,45,46,53,69,78,84,102,103]. Water-soluble Qdots can bear various organic functional groups (-COOH, $-NH_2$, or -SH), that can be coupled with most biological molecules that have other reactive groups. For example, EDC, 1-ethyl-3-(3-dimethylaminopropyl) carbodiimide, is a very common cross-linker used to link carboxylic (or phosphate) groups to primary amines ($-NH_2$), whereas SMCC, 4-(N-maleimidomethyl)-cyclohexanecarboxylic acid N-hydroxysuccinimide ester, is used as a bifunctional cross-linker for the linkage between thiol and amino groups [104]. A huge variety of suitable cross-linking reactions can be found in textbooks [105]. However, as most of these cross-linking protocols have been developed for coupling of proteins, the reaction conditions have to be adjusted most of the times when Qdots are involved.

Instead of directly conjugating the biological molecules of interest to the hydrophilic shell of Qdots, another strategy involving the streptavidin-biotin system can also be realized. Many biological molecules are available with biotin tags or can easily be synthesized with such a tag. Biotin-tagged biomolecules can spontaneously attach to streptavidin coated Qdots. This approach is very universal and streptavidin-coated Qdots are even commercial available, so that they can be readily conjugated with biotinylated proteins and antibodies [78,106-109].

We can conclude that bioconjugation of Qdots is well established and there have been numerous reports of conjugating nanocrystals with various biological molecules, including small molecules like biotin [42], folic acid [103], or neurotransmitter serotonin [95], peptides [46,94], proteins like avidin or streptavidin [52,78,110,111], albumin [49], transferrin [43], trichosanthin [45], lectin wheat germ agglutinin [102], or antibodies [46,52,53,78,100], and DNA [57,69,84,111-113]. However, there is still a lot of improvement possible in the future. So far no general universal protocol exists that works for all types of Qdots and for all types of biological molecules. Furthermore the attachment still has to be controlled in a better way, in particular by reducing nonspecific interactions.

2.4. Synthesis of 'Greener' Quantum Dots (GQDs)

Since the first examples from Alivisatos' [42] and Nie's [43] groups the use of Qdots as luminescence probes in biological and biomedical applications has been a fast growing field. In the same vein an ongoing controversy started regarding safety issues about the biomedical use or potential clinical applications of Qdots. Nowadays most Qdots are made of cadmium-based or lead-based materials (see Section 2.1), which contain toxic heavy metals and are therefore potentially harmful. Still, the actual risk of the hydrophilic coated nanocrystals that typically are used in biological applications remains a matter of discussion [1-3] (see also Section 3.2). Some countries such as Japan and several European countries have already started to restrict the import and use of cadmium containing materials in some occasions because of environmental safety concerns [114] due to the dangers of these metals and incidents of cadmium poisoning. Despite their apparent advantages compared to organic dyes, the intrinsic toxicity of cadmium based materials brings doubts about the future of this promising field. Therefore it is paramount to investigate alternative materials that do not contain cadmium. The development of new synthesis of cadmium-free Qdots therefore has currently a scientific priority regarding the long term development of this field. For this purpose already some promising candidates of 'greener' quantum dots (GQDs), defined as artificial fluorescent nanoparticles that possess environmental safety, biocompatibility, non-toxic composition, and tunable photoluminescence, have been investigated recently, i.e. alternative semiconducting [115-117], metallic [44,118-126], carbon-based [127] and silicon nanoparticles [128-130].

The traditional ZnS shell on top of the CdSe nanoparticles serves two purposes, to improve the quantum yield by passivating non-radiative recombination sites on the particle surface and to act as a barrier so that Cd ions from the CdSe core do not come into contact with the surrounding solvent [117]. A direct strategy for the synthesis of GQDs is to substitute cadmium-based by zinc-based materials. Precursors for zinc-based Qdots are for example diethylzinc [38,131,132], zinc oxide [133], or zinc stearate [115]. Most of zinc-based Qdots such as ZnO [134], ZnSe [131,133], ZnTe [135], or ZnSe/ZnS [132,133] have photoluminescence in the UV-blue range of the optical spectrum, due to the higher band gap of these materials. However for biological applications also particles emitting in the IR-to-green range of the spectrum would be needed. In order to improve the tunability of Zn-based particles, they can be overgrown with a shell of another semiconductor material. However Zn(Se,Te)/CdE(E=S,Se,Te) type-II core/shell Qdots again involve cadmium, now instead of in their core in their shell [37,38]. Again the intrinsic toxicity of these shell materials is putting a severe limitation to their broad potential use in bioimaging. Recently, scientists have found that the wide band gap of zinc chalcogenide nanocrystals (without shell) doped with transition metal ions [115,136-142] might probably overcome this concern and yet maintain the advantage of tunability. Doping, the intentional introduction of impurities into a material, has been widely used to control the properties of bulk semiconductors and these principles now are also used to dope semiconductor nanocrystals [142]. Norris *et al.*, for example, have demonstrated high-quality high fluorescent ZnSe colloidal nanocrystals that are doped with paramagnetic Mn^{2+} impurities. Peng and his colleagues then have developed color-tunable ZnSe Qdots that cover most of the visible spectrum by changing the transition metal ions (Cu2+ or Mn2+) and core/shell combination [115] (this is in contrast to the traditional CdSe particles, for which the spectrum of emission is simply changed by varying the size of the particles). The results from this alternative method are one step towards green synthetic chemistry [22] of GQDs, what the inventers of these particles call "from greener methods to greener products".

Another promising strategy for GQDs synthesis might be the use of III-V semiconductors instead of II-VI materials. For example, pure indium in metal form is considered non-toxic in several investigations. The main advantages offered by III-V semiconductor nanocrystals lie in the robustness of the covalent bond in III-V semiconductors versus the ionic bond in II-VI semiconductors, which might make them less cytotoxic as these materials are less likely to corrode and release toxic ions into solution [116,117,143]. In the reports of Yamazaki and Omura *et al.*, InP particles (mean diameter 1.06 μm) were injected into Syrian golden hamsters twice a week for 8 weeks and they survived throughout a two-year observation period [144,145]. This also makes III-V Qdots (e.g., InP) potentially better candidates than II-VI Qdots (e.g., CdSe) for biological applications although a full toxicological study of III-V nanocrystals is still missing. However, some recent reports have already revealed their ongoing potentials. InP nanoparticles, for example, can be synthesized with mean sizes from 1.7 nm to 6.5 nm, resulting in monodisperse colloids with a band-edge luminescence that is tunable from green to near-IR and room-temperature quantum yields in the range of 20~40% [116,146]. Recently, Bharali *et al.* have already applied hydrophilic InP/ZnS core/shell nanocrystals as luminescence probes for live cell imaging utilizing confocal and two-photon microscopy, which is also potentially useful for deep tissue imaging for future *in vivo* studies [117].

Metallic Qdots such as gold or silver Qdots are also promising candidates of GQDs as potentially complementary imaging tools. After Dickson's group [118] demonstrated that silver oxide (Ag_2O) reveals strong photo-activated red emission for green excitation wavelengths shorter than 520 nanometers, fluorescent metallic nanoparticles have started to be realized [44,119-126]. Gold Qdots, the brightest candidates so far, have also been recently described by Robert Dickson and his colleagues [120]. These particles are strongly fluorescent and very small in size. Each nanoparticle consists of only a handful of metal atoms that are encapsulated by dendrimers. The gold clusters exhibit the same size tunability as semiconductor Qdots and can further provide the "missing link" between atomic and nanoparticle behavior in noble metals. Scientists are now exploring the biological potential of metallic Qdots, which open new opportunities for biological labels, energy transfer pairs, and light emitting sources in nanoscale optoelectronics [120].

The next promising materials of GQDs are silicon-based nanomaterials, so called silicon Qdots. Silicon is an indirect gap semiconductor and its efficiency to emit photons upon electronic excitation or charge-carrier injection is extremely low [147]. Since the discovery of the intense red photoluminescence of porous silicon in 1990 [148,149], the optical properties of this material [150,151] and also of silicon nanocrystals [152,153] has been investigated. The photoluminescence of silicon nanocrystals and their yield were recently found as a function of their size, following very closely the quantum-confinement model [147,154-157]. For instance, the discrete sizes of 1.0, 1.67, 2.15, and 2.9 nm in diameter correspond to UV/ blue, green, yellow and red luminescence of silicon nanoparticles [158]. For biological applications the particles have to be functionalized. Hua *et al.* [159,160] suggested a method for efficient grafting of organic molecules onto photoluminescent silicon nanoparticles. Rogozhina *et al.* [129] prepared strongly luminescent, water stable, carboxyl functionalized nanoparticles by thermal hydrosilylation between a bi-functional alkene and ultrasmall (around 1 nm) H-passivated silicon nanoparticles. Li *et al.* [161] demonstrated stable and water-soluble poly(acrylic acid) grafted luminescent silicon nanoparticles. Sato *et al.* [128] introduced acrylic acid onto the surface of photoluminescent silicon nanoparticles by photo-initiated hydrosilylation, thereby producing water-dispersible, propionic acid-terminated particles. Other surface chemistry [162,163] and some early-stage biological applications such as cell imaging [161] or DNA labeling [164] have also been realized. Current advances of silicon Qdots have revealed their potential use for biomolecular labeling and biological imaging.

The final examples of GQDs are carbon-based Qdots, which were recently proclaimed as a "greener" and safer technology in medicine and biology. Some recent studies have shown the potential luminescence of carbon materials such as carbon nanotubes [165,166] or nanodiamonds [167]. Crude carbon nanotubes, for example, contain a variety of impurities that vary by the method of synthesis. During the electrophoretic purification of single-walled carbon nanotubes (SWNTs) derived from arc-discharge soot, Xu *et al.* found a new class of fluorescent nanoparticles derived from SWNTs. The eluted materials from different fractions were fluorescent in green-blue, yellow, and orange but still with low quantum yield. Ya-Pin Sun and his colleagues reported their new finding on quantum-sized carbon particles or 'carbon dots' that also strongly photoluminesce in both solution and the solid state with tunable spectral features and properties comparable to those of surface-oxidized silicon nanoparticles [127]. The as-prepared carbon dots produced via laser ablation of a carbon target in the presence of water vapor with argon as carrier gas emit no detectable photoluminescence from themselves. However, upon surface passivation by attaching simple organic species, such as diamine terminated PEG or poly(propionylethyleneimine - co-ethyleneimine), to the acid-treated carbon particles, bright luminescence emission that covers the visible to infrared wavelength range was observed. Their versatile surface functionalities that are required for their passivation will also be very useful for their use in bioimaging applications. Some preliminary results on the optical imaging of biological species have also been demonstrated [127].

Current trends for the preparation of "greener" quantum dots are becoming more and more important as for cadmium-based Qdots the release of toxic ions has been reported, which is a major obstacle for their use in medical diagnosis and treatment. Many alternatives of GQD families have been discovered recently, i.e. semiconductors (zinc- or indium-based), metallic (gold and silver), silicon, and carbon-based Qdots, but so far these materials cannot fully replace the traditional cadmium-based Qdots from a photophysical point of view. However, recent results also hint that many materials exit quantum confinement effects when their dimensions are reduced to nanometer. Undoubtedly, there will be more "greener" quantum dots (GQDs) available in the future.

3. Properties of Quantum Dots

3.1. Some Basic Photo-Physical Properties

Semiconductor nanocrystals can be physically described as quantum dots [168], i.e. they exhibit atom-like size-dependent energy states due to the confinement of the charge carriers (electrons, holes) in three dimensions [12,169-172]. In semiconductor systems, the energy difference between the valence (ground state) and conduction (excited state) band is called the band gap energy. The energy band gap of semiconductor nanocrystals (Figure 3) changes as a function of size when the size of the nanostructure is in the order of the exciton Bohr radius [173,174]. The smaller the diameter of the Qdot then is, the bigger the energy gap becomes. Upon optical excitation an electron is excited from the valence to the conduction band leaving behind a hole, which is analogous to exciting electrons from the highest occupied molecular orbital (HOMO) to the lowest unoccupied molecular orbital (LUMO) in organic fluorophores. Radiative recombination of electron-hole pairs results in the emission of fluorescence in which the wavelength depends on the band gap and thus on the size of the Qdots [175-179]. Small CdSe nanocrystals (about 2 nm diameter) for example fluoresce in the green, whereas bigger ones (about 5 nm diameter) emit in the red. The growth of a second shell of a semiconductor material with higher bandgap around the semiconductor core [20,32,33] reduces the non-radiative relaxation of electron hole pairs on the particle surface. Non-radiative recombination, i.e. recombination of electron-hole pairs without emission of fluorescent light, reduces the quantum yield and thus has to be reduced as much as possible.

Semiconductor nanocrystals provide many attractive optical properties for biomedical research [44] including a) the long fluorescence lifetime that allows Qdots to be used for time-resolved fluorescence bioimaging [180]; b) the broad spectral window of absorption that can be used to excite the Qdots, as this permits different Qdots to be excited simultaneously using a single wavelength [42]; c) the high stability against photobleaching that allows Qdots to be used in long-term cellular monitoring, such as protein tracking [52]; d) the large Stokes shift (i.e. the separation between the maxima in the absorption and luminescence spectra) that prevents crosstalk and thus enhances the signal-to-noise ratio; d) the high quantum yield that permits single Qdot imaging; e) the two-photon cross section that is significantly bigger than the one of organic fluorophores and thus allows for imaging structures deep inside biological tissues [181]. Conversely, organic fluorophores such as fluorescein, rhodamine, or Cy-dyes typically can only be optically excited within a narrow spectral window which depends on the particular fluorophore. Furthermore, organic dyes suffer from fast photobleaching and relatively broad, overlapping emission lines, and therefore have limitations in biological applications involving long-term imaging and multicolor detection. On the other hand, Qdots are significantly bigger than many organic fluorophores. For further comparison, several excellent reviews [12,175,182-186] provide more details.

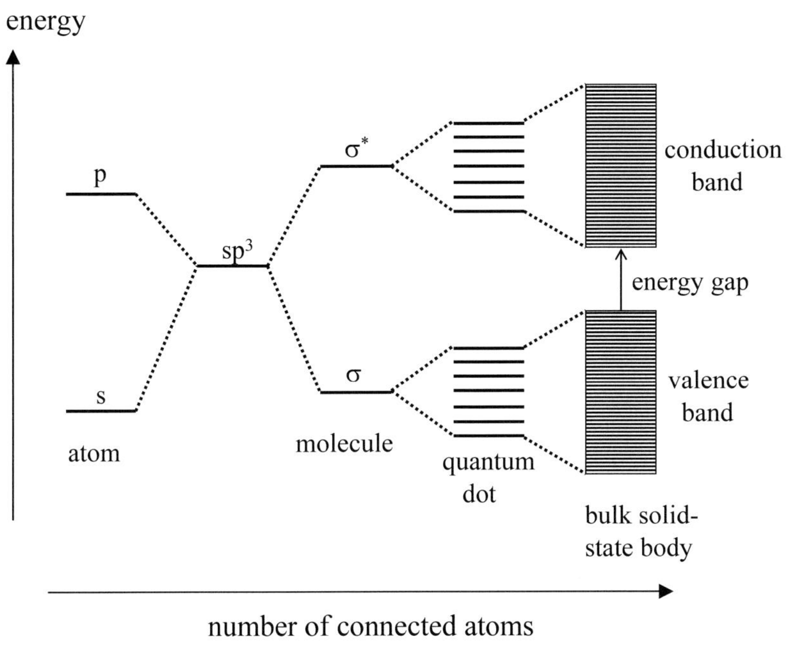

Figure 3. Electronic energy levels depend on the number of bound atoms. By binding more and more atoms together, the discrete energy levels of the atomic orbitals merge into energy bands (here shown for semiconducting materials). Therefore semiconducting nanocrystals (quantum dots) can be regarded as hybrid between small molecules and bulk materials. This figure has been adopted from [Ref. 168].

Since the fluorescence emission spectra of Qdots are relative narrow, many different colors can be distinguished without spectral overlap. For biological fluorescence labeling, this means that more colors can be resolved, and thus more different compartments/structures/processes can be labeled simultaneously, each with a different color. A typical example of high quality multi-color labeling of different cellular structures has been demonstrated by Wu *et al.* [78] and by Mattheakis *et al.*[187] If several Qdots of different colors are decorated to one entity such as microspheres, each microsphere with a different ratio of Qdots of different color can be one spectra code [188-191]. The group of Shuming Nie has embedded Qdots of six different colors with ten intensity levels in polystyrene microspheres. They have demonstrated the tagging ability of biological molecules with these objects and their potential use to code theoretically one million nucleic acids or protein sequences [189].

The most important advantage of Qdots is their reduced vulnerability to photobleaching, as we have mentioned above. In contrast to organic fluorophores that undergo irreversible light-induced reactions Qdots suffer much less from photobleaching due to their inorganic nature. Many practical labeling experiments have demonstrated their stability against photobleaching, especially the performance of Qdots compared with fluorophores typically used in cell biology [43,49,52,53,78,84,106,107,192]. However, another phenomenon known as photo-brightening [64,102,193,194] has been reported, in which upon optical excitation the fluorescence of water soluble Qdots even increases in the beginning. These changes in intensity can lead to false interpretation of results in assays requiring quantitative analysis. Nevertheless, the reduced photobleaching is of particular importance for experiments which involve long term imaging, such as fluorescence labeling of transport processes in cells, or tracking the path of single membrane-bound molecules [108]. Reduced photobleaching is in the same way an advantage for recording z-stacks of high-resolution three-dimensional reconstruction images [107] over time in live-cell imaging [45,192]. Unfortunately, colloidal Qdots also suffer from intermittent blinking whereby the luminescence switches on and off, very similar to organic fluorophores [195-200]. It has been speculated that blinking of colloidal Qdots might result only from the experiment conditions, where Qdots are in the proximity of surfaces and that there might be no blinking of Qdots freely suspended in solution far away from any surface [181]. However, these speculations never have been verified by other groups and therefore are likely not to be true. For bulk measurements blinking is not a big problem, since the camera acquisition times can be longer than the off time of the Qdots while blinking. There is on the other hand a severe problem for single-molecule detection involving the fast imaging rates of biological events that are faster than the times scales of blinking. There is evidence, that upon changing the geometry of Qdots from spherical to rod-shaped objects the blinking is reduced.

3.2. Cytotoxicity / Biocompatibility

Since Qdots are composed of individual ions / atoms (Cd^{2+}, Se^{2+}, Te^{2+}) [201,202] that show an inherent toxicity, harmful effects can be expected, especially when Qdots are used to study live cells and animals. The main concern is the release of toxic ions from the Qdots upon corrosion, which to a large degree will depend on the question if the hydrophilic shell around the Qdot is stable or not.

Colorimetric assays such as the MTT assay (3-(4,5-dimethylthiazol-2-yl)-2,5-diphenyl tetrazolium bromide) have been used to measure cell viability after incubation with Qdots in a quantitative way. They allow measuring absolute numbers for the survival rate of cells upon incubation with toxins. The first comprehensive cytotoxicity study by Derfus *et al.* has used the MTT assay to test cytotoxic effects of different types of Qdots. Primary hepatocyte cultures were examined because the liver is the major target of cadmium injury [2]. The 'worst case' particles used in their study are CdSe Qdots coated with mercaptoacetic acid. In particular after UV exposure or exposure to air significant levels of free Cd^{2+} ions were found in the Qdots solutions, which also correlates with the viability of the cells incubated with these samples [2]. Based on these results cell viability is strongly affected by the liberation of free Cd^{2+} ions from the CdSe lattice. The CdSe corrosion is facilitated by air-induced oxidation of the Qdot surface [20] and UV-radiation-catalyzed oxidation of the Qdot surface [63]. Free Cd^{2+} ions are known to interfere with the function of mitochondria [203]. Furthermore, it was also demonstrated that the quantitative values for the onset of cytotoxic effects in serum-free culture media correlate with the uptake of the particles by the cell as observed by microscope images [3]. In addition to the release of toxic Cd^{2+} ions from the particles, their surface chemistry plays an important role for cytotoxic effects, such as their stability toward aggregation [3]. Reduced cytotoxicity can be achieved by coating CdSe Qdots with appropriate shells such as ZnS, which dramatically reduces the Cd^{2+} release [2]. With the MTT assay Mattheakis *et al.* claim that no toxicity of highly purified polymer-coated CdSe/ZnS Qdots can be observed for typical labeling concentrations of 10 nM [204]. Another system, living embryos, is a very sensitive environment in which to test Qdot toxicity, because even slight cellular perturbations are manifested as different biological phenotypes, i.e. as deformation of the embryos. Larson *et al.* [84] have examined the toxicity of Qdots in Xenopus embryos. They report no negative effects on embryo development upon injection of low concentrations of Qdots (CdSe/ZnS). For higher concentrations some abnormalities were described.

In summary, Qdots of CdSe capped by both ZnS and tight hydrophilic shells (in particular surface silanization [76,205,206]) have been found to lead to no acute and obvious toxicity in studies of cell proliferation and viability in live cells [2,3,46,52,192] and animal models [94,181], as long as the cells are exposed to the Qdots only at low concentrations and for short periods of time. These results suggest that at least for experiments with cell cultures the release of Cd^{2+} can be reduced by appropriate encapsulation of the CdSe cores in a way that experiments with low Qdots concentrations are possible. However, there remains a big need for further systematic investigation into the toxicity of semiconductor nanocrystals in live animals before they potentially could be safely used for diagnostic application. To point is out clearly again: cytotoxicity has certainly to be considered from a different perspective for labeling cell cultures compared to labeling structures in animals or humans [207]. For cell cultures, inert Qdots that do not release any toxic ions are definitely suited. On a short time scale of days Qdots can be inert enough to prevent acute damage of the tissue. On animal studies involving long time scales however, Qdots ingested by the cells might remain in the living tissue for months and presumably even for years [81]. No encapsulation can be as inert as to withstand degradation forever, especially in such complex physiological environments. Eventually, toxic ions might be slowly released in the time course of years. Therefore, alternative Qdot materials, i.e. greener quantum dots (as described in Section 2.4) are required, preferentially composed out of non-toxic materials which are biodegradable. For example, nanoparticles such as iron oxides have been used as contrast agents for magnetic resonance imaging in humans. These magnetic particles are ingested by living cells

[208] but they are degradable into non-toxic elements. It has been shown, that free iron resulting from the degradation of the particles is incorporated into haemoglobin [209] and after some months no residues of the particles remained in the body. There is hope that similar systems can be also found for fluorescent nanoparticles.

4. Quantum Dots as a Cellular Probe

4.1. Labeling of Cellular Structures and Receptors

To investigate cells and cellular processes fluorescence immunolabeling [210-212] is widely used, as for example for probing structures, visualizing compartments within cells, and locating signal transduction-related molecules. Qdots are regarded as new alternative probe in this area because of some of their optical properties, including stability against photobleaching, broad excitation and narrow emission wavelength etc. (see also Section 3.1). Since the first experiment where CdSe/ZnS Qdots were used to stain the nucleus and F-actin in fixed cells [42], there has been an ongoing development in the field that lead to the high quality multi-color labeling images that can be obtained nowadays with standardized procedures [78,213]. In a typical two-step labeling a primary antibody is first reacted with the target and then Qdots conjugated with secondary antibodies, which target the primary antibody, are added [53,54,78]. In this way the fluorescence signal is amplified as more than one secondary antibody can bind to each primary antibody. However the direct conjugation of antibodies to Qdots is still laborious and therefore often biotinylated secondary antibodies are used, which can be finally labeled by streptavidin-Qdot conjugates via the biotin-avidin interaction [78,106,107,214]. Kaul and coworkers, for example, used Qdots as optical reporter to show different staining patterns in normal and transformed cells by the heat shock protein, mortalin, which was immuno-fluorescence labeled [215]. Tokumasu and Dvorak demonstrated the capability to reconstruct a three dimensional high-magnification image of the whole membrane domain band 3 of erythrocytes [107], which would have been difficult to achieve with organic fluorophores. Wu *et al.* developed reliable and specific Qdot probes to stain breast cancer cells that express the surface marker Her2, cytoskeleton fibers, and nuclear antigens in fixed cells, living cells, and tissue sections [78]. The strategy is advantageous because it involves only one universal conjugation of Qdots with streptavidin, and most antibodies are commercial available with an attached biotin group. Further signal amplification can also be achieved by the tyramide signal amplification (TSA) technique [106]. Nowadays Qdots have been used for labeling a large variety of structures in fixed cells. Fixed cells are "dead" because they have been stabilized by cross-linking their sugar skeleton on the cell membrane (e.g. by glutaraldehyde) and their membrane has been permeabilized with appropriate reagents (e.g. by Triton). For living cell images, labeling can be performed by several strategies which will be described in Section 4.2.

If structures on the surface of cells are to be labeled, fluorescent labels do not need to transverse the cell membrane. This means that outside labeling without fixation can be performed also with living cells [55,78,108,109,216-218]. In this way even the dynamic arrangement of single proteins within the cell membrane could be traced [108,109]. Arguably Qdots will be used in the future in particular for time-resolved studies for single molecule tracing, similar to the first experiments demonstrated by Dahan *et al.* [108,219], Lidke *et al.* [109], and Muthukrishnan *et al.* [220] In such single-molecule experiments colloidal Qdots offer significant advantage compared to those methods that involve conventional organic fluorophores. Their stability against photobleaching prolongs the observation times. Fluorescent latex bead have been alternatively used as labels, for example to track surface glycine receptors (GlyR) with/without the GlyR stabilizing protein gephyrin at synapses [221], but the extended size of the beads is suspected to influence the dynamics of the labeled proteins. Qdots would be a compromise, though larger than organic fluorophores they are still smaller that fluorescent latex beads, whereas similar to the latex beads they suffer less from photobleaching. It is likely that colloidal Qdots will never completely replace organic fluorophores as fluorescence labels, but there are niches as long-term imaging for which their advantages are obvious and therefore widely spread use can be expected.

4.2. Incorporation of Quantum Dots by Living Cells

Since the incorporation of Qdots by living cells has first been observed by Nie *et al.* [43], many groups have reported uptake of Qdots by individual cells or by animal tissue. The uptake of colloidal nanoparticles involves similar mechanism as the uptake and processing of nutrients or even the uptake of a virus. Aoyama and coworkers recently used Qdots conjugated sugar balls to mimic their artificial glycoviral gene delivery system [222] by monitoring the size effects of endocytosis in the subviral region [92] for future strategies of artificial molecular delivery machines. Several cellular uptake mechanisms [223-225] have been harnessed to introduce nanoparticles into cells for a variety of applications including drug, nucleic acid and gene delivery as well as cell marking. In general, these transport pathways are part of the whole network of the cellular membrane traffic, which generally involves membrane-surrounded vesicles around the incorporated materials. Natural uptakes by cells is called endocytosis (details can be found in textbooks of cell biology and molecular biology and in some recent reviews [226,227]). Several types of endocytosis are discriminated in accordance with the size of the endocytosed materials and the detailed uptake mechanisms, e.g. pinocytosis for small particles and dissolved macromolecules, phagocytosis for larger structures such as whole cells, debris or bacteria. However, the nomenclature is not always exactly defined. Some membrane proteins such as clathrin can be also involved in endocytosis as surface receptors and therefore clathrin-dependent and clathrin-independent processes are discriminated. Clathrin is a protein cage that assembles around invaginations during endocytosis, forming the so-called clathrin-coated pits [226]. Other mechanism can dominate for other types of cells. Macrophages, for example, are cells specialized in clearing large structures such as debris of dead cells or foreign intruders such as bacteria or colloidal nanoparticles, what is typically referred to as phagocytosis. Endocytotic uptake processes can be specific or nonspecific. Receptor-mediated endocytosis is regarded as specific because it is triggered by a specific lock/key type receptor-ligand binding interaction. A typical example is the uptake of transferrin via transferrin receptor mediated endocytosis, which allows cells to 'engulf' and capture the iron-protein complex. The transferrin / transferrin receptor system has been widely used in drug and gene delivery [228,229]. But also uptake into cells without specific receptor-ligand binding events has been reported. Such nonspecific binding is easily accomplished with cationic particles by simple electrostatic attraction to the outside of the cell, which is exposing negatively charged glycosaminoglycans in the extracellular matrix. In fact most cell labeling experiments with Qdots have been carried out with anionic particles as typically negatively charged carboxyl groups are used to stabilize the particles, but their uptake is less efficient than the one for cationic particles.

For exploiting cellular transport mechanisms such as endocytosis for Qdots labeling, it is highly instructive to consider the vast amount of literature available from the fields of drug and gene delivery [230-234]. Usually, uptaken Qdots will be stored in membrane-surrounded compartments which are not accessible to solutes of the cytoplasm. For experiments involving imaging cellular of reactions in the cytoplasm or the nucleus, more sophisticated tools are required. Derfus *et al.*, for example, have introduced Qdots into the cytoplasm by using three kinds of bio-techniques, transfection, electroporation and microinjection, and only the last one resulted in a homogeneous distribution of particles in the cytoplasm [213]. In the field of nucleic acid/gene delivery, for transport across cellular membranes nucleic acids are self-assembled with polycations or cationic lipids to form charged nanoparticles [230] that can bind to the cell surface by electrostatic interaction (nonspecific) or ligand-receptor interaction (specific) followed by endocytosis. The strategy to release the nucleic acids from the vesicles is to incorporating fusogenic components into the complex such as membrane-disrupting peptides that unfold their activity during the internalization process [232-234]. Otherwise, a cationic moiety such as the polycation polyethylenimine (PEI) can be chosen in a way to comprise an inherent chemical structure that would mimic the 'standard' degradation of internalized materials. Such a polycation that comprises secondary and tertiary amino groups can buffer the natural acidification process within endosomes (the so-called 'proton sponge effect' [235]), causing osmotic destabilization in the endosome that can lead to disruption [236]. Other chemical structures such as several lipids also promote membrane reorganization and release internalized materials such as nucleic acids into the cytoplasm [237]. Much of exploiting cellular transport processes for 'invasion' has been learned from viruses, which involve mechanisms of tricking cells for their own purpose of infection. This suggests that natural transport processes should be exploited for Qdots delivery to cells in an analogous manner. Therefore, in the future similar processes will be utilized to transfer Qdots into the cytoplasm of cells, which is a prerequisite for intracellular staining of living cells.

Although Qdots can be modified with similar surface chemistry to the "objects" studied in the field of drug delivery the efficacy of the uptake as well as the exact mechanism certainly will depend on a variety of parameters including particle size, surface chemistry, and also the type of cells under consideration (Figure 4).

In general as well receptor mediated [43,45,47,55,103,203] as unspecific/adsorptive [49,52,180,192,238,239] endocytotic uptake of Qdots has been reported. Polyethylene glycol (PEG) on the Qdot surface reduces the adsorption of the particles to the membrane and in this way also reduces uptake by cells [55,81,240]. Hanaki *et al.* verified that ingested Qdots are stored in the endosome by making a colocalization experiment with an endosome marker: fluorescein-labeled dextrane as marker for endosomes/lysosomes [49] colocalized with ingested Qdots. Jaiswal *et al.* not only found colocalization of Qdots with the endosome specific marker pECFP [52] but further demonstrated that Qdots uptake can be blocked by cooling the cells to 4 °C, a technique known to block endocytosis. Most of Qdots accumulate in vesicles in the perinuclear region [45,49,52,192] without penetrating the nucleus after entering the cells. Hanaki found that after cell division Qdots can be observed not only in the perinuclear region but also in the neighborhood of the cell membrane. Another general agreement is the fact that ingested Qdots are distributed between both daughter cells upon cell division [49,84,187,192]. This is again in analogy to what is observed in nucleic acid delivery. Only Qdots with specific peptide-modification on their surface, such as nuclear localization signals (NLS) [241-243] and mitochondrial signal peptides [243], can be delivered into cell nucleus and mitochondria, respectively. Ligand-modified Qdots are internalized together with the receptor molecules [55,109]. Lidke *et al.* demonstrated that epidermal-growth-factor- (EGF-) conjugated Qdots bind to erbBX transmembrane receptors, which are responsible for mediating cellular responses to EFG, followed by the internalization of the Qdot-receptor complex.

Some other peptide ligands can also be used for specific cell targeting and delivery of payloads into the target cell. Kim *et al.* found that the cyclic 13-mer Pep42, CTVALPGGYVRVC, can be used to specifically target cancer cells which are expressing GRP78, a member of the heat shock protein family and a marker on malignant cancer cells [244]. Pep42-Qdots were utilized to monitor cellular uptake by confocal laser microscopy, it was shown that the constructs were transported finally to the endoplasmic reticulum. Such a translocation is a receptor-mediated process that did not involve transfer of the particles to the lysosomal compartment. Another bifunctional oligoarginine cell penetrating peptide (based on the HIV-1 Tat protein motif) bearing a terminal polyhistidine tag was synthesized and used to facilitate transmembrane delivery of Qdots bioconjugates [242]. Upon internalization of this construct not all of the Qdots were colocalized within endosome. More recently, Courty *et al.* [219] and Nan *et al.* [245] have observed microtubule-assisted transport of Qdots in cells. Rajan *et al.* even demonstrated how to directly visualize endocytosis, redistribution, and shuttling of Qdots bound to TrKA receptors of PC12 neural processes [246]. Recently, Zhang *et al.* used enzymes to modulate the cellular uptake of Qdots by modifying their surface with peptide ligands that consist of (a) a transporting group (transporter) that facilitates Qdots transport into cells, (b) a blocking group (blocker) that suppresses cellular uptake of Qdots, and (c) a sensing group, linked between the transporter and blocker, that could be cleaved by an extracellular enzyme [247]. Cellular uptake of these peptide-modified Qdots could be actively controlled.

We conclude that there is great potential in further adopting the sophisticated chemistry and biotechnology of advanced drug delivery and gene therapy for their use with Qdots, which eventually will lead to enhanced efficacy and specificity of intracellular labeling and numerous novel applications.

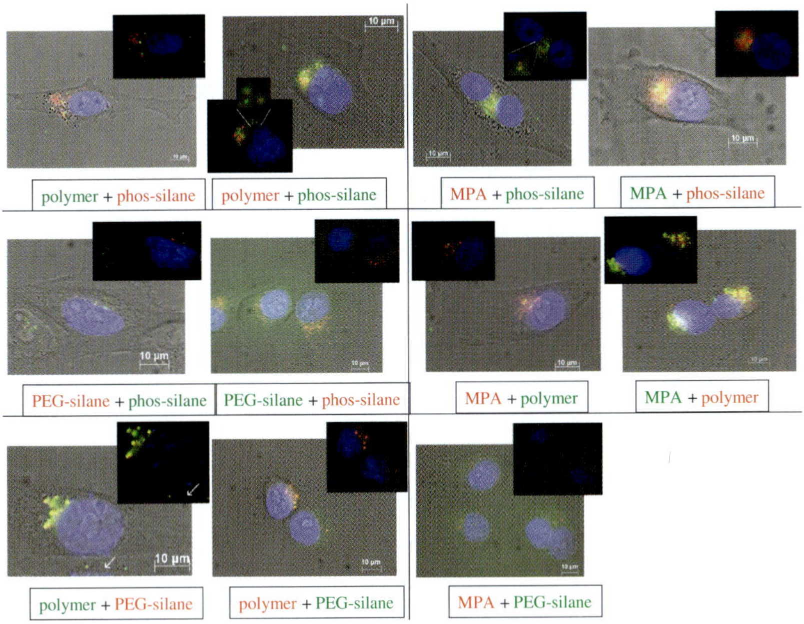

Figure 4. Red and green fluorescent CdSe/ZnS Qdots with different surface coatings have been ingested by MDA-MB-453S cells. Images of the cells were recorded with differential interference contrast (DIC) microscopy, and the nanocrystals and the DAPI-stained nuclei were recorded with fluorescence microscopy. For each experiment, always red and green fluorescent nanocrystals with different surface coatings, i.e. ligand exchange (mercaptopropionic acid, MPA), surface silanization, amphiphilic polymer coating, were used so that their position within the cells can be resolved by their color of fluorescence. The DAPI-stained nuclei appear in blue. The big images contain the merged DIC and fluorescence images, the smaller images show only the fluorescence. All scale bars represent 10 μm. The figure is taken from [Ref. 1-3].

4.3. Tracking the Path and the Fate of Individual Cells with Quantum Dot Labels

In the previous section we have described that Qdots with suitable surface coatings can bind to cells non-specifically or specifically by molecular recognition, can be ingested into cells by natural transport processes, and can be stored within cells in membrane-surrounded structures. Here we will discuss how to track the path and the fate of individual cells upon loading them with Qdots, which is important for many biological, but also medical questions [248].

One important area is embryonic development which has been intensely studied in classic biology, molecular cell biology and stem cell research [249]. Important questions in this area are how a complete organism can develop from one single cell or which cells in an early stage embryo will finally develop to which part of the final creature or which signals between cells during the early stages of development induce the differentiation of individual cells. Such questions are also related to where cells would migrate, how they would differentiate and what their offspring cells would do if implanted into an adult organism. Similarly, for another issue related to cancer research, the spreading of cancer cells (metastasis), it is also required to track cells. Cell labeling is a useful tool in these research areas, and common methods include microinjection of organic fluorophores as makers in embryonic development studies [248], fluorescence labeling of cells in fate-mapping studies [250], or transfection of cells with reporter genes either coding for fluorescent proteins like GFP [251] or coding for some selectable functional marker [252]. More recently, Qdots have been demonstrated to be highly stable tracking and fate mapping markers, owing to their peculiar optical properties, such as reduced photobleaching and long term stability (due to their inert nature) in living cells. From a technical point of view, one has to discriminate between inside and outside labeling. As internal label, Qdots can be either microinjected in cells [84] or ingested by cells [52,187,213]; as external label they can be attached to the cell membrane [47]. Dubertret *et al.* micro-injected Qdot micelles in individual cells of Xenopus blastomeres and monitored the distribution of the label in the growing embryo. Several groups have also used Qdot labels to monitor the fate of the labeled cells within cell cultures [52,187,213]. Cells of different lineage, for example, can be incubated with Qdots of different color and after an appropriate incubation time a sufficient amount of the Qdots has been ingested by the cells and free, non-ingested Qdots are removed by exchanging the culture medium. The labeled cells are then seeded as a co-culture on a new substrate. Different lineage of cells can be then determined by the color of fluorescence of each cell by optical microscopy. This noninvasive labeling with Qdots is maybe the easiest way to distinguish cell lines in co-culture which is hard to be determined by the morphology of the cell. By varying the ratio of the different Qdot colors with which cells are labeled, many different codes can be obtained, and in this way many different cell lines can be tagged with a unique label [187]. Mattheakis *et al.* have used different Qdots to label five different subpopulations of Chinese hamster ovary carcinoma (CHO) cells on one substrate [187]. Jaiswal *et al.* could follow the fate of three subpopulations of Dictyosteluim discoideum cells in one cell colony [52]. Derfus *et al.* could monitor the migration of primary hepatocyte cells in a culture with 3T3 fibroblasts [213] by using similar strategies of cell labeling. Recent progresses had been also made regarding the quantification of the subpopulation with different labels, such as flow cytometry [253,254], confocal laser scanning microscopy [255], and multiphoton microscopy [192].

Qdots can also be used as markers for cell adhesion and cell migration, processes which play essential roles in cancer development and in designing biocompatible surfaces of medical implants. Already 1976 Albrecht-Buehler introduced a method called phagokinetic tracks to follow cellular migration on cell culture substrates [256-258]. Upon migration along a surface coated with markers cells internalized and remove the markers, leaving a marker-free area behind as a blueprint of their pathway. Originally gold colloids were used as markers that could be visualized by dark-field microscopy or by transmission electron microscopy. Later with improved methods a variety of mobility/migration studies have been reported [259,260]. Recently, also Qdots have been used successfully for recording phagokinetic tracks [192,239]. Similarly to the previously used markers, cells nonspecifically incorporate Qdots as they crawl over a Qdot-coated surface, leaving behind Qdot-free zones which are representing the pattern of phagokinetic uptake of Qdots. It has been found that the quantity of incorporated Qdots varies among different cell types. It has been demonstrated that MDA-MB-231 tumor cells uptake more Qdots than non-tumorgenic MCF 10A cells [192]. Unlike the traditional assay, such as Boyden chamber assay [239], the Qdot-based assay can be used with living cells without the need for fixation. These results that Qdots might be used as new tool for cell motility studies.

5. Quantum Dots as Biosensors

5.1 Quantum Dots as FRET Donor

Fluorescence resonance energy transfer (FRET) is a process in which energy is transferred nonradiatively from an excited state donor D (usually a fluorophore) to a proximal ground state acceptor A via resonant, near-field dipole-dipole interaction [261-263]. FRET as a spectroscopic technique is applied in practice for more than 50 years, covering fields from various areas of research such as structural elucidation of biological molecules and their interaction, *in vitro* assays, *in vivo* monitoring in cellular research, nucleic acid analysis, signal transduction, light harvesting and nanomaterials [263]. Because FRET is sensitive to molecular rearrangements on the 1-10 nm range correlated to the size of biological macromolecules in which the distance between the donor and acceptor is changed, scientists have widely used this photophysical process to monitor intracellular interaction and binding events. There are some technical problems associated with using conventional organic dyes for FRET (e.g. fast photobleaching and significant emission overlap between donor and acceptor), which complicate the development of robust, sensitive and reusable optical FRET-based sensors, which could be applied to the wide range of biological sensing applications. One aim of many laboratories [264-267] is the production of nanoscale assemblies that are capable of continuously monitoring concentrations of target species in a reliable manner.

Undoubtedly, Qdots provide a potential solution to these problems. Although several reports of Qdots as FRET donors in a biological context appeared quickly after Qdots had become available [93,268,269], the full potential has only been demonstrated recently [270,271]. There are several examples of Qdots as FRET donors in biological systems. Willard *et al.* reported the first FRET experiment using Qdots as FRET donors in a protein-protein binding assay. The specific binding of biotinylated bovine serum albumin (bBSA) and tetramethylrhodamine-labeled streptavdin (SAv-TMR) was observed by conjugating bBSA to Qdots and observing enhanced TMR fluorescence caused by FRET from the Qdot donors to the TMR acceptors [93]. Medintz *et al.* engineered a histidine tag on a maltose binding protein (MBP) that was finally attached to Qdots, which serve as FRET donor. A beta-cyclodextrin-QSY9 dark quencher conjugate bound in the MBP saccharide binding site resulted in FRET quenching of the Qdot photoluminescence, but the photoluminescence recovered by adding maltose to displace the quencher [270]. The second maltose sensor assembly of this group consists of Qdots coupled with Cy3-labelled MBP bound to beta-cyclodextrin-Cy3.5 that drives the sensor function through a two-step FRET mechanism. In 2004, an in-depth study of FRET between Qdot donors and dye-labeled protein acceptors was reported. This Qdot-MBP conjugate presents two advantages: (a) it permits to tune the degree of spectral overlap between donor and acceptor and (b) provides an unique configuration where a single donor can interact with several acceptors simultaneously [272]. This study demonstrated that Qdots can be used as efficient energy donors in FRET system either by tuning the Qdots emission or by adjusting the number of dye-labeled proteins immobilized on the Qdots. Qdot-based FRET sensors can also be designed in a way that they are capable of switching. The photochromic molecule BIPS (1,2,3-dihydro-1 2-(2-carboxyethyl)-3,3-dimethyl-6-nitrospiro-(2H-1-benzopyran-2,2,2-(2H)-indoline)) is converted into colored merocyanine when it is exposed to UV light and can converted back to colorless spiropyran. Based on this special optical property, Medintz *et al.* again demonstrated reversible modulation of Qdots fluorescence using FRET with the photochromic molecule BIPS [273]. Such a design has become a prototype system as a nanosensor device for sensing chemical, photochromical changes. Their first structural model of a hybrid luminescent Qdot-protein receptor assembly elucidated by using spectroscopic measurement [274] may be also useful in determining the orientation of proteins in other hybrid protein-nanoparticle materials. Extensive reviews can be found in recent References [263,275,276]. Current state of art is the use of peptide-coated Qdots to monitor enzymatic activity within a cell. The Mattoussi group [242,271] has developed a general method for creating self-assembled peptide-Qdot conjugates, which involve tightly binding between poly(histidine)-ended peptide and dihydrolipoic acid (DHLA) molecules on Qdots, for their use in a variety of biomedical research applications. They used several different peptides to detect and quantify the activity of four different proteases (using FRET) that are used to identify some types of malignant cells. The strategy to create these FRET sensors is to attach peptides containing cleavage sites for the various enzymes and a dye molecule that quenches the light emission of the Qdot. When a particular protease is present in an assay mixture, the Qdots fluoresces brightly by releasing the quencher molecule. With the multiplex detecting by using Qdots of four different colors, it is possible to create an assay capable of simultaneously measuring the activity of each of the four enzymes present in a single assay solution [271]. Other protease sensors with organic fluorophores as acceptor have also been reported recently [277,278].

Some researchers have also started to fabricate Qdot biosensors involving the molecular recognition properties of DNA. Patolsky *et al.* demonstrated the dynamics of telomerization and DNA replication by a Qdot-based FRET experiment [279]. Qdots were conjugated to a DNA template molecule and they were mixed with dNTP (N=A, C, G) and Texas-Red-dUTP. After adding the telomerase, DNA replication could be directly resolved in chances of the FRET between the Qdot donor and Texas-Red acceptors upon their attachment. These results demonstrate the potential of Qdot based FRET techniques for their use in fast and sensitive DNA detection and DNA array analysis. Recently, nucleic acids have also been detected by using similar strategies such as two color Qdots [280], molecular beacons [281], gold quenchers [282,283], and small-interfering RNA (siRNA) [284]. Current progress in DNA-based biosensors has enabled the detection of specific biomolecules by utilizing the unique properties of aptamers, which are nucleic acid species that can mimic the structure / binding properties of proteins [285-289]. Levy *et al.* designed a Qdot based aptamer biosensor that is sensitive to bind the blood-clotting protein thrombin as a model system [286]. First, quantum-dot aptamer beacon was synthesized, consisting of organic quenchers bound to an oligonucleotide structure that hybridized to the aptamer structure attached to the Qdot surface (this is the "off" state with FRET quenching). In the presence of the target protein, thrombin, the quadruplex conformation of the aptamer resulted in the displacement of the antisense oligonucleotide quencher and a concomitant increase in fluorescence ("on" state) [286].

5.2 Quantum Dots as FRET Acceptor

In 1996, Kagan *et al.* observed resonance energy transfer between closely packed CdSe Qdots [290,291]. The quenching of the luminescence of the small Qdots was accompanied by an enhancement of the luminescence of the large Qdots. This example of resonance energy transfer from small Qdots to large Qdots proved that Qdots can not only play the role of a donor but also can be used as acceptor in FRET pairs. However, there are only few examples of Qdots as acceptors in biologically motivated experiments due to their broad absorbance profile that overlaps with the absorption of the donor. Clapp *et al.* assessed the ability of luminescent Qdots as energy acceptors in FRET assays with organic dyes serving as donors [292]. Steady-state and time resolved fluorescence measurements on this system showed no apparent FRET from dyes to Qdots. This has been attributed to the long exciton lifetime of the Qdot acceptor compared with that of the dye donor and substantial direct excitation of Qdots [274].

In 2006, biosensors with Qdots as acceptors came to a revival because of the development of bioluminescent Qdots. So *et al.* reported the idea [293] that Qdots can emit light without external optical excitation based on the principle of bioluminescence resonance energy transfer (BRET). BRET is a naturally occurring phenomenon of non-radiative energy transfer from a light emitting protein (the donor, such as R. Reniformis luciferase) whose luminescence is excited by a chemical stimulus to a fluorescence protein (the acceptor, such as green fluorescent protein, GFP) in close proximity [294,295]. In other words: The protein donor (that is bioluminescent) is excited by a non-optical stimulus and this energy is transferred in a non-radiative way to an acceptor, which in turn emits a photon. By conjugating luciferase proteins to Qdots, BRET, analogous to FRET, was demonstrated, whereby the Qdots had the role as acceptor [296-298]. More recently, Huang *et al.* reported another resonance energy transfer between chemiluminescent donors and Qdots as acceptors, which is called chemiluminescence resonance energy transfer (CRET), and is very similar to BRET [299]. CRET had been triggered by the oxidation of a luminescent substrate (without any optical excitation source), the energy had been transferred in a non-radiative way to the Qdot acceptor, and upon relaxation photos were emitted. Qdots are very useful as acceptors for luminescent donors, since the emission spectra of bioluminescent molecules are not sharp and therefore the broad absorption spectra of the Qdots are advantageous.

6. Quantum Dots as *in vivo* Probes

New challenges arise with the application of Qdots as contrast agents for *in vivo* imaging. Unlike studying the monolayer of cultured cells, biological thick tissue attenuates most signals used for imaging. Organic fluorophores and chemiluminescence are currently the most common used optical probes for animal imaging [300-302]. Comparing whole animal imaging with optical probes to other techniques that rely on signals transmitted through thick-tissue samples, such as ultrasonic waves (ultrasound imaging), X-ray (computed X-ray tomography, CT), gamma rays (Positron emission tomography, PET) or radio waves (magnetic resonance imaging, MRI) points out, that optical imaging is still limited by the poor transmission of visible light through biological tissue. The use of near-infrared (NIR) wavelengths (typically 700-900 nm) for excitation or emission can be used to improve depth sensitivity because of the low tissue absorption and scattering effects in this emission range (as described in some recent reviews [303-305]). Cyanine-dye series are the most popular organic fluorophores emitting in the NIR, but there is still only limited commercial availability. Since Qdots can be synthesized with NIR emission [40,306,307] they are potential candidates for *in vivo* imaging.

The first two *in vivo* experiments with Qdots were reported in 2002. Akerman and colleagues showed that CdSe/ZnS Qdots coated with a lung-targeting peptide accumulate in the lungs of mice after intravenous injection, whereas two other peptides specifically direct Qdots to blood vessels or lymphatic vessels in tumors [94]. They investigated tumor-tissue sections of the animals that had been injected with Qdots with epifluorescence microscopy. Dubertret *et al.* studied the behavior of specific cells during embryogenesis by microinjecting phospholipids-coated Qdots into early-stage *Xenopus* embryos. Both cases demonstrated that Qdots provide a stable, robust fluorescent probe for *in vivo* experiments, although both of them still were done with particles that were fluorescent in the visible. Live animal imaging using Qdots with multiphoton microscopy was later achieved by Larson and his colleagues [181] in 2003. Colloidal semiconductor nanocrystals have larger two-photon cross-sections than organic dyes. After intravenous injection into mice the Qdots were visualized dynamically through the skin of the living animals, in capillaries hundreds of micrometers deep inside the tissue by using an excitation wavelength of 900 nm. In 2004 Ballou *et al.* presented a detailed study of *in vivo* in-depth imaging of Qdots with different surface coatings by fluorescence imaging of living animals, by necropsy, by frozen tissue sections for optical microscopy, and by electron microscopy [81]. Gao *et al.* demonstrated whole-animal imaging of prostate cancer in mice using Qdots conjugated to antibodies. *In vivo* imaging of Qdots bound at the tumor site were much brighter and sensitive than comparable images with GFP, which were buried in autofluorescence and background [83]. Kim and coworkers demonstrated the use of Qdots during surgical procedures by mapping sentinel-lymph-nodes (SLN) with NIR fluorescent Qdots in mice and pigs [306]. After Qdots entered the lymphatic system by intradermal injection the surgeon could follow the guidance of the Qdots flow in real time using an intraoperative imaging system which allowed for rapid identifying the position of the SLN in a precise surgical procedure. This study has to be seen as breakthrough, as Qdots were for the first time effectively used as contrast agent for surgery. In 2006 Cai *et al.* reported another study of *in vivo* targeting and imaging of tumor vasculature using RGD peptide-labeled Qdots. This study opened up new perspectives for integrin-targeted near-infrared optical imaging. Recently, Rao and coworkers demonstrated the use of bioluminescence to excite Qdots for *in vivo* imaging without the use of a laser excitation source [293] (see also in

Section 5.2). Bioluminescence is widely used for *in vivo* imaging of nude mice. Although Qdots can emit light in the NIR region of the spectrum, and thus have demonstrated good tissue penetration of the emitted light in previous experiments, they all required efficient excitation with blue light that, however, does not penetrate well the tissue and also produces high background owing to excitation of endogenous fluorophores. Compared to existing Qdots, self-illuminating Qdot conjugates have greatly enhanced the sensitivity in small animal imaging, with an *in vivo* signal-to-background ratio of $> 10^3$ for 5 pmol of conjugate [293].

There are still many challenges waiting ahead for establishing Qdots for clinical *in vivo* applications. In the long-term study experiments from Ballou *et al.* [81] for example it has been demonstrated that Qdots remain fluorescent after at least four months *in vivo* without acute toxicity, but their potentially deleterious effects still need to be investigated in detail prior to long-term usage in higher organism. Another severe problem is the limited retention time. Even PEG-coated Qdots are cleared from the blood stream by the liver [81]. Without higher retention times they are no good transport system for drug delivery. Klostranec and Chan pointed out in their recent review [44] several fundamental questions that have to be addressed: A) What are the proper doses of Qdots for optimal *in vivo* imaging? B) What is the kinetics of Qdots *in vivo*? C) How to prevent the uptake of Qdots by the reticuloendothelial cells? D) Are Qdots toxic to the animal? For a more detailed discussion we refer the interested reader to this review [44]. It is worth noting that the metabolism and degradation of Qdots in the body remains to be investigated and several reports have shown that injected Qdots can accumulate in the kidney, liver, spleen or bone marrow (organs that are part of the reticuloendothelial system, RES) [83,94,308]. It is not known whether Qdots are ultimately secreted from the body. The above studies have shown the great potential to use Qdots as *in vivo* probes for cancer studies, drug delivery and non-invasive whole body imaging. However, a lot of research is still to be completed before Qdots can be used safely as probes in the human body.

7. Perspectives

The development and maturity of Qdots in the last twenty years is a good example for the general paradigm underlying nanobiotechnology, in which researchers develop and characterize nanostructures, understand and manipulate their surface chemistry for biological application, and demonstrate their ability to solve biological problems or integrate them into systems or devices for clinical diagnostics [44]. New crisis and challenges arose in this era, especially cytotoxicity, that were not initially considered in the first demonstration experiments. Although recent studies have shown the potential for the use of Qdots in cancer diagnosis and therapy, it is an urgent need to understand Qdot toxicity and metabolism in the body. Fortunately, there is an increasing number of new reports about "greener" quantum dots (GQDs), which will hopefully be the first step in resolving the current crisis. In order to advance in the field of nanobiotechnology, fully investigating and learning the fundamental properties of nanostructures in biological systems is still required. This will be also a key to create future biosensors or nano machines such as artificial viruses [92,222,309,310]. Some recent reports have described how to assemble nanostructures [311], and we give a particular importance to the appearance of mono-functional techniques [312-316]. Simple Qdots could be combined with other nanomaterials to assemble small but complex multifunctional machines with dimensions similar to a standard virus (<150nm). Additionally, multimodal imaging probes and detection systems will be areas of increased interests. Multilevel imaging from molecular to medical scales demonstrates the need for the development of nanoparticles that can be used in all kinds of imaging techniques, such as MRI, PET or SPECT, optical confocal microscopy, CT, and TEM [317]. Qdots as bimodal imaging probes have been introduced by pairing their optical capabilities with electron microscopy (EM) [318] or MRI [77,86,218] modalities. We conclude that the recent progress of Qdots in biotechnology, offers perspective for the use of Qdots in fields ranging from material chemistry, optical physics, cellular biology and medical applications.

Acknowledgements

Cheng-An Lin thanks the sandwich program Deutscher Akademischer Austauschdienst (DAAD-Germany) and National Science Council (NSC-Taiwan) for funding as well as Stefan Kudera & Marco Zanella who helped him a lot in Qdots synthesis. He also thanks his parents and wife, Pei-Yun Li, for long-term encouragement. This work was also supported, in part, by NSC National Program of Nanoscience and Nanotechnology Grant NSC 94-2120-M-033, by the Center-of-Excellence Program on Membrane Technology (Ministry of Education, Taiwan, R.O.C.) and by the German Research Foundation (DFG, Emmy Noether program).

References

1. Michalet, X. et al. Quantum dots for live cells, in vivo imaging, and diagnostics. Science 307, 538-44 (2005).
2. Derfus, A. M., Chan, W. C. W. & Bhatia, S. N. Probing the cytotoxicity of semiconductor quantum dots. Nano Letters 4, 11-18 (2004).
3. Kirchner, C. et al. Cytotoxicity of colloidal CdSe and CdSe/ZnS nanoparticles. Nano Letters 5, 331-338 (2005).
4. Stephens, D. J. & Allan, V. J. Light Microscopy Techniques for Live Cell Imaging. Science 300, 82-86 (2003).
5. Waggoner, A. Fluorescent labels for proteomics and genomics. Current Opinion in Chemical Biology 10, 62-66 (2006).
6. Giepmans, B. N. G., Adams, S. R., Ellisman, M. H. & Tsien, R. Y. The fluorescent toolbox for assessing protein location and function. Science 312, 217-224 (2006).
7. Weijer, C. J. Visualizing Signals Moving in Cells. Science 300, 96-100 (2003).
8. Miyawaki, A., Sawano, A. & Kogure, T. Lighting up cells: labelling proteins with fluorophores. Nature Cell Biology 5, S1-S7 (2003).
9. Pawley, J. B. & Centonze, V. E. in Cell Biology - A Laboratory Handbook (ed. Celis, J. E.) 44-64 (Academic Press, San Diego, California, 1994).
10. Kulzer, F., Kummer, S., Matzke, R., Bräuchle, C. & Basche, T. Single-molecule optical switching of terrylene in p-terphenyl. Nature 387, 688-691 (1997).
11. Dickson, R. M., Cubitt, A. B., Tsien, R. Y. & Moerner, W. E. On/off blinking and switching behavior of single molecules of green fluorescent protein. Nature 388, 355-358 (1997).
12. Alivisatos, A. P. Semiconductor Clusters, Nanocrystals, and Quantum Dots. Science 271, 933-937 (1996).

13. Brus, L. E. Electron-electron and electron-hole interactions in small semiconductor crystallites: the size dependence of the lowest excited electronic state. J. Chem. Phys. 80, 4403-4409 (1984).
14. Brus, L. Electronic Wave Functions in Semiconductor Clusters: Experiment and Theory. J. Phys. Chem 90, 2555-2560 (1986).
15. Bruchez, M. J., Moronne, M., Gin, P., Weiss, S. & Alivisatos, A. P. Semiconductor Nanocrystals as Fluorescent Biological Labels. Science 281, 2013-2016 (1998).
16. Chan, W. C. W. & Nie, S. Quantum Dot Bioconjugates for Ultrasensitive Nonisotopic Detection. Science 281, 2016-2018 (1998).
17. Jana, N. R. & Peng, X. G. Single-phase and gram-scale routes toward nearly monodisperse Au and other noble metal nanocrystals. Journal of the American Chemical Society 125, 14280-14281 (2003).
18. Murray, C. B., Norris, D. J. & Bawendi, M. G. Synthesis and Characterization of Nearly Monodisperse CdE (E = S, Se, Te) Semiconductor Nanocrystallites. Journal of the American Chemical Society 115, 8706-8715 (1993).
19. Hyeon, T., Lee, S. S., Park, J., Chung, Y. & Bin Na, H. Synthesis of highly crystalline and monodisperse maghemite nanocrystallites without a size-selection process. Journal of the American Chemical Society 123, 12798-12801 (2001).
20. Dabbousi, B. O. et al. (CdSe)ZnS Core-Shell Quantum Dots: Synthesis and Characterization of a Size Series of Highly Luminescent Nanocrystallites. Journal of Physical Chemistry B 101, 9463-9475 (1997).
21. Talapin, D. V., Rogach, A. L., Kornowski, A., Haase, M. & Weller, H. Highly luminescent monodisperse CdSe and CdSe/ZnS nanocrystals synthesized in a hexadecylamine-trioctylphosphine oxide-trioctylphospine mixture. Nano Letters 1, 207-211 (2001).
22. Peng, Z. A. & Peng, X. G. Formation of high-quality CdTe, CdSe, and CdS nanocrystals using CdO as precursor. Journal of the American Chemical Society 123, 183-184 (2001).
23. Reiss, P., Bleuse, J. & Pron, A. Highly luminescent CdSe/ZnSe core/shell nanocrystals of low size dispersion. Nano Letters 2, 781-784 (2002).
24. Spanhel, L., Haase, M., Weller, H. & Henglein, A. Photochemistry of Colloidal Semiconductors. 20. Surface Modification and Stability of Strong Luminescing CdS Particles. Journal of the American Chemical Society 109, 5649-5655 (1987).
25. Steigerwald, M. L. et al. Surface Derivatization and Isolation of Semiconductor Cluster Molecules. Journal of the American Chemical Society 110, 3046-3050 (1988).
26. Ekimov, A. I. & Onushchenko, A. A. Quantum Size Effect in the Optical-Spectra of Semiconductor Micro-Crystals. SOVIET PHYSICS SEMICONDUCTORS-USSR 16, 775-778 (1982).
27. Efros, A. L. & L.Efros, A. Interband Absorption of Light in a Semiconductor Sphere. SOVIET PHYSICS SEMICONDUCTORS-USSR 16, 772-775 (1982).

28. Kortan, A. R. et al. Nucleation and Growth of CdSe on ZnS Quantum Crystallite Seeds, and Viece Verca, in Inverse Micelle Media. Journal of the American Chemical Society 112, 1327-1332 (1990).
29. Murray, C. B., Kagan, C. R. & Bawendi, M. G. Synthesis and Characterization of Monodisperse Nanocrystals and Close-Packed Nanocrystals Assembles. Annu. Rev. Mater. Sci. 30, 545-610 (2000).
30. Pellegrino, T. et al. On the Development of Colloidal Nanoparticles towards Multifunctional Structures and their Possible Use for Biological Applications. Small 1, 48-63 (2005).
31. Peng, X., Wickham, J. & Alivisatos, A. Kinetics of II-VI and III-V colloidal semiconductor nanocrystal growth: "Focusing" of size distributions. Journal of the American Chemical Society 120, 5343-5344 (1998).
32. Hines, M. A. & Guyot-Sionnest, P. Synthesis and Characterization of Strongly Luminescing ZnS-Capped CdSe Nanocrystals. Journal of Physical Chemistry B 100, 468-471 (1996).
33. Peng, X., Schlamp, M., Kadavanich, A. & Alivisatos, A. Epitaxial growth of highly luminescent CdSe/CdS core/shell nanocrystals with photostability and electronic accessibility. Journal of the American Chemical Society 119, 7019-7029 (1997).
34. van Sark, W. G. J. H. M. et al. Photooxidation and photobleaching of single CdSe/ZnS quantum dots probed by room-temperature time-resolved spectroscopy. Journal of Physical Chemistry B 105, 8281-8284 (2001).
35. Li, J. J. et al. Large-scale synthesis of nearly monodisperse CdSe/CdS core/shell nanocrystals using air-stable reagents via successive ion layer adsorption and reaction. Journal of the American Chemical Society 125, 12567-12575 (2003).
36. Battaglia, D., Li, J. J., Wang, Y. J. & Peng, X. G. Colloidal two-dimensional systems: CdSe quantum shells and wells. Angewandte Chemie-International Edition 42, 5035-5039 (2003).
37. Zhong, X. H., Xie, R. G., Zhang, Y., Basche, T. & Knoll, W. High-quality violet-to red-emitting ZnSe/CdSe core/shell nanocrystals. Chemistry of Materials 17, 4038-4042 (2005).
38. Xie, R. G., Zhong, X. H. & Basche, T. Synthesis, characterization, and spectroscopy of type-II core/shell semiconductor nanocrystals with ZnTe cores. Advanced Materials 17, 2741-+ (2005).
39. Talapin, D. V. et al. CdSe/CdS/ZnS and CdSe/ZnSe/ZnS core-shell-shell nanocrystals. Journal of Physical Chemistry B 108, 18826-18831 (2004).
40. Aharoni, A., Mokari, T., Popov, I. & Banin, U. Synthesis of InAs/CdSe/ZnSe core/shell1/shell2 structures with bright and stable near-infrared fluorescence. Journal of the American Chemical Society 128, 257-264 (2006).
41. Pan, D. C. et al. Semiconductor "nano-onions" with multifold alternating CdS/CdSe or CdSe/CdS structure. Chemistry of Materials 18, 4253-4258 (2006).

42. Bruchez, M., Jr., Moronne, M., Gin, P., Weiss, S. & Alivisatos, A. P. Semiconductor nanocrystals as fluorescent biological labels. Science 281, 2013-6 (1998).
43. Chan, W. C. & Nie, S. Quantum dot bioconjugates for ultrasensitive nonisotopic detection. Science 281, 2016-8 (1998).
44. Klostranec, J. M. & Chan, W. C. W. Quantum Dots in Biological and Biomedical Research: Recent Progress and Present Challenges. Advanced Materials 18, 1953-1964 (2006).
45. Zhang, C. Y. et al. Quantum dot-labeled trichosanthin. Analyst 125, 1029-1031 (2000).
46. Winter, J. O., Liu, T. Y., Korgel, B. A. & Schmidt, C. E. Recognition molecule directed interfacing between semiconductor quantum dots and nerve cells. Advanced Materials 13, 1673-1677 (2001).
47. Kloepfer, J. A. et al. Quantum Dots as Strain- and Metabolism-Specific Microbiological Lables. Applied and Environmental Microbiology 69, 4205-4213 (2003).
48. Mitchell, G. P., Mirkin, C. A. & Letsinger, R. L. Programmed Assembly of DNA Functionalized Quantum Dots. Journal of the American Chemical Society 121, 8122-8123 (1999).
49. Hanaki, K. et al. Semiconductor quantum dot/albumin complex is a long-life and highly photostable endosome marker. Biochemical and Biophysical Research Communications 302, 496-501 (2003).
50. Chen, C.-C., Yet, C.-P., Wang, H.-N. & Chao, C.-Y. Self-Assembly of Monolayers of Cadmium Selenide Nanocrystals with Dual Color Emission. Langmuir 15, 6845-6850 (1999).
51. Mattoussi, H. et al. Self-Assembly of CeSe-ZnS Quantum Dot Bioconjugates Using an Engineered Recombinant Protein. Journal of the American Chemical Society 122, 12142-12150 (2000).
52. Jaiswal, J. K., Mattoussi, H., Mauro, J. M. & Simon, S. M. Long-term multiple color imaging of live cells using quantum dot bioconjugates. Nature Biotechnology 21, 47-51 (2003).
53. Sukhanova, A. et al. Highly stable fluorescent nanocrystals as a novel class of labels for immunohistochemical analysis of paraffin-embedded tissue sections. Laboratory Investigation 82, 1259-1261 (2002).
54. Sukhanova, A. et al. Biocompatible fluorescent nanocrystals for immunolabeling of membrane proteins and cells. Analytical Biochemistry 324, 60-67 (2004).
55. Pinaud, F., King, D., Moore, H. P. & Weiss, S. Bioactivation and cell targeting of semiconductor CdSe/ZnS nanocrystals with phytochelatin-related peptides. Journal of the American Chemical Society 126, 6115-6123 (2004).
56. Tsay, J. M., Doose, S., Pinaud, F. & Weiss, S. Enhancing the photoluminescence of peptide-coated nanocrystals with shell composition and UV irradiation. Journal of Physical Chemistry B 109, 1669-1674 (2005).

57. Pathak, S., Choi, S. K., Arnheim, N. & Thompson, M. E. Hydroxylated quantum dots as luminescent probes for in situ hybridization. Journal of the American Chemical Society 123, 4103-4104 (2001).
58. Wang, Y. A., Li, J. J., Chen, H. Y. & Peng, X. G. Stabilization of inorganic nanocrystals by organic dendrons. Journal of the American Chemical Society 124, 2293-2298 (2002).
59. Liu, Y. C., Kim, M., Wang, Y. J., Wang, Y. A. & Peng, X. G. Highly luminescent, stable, and water-soluble CdSe/CdS core - Shell dendron nanocrystals with carboxylate anchoring groups. Langmuir 22, 6341-6345 (2006).
60. Wisher, A. C., Bronstein, I. & Chechik, V. Thiolated PAMAM dendrimer-coated CdSe/ZnSe nanoparticles as protein transfection agents. Chemical Communications, 1637-1639 (2006).
61. Palaniappan, K., Xue, C. H., Arumugam, G., Hackney, S. A. & Liu, J. Water-soluble, cyclodextrin-modified CdSe-CdS core-shell structured quantum dots. Chemistry of Materials 18, 1275-1280 (2006).
62. Skaff, H. & Emrick, T. The use of 4-substituted pyridines to afford amphiphilic, pegylated cadmium selenide nanoparticles. Chemical Communications, 52-53 (2003).
63. Aldana, J., Wang, Y. A. & Peng, X. G. Photochemical instability of CdSe nanocrystals coated by hydrophilic thiols. Journal of the American Chemical Society 123, 8844-8850 (2001).
64. Gerion, D. et al. Synthesis and Properties of Biocompatible Water-Soluble Silica-Coated CdSe/ZnS Semiconductor Quantum Dots. Journal of Physical Chemistry B 105, 8861-8871 (2001).
65. Uyeda, H. T., Medintz, I. L., Jaiswal, J. K., Simon, S. M. & Mattoussi, H. Synthesis of compact multidentate ligands to prepare stable hydrophilic quantum dot fluorophores. Journal of the American Chemical Society 127, 3870-8 (2005).
66. Jiang, W., Mardyani, S., Fischer, H. & Chan, W. C. W. Design and characterization of lysine cross-linked mereapto-acid biocompatible quantum dots. Chemistry of Materials 18, 872-878 (2006).
67. Giersig, M., Liz-Marzan, L. M., Ung, T., Su, D. & Mulvaney, P. Chemistry of Nanosized Silica-Coated metal Particles-EM-Study. Ber. Bunsen. Phys. Chem. 101, 1617-1620 (1997).
68. Correa-Duarte, M. A., Giersig, M. & Liz-Marzán, L. M. Stabilization of CdS semiconductor nanoparticles against photodegradation by a silica coating procedure. Chemical Physics Letters 286, 497-501 (1998).
69. Parak, W. J. et al. Conjugation of DNA to silanized colloidal semiconductor nanocrystalline quantum dots. Chemistry of Materials 14, 2113-2119 (2002).
70. Nann, T. & Mulvaney, P. Single quantum dots in spherical silica particles. Angewandte Chemie-International Edition 43, 5393-5396 (2004).

71. Mulvaney, P., Liz-Marzan, L. M., Giersig, M. & Ung, T. Silica encapsulation of quantum dots and metal clusters. Journal of Materials Chemistry 10, 1259-1270 (2000).
72. Rogach, A. L., Nagesha, D., Ostrander, J. W., Giersig, M. & Kotov, N. A. "Raisin bun"-type composite spheres of silica and semiconductor nanocrystals. Chemistry of Materials 12, 2676-2685 (2000).
73. Farmer, S. C. & Patten, T. E. Photoluminescent polymer/quantum dot composite nanoparticles. Chemistry of Materials 13, 3920-3926 (2001).
74. Yang, H. S., Holloway, P. H. & Santra, S. Water-soluble silica-overcoated CdS: Mn/ZnS semiconductor quantum dots. Journal of Chemical Physics 121, 7421-7426 (2004).
75. Santra, S., Yang, H. S., Holloway, P. H., Stanley, J. T. & Mericle, R. A. Synthesis of water-dispersible fluorescent, radio-opaque, and paramagnetic CdS: Mn/ZnS quantum dots: A multifunctional probe for bioimaging. Journal of the American Chemical Society 127, 1656-1657 (2005).
76. Zhelev, Z., Ohba, H. & Bakalova, R. Single quantum dot-micelles coated with silica shell as potentially non-cytotoxic fluorescent cell tracers. Journal of the American Chemical Society 128, 6324-6325 (2006).
77. Bakalova, R. et al. Silica-shelled single quantum dot micelles as imaging probes with dual or multimodality. Analytical Chemistry 78, 5925-5932 (2006).
78. Wu, X. Y. et al. Immunofluorescent labeling of cancer marker Her2 and other cellular targets with semiconductor quantum dots. Nature Biotechnology 21, 41-46 (2003).
79. Pellegrino, T. et al. Hydrophobic nanocrystals coated with an amphiphilic polymer shell: A general route to water soluble nanocrystals. Nano Letters 4, 703-707 (2004).
80. Petruska, M. A., Bartko, A. P. & Klimov, V. I. An amphiphilic approach to nanocrystal quantum dot-titania nanocomposites. Journal of the American Chemical Society 126, 714-715 (2004).
81. Ballou, B., Lagerholm, B. C., Ernst, L. A., Bruchez, M. P. & Waggoner, A. S. Noninvasive imaging of quantum dots in mice. Bioconjugate Chemistry 15, 79-86 (2004).
82. Luccardini, C., Tribet, C., Vial, F., Marchi-Artzner, V. & Dahan, M. Size, charge, and interactions with giant lipid vesicles of quantum dots coated with an amphiphilic macromolecule. Langmuir 22, 2304-2310 (2006).
83. Gao, X. H., Cui, Y. Y., Levenson, R. M., Chung, L. W. K. & Nie, S. M. In vivo cancer targeting and imaging with semiconductor quantum dots. Nature Biotechnology 22, 969-976 (2004).
84. Dubertret, B. et al. In vivo imaging of quantum dots encapsulated in phospholipid micelles. Science 298, 1759-1762 (2002).
85. Fan, H. Y. et al. Surfactant-assisted synthesis of water-soluble and biocompatible semiconductor quantum dot micelles. Nano Letters 5, 645-648 (2005).

86. Mulder, W. J. M. et al. Quantum dots with a paramagnetic coating as a bimodal molecular imaging probe. Nano Letters 6, 1-6 (2006).
87. Jin, T., Fujii, F., Sakata, H., Tamura, M. & Kinjo, M. Calixarene-coated water-soluble CdSe-ZnS semiconductor quantum dots that are highly fluorescent and stable in aqueous solution. Chemical Communications, 2829-2831 (2005).
88. Jin, T., Fujii, F., Sakata, H., Tamura, M. & Kinjo, M. Amphiphilic p-sulfonatocalix[4]arene-coated CdSe/ZnS quantum dots for the optical detection of the neurotransmitter acetylcholine. Chemical Communications, 4300-4302 (2005).
89. Jin, T., Fujii, F., Yamada, E., Nodasaka, Y. & Kinjo, M. Control of the optical properties of quantum dots by surface coating with calix[n]arene carboxylic acids. Journal of the American Chemical Society 128, 9288-9289 (2006).
90. Wang, Y., Wong, J. F., Teng, X. W., Lin, X. Z. & Yang, H. "Pulling" nanoparticles into water: Phase transfer of oleic acid stabilized monodisperse nanoparticles into aqueous solutions of alpha-cyclodextrin. Nano Letters 3, 1555-1559 (2003).
91. Feng, J. et al. Cyclodextrin driven hydrophobic/hydrophilic transformation of semiconductor nanoparticles. Applied Physics Letters 86, - (2005).
92. Osaki, F., Kanamori, T., Sando, S., Sera, T. & Aoyama, Y. A quantum dot conjugated sugar ball and its cellular uptake on the size effects of endocytosis in the subviral region. Journal of the American Chemical Society 126, 6520-6521 (2004).
93. Willard, D. M., Carillo, L. L., Jung, J. & Van Orden, A. CdSe-ZnS quantum dots as resonance energy transfer donors in a model protein-protein binding assay. Nano Letters 1, 469-474 (2001).
94. Akerman, M. E., Chan, W. C. W., Laakkonen, P., Bhatia, S. N. & Ruoslahti, E. Nanocrystal targeting in vivo. Proceedings of the National Academy of Sciences of the United States of America 99, 12617-12621 (2002).
95. Rosenthal, S. J. et al. Targeting Cell Surface Receptors with Ligand-Conjugated Nanocrystals. Journal of the American Chemical Society 124, 4586-4594 (2002).
96. Lakowicz, J. R., Gryczynski, I., Gryczynski, Z., Nowaczyk, K. & Murphy, C. J. Time-resolved spectral observations of cadmium-enriched cadmium sulfide nanoparticles and the effects of DNA oligomer binding. Analytical Biochemistry 280, 128-136 (2000).
97. Mahtab, R., Harden, H. H. & Murphy, C. J. Temperature- and salt-dependent binding of long DNA to protein-sized quantum dots: Thermodynamics of "inorganic protein"-DNA interactions. Journal of the American Chemical Society 122, 14-17 (2000).
98. Mattoussi, H. et al. Self-assembly of CdSe-ZnS quantum dot bioconjugates using an engineered recombinant protein. Journal of the American Chemical Society 122, 12142-12150 (2000).

99. Mattoussi, H. et al. Bioconjugation of highly luminescent colloidal CdSe-ZnS quantum dots with an engineered two-domain recombinant protein. Physica Status Solidi B-Basic Research 224, 277-283 (2001).
100. Goldman, E. R. et al. Conjugation of luminescent quantum dots with antibodies using an engineered adaptor protein to provide new reagents for fluoroimmunoassays. Analytical Chemistry 74, 841-847 (2002).
101. Goldman, E. R. et al. Luminescent quantum dot-adaptor protein-antibody conjugates for use in fluoroimmunoassays. Physica Status Solidi B-Basic Research 229, 407-414 (2002).
102. Kloepfer, J. A. et al. Quantum dots as strain- and metabolism-specific microbiological labels. Applied and Environmental Microbiology 69, 4205-4213 (2003).
103. Chan, W. C. W. et al. Luminescent quantum dots for multiplexed biological detection and imaging. Current Opinion in Biotechnology 13, 40-46 (2002).
104. Aslam, M. & Dent, A. Bioconjugation: Protein Coupling Techniques for the Biomedical Sciences (Macmillan Ref., London, 2000).
105. Hermanson, G. T. Bioconjugate Techniques (Academic Press, San Diego, 1996).
106. Ness, J. M., Akhtar, R. S., Latham, C. B. & Roth, K. A. Combined tyramide signal amplification and quantum dots for sensitive and photostable immunofluorescence detection. Journal of Histochemistry & Cytochemistry 51, 981-987 (2003).
107. Tokumasu, F. & Dvorak, J. Development and application of quantum dots for immunocytochemistry of human erythrocytes. Journal of Microscopy-Oxford 211, 256-261 (2003).
108. Dahan, M. et al. Diffusion dynamics of glycine receptors revealed by single-quantum dot tracking. Science 302, 442-445 (2003).
109. Lidke, D. S. et al. Quantum dot ligands provide new insights into erbB/HER receptor-mediated signal transduction. Nature Biotechnology 22, 198-203 (2004).
110. Goldman, E. R. et al. Avidin: A natural bridge for quantum dot-antibody conjugates. Journal of the American Chemical Society 124, 6378-6382 (2002).
111. Schroedter, A., Weller, H., Eritja, R., Ford, W. E. & Wessels, J. M. Biofunctionalization of silica-coated CdTe and gold nanocrystals. Nano Letters 2, 1363-1367 (2002).
112. Gerion, D. et al. Sorting fluorescent nanocrystals with DNA. Journal of the American Chemical Society 124, 7070-7074 (2002).
113. Gerion, D. et al. Room-temperature single-nucleotide polymorphism and multiallele DNA detection using fluorescent nanocrystals and microarrays. Analytical Chemistry 75, 4766-4772 (2003).
114. Eisenstein, M. Helping cells to tell a colorful tale. Nature Methods 3, 647-654 (2006).

115. Pradhan, N., Goorskey, D., Thessing, J. & Peng, X. G. An alternative of CdSe nanocrystal emitters: Pure and tunable impurity emissions in ZnSe nanocrystals. Journal of the American Chemical Society 127, 17586-17587 (2005).
116. Talapin, D. V. et al. Etching of colloidal InP nanocrystals with fluorides: Photochemical nature of the process resulting in high photoluminescence efficiency. Journal of Physical Chemistry B 106, 12659-12663 (2002).
117. Bharali, D. J., Lucey, D. W., Jayakumar, H., Pudavar, H. E. & Prasad, P. N. Folate-receptor-mediated delivery of InP quantum dots for bioimaging using confocal and two-photon microscopy. Journal of the American Chemical Society 127, 11364-11371 (2005).
118. Peyser, L. A., Vinson, A. E., Bartko, A. P. & Dickson, R. M. Photoactivated fluorescence from individual silver nanoclusters. Science 291, 103-106 (2001).
119. Zheng, J., Petty, J. T. & Dickson, R. M. High quantum yield blue emission from water-soluble Au-8 nanodots. Journal of the American Chemical Society 125, 7780-7781 (2003).
120. Zheng, J., Zhang, C. W. & Dickson, R. M. Highly fluorescent, water-soluble, size-tunable gold quantum dots. Physical Review Letters 93, - (2004).
121. Zheng, J. & Dickson, R. M. Individual water-soluble dendrimer-encapsulated silver nanodot fluorescence. Journal of the American Chemical Society 124, 13982-13983 (2002).
122. Petty, J. T., Zheng, J., Hud, N. V. & Dickson, R. M. DNA-templated Ag nanocluster formation. Journal of the American Chemical Society 126, 5207-5212 (2004).
123. Zhang, J. G., Xu, S. Q. & Kumacheva, E. Photogeneration of fluorescent silver nanoclusters in polymer microgels. Advanced Materials 17, 2336-+ (2005).
124. Negishi, Y., Nobusada, K. & Tsukuda, T. Glutathione-protected gold clusters revisited: Bridging the gap between gold(I)-thiolate complexes and thiolate-protected gold nanocrystals. Journal of the American Chemical Society 127, 5261-5270 (2005).
125. Liu, Z. L., Peng, L. X. & Yao, K. L. Intense blue luminescence from self-assembled Au-thiolate clusters. Materials Letters 60, 2362-2365 (2006).
126. Shi, X. Y., Ganser, T. R., Sun, K., Balogh, L. P. & Baker, J. R. Characterization of crystalline dendrimer-stabilized gold nanoparticles. Nanotechnology 17, 1072-1078 (2006).
127. Sun, Y. P. et al. Quantum-sized carbon dots for bright and colorful photoluminescence. Journal of the American Chemical Society 128, 7756-7757 (2006).
128. Sato, S. & Swihart, M. T. Propionic-acid-terminated silicon nanoparticles: Synthesis and optical characterization. Chemistry of Materials 18, 4083 - 4088 (2006).

129. Rogozhina, E. V., Eckhoff, D. A., Gratton, E. & Braun, P. V. Carboxyl functionalization of ultrasmall luminescent silicon nanoparticles through thermal hydrosilylation. Journal of Materials Chemistry 16, 1421-1430 (2006).
130. Jurbergs, D., Rogojina, E., Mangolini, L. & Kortshagen, U. Silicon nanocrystals with ensemble quantum yields exceeding 60%. Applied Physics Letters 88, - (2006).
131. Hines, M. A. & Guyot-Sionnest, P. Bright UV-blue luminescent colloidal ZnSe nanocrystals. Journal of Physical Chemistry B 102, 3655-3657 (1998).
132. Hwang, C. S. & Cho, I. H. Characterization of the ZnSe/ZnS core shell quantum dots synthesized at various temperature conditions and the water soluble ZnSe/ZnS quantum dot. Bulletin of the Korean Chemical Society 26, 1776-1782 (2005).
133. Chen, H. S. et al. Colloidal ZnSe, ZnSe/ZnS, and ZnSe/ZnSeS quantum dots synthesized from ZnO. Journal of Physical Chemistry B 108, 17119-17123 (2004).
134. Demir, M. M., Munoz-Espi, R., Lieberwirth, I. & Wegner, G. Precipitation of monodisperse ZnO nanocrystals via acid-catalyzed esterification of zinc acetate. Journal of Materials Chemistry 16, 2940-2947 (2006).
135. Jun, Y. W., Choi, C. S. & Cheon, J. Size and shape controlled ZnTe nanocrystals with quantum confinement effect. Chemical Communications, 101-102 (2001).
136. Hefetz, Y. et al. Optical properties of ZnSe/(Zn,Mn)Se multiquantum wells. Applied Physics Letters 47, 989-991 (1985).
137. Holtz, P. O., Monemar, B. & Loykowski, H. J. Optical properties of Ag-related centers in bulk ZnSe. Physical Review B 32, 986-996 (1985).
138. Bhargava, R. N., Gallagher, D., Hong, X. & Nurmikko, A. Optical properties of manganese-doped nanocrystals of ZnS. Physical Review Letters 72, 416-419 (1994).
139. Bol, A. A. & Meijerink, A. Luminescence quantum efficiency of nanocrystalline ZnS: Mn2+. 1. Surface passivation and Mn2+ concentration. Journal of Physical Chemistry B 105, 10197-10202 (2001).
140. Bol, A. A. & Meijerink, A. Luminescence quantum efficiency of nanocrystalline ZnS: Mn2+. 2. Enhancement by UV irradiation. Journal of Physical Chemistry B 105, 10203-10209 (2001).
141. Schwartz, D. A., Norberg, N. S., Nguyen, Q. P., Parker, J. M. & Gamelin, D. R. Magnetic quantum dots: Synthesis, spectroscopy, and magnetism of CO2+- and Ni2+-doped ZnO nanocrystals. Journal of the American Chemical Society 125, 13205-13218 (2003).
142. Erwin, S. C. et al. Doping semiconductor nanocrystals. Nature 436, 91-94 (2005).
143. Kumar, S. (Albert Ludwigs-Universitat, Freiburg im Breisgan, 2004).
144. Yamazaki, K. et al. Long term pulmonary toxicity of indium arsenide and indium phosphide instilled intratracheally in hamsters. Journal of Occupational Health 42, 169-178 (2000).

145. Omura, M. et al. Changes in the testicular damage caused by indium arsenide and indium phosphide in hamsters during two years after intratracheal instillations. Journal of Occupational Health 42, 196-204 (2000).
146. Lucey, D. W. et al. Monodispersed InP quantum dots prepared by colloidal chemistry in a noncoordinating solvent. Chemistry of Materials 17, 3754-3762 (2005).
147. Huisken, F., Ledoux, G., Guillois, O. & Reynaud, C. Light-emitting silicon nanocrystals from laser pyrolysis. Advanced Materials 14, 1861-1865 (2002).
148. Canham, L. T. Silicon quantum wire array fabrication by electrochemical and chemical dissolution of wafers. Applied Physics Letters 57, 1046-1048 (1990).
149. Lehmann, V. & Gösele, U. Porous silicon formation: A quantum wire effect. Applied Physics Letters 58, 856-858 (1991).
150. Fauchet, P. M. Photoluminescence and electroluminescence from porous silicon. Journal of Luminescence 70, 294-309 (1996).
151. Cullis, A. G., Canham, L. T. & Calcott, P. D. J. The structural and luminescence properties of porous silicon. Journal of Applied Physics 82, 909 (1997).
152. Brus, L. E. et al. Electronic Spectroscopy and Photophysics of Si Nanocrystals: Relationship to Bulk c-Si and Porous Si. Journal of the American Chemical Society 117, 2915 - 2922 (1995).
153. Ehbrecht, M., Kohn, B., Huisken, F., Laguna, M. A. & Paillard, V. Photoluminescence and resonant Raman spectra of silicon films produced by size-selected cluster beam deposition. Physical Review B 56, 6958-6964 (1997).
154. Ledoux, G. et al. Photoluminescence properties of silicon nanocrystals as a function of their size. Physical Review B 62, 15942-15951 (2000).
155. Ledoux, G., Gong, J., Huisken, F., Guillois, O. & Reynaud, C. Photoluminescence of size-separated silicon nanocrystals: Confirmation of quantum confinement. Applied Physics Letters 80, 4834-4836 (2002).
156. Yamani, Z., Ashhab, S., Nayfeh, A., Thompson, W. H. & Nayfeh, M. Red to green rainbow photoluminescence from unoxidized silicon nanocrystallites. Journal of Applied Physics 83, 3929-3931 (1998).
157. Li, X. G., He, Y. Q., Talukdar, S. S. & Swihart, M. T. Process for preparing macroscopic quantities of brightly photoluminescent silicon nanoparticles with emission spanning the visible spectrum. Langmuir 19, 8490-8496 (2003).
158. Belomoin, G. et al. Observation of a magic discrete family of ultrabright Si nanoparticles. Applied Physics Letters 80, 841-843 (2002).
159. Hua, F. J., Swihart, M. T. & Ruckenstein, E. Efficient surface grafting of luminescent silicon quantum dots by photoinitiated hydrosilylation. Langmuir 21, 6054-6062 (2005).
160. Hua, F. J., Erogbogbo, F., Swihart, M. T. & Ruckenstein, E. Organically capped silicon nanoparticles with blue photoluminescence prepared by hydrosilylation followed by oxidation. Langmuir 22, 4363-4370 (2006).

161. Li, Z. F. & Ruckenstein, E. Water-soluble poly(acrylic acid) grafted luminescent silicon nanoparticles and their use as fluorescent biological staining labels. Nano Letters 4, 1463-1467 (2004).
162. Pettigrew, K. A., Liu, Q., Power, P. P. & Kauzlarich, S. M. Solution synthesis of alkyl- and alkyl/alkoxy-capped silicon nanoparticles via oxidation of Mg2Si. Chemistry of Materials 15, 4005-4011 (2003).
163. Fojtik, A. & Henglein, A. Surface chemistry of luminescent colloidal silicon nanoparticles. Journal of Physical Chemistry B 110, 1994-1998 (2006).
164. Wang, L., Reipa, V. & Blasic, J. Silicon nanoparticles as a luminescent label to DNA. Bioconjugate Chemistry 15, 409-412 (2004).
165. Xu, X. Y. et al. Electrophoretic analysis and purification of fluorescent single-walled carbon nanotube fragments. Journal of the American Chemical Society 126, 12736-12737 (2004).
166. Peng, H. Q., Alvarez, N. T., Kittrell, C., Hauge, R. H. & Schmidt, H. K. Dielectrophoresis field flow fractionation of single-walled carbon nanotubes. Journal of the American Chemical Society 128, 8396-8397 (2006).
167. Yu, S. J., Kang, M. W., Chang, H. C., Chen, K. M. & Yu, Y. C. Bright fluorescent nanodiamonds: No photobleaching and low cytotoxicity. Journal of the American Chemical Society 127, 17604-17605 (2005).
168. Parak, W. J., Manna, L., Simmel, F. C., Gerion, D. & Alivisatos, P. in Nanoparticles - From Theory to Application (ed. Schmid, G.) 4-49 (Wiley-VCH, Weinheim, 2004).
169. Bawendi, M. G., Steigerwald, M. L. & Brus, L. E. The quantum mechanics of large semiconductor clusters ("quantum dots"). Annual Review of Physical Chemistry 41, 477-496 (1990).
170. Alivisatos, A. P. Nanocrystals: building blocks for modern materials design. Endeavour 21, 56-60 (1997).
171. Efros, A. L. & Rosen, M. The electronic structure of semiconductor nanocrystals. Annual Review of Materials Science 30, 475-521 (2000).
172. Sapra, S. & Sarma, D. D. Evolution of the electronic structure with size in II-VI semiconductor nanocrystals. Physical Review B 69, - (2004).
173. Brus, L. E. Electron-electron and electron-hole interactions in small semiconductor crystallites: the size dependence of the lowest excited electronic state. Journal of Chemical Physics 80, 4403-4409 (1984).
174. Brus, L. Electronic Wave Functions in Semiconductor Clusters: Experiment and Theory. Journal of Physical Chemistry 90, 2555-2560 (1986).
175. Murray, C. B., Kagan, C. R. & Bawendi, M. G. Synthesis and characterization of monodisperse nanocrystals and close-packed nanocrystal assemblies. Annual Review of Materials Science 30, 545-610 (2000).
176. Qu, L. H. & Peng, X. G. Control of photoluminescence properties of CdSe nanocrystals in growth. Journal of the American Chemical Society 124, 2049-2055 (2002).

177. Kippeny, T., Swafford, L. A. & Rosenthal, S. J. Semiconductor nanocrystals: A powerful visual aid for introducing the particle in a box. Journal of Chemical Education 79, 1094-1100 (2002).
178. Yu, W. W., Qu, L. H., Guo, W. Z. & Peng, X. G. Experimental determination of the extinction coefficient of CdTe, CdSe, and CdS nanocrystals. Chemistry of Materials 15, 2854-2860 (2003).
179. Cademartiri, L. et al. Size-dependent extinction coefficients of PbS quantum dots. Journal of the American Chemical Society 128, 10337-10346 (2006).
180. Dahan, M. et al. Time-gated biological imaging by use of colloidal quantum dots. Opt. Letters 26, 825-827 (2001).
181. Larson, D. R. et al. Water-soluble quantum dots for multiphoton fluorescence imaging in vivo. Science 300, 1434-1436 (2003).
182. Henglein, A. Small-particle research: physicochemical properties of extremely small colloidal metal and semiconductor particles. Chemical Reviews 89, 1861 - 1873 (1989).
183. Nirmal, M. & Brus, L. Luminescence Photophysics in Semiconductor Nanocrystals. Accounts of Chemical Research 32, 407-414 (1999).
184. Alivisatos, A. P. Perspectives on the Physical Chemistry of Semiconductor Nanocrystals. Journal of Physical Chemistry A 100, 13226-13239 (1996).
185. Weller, H. Quantized semiconductor particles: a novel state of matter for materials science. Advanced Materials 5, 88-95 (1993).
186. Weller, H. Colloidal Semiconductor Q-Particles: Chemistry in the Transition Region Between Solid State and Molecules. Angewandte Chemie International Edition 32, 41-53 (1993).
187. Mattheakis, L. C. et al. Optical coding of mammalian cells using semiconductor quantum dots. Analytical Biochemistry 327, 200-208 (2004).
188. Rosenthal, S. J. Bar-coding biomolecules with fluorescent nanocrystals. Nature Biotechnology 19, 621-622 (2001).
189. Han, M., Gao, X., Su, J. Z. & Nie, S. Quantum-dot-tagged microbeads for multiplexed optical coding of biomolecules. Nature Biotechnology 19, 631-635 (2001).
190. Cai, W. B. et al. Peptide-labeled near-infrared quantum dots for imaging tumor vasculature in living subjects. Nano Letters 6, 669-676 (2006).
191. Joumaa, N., Lansalot, M., Theretz, A. & Elaissari, A. Synthesis of quantum dot-tagged submicrometer polystyrene particles by miniemulsion polymerization. Langmuir 22, 1810-1816 (2006).
192. Parak, W. J. et al. Cell motility and metastatic potential studies based on quantum dot imaging of phagokinetic tracks. Advanced Materials 14, 882-885 (2002).
193. Cordero, S. R., Carson, P. J., Estabrook, R. A., Strouse, G. F. & Buratto, S. K. Photo-activated luminescence of CdSe quantum dot monolayers. Journal of Physical Chemistry B 104, 12137-12142 (2000).

194. Jones, M., Nedeljkovic, J., Ellingson, R. J., Nozik, A. J. & Rumbles, G. Photoenhancement of luminescence in colloidal CdSe quantum dot solutions. Journal of Physical Chemistry B 107, 11346-11352 (2003).
195. Michalet, X. et al. Properties of fluorescent semiconductor nanocrystals and their application to biological labeling. Single Molecules 2, 261-276 (2001).
196. Nirmal, M. et al. Fluorescence Intermittency in Single Cadmium Selenide Nanocrystals. Nature 383, 802-804 (1996).
197. Efros, A. L. & Rosen, M. Random Telegraph Signal in the Photoluminescence Intensity of a Single Quantum Dot. Physical Review Letters 78, 1110-1113 (1997).
198. Kuno, M., Fromm, D. P., Hamann, H. F., Gallagher, A. & Nesbitt, D. J. Nonexponential "blinking" kinetics of single CdSe quantum dots: A universal power law behavior. Journal of Chemical Physics 112, 3117-3120 (2000).
199. Koberling, F., Mews, A. & Basche, T. Oxygen-induced blinking of single CdSe nanocrystals. Advanced Materials 13, 672-676 (2001).
200. van Sark, W. G. J. H. M., Frederix, P. L. T. M., Bol, A. A., Gerritsen, H. C. & Meijerink, A. Blueing, bleaching, and blinking of single CdSe/ZnS quantum dots. Chemphyschem 3, 871-879 (2002).
201. Kondoh, M. et al. Cadmium induces apoptosis partly via caspase-9 activation in HL-60 cells. Toxicology 170, 111-117 (2002).
202. Shen, H.-M., Yang, C.-F. & Ong, C.-N. Sodium selenite-induced oxidative stress and apoptosis in human hepatoma HepG2 cells. International Journal of Cancer 81, 820-828 (1999).
203. Rikans, L. E. & Yamano, T. Mechanisms of cadmium-mediated acute hepatotoxicity. Journal of Biochemical and Molecular Toxicology 14, 110-117 (2000).
204. Mattheakis, L. C. et al. Optical coding of mammalian cells using semiconductor quantum dots. Analytical Biochemistry 327, 200-208 (2004).
205. Selvan, S. T., Tan, T. T. & Ying, J. Y. Robust, non-cytotoxic, silica-coated CdSe quantum dots with efficient photoluminescence. Advanced Materials 17, 1620-1625 (2005).
206. Zhang, T. T. et al. Cellular effect of high doses of silica-coated quantum dot profiled with high throughput gene expression analysis and high content cellomics measurements. Nano Letters 6, 800-808 (2006).
207. Parak, W. J., Pellegrino, T. & Plank, C. Labelling of cells with quantum dots. Nanotechnology 16, R9-R25 (2005).
208. Schulze, E. et al. Cellular uptake and trafficking of a prototypical magnetic iron oxide label in vitro. Investigative Radiology 30, 604-610 (1995).
209. Weissleder, R. et al. Superparamagnetic iron oxide: Pharmacokinetics and toxicity. American Journal of Roentgenology 152, 167-173 (1989).
210. Whitaker, M. in Cell Biology - A Laboratory Handbook (ed. Celis, J. E.) 37-43 (Academic Press, San Diego, California, 1994).

211. Osborn, M. in Cell Biology - A Laboratory Handbook (ed. Celis, J. E.) 347-354 (Academic Press, San Diego, California, 1994).
212. Herzog, M., Draeger, A., Ehler, E. & Small, J. V. in Cell Biology - A Laboratory Handbook (ed. Celis, J. E.) 355-360 (Academic Press, San Diego, California, 1994).
213. Derfus, A. M., Chan, W. C. W. & Bhatia, S. N. Intracellular delivery of quantum dots for live cell labeling and organelle tracking. Advanced Materials 16, 961-+ (2004).
214. Minet, O., Dressler, C. & Beuthan, J. Heat stress induced redistribution of fluorescent quantum dots in breast tumor cells. Journal of Fluorescence 14, 241-247 (2004).
215. Kaul, Z. et al. Mortalin imaging in normal and cancer cells with quantum dot immuno-conjugates. Cell Research 13, 503-507 (2003).
216. Rosenthal, S. J. et al. Targeting cell surface receptors with ligand-conjugated nanocrystals. Journal of the American Chemical Society 124, 4586-4594 (2002).
217. Howarth, M., Takao, K., Hayashi, Y. & Ting, A. Y. Targeting quantum dots to surface proteins in living cells with biotin ligase. Proceedings of the National Academy of Sciences of the United States of America 102, 7583-7588 (2005).
218. van Tilborg, G. A. F. et al. Annexin A5-conjugated quantum dots with a paramagnetic lipidic coating for the multimodal detection of apoptotic cells. Bioconjugate Chemistry 17, 865-868 (2006).
219. Courty, S., Luccardini, C., Bellaiche, Y., Cappello, G. & Dahan, M. Tracking individual kinesin motors in living cells using single quantum-dot imaging. Nano Letters 6, 1491-1495 (2006).
220. Muthukrishnan, G., Hutchins, B. M., Williams, M. E. & Hancock, W. O. Transport of semiconductor nanocrystals by kinesin molecular motors. Small 2, 626-630 (2006).
221. Meier, J., Vannier, C., Serge, A., Triller, A. & Choquet, D. Fast and reversible trapping of surface glycine receptors by gephyrin. Nature Neuroscience 4, 253-260 (2001).
222. Aoyama, Y. et al. Artificial viruses and their application to gene delivery. size-controlled gene coating with glycocluster nanoparticles. Journal of the American Chemical Society 125, 3455-3457 (2003).
223. Mellman, I. & Warren, G. The road taken: Past and future foundations of membrane traffic. Cell 100, 99-112 (2000).
224. Munro, S. Organelle identity and the organization of membrane traffic. Nature Cell Biology 6, 469-472 (2004).
225. Behnia, R. & Munro, S. Organelle identity and the signposts for membrane traffic. Nature 438, 597-604 (2005).
226. Mousavi, S. A., Malerod, L., Berg, T. & Kjeken, R. Clathrin-dependent endocytosis. Biochemical Journal 377, 1-16 (2004).

227. Pelkmans, L. & Helenius, A. Insider information: what viruses tell us about endocytosis. Current Opinion in Cell Biology 15, 414-422 (2003).
228. Wagner, E., Curiel, D. & Cotten, M. Delivery Of Drugs, Proteins And Genes Into Cells Using Transferrin As A Ligand For Receptor-Mediated Endocytosis. Advanced Drug Delivery Reviews 14, 113-135 (1994).
229. Widera, A., Norouziyan, F. & Shen, W. C. Mechanisms of TfR-mediated transcytosis and sorting in epithelial cells and applications toward drug delivery. Advanced Drug Delivery Reviews 55, 1439-66 (2003).
230. Niidome, T. & Huang, L. Gene therapy progress and prospects: nonviral vectors. Gene Ther 9, 1647-52 (2002).
231. Cotten, M. & Wagner, E. Non-Viral Approaches to Gene Therapy. Current Opinion in Biotechnology 4, 705-710 (1993).
232. Wagner, E., Plank, C., Zatloukal, K., Cotten, M. & Birnstiel, M. L. Influenza virus hemagglutinin HA-2 N-terminal fusogenic peptides augment gene transfer by transferrin-polylysine-DNA complexes: toward a synthetic virus-like gene-transfer vehicle. Proc Natl Acad Sci U S A 89, 7934-8 (1992).
233. Plank, C., Oberhauser, B., Mechtler, K., Koch, C. & Wagner, E. The Influence of Endosome-disruptive peptides on Gene Transfer Using Synthetic Virus-like Gene Transfer Systems. Journal of Biological Chemistry 269, 12918-12924 (1994).
234. Plank, C., Zauner, W. & Wagner, E. Application of membrane-active peptides for drug and gene delivery across cellular membranes. Advanced Drug Delivery Reviews 34, 21-35 (1998).
235. Boussif, O. et al. A versatile vector for gene and oligonucleotide transfer into cells in culture and in vivo: polyethylenimine. Proc Natl Acad Sci U S A 92, 7297-301. (1995).
236. Sonawane, N. D., Szoka, F. C., Jr. & Verkman, A. S. Chloride accumulation and swelling in endosomes enhances DNA transfer by polyamine-DNA polyplexes. J Biol Chem 278, 44826-31 (2003).
237. Xu, Y. & Szoka, F. C., Jr. Mechanism of DNA release from cationic liposome/DNA complexes used in cell transfection. Biochemistry 35, 5616-23 (1996).
238. Parak, W. et al. Biological applications of colloidal nanocrystals. Nanotechnology 14, R15-R27 (2003).
239. Pellegrino, T. et al. Quantum dot-based cell motility assay. Differentiation 71, 542-548 (2003).
240. Bentzen, E. L. et al. Surface modification to reduce nonspecific binding of quantum dots in live cell assays. Bioconjugate Chemistry 16, 1488-1494 (2005).
241. Chen, F. Q. & Gerion, D. Fluorescent CdSe/ZnS nanocrystal-peptide conjugates for long-term, nontoxic imaging and nuclear targeting in living cells. Nano Letters 4, 1827-1832 (2004).
242. Delehanty, J. B. et al. Self-assembled quantum dot-peptide bioconjugates for selective intracellular delivery. Bioconjugate Chemistry 17, 920-927 (2006).

243. Hoshino, A. et al. Quantum dots targeted to the assigned organelle in living cells. Microbiology and Immunology 48, 985-994 (2004).
244. Kim, Y. et al. Targeting heat shock proteins on cancer cells: Selection, characterization, and cell-penetrating properties of a peptidic GRP78 ligand. Biochemistry 45, 9434-9444 (2006).
245. Nan, X. L., Sims, P. A., Chen, P. & Xie, X. S. Observation of individual microtubule motor steps in living cells with endocytosed quantum dots. Journal of Physical Chemistry B 109, 24220-24224 (2005).
246. Rajan, S. S. & Vu, T. Q. Quantum Dots Monitor TrkA Receptor Dynamics in the Interior of Neural PC12 Cells. Nano Letters 6, 2049 - 2059 (2006).
247. Zhang, Y., So, M. K. & Rao, J. Protease-Modulated Cellular Uptake of Quantum Dots. Nano Letters 6, 1988-1992 (2006).
248. Pearson, H. Developmental biology: Your destiny, from day one. Nature 417, 14-15 (2002).
249. Kubo, A. et al. Development of definitive endoderm from embryonic stem cells in culture. Development 131, 1651-62 (2004).
250. Ferrari, A. et al. In vivo tracking of bone marrow fibroblasts with fluorescent carbocyanine dye. Journal of Biomedical Materials Research 56, 361-367 (2001).
251. Tombolini, R. & Jansson, J. K. in Bioluminescence Methods and Protocols (ed. LaRossa, R. A.) 285-298 (Humana Press, Totowa, New Jersey, 1998).
252. Brenner, M. K. et al. Gene-marking to trace origin of relapse after autologous bone-marrow transplantation. Lancet 341, 85-86 (1993).
253. Chattopadhyay, P. K. et al. Quantum dot semiconductor nanocrystals for immunophenotyping by polychromatic flow cytometry. Nature Medicine 12, 972-977 (2006).
254. Zhelev, Z. et al. Fabrication of quantum dot-lectin conjugates as novel fluorescent probes for microscopic and flow cytometric identification of leukemia cells from normal lymphocytes. Chemical Communications, 1980-1982 (2005).
255. Voura, E. B., Jaiswal, J. K., Mattoussi, H. & Simon, S. M. Tracking metastatic tumor cell extravasation with quantum dot nanocrystals and fluorescence emission-scanning microscopy. Nature Medicine 10, 993-998 (2004).
256. Albrecht-Buehler, G. & Lancaster, R. M. A quantitative description of the extension and retraction of surface protrusions in spreading 3T3 mouse fibroblasts. The Journal of Cell Biology 71, 370-382 (1976).
257. Albrecht-Buehler, G. The Phagokinetic Tracks of 3T3 Cells. Cell 11, 395-404 (1977).
258. Albrecht-Buehler, G. Phagokinetic Tracks of 3T3 Cells: Parallels between the Orientation of Track Segments and of Cellular Structures Which Contain Actin or Tubulin. Cell 12, 333-339 (1977).
259. Chen, J. D., Helmold, M., Kim, J. P., Wynn, K. C. & Woodley, D. T. Human Keratincytes Make Uniquely Linear Phagokinetic Tracks. Dermatology 188, 6-12 (1994).

260. Stokes, C. L., Lauffenburger, D. A. & Williams, S. K. Migration of individual microvessel endothelial cells: stochastic model and parameter measurement. Journal of Cell Science 99, 41-430 (1991).
261. Forster, T. Transfer mechanisms of electronic excitation (Discuss. Faraday Soc., 1959).
262. Lakowicz, J. R. Principles of Fluorescence Spectroscopy (Kluwer/Plenum, New York, 1999).
263. Sapsford, K. E., Berti, L. & Medintz, I. L. Materials for fluorescence resonance energy transfer analysis: Beyond traditional donor-acceptor combinations. Angewandte Chemie-International Edition 45, 4562-4588 (2006).
264. Iqbal, S. S. et al. A review of molecular recognition technologies for detection of biological threat agents. Biosensors and Bioelectronics 15, 549 (2000).
265. O'Connell, P. J. & Guilbault, G. G. Future trends in biosensor research. Analytical Letters 34, 1063-1078 (2001).
266. De Lorimier, R. M. et al. Construction of a fluorescent biosensor family. Protein Science 11, 2655 (2002).
267. Scheller, F. W., Wollenberger, U., Warsinke, A. & Lisdat, F. Research and development in biosensors. Current Opinion in Biotechnology 12, 35-40 (2001).
268. Mamedova, N. N., Kotov, N. A., Rogach, A. L. & Studer, J. Albumin-CdTe nanoparticle bioconjugates: Preparation, structure, and interunit energy transfer with antenna effect. Nano Letters 1, 281-286 (2001).
269. Tran, P. T., Goldman, E. R., Anderson, G. P., Mauro, J. M. & Mattoussi, H. Use of luminescent CdSe-ZnS nanocrystal bioconjugates in quantum dot-based nanosensors. Physica Status Solidi B-Basic Research 229, 427-432 (2002).
270. Medintz, I. L. et al. Self-assembled nanoscale biosensors based on quantum dot FRET donors. Nature Materials 2, 630-638 (2003).
271. Medintz, I. L. et al. Proteolytic activity monitored by fluorescence resonance energy transfer through quantum-dot-peptide conjugates. Nature Materials 5, 581-589 (2006).
272. Clapp, A. R. et al. Fluorescence resonance energy transfer between quantum dot donors and dye-labeled protein acceptors. Journal of the American Chemical Society 126, 301-310 (2004).
273. Medintz, I. L., Trammell, S. A., Mattoussi, H. & Mauro, J. M. Reversible modulation of quantum dot photoluminescence using a protein-bound photochromic fluorescence resonance energy transfer acceptor. Journal of the American Chemical Society 126, 30-31 (2004).
274. Medintz, I. L. et al. A fluorescence resonance energy transfer-derived structure of a quantum dot-protein bioconjugate nanoassembly. Proceedings of the National Academy of Sciences of the United States of America 101, 9612-9617 (2004).
275. Medintz, I. L., Uyeda, H. T., Goldman, E. R. & Mattoussi, H. Quantum dot bioconjugates for imaging, labelling and sensing. Nature Materials 4, 435-446 (2005).

276. Clapp, A. R., Medintz, I. L. & Mattoussi, H. Forster resonance energy transfer investigations using quantum-dot fluorophores. Chemphyschem 7, 47-57 (2006).
277. Nie, Q. L., Tan, W. B. & Zhang, Y. Synthesis and characterization of monodisperse chitosan nanoparticles with embedded quantum dots. Nanotechnology 17, 140-144 (2006).
278. Shi, L. F., De Paoli, V., Rosenzweig, N. & Rosenzweig, Z. Synthesis and application of quantum dots FRET-based protease sensors. Journal of the American Chemical Society 128, 10378-10379 (2006).
279. Patolsky, F. et al. Lighting-up the dynamics of telomerization and DNA replication by CdSe-ZnS quantum dots. Journal of the American Chemical Society 125, 13918-13919 (2003).
280. Zhang, C. Y. & Johnson, L. W. Homogenous rapid detection of nucleic acids using two-color quantum dots. Analyst 131, 484-488 (2006).
281. Kim, J. H., Morikis, D. & Ozkan, M. Adaptation of inorganic quantum dots for stable molecular beacons. Sensors and Actuators B-Chemical 102, 315-319 (2004).
282. Dyadyusha, L. et al. Quenching of CdSe quantum dot emission, a new approach for biosensing. Chemical Communications, 3201-3203 (2005).
283. Gill, R., Willner, I., Shweky, I. & Banin, U. Fluorescence resonance energy transfer in CdSe/ZnS-DNA conjugates: Probing hybridization and DNA cleavage. Journal of Physical Chemistry B 109, 23715-23719 (2005).
284. Bakalova, R., Zhelev, Z., Ohba, H. & Baba, Y. Quantum dot-conjugated hybridization probes for preliminary screening of siRNA sequences. Journal of the American Chemical Society 127, 11328-11335 (2005).
285. Dwarakanath, S. et al. Quantum dot-antibody and aptamer conjugates shift fluorescence upon binding bacteria. Biochemical and Biophysical Research Communications 325, 739-743 (2004).
286. Levy, M., Cater, S. F. & Ellington, A. D. Quantum-dot aptamer beacons for the detection of proteins. Chembiochem 6, 2163-2166 (2005).
287. Hansen, J. A. et al. Quantum-dot/aptamer-based ultrasensitive multi-analyte electrochemical biosensor. Journal of the American Chemical Society 128, 2228-2229 (2006).
288. Famulok, M. & Mayer, G. Chemical biology - Aptamers in nanoland. Nature 439, 666-669 (2006).
289. Liu, J. W. & Lu, Y. Fast colorimetric sensing of adenosine and cocaine based on a general sensor design involving aptamers and nanoparticles. Angewandte Chemie-International Edition 45, 90-94 (2006).
290. Kagan, C. R., Murray, C. B., Nirmal, M. & Bawendi, M. G. Electronic energy transfer in CdSe quantum dot solids. Physical Review Letters 76, 1517-1520 (1996).

291. Kagan, C. R., Murray, C. B. & Bawendi, M. G. Long-range resonance transfer of electronic excitations in close-packed CdSe quantum-dot solids. PHYSICAL REVIEW B 54, 8633-8643 (1996).
292. Clapp, A. R., Medintz, I. L., Fisher, B. R., Anderson, G. P. & Mattoussi, H. Can luminescent quantum dots be efficient energy acceptors with organic dye donors? Journal of the American Chemical Society 127, 1242-1250 (2005).
293. So, M. K., Xu, C. J., Loening, A. M., Gambhir, S. S. & Rao, J. H. Self-illuminating quantum dot conjugates for in vivo imaging. Nature Biotechnology 24, 339-343 (2006).
294. Wilson, T. & Hastings, J. W. Bioluminescence. Annu Rev Cell Dev Biol 14, 197-230 (1998).
295. De, A. & Gambhir, S. S. Noninvasive imaging of protein-protein interactions from live cells and living subjects using bioluminescence resonance energy transfer. Faseb Journal 19, - (2005).
296. Evanko, D. Bioluminescent quantum dots. Nature Methods 3, 240-241 (2006).
297. Frangioni, J. V. Self-illuminating quantum dots light the way. Nature Biotechnology 24, 326-328 (2006).
298. Zhang, Y. et al. HaloTag protein-mediated site-specific conjugation of bioluminescent proteins to quantum dots. Angewandte Chemie-International Edition 45, 4936-4940 (2006).
299. Huang, X. Y., Li, L., Qian, H. F., Dong, C. Q. & Ren, J. C. A resonance energy transfer between chemiluminescent donors and luminescent quantum-dots as acceptors (CRET). Angewandte Chemie-International Edition 45, 5140-5143 (2006).
300. Weissleder, R. Scaling down imaging: molecular mapping of cancer in mice. Nat Rev Cancer 2, 11-8 (2002).
301. Weissleder, R., Tung, C. H., Mahmood, U. & Bogdanov, A., Jr. In vivo imaging of tumors with protease-activated near-infrared fluorescent probes. Nat Biotechnol 17, 375-8 (1999).
302. Contag, C. H. & Ross, B. D. It's not just about anatomy: in vivo bioluminescence imaging as an eyepiece into biology. J Magn Reson Imaging 16, 378-87 (2002).
303. Sevick-Muraca, E. M., Houston, J. P. & Gurfinkel, M. Fluorescence-enhanced, near infrared diagnostic imaging with contrast agents. Curr Opin Chem Biol 6, 642-50 (2002).
304. Frangioni, J. V. In vivo near-infrared fluorescence imaging. Curr Opin Chem Biol 7, 626-34 (2003).
305. Ntziachristos, V., Bremer, C. & Weissleder, R. Fluorescence imaging with near-infrared light: new technological advances that enable in vivo molecular imaging. Eur Radiol 13, 195-208 (2003).
306. Kim, S. et al. Near-infrared fluorescent type II quantum dots for sentinel lymph node mapping. Nature Biotechnology 22, 93-97 (2004).

307. Bakueva, L. et al. PbS quantum dots with stable efficient luminescence in the near-IR spectral range. Advanced Materials 16, 926-929 (2004).
308. Fischer, H. C., Liu, L., Pang, K. S. & Chan, W. C. W. Pharmacokinetics of Nanoscale Quantum Dots: In Vivo Distribution, Sequestration, and Clearance in the Rat. Advanced Functional Materials 16, 1299-1305 (2006).
309. Everts, M. et al. Covalently linked au nanoparticles to a viral vector: Potential for combined photothermal and gene cancer therapy. Nano Letters 6, 587-591 (2006).
310. Chen, C. et al. Nanoparticle-templated assembly of viral protein cages. Nano Letters 6, 611-615 (2006).
311. Niemeyer, C. M. Nanoparticles, proteins, and nucleic acids: Biotechnology meets materials science. Angewandte Chemie-International Edition 40, 4128-4158 (2001).
312. Sperling, R. A., Pellegrino, T., Li, J. K., Chang, W. H. & Parak, W. J. Electrophoretic separation of nanoparticles with a discrete number of functional groups. Advanced Functional Materials 16, 943-948 (2006).
313. Worden, J. G., Dai, Q., Shaffer, A. W. & Huo, Q. Monofunctional group-modified gold nanoparticles from solid phase synthesis approach: Solid support and experimental condition effect. Chemistry of Materials 16, 3746-3755 (2004).
314. Worden, J. G., Shaffer, A. W. & Huo, Q. Controlled functionalization of gold nanoparticles through a solid phase synthesis approach. Chemical Communications, 518-519 (2004).
315. Liu, X. et al. Monofunctional gold nanoparticles prepared via a noncovalent-interaction-based solid-phase modification approach. Small 2, 1126-1129 (2006).
316. Zanchet, D. et al. Electrophoretic and Structural Studies of DNA-Directed Au Nanoparticle Groupings. Journal of Physical Chemistry B 106, 11758-11763 (2002).
317. Smith, A. M., Ruan, G., Rhyner, M. N. & Nie, S. M. Engineering luminescent quantum dots for In vivo molecular and cellular imaging. Annals of Biomedical Engineering 34, 3-14 (2006).
318. Giepmans, B. N. G., Deerinck, T. J., Smarr, B. L., Jones, Y. Z. & Ellisman, M. H. Correlated light and electron microscopic imaging of multiple endogenous proteins using Quantum dots. Nature Methods 2, 743-749 (2005).

CHAPTER 12

DNA-BASED ARTIFICIAL NANOSTRUCTURES

Giampaolo Zuccheri, Marco Brucale, Alessandra Vinelli, Bruno Samorì*

Department of Biochemistry, National Institute for the Physics of the Matter, Italian Interuniversity Consortium for Materials Science and Technology at the University of Bologna Via Irnerio, 48, Bologna,40126 - Italy

**E-mail: samori@alma.unibo.it*

A multitude of DNA-based nanostructures of different size and dimensionality have appeared in the literature in the last few years. The programmed self-assembly of oligodeoxynucleotides can be employed to build DNA nanostructures by design in a very successful approach towards bottom-up nanofabrication. This review paper will focus on the developments in this area of research and on some of the properties of DNA that elicited the birth of this new area of research.

1. Introduction

More than 50 years after the discovery of the structure of the DNA double-helix [1], our gratitude towards J. D. Watson and F. H. C. Crick is renewed daily, as the many wonders of this molecule never stop to suggest new pieces of research. On analyzing DNA fiber diffraction data, they described the canonical B structure, the most common in living organisms and in normal solution conditions. They had already envisaged that "the specific base pairing immediately suggests a possible copying mechanism for the genetic material" laying the foundations for the comprehension of the code underlying the functioning and heredity of all living organisms. DNA itself also controls the expression of codes written in its base sequence, for instance through the control of protein-recognition mechanisms that are based on the modulation of its structure and dynamics along the chain. Ultrastructural characterization techniques methods led towards the discovery that the codes contained in the DNA base-sequence rule these interactive processes from the atomic scale of the single base-pair level to the nanometer and micrometer scale-lengths of its superstructures.

The term “code” was defined by Trifonov as “any pattern or bias in the sequence which corresponds to one or another specific biological (biomolecular) function or interaction” [2]. The codes of DNA are generally chemical in nature, mostly structural: it is the stereochemistry of an interaction between a couple of aromatic systems that determines the base-pairing specificity. On a larger scale, the composition in space of many local chain deformations drives, for instance, the DNA wrapping around the histone proteins in nucleosomes in chromatin.

The informational codes of DNA can provide also serve as a toolbox of assembly information that can be used to switch self-organization among different length and energy scales. The field of DNA nanotechnology has been so far relying heavily on the Watson-Crick base-pairing code but there is room for much more, as other informational codes are also available to DNA.

In this review paper, we will work to the explanation of some of the properties of DNA structures along two directions: in the first part we will give a brief overview on some of the nanometer size properties of DNA and on some of the informational codes that can be written in it. In the second, more conspicuous section, we will describe a number of DNA-based nanostructures that have been recently presented, hopefully classifying them to help the reader in the comprehension of such diverse artificial constructs.

2. Affinity vs. Specificity in DNA Interactions

For all the applications of the DNA base-pairing, maximizing the affinity and the specificity of the Watson-Crick interaction code is particularly important. In those biomolecular interactions and recognitions that are based on shape complementarities, or steric fit between the two counterparts (enzyme-substrate, antigen-antibody, aptamer-small molecule complexes) both high specificity and high affinity can be achieved at the same time. A non-precise steric fit between two surfaces results in significant energetic penalties. The recognition mode and the association between two nucleic acid chains are based on a 1D nucleation-zipping mechanism instead [3,4]. Steric fit and nucleation zipping differ substantially. A strong zip with one irregular or missing link can still be fastened with high affinity leaving out the small mismatched part. The free energy loss is in this case very small and an increase in affinity does not affect this loss because the mismatched complex will be stabilized comparably. This mechanism results in a gradual decrease of nucleic acid hybridization specificity with increasing the binding affinity. A scientist that is designing a nucleic-acid probe or a molecular construction must bear in mind that a longer oligonucleotide is not always better, and that there might be alternative strategies for increasing the specificity and the affinity at the same time as brilliantly reviewed by Demidov and coworkers [3], such as using nucleic acid homologues or introducing additional energy penalties for a mismatched pairing. The first countermeasure usually taken when designing DNA molecules that need to pair efficiently is reducing the symmetrical and repetitive elements of the sequences. Computer software is available for this task, and it is commonly used for the design of DNA nanostructures.

3. Structural Codes for DNA in the Nanoscale: Shape and Dynamics

The base sequence of a DNA segment encodes for the average shape of DNA molecules and also for the dynamics of the chain. DNA is continuously morphing into shapes and structures that are slight modifications of the canonical B-form. Atomic Force Microscopy (AFM) micrographs can give hint of the apparently chaotic movements of single DNA molecules. A spread of DNA molecules on a surface yields images similar to the micrograph in Figure 1. No two macromolecules have the same shape and conformation, in spite of having the identical chemical composition. Contrary to the first impression, the dynamics that leads to such a variety of shapes is not random.

3.1. *The DNA Shape Code: How Local Deformations Can Affect the Average Molecular Shape*

A DNA molecule assumes a shape that is the result of the superimposition of the thermal fluctuations upon its intrinsic, lowest energy, structure, depending from its base sequence [5-7]. The average structure of dsDNA is a result of the sequence since the chemical inhomogeneities imparted by the different base pairs along the chain give rise to modulations of the orientations of the average planes of the base pairs. These orientations are commonly expressed in terms of the base-step orientational parameters: roll, tilt, and twist (see Figure 2). Considerable effort has gone into defining sets of these parameters corresponding to the lowest energy structures directed by the sequence. Donald Crothers has critically reviewed the recent achievements on the matter [8].

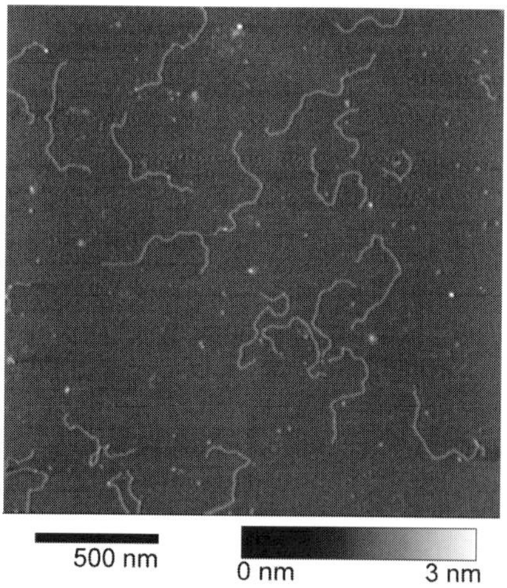

Figure 1. AFM image of a specimen of double-stranded DNA molecules. The heights of the features on the surface are coded in shades of color according to the attached look-up table. All the molecules (besides some obvious fragments) have the same base sequence and length, but they display a different shape due to the intrinsic flexibility of the polymers.

Variations in the roll or tilt angles give rise to bending of the double-helical axis. These local bends might just lead to a local zigzag pattern of the chain axis, which would remain globally straight on a larger scale, unless the deformations are composed in phase with the repeat of the helical winding. In this latter case, they might give rise to extended curvatures that propagate from the Ångström to the nanometer scale [9]. A notable natural example is the most highly curved DNA segment known in nature: the 211 bp segment from the kinetoplast DNA of the Trypanosomatidae Protozoan Crithidia fasciculata. Its sequence (reported in Figure 3) is characterized by a periodical recurrence of tracts of 3 to 6 adenines; spaced by 10 or 11 bp, i.e. the average helical repeat of B-DNA. This distribution of the A-tracts, perfectly phased with the helical winding, makes this short DNA segment have its lowest conformational energy when wrapped in a circle (see Figure 3). Electron microscopy evidence of such a large curvature was first presented by Jack Griffith [10].

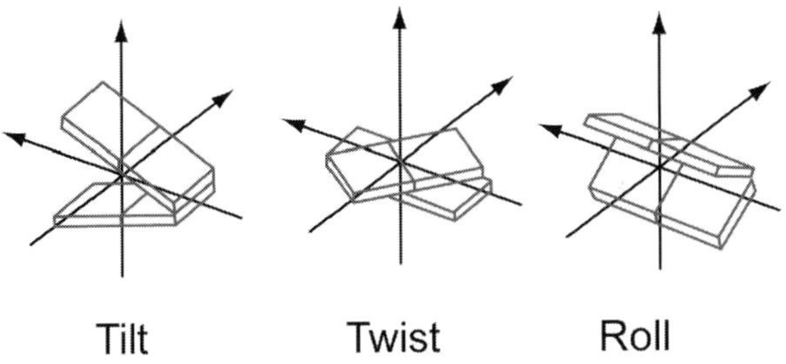

Figure 2. The most important dinucleotide-step orientational angles: combinations of values of these along the helix will determine the average shape of a DNA molecule.

Intrinsic curvatures have been monitored and studied by X-ray crystallography on very short double-stranded oligonucleotides [11] on longer DNA molecules by gel retardation [12,13], circularization kinetic [14-16], electron microscopy (EM) [17,18], atomic force microscopy (AFM) [19,20], and have been simulated by molecular dynamics [21]. Often, these experiments were carried out with peculiar dsDNA constructs, i.e. on constructs with (i) anomalous flexibility sites, like single-stranded stretches [20], internal loops due to mismatches [13], a single nick [22], a double-stranded linker connecting triple-helix tracts [23]; (ii) segments with very accurate phasing [20] or unphasing [6,14] of the adenine tracts with the helical periodicity. These experimental efforts usually derived conclusions on the local helical curvature from analyzing global parameters of the whole chain under investigation, like its persistence length, its end-to-end distance, or the cyclization J factor. The effect of defined sequence variants on the curvature was usually inferred from comparisons of the global parameters among sequences. A combinatorial approach has also been proposed [14].

GATCCCGCCT **AAAA**TTCCAA CCG**AAAA**TCG CGAGGTTACT
TTTTTGGAGC CCG**AAAA**CCA CCC**AAAA**TCA AGG**AAAAA**TG
GCC**AAAAAA**T GCC**AAAAAA**T AGCG**AAAA**TA CCCCG**AAAA**T
TGGC**AAAAA**T TAAC**AAAAAA** TAGCGAATTT CCCTGAATTT
TAGGCG**AAAA** AACCCCCGAA AATGGCCAAA AACGCACTGA
AAATC**AAAA**T CTGAACGTCT CGG

Figure 3. Below, the base sequence of the curved section of the kinetoplast DNA of Crithidia fasciculate, characterized by the phased repetitions of A-tracts. Above, the depiction of a computer model of its lowest energy structure. It can be appreciated that the A-tracts (red portions) are segregated on the face of the molecule looking towards the observer.

The trajectory of the double-helical axis of individual dsDNA chains deposited on a substrate can be traced with the aid of the EM or the AFM. It is possible to set up methods to map the intrinsic curvature along the chain of a natural DNA of any sequence, just by gathering a collection of single-molecule data from high-resolution microscopy imaging. The curvature can be calculated from the local angular chain deflections along a large number of profiles, averaging all the values over the ensemble of profiles [19].

The intrinsic DNA curvatures can be theoretically predicted and experimentally evaluated, as described above. Nearest-neighbor methods for the computation of axial curvature of DNA are currently available and the different sets of parameters (results of different methods and optimizations) are in good agreement for their general results [8,24-29].

Despite many efforts have been spent in this direction, the description of the origin of intrinsic curvatures at the atomic level remains somewhat disputed. Curvature is a long-range superstructural property that is more determined by the way the double helix composes and phases the local bends over different spatial scales, than by their individual values [30]. Certain flexibility in the sequence is allowed without producing serious changes in the average shape of the molecule.

3.2. DNA Flexibility: Curvature is Only Half of the Story (but the Story is Not Complete Yet)

The base sequence determines not only the global and local average shape of a dsDNA molecule but also its response to the thermal fluctuations. In this way, the base sequence controls the formation of conformers. A conformation, even if poorly populated, can play an important role in a biological function: it can be recognized and selected to switch on processes that the most stable structures might not be able to activate. One of the experimental observables that gives insight on the accessible conformational space of a chain is its local axial flexibility, (i.e., the tendency of the long axis of the double helix to deviate both locally and globally from a straight trajectory).

While a significant agreement is found in the literature on the origin and determinants of DNA curvature, the issue of DNA flexibility is still under debate. Debate could be originating from the different experimental methods and so the different viewpoints. Evidence gathered at atomic resolution might give information on the atomic determinants of some of the possible chain motions, while evidence collected at a larger size-scale might give more information on the global behavior of a molecule, not being able to interpret fully the high-resolution determinants of the resultant flexibility [30].

3.3. Surface-DNA Interactions can be Sequence-Dependent

Macromolecules can exert an exceptional degree of control over nucleation, phase stabilization, assembly, and pattern formation in inorganic structures [31-33]. Peptides that can show selectivity for binding to metal and metal oxide surfaces or that can recognize and control the growth of an inorganic semiconductor surface like that of GaAs have been selected [34-36]. Much interest in now attracted by the interfacing of inorganic and organic/biological materials and the full understanding of the rules for their interaction.

At first sight, a straight DNA chain can rotate around its axis on the surface, so many azimuthal orientations are expected to be equally probable and the chemical interactions with the crystal surface is averaged to a cylindrical symmetry. On the basis of this consideration, it is not expected that DNA should exhibit any sequence preference in its binding to inorganic surfaces, nor that any azimuthal orientation should be preferred. On a more careful analysis, it can become evident that the surface density of charges on the outer surface of DNA can be modulated by the nucleobases hidden in the interior of the double-helix [37]. On this ground, it is possible that, on interacting with a charged surface, there could be preferred azimuthal orientations of a straight DNA helix.

When a DNA helix is intrinsically curved, the curvature plane defines two faces for a DNA section. The chemical inhomogeneities that yield the curvature act also in such a way that the two resulting faces of DNA will expose sides of the molecule with a different chemical composition. If DNA curvature is induced by regular phasing of Adenine tracts (A-tracts) with the helical periodicity, then the two defined faces will vary in the richness of adenines that they could expose to a surface adsorbed on that face. One face will turn out to be rich in adenines and the opposite one will turn out to be rich in thymines. By properly exploiting the internal symmetry of tailor-made DNA molecules, we could show that the propensity in the adsorption of the different faces of highly curved DNA molecules can be significantly different, so that one face (the thymine rich, in our molecules) will adsorb up to 10 times more frequently that the other face on the surface of freshly-cleaved muscovite mica [38]. This recognition effect seems to be directly dependent on the

DNA curvature, itself related to the base sequence. We can hypothesize that a recognition process of this kind might have been relevant in pre-cellular stages of life evolution. Inorganic surfaces have served as catalysts of prebiotic syntheses [39] and also as templates for the self-organization of increasingly more complex bio-structures [40-44]. We expect that specific surface-biomolecule recognition processes, such as the ones mentioned above, might play a role in the bottom-up assembly of more and more complex nanostructures [45].

4. A Practical Application of the Watson-Crick DNA Code: DNA Chips and DNA Detection

One of the most established technical applications in which the base-pairing codes of DNA are exploited is genetic analysis. Here the most basic assembly, hybridization through Watson-Crick base pairing, is exploited for practical reasons. For research and diagnostic purposes, cells are scanned for the presence of genes, or for the level of expression of peculiar genetic products [46,47]. Parallel genetic testing of a specimen for many genes is currently performed on devices termed "sensing arrays" with multiple spots on a surface. Each spot exposes a different probe oligonucleotide that can hybridize with the DNA or RNA target. The read-out of the hybridization can be performed thanks to the introduction of fluorescent [46], electroactive [48-50], or nanoparticulate labels [51,52]. Several strategies nowadays go towards the fully electronic detection of (labeled or unlabeled) nucleic acids and proteins [53] but the sensitivity still needs improvement (more than 104 molecule of analyte required) to help the technology for simple-to-use, low-cost, point-of-care diagnostics. The market is hungry for lab-on-chip devices for applications to environmental, biohazard and bio-terrorism testing and also for the detection of pathogens in human patients.

It is nowadays possible to immobilize nucleic acids at surfaces through many strategies, like electropolymerization, streptavidin-biotin interactions, gold-thiol chemistry and other techniques [54]. There certainly is a drive towards PCR-free DNA arrays that could include fully-electronic detection methods for revealing unlabelled DNA molecules. This very important goal seems to strongly depend on the development (i) of nanoscience-based strategies for signal enhancement (ii) of methodologies to increase both the affinity and the specificity of the base pairing recognition processes. In the section about nanoparticle-based assemblies some is said about the research undergoing for the first theme, while much has already been done for the latter. A particularly clear overview about the very special interplay between affinity and selectivity in nucleic acid interaction can be found in the already cited paper by Demidov and Frank-Kamenetskii [3]. Some of the approaches towards increasing both affinity and specificity at the same time use oligonucleotides in novel topologies (e.g. circular, dendrimeric, nanoparticle-bound), while others employ newer types of oligonucleotide analogs (locked-nucleic acids, peptide nucleic acids).

5. Base-Pairing for Nanoscience and Nanotechnology

Nanoscience and nanotechnology are more and more extensively using the informational codes of DNA to create tools and molecular constructions. The controlled assembly of matter on the nanoscale is currently a foremost goal in several fields, including materials science, electronic engineering, and biosensors development. Self-assembled DNA nanoarchitectures provide a versatile tool to this end. Among the available materials, DNA offers possibilities unmatched by other molecules, due to the peculiar characteristics that make it work so well in the cell.

The specific recognition between complementary stretches of nucleotides to form a Watson-Crick double-chain has been used more and more extensively in nanoscience to organize specific recognition processes that are used to assemble nanoscale constructions made of DNA or of DNA and other components in which the DNA is the assembly tool. DNA enables the creation of constructions with a predictable structure. The constructs can reach an impressive level of complexity and efficiency: they can be made of up to hundreds different oligonucleotides that auto-assemble precisely.

The key feature of DNA is its ability to form bonds following known rules. Watson and Crick sequence-dependent complimentarity can be exploited to drive the intra- or inter-molecular self-assembly of DNA molecules into well-defined shapes. These shapes are potentially fairly persistent, since DNA is a rather rigid polymer [55]. DNA can be easily modified with extreme precision and versatility by synthetic chemistry and by a rich variety of modifying enzymes borrowed from living organisms. DNA molecules can be functionalized with different species such as metal nanoparticles, proteins, carbon nanotubes, organic dyes [56-64], and still retain their self-assembly abilities. The upper limit for the formation of different DNA-based nano-objects is only the researchers' fantasy.

Inspired by the intermediate forms that nucleic acids take during their functional lifetime inside cells, in 1982, Ned Seeman started to investigate the possibility of obtaining topologically and geometrically defined DNA branched nanostructures [65]. In particular, he employed the blocked Holliday junction [66] (Figure 4), and other junctions with three to six double-helical arms [67] to introduce stable branching points in DNA. This "escape" from the boundary of the intrinsic linearity of DNA means the possibility of obtaining arbitrary shapes from DNA-mediated self-assembly. The mechanical coupling of DNA strands connected by branching points allows the formation of frames with a rigidity surpassing by far that of an isolated double stranded chain, leading to the formation of structures with a deterministic control of the geometric shape.

Once the desired set of strands is appropriately designed (a feat nowadays achieved thanks to computer software) and synthesized, the actual assembly of the complete structure can be as simple as mixing all the components at high temperature and then letting them cool down in a near-equilibrium regime. Most of the ongoing development of DNA-based nanostructures aims at expanding their complexity over spatial and temporal dimensions through hierarchical integrations of elementary structural and functional units.

It may come naturally to the mind of the reader that errors might soon emerge in DNA-based nanostructures as their complexity increases. These can become so serious as to impair the structural integrity of the assembled object. Devising countermeasures to this inconvenience is an active field of research [68]. Mismatches occur when binding of components (oligonucleotides or a number of partial assemblies) takes place in an energetically sub-optimal configuration. It has been shown that arbitrarily low error rates can be achieved by appropriate control the assembly rate, although this necessarily occurs at the cost of a significant slow-down [69]. Decreasing error rates by a factor of 10 entails slowing down the self-assembly process by a factor of 100. Erik Winfree and co-workers have proposed several strategies to decrease the error rate in the formation of 2D assemblies without slowing down the process. These strategies aim at making the incorporation of erroneous components a much slower or energetically less favored process, leaving time for the mistakes to self-correct via dissociation [69,70].

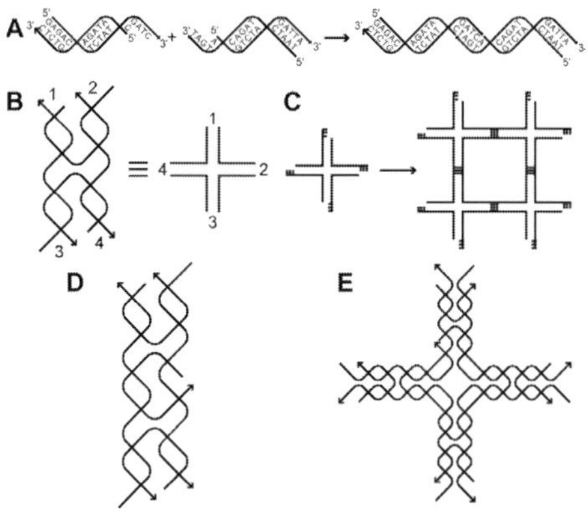

Figure 4. Schematic representation of some of the main assembly features used in structural DNA nanotechnology. Cohesion (A) of two double-helical fragments via "sticky-ends," single-stranded overhangs protruding from a double helical segment. Adjoined sticky ends (with two nicks) can be further stabilized by chemical or enzymatic ligation to make the phosphodiester backbones continuous. Schematic illustration (B) of a stable Holliday Junction. Four oligonucleotides form a branched junction in which the sequences flanking the center are not symmetrical. The branching point cannot thus migrate as it does in the naturally occurring symmetric junctions. The angles and lengths in the scheme do not correspond to the actual structure. Sticky-end cohesion (C) of branched structures. Four-arm junctions bearing sticky-ends on the arms can form superstructures by binding to other similar junctions according to the sequences of the involved sticky ends. The shape depicted on the right, formed by just four junctions is often called 'DNA parallelogram' or 'DNA rhombus motif.' Further junctions can bind to the quadrilateral and eventually form a continuous lattice. Scheme (D) of a DNA double crossover (DX) motif. The depicted motif can be seen as two linked four-way junctions. Several possible DX conformational isomers exist in addition to that depicted here. Scheme (E) of a DNA four-by-four (4x4) motif. The motif can be seen as four distinct four-way junctions joined through the central loop.

5.1. An Evolving Fauna of DNA-Based Molecular Nanostructures

The laboratories of Ned Seeman, Eric Winfree and John Reif have been some of the most notable pioneers of the implementation of the self-assembly of DNA oligonucleotides into complex nanostructures [71-80]. The blocked Holliday junction can be the building block for small or large polymeric 1D or 2D nano objects [81]. Computerized methods have been developed to design the oligonucleotide sequences that can auto-assemble specifically, minimizing the possible frame-shift errors or the mismatches. Atomic force microscopy and several classical biochemical techniques, such as electrophoresis and fluorescence spectroscopy, are used to characterize the building blocks and the constructed nano-objects. One of the peculiar properties of the DNA-based constructs born in Seeman's lab is their rigidity. Braided DNA structures are at least twice as stiff as the already fairly rigid dsDNA, i.e. they have a persistence length of up to 100 nm [71].

DNA does not "simply" have coding and structural functions in this type of structures, it can also have functional properties. For example, DNA can undergo structural transitions in response to changes in its environment. These can be employed towards the construction of true DNA-based nanomotors, described later in this paper.

Desired functionality of DNA nanostructures can also be achieved thanks to the decoration with other functional elements, such as organic molecules, proteins or other biomolecules. Due to the contemporary development of a huge library of methods and technologies, nanoparticles have been used in conjunction with DNA to harness the properties of both systems towards the formation of more and more complex nanostructures.

5.2. Hybrid Nanostructures Based on DNA Assembly: Metal Nanoparticles Plus DNA as an Example

The integration of metal or semiconductor nanoparticles with DNA can match the unique electronic, photonic, and catalytic properties of nanoparticles with the structural and recognition properties of DNA to create novel hybrid nanobiomaterials. Of particular interest for researchers are the use of biomolecule–nanoparticle assemblies for bioanalytical applications and for the fabrication of bioelectronic devices.

The groups of Chad Mirkin at Northwestern University and Paul Alivisatos at the University of California at Berkeley opened the way to the DNA-mediated nanoparticles assembly. Both groups work on the synthesis, characterization and assembly of DNA/AuNPs conjugates. AuNPs present fascinating aspects such as their size-related electronic, magnetic and optical properties and their possible applications to catalysis and biology [82].

The first example of assembly of gold nanoparticles in big aggregates with DNA was reported in 1996 [83] and a highly selective, colorimetric polynucleotide detection method based on optical properties of mercaptoalkyl oligonucleotide-modified gold nanoparticle probes was reported in 1997 [84]. Introduction of a single-stranded target oligonucleotide (30 bases) into a solution containing the appropriate probes resulted in the formation of a polymeric network of nanoparticles with a concomitant red-to-pinkish/purple color change. A temperature increase over the duplex melting temperature causes the disassembly of the structure and the change of color from purple to red. The optical properties of macroscopic DNA-linked Au nanoparticle aggregates can be controlled through choice of DNA linker length and the differences in the optical properties observed for the DNA-linked aggregates formed with the oligonucleotide linkers of different length are due not only to the interparticle distance but also to aggregate size [85]. These evidences have important implications for the development of colorimetric detection methods based on gold nanoparticles. Detection methods that rely on these materials show promise with respect to increased selectivity and sensitivity as compared with many conventional assays that rely on molecular probes [86,87]. In the case of target selectivity, the nanoparticle probes can be used to differentiate perfectly complementary targets from those with single-base mismatches, whereas the analogous assays based upon organic fluorophores do not offer such selectivity. The origin of this selectivity derives, in part, from the extraordinarily sharp melting profiles exhibited by duplex DNA structures formed between target strands of DNA and the nanoparticle probes [84]. The melting properties of DNA linked nanoparticle aggregates are affected by a number of factors, including DNA surface density, nanoparticle size, interparticle distance, and salt concentration [88] that should be combined in order to obtain the maximum level of selectivity.

Paul Alivisatos and coworkers have showed that nanocrystals modified with ssDNA can be arranged into homodimeric and homotrimeric assemblies [90] and also in heterodimeric and heterotrimeric "nanocrystal molecules" [91]. Phosphine stabilized gold nanoparticles can be isolated in an electrophoresis gel and they are stable for several cycles of separation and recovery. Thiol-modified single-stranded oligonucleotides can be incorporated into the protective phosphine shell and can react directly with the gold surface. ssDNA/gold conjugates are assembled with different strategies. The first involved the use of two complementary ssDNA/conjugates to form double stranded nanocrystal structures. In the second approach, the DNA/nanoparticles conjugate are later assembled with template strands [91]. The nanocrystals and the different hybrid conjugates can be characterized with TEM and gel electrophoresis. A body of experiments for the investigation of the surface coverage of Au nanoparticles with DNA, the conformation of bound DNA [92] and the role of nonspecific adsorption on the surface of gold nanoparticles is already available [93].

Starting from the early studies on DNA modified gold nanoparticle several applications have been reported on literature, in particular in the fields of analytical chemistry and biodiagnostic. Christof Niemeyer and colleagues have reported the preparation and characterization of oligofunctional gold nanoparticle conjugates containing different DNA sequences: the bottom-up assembly of complex biomolecular functionalized nanoparticles is thus possible [94]. A method based on strand displacement for the reversible sequence-specific switching of DNA/gold nanoparticles aggregation has also been reported by Miemeyer [95]. This strategy takes advantage of linker oligonucleotide whose sequence is divided in three sections, two complementary to that of the oligonucleotides bound to 23 nm Au nanoparticles, and one which forms a dangling end in the nanoparticles aggregate. Linker introduction into a solution containing the appropriate conjugates results in the formation of a polymeric network of nanoparticles with the concomitant color change. The dangling end can later serve as a nucleation section to promote hybridization with a DNA sequence fully complementary to the linker: the complete pairing of this molecule induces aggregate disassembly.

Recently, numerous applications of modified gold nanoparticles in colorimetric sensors have hit the literature. Pb(II) detection can be realized with DNazyme driven disintegration of DNA/gold nanoparticles aggregates [96]. A DNazyme (called also deoxyribozyme) is a catalytic DNA molecule of a particular sequence and 3-dimensional structure that can carry out specific chemical reactions, often with an efficiency comparable to that of protein enzymes. The DNA enzyme used by Liu and Lu is a RNA cleaving DNazyme activated by the presence of Pb(II) that was obtained through an in vitro selection process [89,97]. In the presence of Pb(II), this DNazyme can cleave a target molecule: if this cleaved molecule served as the linker between Au-oligonucleotides conjugates, then disintegration of the aggregate takes place and a change in color from blue to red can be observed (see Figure 5 for a scheme).

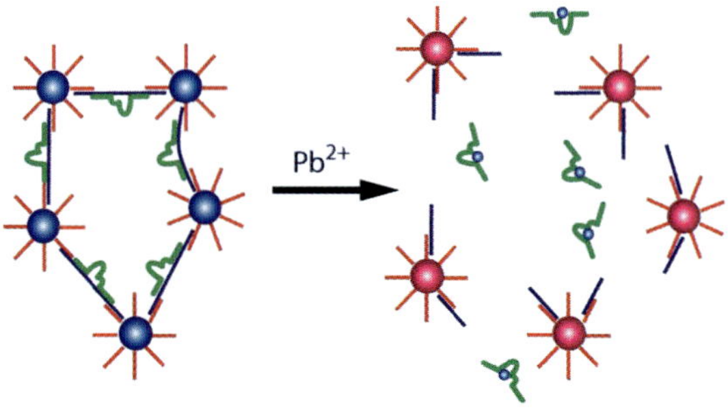

Figure 5. Scheme of the DNAzyme-mediated disintegration of an assembly made of oligonucleotide-nanoparticle conjugates. The presence of Pb(II) in solution triggers the enzymatic cleavage of the RNA linkers that hold the nanoparticles connected [89]. The solution color shifts from blue/purple to red upon disassembly of the nanoparticles.

Notable diagnostic applications are not lacking. Oligonucleotides have been successfully used as biochemical barcodes to measure the concentration of amyloid-α-derived diffusible ligands, a potential soluble pathogenic marker for Alzheimer's disease [98]. The key to the bio-barcode assay is the homogeneous isolation of specific antigens by means of a sandwich process involving oligonucleotide-modified Au nanoparticles (NPs with biobarcodes) and magnetic microparticles, both functionalized with specific antibodies to the antigen of interest. This system shows an extraordinary increase on sensitivity respect to conventional assay for early disease markers detection thanks to the very effective sequestration of antigen and in particular to the amplification process that occurs as a result of the large number of barcode DNA strands released for each antigen recognition and binding event. A similar bio-barcode assay has been previously employed towards the detection of the prostate-specific antigen reaching a sensitivity six orders of magnitude greater than the conventional ELISA assay for the same target [99].

The optical properties of colloidal gold have been recently harnessed towards the implementation of a molecular ruler [100]. When two nanoparticles are brought into proximity, the measured resonance in the wavelength of the plasmon resonance depends on the inter-particle separation. This effect has been applied to dynamics of DNA hybridization on the single-molecule level. Sonnichsen and coworkers have used plasmon coupling to monitor the directed assembly of functionalized 40 nm particle pairs through a 33-nucleotide ssDNA molecule. Liu and coworkers have designed a Au/DNA nanoplasmonic molecular ruler that is able to measure length changes in the DNA molecules anchored to the nanoparticles by measuring shifts in the scattering spectrum of a single nanoparticle [101]. Endonuclease-mediated shortening of DNA can be measured with base-pair resolution due to a reported wavelength change of 1.23 nm/bp.

5.3. Nature and Nanotechnology are a Matter of Hierarchy (and Topology)

The large and complex functional aggregates that fill the cells of living organisms are usually made from building blocks that are orders of magnitude smaller than the final assembly. The huge gap between the basic components and the functional structure is not usually crossed with one step. Subunits combine into higher-order constructs that will on their turn often combine into even higher order structures.

The reader should find it easy to rationalize the artificial nanostructures that we have presented so far in terms of the hierarchical level that are present. The lowest level is represented by the most basic components: the synthetic oligonucleotides concurring to the formation of the desired structure. The next level up is obtained by the pairing of the oligonucleotides through relatively long stretches of bases. Sometimes, this can be the highest hierarchical level present. A further level of hierarchy can be the combination of these objects into larger superstructures. Typically, the constituent sub-structures are reciprocally bound together by means of cohesive "sticky ends" [102] (see Figure 4) that entail the pairing of short (4-7 nucleotides long) complementary oligonucleotides protruding from paired double-strands. This, or other types of interactions, do not disrupt the pre-existing levels and need them to settle. A further level up in the hierarchy can also be the functionalization of a pre-formed structure with elements of a different nature, which can endow the nanostructure with a new function.

Even the most complex DNA structures could, in principle, be assembled in a single step, just by mixing all the constituent oligonucleotides at high temperature and then cooling the mix very slowly to allow the maximization of the number of interactions so that the most thermodynamically stable result is obtained (i.e. what was planned). However, it is likely that there could be an inherent complexity threshold in the successful self-assembly of a 'one-pot' mix of many oligonucleotides. The energy landscape governing such multi-component assembly will be increasingly rugged, and successful assembly of such complex system would require a prohibitively long assembly time so that trappings to sub-optimal structures will inevitably occur.

A stepwise procedure can be adopted in order to reach high levels of complexity, and the structures need to be designed accordingly. If the design can be organized so to be separated in domains of thermal stability for the different substructures, then each assembly step can be performed in different vessels, and the resulting assemblies united only before the next level up the complexity ladder. During each step, only the sequences actually directing the assembly at this level (e.g., the appropriate sticky ends) are relevant to the process, while the sequences already included in the lower level (higher stability) structures are not disassembled. This also saves for the need to have the base sequences of all the oligonucleotides compatible with each others as, once assembled, the paired sections of the oligonucleotides can be effectively screened from antagonistic reactions. One evident example of such a multiple pot strategy has been implemented by Jaeger and coworkers [103].

The topological dimension of an object is a measure of its space-covering properties. Roughly speaking, it is the number of coordinates needed to univocally specify a point in the object. For example, a line is topologically mono-dimensional and a surface is two-dimensional: both can be morphed into geometrically three-dimensional objects without affecting their topological state. When an object is made of multiple parts, their connectedness defines the topological dimension of the object, regardless of the geometrical dimension of the parts. In the next sections, we will attempt to rationalize most of the different artificial DNA nanostructures so far reported on the basis of the number of their topological dimensions.

5.3.1. Zero-Dimensional Topologies in DNA Artificial Nanostructures: Discrete DNA Constructs

Multiple branched junctions were first utilized to design and create discrete artificial nanostructures such as a DNA cube, a DNA truncated octahedron or DNA Borromean rings [67]. These were not truly thought to have a precise persistent geometrical shape (as these type of structures are more commonly designed nowadays) but to show efficiency in the assembly of complex nanostructures based on branched joints [106]. The topological control was, anyway, very high. As later demonstrated, a rigid structure can be obtained even from flexible components by means of a careful design implementing the concept of tensegrity as shown in a number of different nanostructures [107-109]. The majority of the DNA-based nanomotors (see below) reported so far are indeed singular, 0D objects not designed to oligomerize or interact, while there are examples of motor elements introduced in higher-dimensional assemblies [110].

5.3.2. Mono-Dimensional Topologies: Linear Arrays of Supramolecularly Connected Components to Make DNA Nano-Objects

A one-dimensional arrangement of rigid elements (or tiles) rigidly connected through supramolecular interactions will make a rigid polymeric nanostructure propagating in just one dimension. Some of the first moduli to be arranged in such a fashion were the DNA rhombus motif [72,111,112], and the DNA double crossover (DX) molecule [113,114] (see Figure 4). Typically, each tile in an 1D array binds to the successive one by means of more than one sticky end. Sticky ends having different sequences also have different thermal stabilities, so that during a thermal annealing they are stabilized in a certain sequence. This observation prompted us to study the assembly behavior of DNA rhombus tiles using pairs of either identical or slightly different sticky ends. We found that it is possible to obtain different assembly geometries with the same tile just by changing the annealing conditions [115]. Recently, a new type of DNA motif, the helical bundle, permitted to obtain 1D linearly arranged arrays reaching contour lengths of several micrometers [116]. Linear arrays do not need to be spatially one-dimensional, as exemplified by the assembly of flat rigid tiles into a super-helix developing in 3D (even if into a flexible chain) [117].

1D arrays of tiles do not need to be simply the repetition of the same modulus. A completely aperiodic assembly can be planned as a way to implement a mathematical algorithm into a DNA assembly: the coding of different connections between the tiles works in such a way as to define a unique sequence of tiles along the construct, defined by the algorithm. It is also possible to design a set of two or more sub-units with different complementarities. This algorithmic self-assembly can be used also as a physical implementation of computations [118-121].

5.3.3. Two-Dimensional Topologies of DNA Tiles

Two-dimensional topologies in the supramolecular connectivity of tiles can be implemented with a variety of tiles, including the double- and triple-crossover (TX) [113,122,123], the parallelogram or rhombus motif [112], the four-by-four structure [60] (see Figure 4), the three-and six-helix bundles [116,124], and the DX triangle [125]. The exact positioning of the cohesive sticky ends on a tile defines the topology of an assembly, so the same type of tile can often be used both for 1D and 2D topologies, only by acting on the sticky-ends for the connections (see Figure 6 for an example of different connectivities on the same tile).

2D lattices are usually better behaved than 1D linear structures. Each tile in a regular 2D lattice is linked to at least three other adjoining tiles, instead of only two as in linear arrays. This implies that the persistence of the whole structure doesn't critically depend on the simultaneous stability of each single tile-tile interaction. On the contrary, in linear arrays, the rupture of even one single interaction between two adjacent tiles results in the breakdown of the whole structure into smaller parts. Conversely, in 2D lattices the rupture of one tile-tile bond is not sufficient to impair the overall structural integrity, and the removal of an internal tile would require the simultaneous rupture of all the interactions with its neighboring tiles. These considerations motivate that 2D arrays are usually better behaved (bigger) than their 1D counterpart.

DNA 2D lattices can be easily used to obtain nano-patterned surfaces which can then be used as information-containing scaffolds upon which the controlled localization of matter on the nanoscale can be implemented [30,57]. Since DNA can be functionalized with an ample set of functional chemical moieties [56-64,126], 2D DNA arrays have several potential applications, including molecular electronics, sensors, and 'smart' materials, thus considerably increasing the attractiveness of DNA tiling assemblies. Making DNA nano-object functional by decoration with other molecules is certainly an expanding field of research, bound to bring DNA-nanotechnology into applicative fields. In Figure 7 are collected a few examples of DNA nanostructures with different dimensionality.

If the basic tile of a periodic array is not perfectly planar, or the tiles are connected so that they do not lie on a plane, the deformations in the propagating structure will sum up to an extent which could ultimately become a limiting factor in determining the extension of the lattice. Many published atomic force microscopy images of 2D lattices show ribbon-like structures rather than indefinitely wide periodic arrays. It has been proposed [104] that this is due to the tendency of the lattices to form curved surfaces that eventually reach a tube-like shape, inhibiting further growth of the lattice. The ribbons seen in the images could be the result of discrete tubes forming in solution and then unrolling themselves on the mica surface during deposition. Recently, this feature has been rationalized and intentionally included in the design of sets of tiles that self-assemble in DNA hollow nanotubes [104,116,124,127,128].

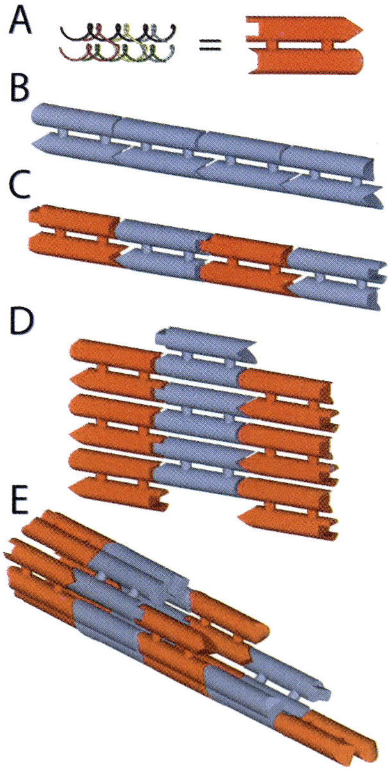

Figure 6. Scheme of some of the possible spatial arrangements of a DNA tile. A DX tile (A) (see also Figure 4) is schematized as two connected cylinders with two different types of joints at the ends. An example of (B) a 1D construct obtained with one type of tile and (C) a 1D construct obtained with two alternating tiles. An example of (D) a 2D array built of DX tiles. A DNA nanotube (E) can be built by connecting flat tiles at an angle as demonstrated for DX [104] or TX (triple crossover) tiles [105].

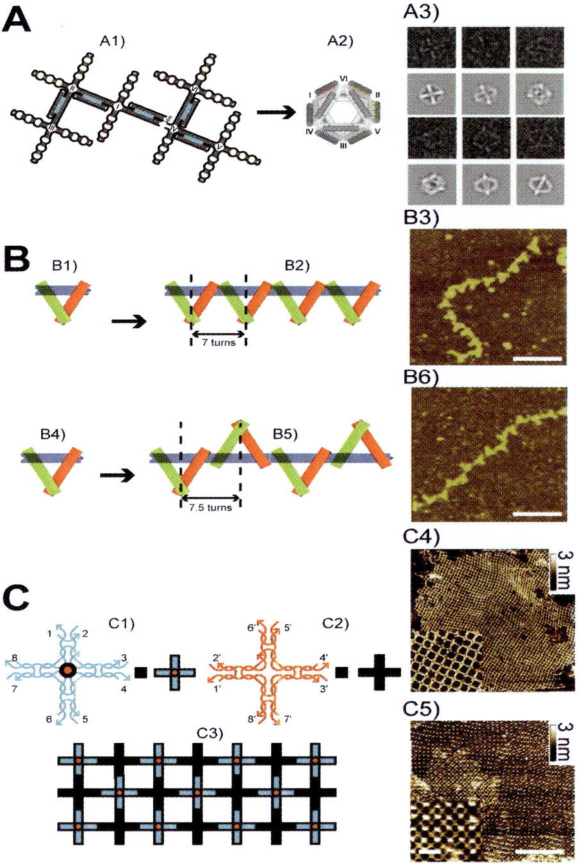

Figure 7. Examples of DNA nanoarchitectures with different topological dimensionalities. A zero-dimensional array (A), i.e., an individual DNA octahedral object made from the directed folding of a 1.7 kilobases long DNA strand [109]. A1 is the 'unfolded' structure, A2 the model of the folded and A3 contains cryo-TEM micrographs: in A3, the first and the third row panels are the raw images of individual particles, and the second and fourth are the result of projecting the experimental data onto a three-dimensional model of the folded octahedron obtained by single particle reconstruction techniques. One-dimensional DNA arrays (B) made of triangular tiles [107]. B1 and B2 are the monomer and the 1D array of tiles spaced by 7 full turns of DNA. B4 and B5 represent the monomer and 1D array of tiles separated by 7.5 turns. B3 and B6 are AFM images of the arrays on their left. Two-dimensional DNA array (C) constituted by two alternating tiles [60]. C1, C2 schematize the tiles and C3 the 2D array. C4 and C5 are the AFM images of the 2D array and of the same type of array decorated with streptavidin molecules at the central points of the tiles. Size bar is 300 nm in the main figure, 100 nm in the inset. Reprinted with permission from [Ref. 117].

5.4. Raising the Size and Complexity: Algorithmic Assembly, DNA Origami and Other Assemblies on Long Template Strands

The ability to build nanoscale scaffolds of arbitrary shapes with a command over their nanometer-resolution structure will probably be a key technological advance towards the construction of nanoscale circuitry, data storage units, sensor arrays or artificial molecular factories [[72,129,130]. Aperiodically patterned, self-assembled 2D or 3D DNA lattices are regarded as extremely promising candidates for playing the role of such nanoscale scaffolds. Three main strategies have been proposed and implemented in the past years for the creation of DNA aperiodic lattices: algorithmic self-assembly, stepwise assembly and directed nucleation assembly [131].

The algorithmic assembly of discrete DNA tiles is potentially a very simple and powerful approach for obtaining arbitrarily complex aperiodic structures [132]. It has been demonstrated that an hypothetical set of tiles with the ability to self-assemble according to an algorithm can perform Turing-universal computation [133]. Accordingly, a set of DNA tiles with the ability of self-assemble through sticky-end mediated, information-driven cohesion (incorporating the algorithm of the assembly) can be designed to obtain potentially any shape [119,123]. However, practical considerations hinder somewhat the seemingly unlimited potential of algorithmic assemblies of DNA tiles. Any complex shape would need a large number of individually synthesized DNA tiles [134-136]. The result of their assembly would then be dictated not only by the sequences of their sticky ends, but also by the subtle interplay of assembly growth kinetics, mismatched pairings incidence, nucleation energies, concentration of the individual components and the temperature at which the assembly is performed [68,69,137-139]. Even prior to that, individual DNA tiles must be flawlessly formed by exact stoichiometric amounts of component strands to avoid competing coupling reactions of incomplete tiles. Due to these practical limitations, successful experimental implementations of the algorithmic self-assembly of complex 2D structures was only recently reported [103,136].

The complications arising from having multiple simultaneous processes competing during the self-assembly of the desired structure can be avoided by performing a stepwise, sequential assembly [134]. This approach entails the formation of subsets of the structure which are then brought together in a stepwise fashion, thus making it possible to remove any excess of unreacted species after each step. As also mentioned above, another potential advantage of this approach in the context of building aperiodic DNA lattices is that the number of different tiles required for building a given structure is lower than that needed following the aforementioned algorithmic assembly strategy. This is because the final positions of the individual tiles in the assembly is not dictated solely by their sticky-ends, but also by the order in which they are assembled. Identical tiles could thus be incorporated in the assembly at different positions. The drawback of this approach is that it needs extensive external input from the operator and its overall attractiveness is thus slightly lower than that of unmediated self-assembly.

The most successful strategy employed so far toward the assembly of complex aperiodic structures is that of the directed nucleation assembly. Proposed by LaBean, Winfree and Reif in 1999 [142] this approach has been used for striking implementations in the following years [109,131,140,141]. The method entails the use of a long ssDNA template strand encoding the pattern information of the complete structure. Several shorter strands are then designed to assemble at specified positions on the template strand, folding into the desired shape and completing the formation of the structure. Although this approach is apparently similar to the algorithmic assembly of individual tiles, its practical advantages are manifold. Since the structure can only form around the template strand, its dimensions are defined with extreme precision and their persistence does not rely on the simultaneous stability of a series of tile-to-tile connective reversible interactions. Moreover, the result of the assembly is less prone to deviations from the desired shape caused by error incorporation. This is because most of the tile-to-tile connections present in an algorithmic array are substituted in a templated array by the irreversible, covalent connection present throughout the template strand backbone.

As proposed by Hao Yan *et al.* in 2003 [131] the directed nucleation approach is applicable to the formation of 2D planar aperiodic patterns. Recently, Paul Rothemund experimentally implemented this design

[70,140] and brought it a step further in its potential by demonstrating the formation of "DNA origami" shapes (see Figure 8). In the DNA origami approach, a long ssDNA template is folded with the help of many short "staple" strands to produce an arbitrary planar shape. The shape is constituted by parallel double helices lying side-to-side in the same plane being connected by strand exchange crossovers formed by the shorter strands. The long template strand traverses the shape from side to side in a raster-filling path, thus participating in each helix.

In Rothemund's experiments, the template strand is not an ad-hoc synthesized single-stranded polynucleotide, but the naturally occurring 7249-nt single-stranded genomic sequence of M13mp18 virus. This simple fact implies the shift of two of the main paradigms in the field of structural DNA nanotechnology: sequence symmetry minimization and exact stoichiometry determination. All previous DNA structural nanotechnology designs relied on the careful optimization of all the strand sequences involved in the assembly, typically (but not always, as exemplified by Mao and coworkers [143-145]) investing particular care in the minimization of sequence symmetry [146,147]; this was done in order to avoid errors in the pairing of the component strands. Furthermore, structures were always formed by as exact stoichiometric ratios of strands as possible, to maximize complete formation. In the origami approach, no sequence optimization is performed and exact stoichiometry determination is unnecessary. Although long-range correlations are present in the sequences of naturally occurring DNA [148], at the scale involved in structural DNA nanotechnology a viral ssDNA strand can be regarded as having an essentially random sequence, and therefore more secondary structure and short-range sequence symmetry than an optimized synthetic strand. Nevertheless, perfectly assembled DNA origami do form with substantial yield, typically around 70 % in Rothemund's experiments. Since the short staple strands are not designed to bind to each other, their relative stoichiometry does not need to be precisely controlled. Most importantly, since complete structures can only form on template strands, it is possible to use a vast stoichiometric excess of staple strands (100-fold). This means than any undesired template-template intra- or inter-molecular interaction is very unlikely to remain stable in the assembly conditions. Likewise, imperfect staple-template interactions should be unlikely because the strand invasion equilibria at work during the assembly will favor the complete, correct pairings. An additional advantage of the origami approach over the other strategies is that a less than ideal purity of the short staple strands does not impair the overall result, since these are used in large excess. This makes the time- and resources-consuming step of synthetic ODNs purification unnecessary. Taken together, these advantages make it possible to use the origami approach to build DNA aperiodic structures that are significantly larger and more complex than previously possible. Due to the low cost of unpurified short synthetic oligonucleotides, and the possibility to amplify the template strand with enzymatic methods, the origami approach also permits to conveniently increase the production scale of the constructs.

A particularly interesting technique demonstrated by Rothemund is using the staple strands to extrude short dsDNA hairpins out of the origami plane at specific points [140]. This allows to emboss arbitrary shapes on the structure with a theoretical resolution limit of 5.4 nm and 6 nm along the two orthogonal axes. Substituting the dsDNA hairpins with sticky ends for directing the assembly of DNA-modified metal nanoparticles, nanotubes, fluorophores or proteins seems like a natural extension of this approach and will definitely represent a major step forward toward the bottom-up construction of functional nanoscale devices.

Blunt-end helix-stacking interactions can be a tool in directing the interaction of DNA assemblies [135,149]. Due to the number of constituent double-helices, the large tile constructs obtained with the directed nucleation and origami approaches are especially prone to this type of edge-to-edge, blunt-end cohesion, so that it can be exploited to design and obtain even larger assemblies (tens of megaDaltons for Rothemund's examples [140]) or periodic lattices of shapes.

Shih and coworkers demonstrated how the long template strategy can be used to obtain a three-dimensional hollow octahedron [109]. In their example, a 1669-nt ssDNA strand folds into a regular octahedron with an outer diameter of 22 nm, upon hybridization with five 40-nt helper strands. The long template strand was constructed by a polymerase chain reaction (PCR)-based assembly, using synthetic oligonucleotides [150].

Their methodology is plausibly extensible to other shapes and is amenable to large-scale production since the bulk of the 3D structures is constituted by the template strand that can be conveniently amplified by polymerases once it is assembled from shorter synthetic oligonucleotides.

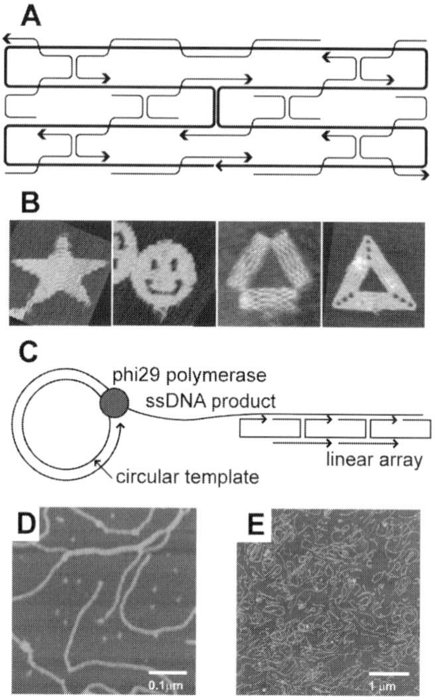

Figure 8. Examples of the use of a long template in DNA nanoarchitectures. Schematic representation (A) of the design strategy of DNA origami. A long ssDNA template with a naturally occurring sequence is folded so that it fills completely the desired 2D figure (in this case simply a rectangle). Several short "staple" strands pin the long strand in place. Each staple strand binds to different domains of the template strands forming multiple crossovers. The spacing between successive crossovers is typically 1.5 full helical turns, so that they lie on the same plane forming a 180° angle. Examples (B) of shapes obtained by Paul Rothemund with the DNA origami design. All the AFM images are 165 nm x 165 nm. Adapted with permission from [Ref. 140]. Schematic representation (C) of the strategy employed by Lubrich *et al.* for the production of long 1D arrays. Rolling Circle Amplification (RCA) of a circularized synthetic oligonucleotide yields a several kilobases long ssDNA template with a repetitive periodic sequence. Shorter synthetic oligonucleotides then assemble on the template yielding the desired periodic array. AFM images (D and E) of the linear arrays obtained with the strategy outlined above. Reprinted from [Ref. 141] (permission requested).

5.5. Building 3D Objects

The quest for building fully 3D objects out of DNA is a very hot topic, since the original proposal of Seeman's of using such 3D DNA lattices to organize matter in an ordered fashion, for instance to make it amenable of X-ray crystallographic studies [65]. Some progress has been made and some fully 3D DNA nano-objects have been presented [151]. Research is still going in the direction of trying to extend some of the concepts devised for 2D arrays in producing 3D arrays [107], and some preliminary results (and macroscopic crystals of DNA nanostructures) are coming out of the lab of Ned Seeman and his co-workers (personal communications).

One of the possible alternatives towards making 3D objects is coiling a 3D helix in space. One way of making a 3D-helix out of linearly arranged flat and rigid DNA tiles is to equip each tile with two sticky ends at a non-180° angle. Also, the joining of the two sticky ends must form a non-integer number of half helical turns, so that two successive tiles form a dihedral angle. The overall architecture of such a super-helix is thus determined by the geometrical features of the tiles and their connections. Unless the joints are made very rigid, for instance by double-connections [152], these objects will not enjoy the geometrical determination that generally characterizes the structures described in this paper. The binding of two tiles through a single sticky end does not confer rigidity to the resulting construct, still, a fully spatial but flexible nanostructure can thus be built [117].

5.6. Strategies to Enhance the Structural Rigidity of the Nanostructures

Since the aim of structural DNA nanotechnology is to provide a facile bottom-up access to the accurate spatial arrangement of matter in the nanoscale, its success relies on the development of structurally well-defined DNA motifs. Apparently, the most straightforward strategy to obtain the highest geometrical control over assemblies of DNA motifs is to remove any potential flexibility from the structures. Much effort has been invested in designing rigid motifs, including double crossovers

[123], triple crossovers [122], double-double crossovers [153], 8- and 12-helix tiles [149], which have been polymerized to form ordered 2D arrays. A strategy proposed by Mao and coworkers to obtain rigid tiles is to implement the concept of tensegrity [107].

In the last years however, new motifs were reported [60,145,154] which have the capability of forming large, ordered 2D arrays even if their structure should be relatively flexible. This prompted He and Mao to investigate the role of tile rigidity in the formation of large, well-ordered planar 2D arrays [155]. The gist of their reasoning is that if the motif is too flexible, it will not preserve any direction of propagation in the plane and no large 2D arrays will form. On the other hand, if the motif is too rigid, any unpredicted distortion of the individual tile could induce a stress on the tile-to-tile connections that could also prevent the formation of large 2D arrays. So a subtle balance of flexibility and stress (and hence rigidity) in DNA nanostructures appears to be an important factor in determining the success of 2D planar assemblies, even if presently there is no design tool to predict how to attain the ideal balance. To prove this assertion, He and Mao synthesized a group of almost identical tiles, but having a range of different flexibilities. They found that only the middle motifs in the series could form 2D arrays, while neither the most flexible nor the most rigid did [155]. Following an orthogonal approach, Seeman and coworkers built a set of new motifs that can assemble into 2D arrays by means of adjoining pairs of sticky ends, and compared the results to these obtained with similar motifs with isolated sticky ends [156]. Just as two joining sticky ends form a nicked double helix junction, double sticky end interactions result in the formation of a nicked DX structure, which appears to be sturdier and thus less susceptible to errors in twist between succeeding tiles. This means that lattices structures unattainable with single sticky-end cohesion are instead accessible with double cohesion tiles designs. Taken together, these results seem to indicate that the best tiles for obtaining large, ordered DNA 2D lattices are those with rigid, strong cohesion points kept together by a moderately flexible core.

5.7. The Enhancement of Symmetry in the Assembly: an Alternative Strategy

Since its outset, structural DNA nanotechnology designs relied on the minimization of sequence symmetry in the component strands [147]. The reasoning is that each section of the construct ought to have an unique sequence that should ideally pair only with its intended targets, hence promoting the formation of the desired structure. The more sequence symmetry is present in the involved strands, the more the chances of intra- or inter-molecular incorrect pairings during the self-assembly.

Recently, Chengde Mao and coworkers started to investigate the possibility of using sequence symmetry as an asset rather than as a limitation. The gist of the idea is to identify the symmetry elements of the desired structure backbone and use them to reduce the complexity of the sequences involved in the assembly. In Mao's first proof of concept [143-145], the structure under examination is a four-by-four DNA motif, formed by nine different strands, roughly having the shape of a four-arms cross with a fourfold rotational symmetry. Two strands are needed to form each arm, and one strand joins them together at the center. Since the nine strands have nine unique sequences, when the DNA bases are taken into account, the structure has in fact no symmetry elements. The same backbone shape, however, can be obtained with just three different sequences if the arms are designed to be identical (thus reusing the same two strands four times around the central strand). It is important to note that sequence symmetry minimization is still employed in the design of each individual arm, but the arm is present in the structure more than once.

Generalizing these concepts, the introduction of a degree of sequence symmetry in the design of DNA nanoarchitectures could bring on several advantages: 1) the sequence space needed for a given structure backbone is minimized, and thus sequence design is simplified; 2) the number of different strands is reduced; and 3) the unpredictable distortions present in most DNA tiles design self-cancel [145]. Liu *et al.* tried to bring the consequences of the reasoning presented above to the limit. Employing sequence symmetry in the design, they were able to build large 2D crystalline arrays using only four different strands [144], and even micrometers-long DNA hollow nanotubes formed by just one short DNA oligonucleotide [143].

5.8. The Temporal Dimensionality

Another goal of structural DNA nanotechnology is to be able to dynamically alter the nanoscale structure of artificial self-assembled constructs. Objects that can modify their structure in response to a specific event or in accord with a predetermined program have the potential of functional utility; dynamic DNA nanostructures are thus promising nanoscale components for building functional nanodevices and machines. Proposed applications of DNA molecular devices include such materials science, nanoelectronics, biosensors, chemical synthesis [157] and molecular therapy. Many DNA self-assembled constructs capable of controlled motion have been reported in the last years [158-160].

Several different principles are employed to obtain controlled movement in these constructs, including DNA conformational transitions [74,161-163], strand-displacement equilibria [75,110,164-169], and protein binding [170]. Motion is either triggered by specific, externally provided stimuli, or is autonomous [166,171-173] (see Figure 8 for a few examples).

For example, under proper conditions, a GC-alternating sequence can undergo the B-Z transition that implies a reversal of handedness of the double-helix. Seeman and coworkers have employed this controlled and induced rotation to change the distance of separation of objects in space [74]. When these objects were fluorescent dyes, the motion could be easily followed studying their photophysical properties. Many other examples of controllable dynamic objects made of DNA are nowadays found in the scientific literature.

One key feature of strand-displacement-based devices (prototyped by Yurke *et al.* [164]) is that the sequence-dependent mechanisms for the actuation of the device allow simultaneous control of different devices in the same environment, or different parts of the same device. A number of ingenious variations of this idea have been reported, including DNA "walkers" that can move along a track step-by-step [167,168], a 2D DNA periodic lattice with tunable cavities [110], and a pair of DNA "molecular gears" that can revolve against each other [169].

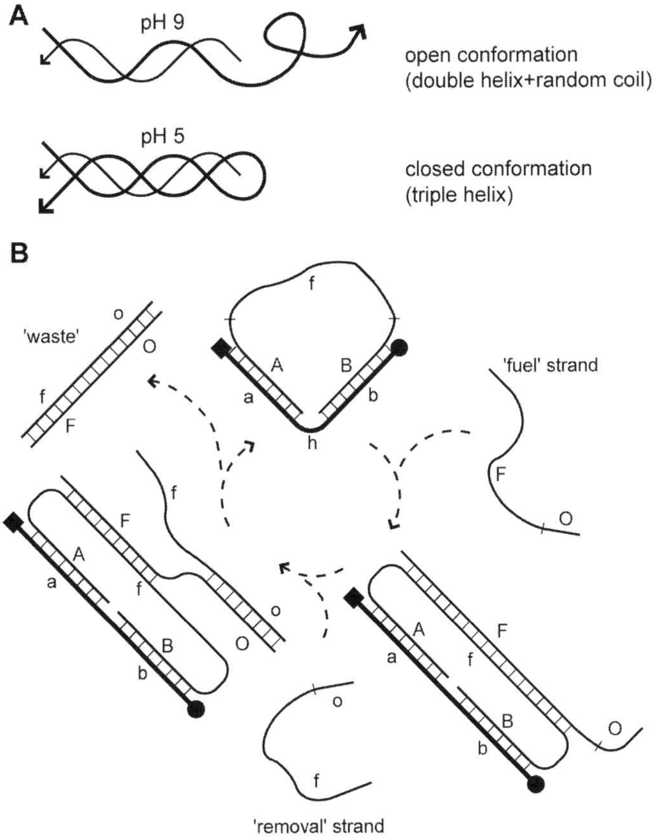

Figure 9. Examples of strategies utilized for obtaining controlled motion of DNA nanostructures. A DNA conformational change (A) triggered by external input. One long and one short oligonucleotide combine to form a structure which comprises a double helical domain and a single-stranded overhang. At acidic pH, the overhang folds on the duplex by forming a cytosine motif triple helix. Cycling the pH between acidic and basic values causes repeated movement of the structure. Sequence-dependant addressing (B) of DNA "molecular tweezers." Letters represent sequence tracts, the uppercase pairs with the lowercase. Addition of the "fuel" strand causes the tweezers to open. Its removal by a nucleation-zipping mechanism restores the system to its original open state. Each cycle produces a "waste" duplex product formed by the joining of one fuel and one removal strand.

In addition to "clocked" devices that respond directly to a change in their environment, autonomously running DNA nanomachines have been built. Yin *et al.* reported a walking device autonomously processing along a track by means of a sequence of enzymatic reactions [171]. Mao and coworkers built a set of molecular tweezers containing an RNA-cleaving DNAzyme that can continuously cycle between its open and closed shape when the appropriate substrate is accessible [172]. The cycling can be stopped by putting a "brake" on the device [173]. Another strategy to obtain free-running devices is to use "catalytic" strands that can unlock kinetically stable loops for the invasion of other strands [166,174,175]. Two further examples of autonomous devices was reported by Simmel and colleagues. In the first case, a pH-sensitive molecular construct is located in a reactor in which a non-equilibrium oscillatory chemical reaction is taking place [176,177]. The reaction produces pH variations that cause the device to cycle between its conformations. The second strategy entails the use of transcription of a designed sequence to control a set of molecular DNA tweezers [178].

Hybridization chain reaction [179] can be exploited to obtain a 'triggered' self-assembly of static DNA nanostructures. Briefly, it is possible to store potential energy in locked conformations such as loops that are kinetically inaccessible at room temperature in the laboratory time scale, and then unlock them via a chain reaction of successive hybridizations initiated by a 'catalyst strand.' During the reaction, all the loops are opened one at a time and incorporated into a growing nanostructure.

In order to build complex functional nanodevices constituted by several different components working in concert, immobilizing all the different portions on some sort of common substrate or surface will probably be convenient. An immediate problem posed by this is that even if the functioning of a component is verified in solution, it might not work with the same efficiency, or even at all, when it is immobilized in a sterically hindered context such as on a surface. The functioning of DNA nanodevices immobilized on a nanoparticle surface was verified by Li and Tan [165]. Another issue is to understand the details of how the synthetic devices work at single molecule level. Most published data regarding DNA nanodevices motion refer to populations of devices, and almost nothing is known about the behavior of the individual in the population. Single-molecule studies of artificial DNA nanomotors are beginning to appear in the literature [180]. Combining the two approaches, the single-molecule behavior of DNA nanodevices immobilized on a glass surface was recently verified by our group (Kolaric *et al.* submitted for publication). Shu *et al.* recently reported an interesting use of DNA nanomotors on surfaces [181]. They immobilized an ensemble of DNA motors on microfabricated silicon cantilevers and verified that the forces exerted by cycling the nanodevices can induce a surface stress capable of bending the cantilevers, demonstrating the translation of biochemical energy into micromechanical work.

6. Conclusions and Outlook

DNA is currently one of the biological molecules that most fascinates the nanoscience and nanotechnology community. Its information-driven self-assembly capabilities and well-characterized structure offer an extremely powerful tool for bottom-up nanofabrication of diverse materials and devices. After more than twenty years from its birth, structural DNA nanotechnology is an increasingly active and exciting field. Consolidated results of DNA nanotechnology include building an almost limitless variety of shapes by self-assembly, producing controlled nanoscale motion within these structures, and functionalizing them with proteins, metal nanoparticles, carbon nanotubes, fluorescent dyes and other species. This nanoscale construction kit is grown to such a vast size and potential that its practical technological implementations seem to be completely within reach. As Lloyd M. Smith remarks in the same issue of Nature on which Paul Rothemund's DNA origami paper was published, "the barrier we have to surmount next is to deploy our knowledge to develop structures and devices that are really useful. Happily, in that endeavor, we are now perhaps limited more by our imagination than by our ability" [182].

Acknowledgements

The authors wish to acknowledge support from EUROCORESSONS program "BIONICS" through funds from the Italian National Research Council (CNR) and EU FP6-STREP program "NUCAN."

References

1. J. D. Watson, F. H. C. Crick, Nature 171, 737 (1953).
2. E. N. Trifonov, Bull. Math. Biol. 51, 417 (1989).
3. V. V. Demidov, M. D. Frank-Kamenetskii, Trends Biochem. Sci. 29, 62 (2004).
4. V. A. Bloomfield, D. M. Crothers, I. Tinoco, Nucleic acids: structures, properties, and functions (University Science Books, Sausalito, Calif., 2000), pp. x.
5. J. A. Schellman, S. C. Harvey, Biophys Chem. 55, 95 (1995).
6. J. Bednar, P. Furrer, V. Katritch, A. Z. Stasiak, J. Dubochet, A. Stasiak, J. Mol. Biol. 254, 579 (1995).
7. C. R. Calladine, H. R. Drew, J. Mol. Biol. 257, 479 (1996).
8. D. M. Crothers, Proc. Natl. Acad. Sci. USA 95, 15163 (1998).
9. M. A. El Hassan, C. R. Calladine, Philosophical transactions-Royal Society of London. Physical sciences and engineering 355, 43 (1997).
10. J. Griffith, M. Bleyman, C. A. Rauch, P. A. Kitchin, P. T. Englund, Cell 46, 717 (1986).
11. R. E. Dickerson, D. Goodsell, M. L. Kopka, J. Mol. Biol. 256, 108 (1996).
12. P. R. Hardwidge, R. B. Den, E. D. Ross, L. J. Maher, 3rd, J. Biomol. Struct. Dyn. 18, 219 (2000).
13. J. D. Kahn, E. Yun, D. M. Crothers, Nature 368, 163 (1994).
14. Y. Zhang, D. M. Crothers, Proc. Natl. Acad. Sci. USA 100, 3161 (2003).
15. Y. Zhang, D. M. Crothers, Biophys. J. 84, 136 (2003).
16. P. De Santis, M. Fuà, M. Savino, C. Anselmi, G. Bocchinfuso, J. Phys. Chem. 100, 9968 (1996).
17. J. A. Cognet, C. Pakleza, D. Cherny, E. Delain, E. Le Cam, J. Mol. Biol. 285, 997 (1999).
18. G. Muzard, B. Theveny, B. Revet, EMBO J. 9, 1289 (1990).
19. G. Zuccheri, A. Scipioni, V. Cavaliere, G. Gargiulo, P. De Santis, B. Samorì, Proc. Natl. Acad. Sci. U S A 98, 3074 (2001).
20. C. Rivetti, C. Walker, C. Bustamante, J. Mol. Biol. 280, 41 (1998).
21. W. K. Olson, V. B. Zhurkin, Curr. Opin. Struct. Biol. 10, 286 (2000).
22. E. Le Cam, F. Fack, J. Menissier-de Murcia, J. A. Cognet, A. Barbin, V. Sarantoglou, B. Revet, E. Delain, G. de Murcia, J. Mol. Biol. 235, 1062 (1994).
23. T. Akiyama, M. E. Hogan, Biochemistry 36, 2307 (1997).
24. P. De Santis, A. Palleschi, M. Savino, A. Scipioni, Biochemistry 29, 9269 (1990).

25. D. Boffelli, P. De Santis, A. Palleschi, G. Risuleo, M. Savino, FEBS Lett. 300, 175 (1992).
26. P. De Santis, A. Palleschi, M. Savino, A. Scipioni, Biophys. Chem. 42, 147 (1992).
27. E. S. Shpigelman, E. N. Trifonov, A. Bolshoy, Comput. Appl. Biosci. 9, 435 (1993).
28. A. Bolshoy, P. McNamara, R. E. Harrington, E. N. Trifonov, Proc. Natl. Acad. Sci. U S A 88, 2312 (1991).
29. A. A. Gorin, V. B. Zhurkin, W. K. Olson, J. Mol. Biol. 247, 34 (1995).
30. B. Samorì, G. Zuccheri, Angew. Chem. Int. Ed. Engl. 44, 1166 (2005).
31. N. C. Seeman, A. M. Belcher, Proc. Natl. Acad. Sci. U S A 99 Suppl 2, 6451 (2002).
32. A. M. Belcher, X. H. Wu, R. J. Christensen, P. K. Hansma, G. D. Stucky, D. E. Morse, Nature 381, 56 (1996).
33. G. Falini, S. Albeck, S. Weiner, L. Addadi, Science 271, 67 (1996).
34. S. Brown, Proc. Natl. Acad. Sci. USA 89, 8651 (1992).
35. S. Brown, Nat. Biotechnol 15, 269 (1997).
36. S. R. Whaley, D. S. English, E. L. Hu, P. F. Barbara, A. M. Belcher, Nature 405, 665 (2000).
37. A. Scipioni, S. Pisano, A. Bergia, M. Savino, S. B., P. De Santis, Chem. Biochem. 9999, NA (2006).
38. B. Sampaolese, A. Bergia, A. Scipioni, G. Zuccheri, M. Savino, B. Samorì, P. De Santis, Proc. Natl. Acad. Sci. USA 99, 13566 (2002).
39. R. Saladino, U. Ciambecchini, C. Crestini, G. Costanzo, R. Negri, E. Di Mauro, Chem. Biochem 4, 514 (2003).
40. N. Lahav, S. Nir, A. C. Elitzur, Prog. Biophys. Mol. Biol. 75, 75 (2001).
41. P. Szabó, I. Scheuring, T. Czárán, E. Szathmáry, Nature 420, 340 (2002).
42. S. J. Sowerby, N. G. Holm, G. B. Petersen, Biosystems 61, 69 (2001).
43. S. J. Sowerby, C. A. Cohn, W. M. Heckl, N. G. Holm, Proc. Natl. Acad. Sci. USA 98, 820 (2001).
44. C. Anselmi, P. De Santis, R. Paparcone, M. Savino, A. Scipioni, Orig. Life Evol. Biosph. 34, 143 (2004).
45. B. Samorì, G. Zuccheri, A. Scipioni, P. De Santis, in Nanotechnology: Science and Computation J. Chen, N. Jonoska, G. Rozenberg, Eds. (Springer, Berlin Heidelberg, 2006) pp. 249.
46. E. M. Southern, Methods. Mol. Biol. 170, 1 (2001).
47. M. J. Heller, Annu. Rev. Biomed. Eng. 4, 129 (2002).
48. E. Palecek, Talanta 56, 809 (2002).
49. E. Palecek, F. Jelen, Crit. Rev. Anal. Chem. 32, 261 (2002).
50. K. Kerman, M. Kobayashi, E. Tamiya, Meas. Sci. Technol. 15, R1 (2004).
51. T. A. Taton, C. A. Mirkin, R. L. Letsinger, Science 289, 1757 (2000).
52. S. J. Park, T. A. Taton, C. A. Mirkin, Science 295, 1503 (2002).

53. C. Guiducci, C. Stagni, G. Zuccheri, A. Bogliolo, L. Benini, B. Samorì, B. Riccò, Biosensors and Bioelectronics 19, 781 (2004).
54. R. Hölzel, N. Gajovic Eichelmann, F. F. Bier, Biosensors and Bioelectronics 18, 555 (2003).
55. C. Bustamante, J. F. Marko, E. D. Siggia, S. Smith, Science 265, 1599 (1994).
56. C. A. Mirkin, R. L. Letsinger, R. C. Mucic, J. J. Storhoff, Nature 382, 581 (1996).
57. C. M. Niemeyer, Curr. Opin. Chem. Biol. 4, 609 (2000).
B. Samorì, M. Brucale, A. Vinelli, G. Zuccheri
58. S. J. Park, A. A. Lazarides, C. A. Mirkin, R. L. Letsinger, Angew. Chem. Int. Ed. Engl. 40, 2909 (2001).
59. K. A. Williams, P. T. M. Veenhuizen, B. G. d. l. Torre, R. Eritja, C. Dekker, Nature 420, 761 (2002).
60. H. Yan, S. H. Park, G. Finkelstein, J. H. Reif, T. H. LaBean, Science 301, 1882 (2003).
61. E. Katz, I. Willner, 43, 6042 (2004).
62. C. M. Niemeyer, Biochem. Soc. Trans. 32, 51 (2004).
63. R. Singh, D. Pantarotto, D. McCarthy, O. Chaloin, J. Hoebeke, C. D. Partidos, J. P. Briand, M. Prato, A. Bianco, K. Kostarelos, J. Am. Chem. Soc. 127, 4388 (2005).
64. G. Zuccheri, M. Brucale, B. Samorì, Small, 1, 590 (2005).
65. N. C. Seeman, J. Theor. Biol. 99, 237 (1982).
66. N. R. Kallenbach, R. I. Ma, N. C. Seeman, Nature 305, 829 (1983).
67. N. C. Seeman, Annu. Rev. Biophys. Biomol. Struct. 27, 225 (1998).
68. S. Roweis, E. Winfree, J. Comput. Biol. 6, 65 (1999).
69. E. Winfree, R. Bekbolatov, DNA Based Computers 9 (2003).
70. E. Winfree. vol. 2006.
71. P. Sa-Ardyen, A. V. Vologodskii, N. C. Seeman, Biophys. J. 84, 3829 (2003).
72. N. C. Seeman, Nature 421, 427 (2003).
73. N. C. Seeman, Annu. Rev. Biophys. Biomol. Struct. 27, 225 (1998).
74. C. Mao, W. Sun, Z. Shen, N. C. Seeman, Nature 397, 144 (1999).
75. H. Yan, X. Zhang, Z. Shen, N. C. Seeman, Nature 415, 62 (2002).
76. E. Winfree, F. Liu, L. A. Wenzler, N. C. Seeman, Nature 394 (1998).
77. L. Feng, S. H. Park, J. H. Reif, H. Yan, Angew. Chem. Int. Ed. Engl. 42, 4342 (2003).
78. H. Yan, T. H. LaBean, L. Feng, J. H. Reif, Proc. Natl. Acad. Sci. USA 100, 8103 (2003).
79. H. Li, S. H. Park, J. H. Reif, T. H. LaBean, H. Yan, J. Am. Chem. Soc. 126, 418 (2004).
80. R. M. Dirks, M. Lin, E. Winfree, N. A. Pierce, Nucleic Acids Res. 32, 1392 (2004).
81. C. Mao, W. Sun, N. C. Seeman, J. Am. Chem. Soc. 121, 5437 (1999).
82. M. C. Daniel, D. Astruc, Chemical Reviews 104, 293 (2004).

83. C. A. Mirkin, R. L. Letsinger, R. C. Mucic, J. J. Storhoff, Nature 382, 607 (1996).
84. R. Elghanian, J. J. Storhoff, R. C. Mucic, R. L. Letsinger, C. A. Mirkin, Science 277, 1078 (1997).
85. J. J. Storhoff, A. A. Lazarides, R. C. Mucic, C. A. Mirkin, R. L. Letsinger, G. C. Schatz, J. Am. Chem. Soc. 122, 4640 (2000).
86. B. W. Kirk, M. Feinsod, R. Favis, R. M. Kliman, F. Barany, Nucleic Acids Res. 30, 3295 (2002).
87. B. Schweitzer, S. Wiltshire, J. Lambert, S. O'Malley, K. Kukanskis, Z. R. Zhu, S. F. Kingsmore, P. M. Lizardi, D. C. Ward, Proc. Natl. Acad. Sci. USA 97, 10113 (2000).
88. R. C. Jin, G. S. Wu, Z. Li, C. A. Mirkin, G. C. Schatz, J. Am. Chem. Soc. 125, 1643 (2003).
89. J. Liu, Y. Lu, J. Amer. Chem. Soc. 127, 12677 (2005).
90. A. P. Alivisatos, K. P. Johnsson, X. G. Peng, T. E. Wilson, C. J. Loweth, M. P. Bruchez, P. G. Schultz, Nature 382, 609 (1996).
91. C. J. Loweth, W. B. Caldwell, X. G. Peng, A. P. Alivisatos, P. G. Schultz, Angew. Chem. Int. Ed. 38, 1808 (1999).
92. W. J. Parak, T. Pellegrino, C. M. Micheel, D. Gerion, S. C. Williams, A. P. Alivisatos, Nano Lett. 3, 33 (2003).
93. J. J. Storhofff, R. Elghanian, C. A. Mirkin, R. L. Letsinger, Langmuir 18, 6666 (2002).
94. C. M. Niemeyer, B. Ceyhan, P. Hazarika, Angew. Chem. Int. Ed. 42, 5766 (2003).
95. P. Hazarika, B. Ceyhan, C. M. Niemeyer, Angew. Chem. Int. Ed. 43, 6469 (2004).
96. J. W. Liu, Y. Lu, J. Am. Chem. Soc. 125, 6642 (2003).
97. S. W. Santoro, G. F. Joyce, Proc. Natl. Acad. Sci. USA 94, 4262 (1997).
98. D. G. Georganopoulou, L. Chang, J. M. Nam, C. S. Thaxton, E. J. Mufson, W. L. Klein, C. A. Mirkin, Proc. Natl. Acad. Sci. USA 102, 2273 (2005).
99. J. M. Nam, C. S. Thaxton, C. A. Mirkin, Science 301, 1884 (2003).
100. C. Sonnichsen, B. M. Reinhard, J. Liphardt, A. P. Alivisatos, Nature Biotechnology 23, 741 (2005).
101. G. Liu, Y. Yin, S. Kunchakarra, B. Mukherjee, D. Gerion, S. D. Jett, D. G. Bear, J. W. Gray, A. P. Alivisatos, L. P. Lee, F. F. Chen, Nature Nanotechnology 1, 47 (2006).
102. H. Qiu, J. C. Dewan, N. C. Seeman, J. Mol. Biol. 267, 881 (1997).
103. A. Chworos, I. Severcan, A. Y. Koyfman, P. Weinkam, E. Oroudjev, H. G. Hansma, L. Jaeger, Science 306, 2068 (2004).
104. P. W. Rothemund, A. Ekani-Nkodo, N. Papadakis, A. Kumar, D. K. Fygenson, E. Winfree, J. Am. Chem. Soc. 126, 16344 (2004).
105. D. Liu, S. H. Park, J. H. Reif, T. H. LaBean, Proc. Natl. Acad. Sci. USA 101, 717 (2004).

106. N. C. Seeman, DNA Cell Biol 10, 475 (1991).
107. D. Liu, M. Wang, Z. Deng, R. Walulu, C. Mao, J. Am. Chem. Soc. 126, 2324 (2004).
108. R. P. Goodman, R. M. Berry, A. J. Turberfield, Chem. Comm., 1372 (2004).
109. W. M. Shih, J. D. Quispe, G. F. Joyce, Nature 427, 618 (2004).
110. L. Feng, S. H. Park, J. H. Reif, H. Yan, Angew. Chem. Int. Ed. Engl. 42, 4342 (2003).
111. N. C. Seeman, Biochemistry 42, 7259 (2003).
112. C. Mao, W. Sun, N. C. Seeman, J. Am. Chem. Soc 121, 5437 (1999).
113. T. J. Fu, N. C. Seeman, Biochemistry 32, 3211 (1993).
114. X. Li, X. Yang, J. Qi, N. C. Seeman, J. Am. Chem. Soc 118, 6131 (1996).
115. M. Brucale, G. Zuccheri, B. Samorì, Org. Biomol. Chem. 4, 3427 (2006).
116. F. Mathieu, S. Liao, J. Kopatsch, T. Wang, C. Mao, N. C. Seeman, Nano Lett. 5, 661 (2005).
117. M. Brucale, G. Zuccheri, B. Samori, Trends Biotechnol 24, 235 (2006).
118. L. M. Adleman, Science 266, 1021 (1994).
119. E. Winfree, in DNA Based Computers: DIMACS Workshop R. J. Lipton, E. B. Baum, Eds. (American Mathematical Society, Providence, RI, 1996), vol. 27, pp. 199.
120. S. Roweis, E. Winfree, R. Burgoyne, N. V. Chelyapov, M. F. Goodman, P. W. Rothemund, L. M. Adleman, J. Comput. Biol. 5, 615 (1998).
121. C. Mao, T. H. LaBean, J. H. Relf, N. C. Seeman, Nature 407, 493 (2000).
122. T. H. LaBean, H. Yan, J. Kopatsch, F. Liu, E. Winfree, J. H. Reif, N. C. Seeman, J. Am. Chem. Soc. 122, 1848 (2000).
123. E. Winfree, F. Liu, L. A. Wenzler, N. C. Seeman, Nature 394, 539 (1998).
124. S. H. Park, R. Barish, H. Li, J. H. Reif, G. Finkelstein, H. Yan, T. H. LaBean, Nano Lett. 5, 693 (2005).
125. B. Ding, R. Sha, N. C. Seeman, J. Am. Chem. Soc. 126, 10230 (2004).
126. S. H. Park, P. Yin, Y. Liu, J. H. Reif, T. H. LaBean, H. Yan, Nano Lett. 5, 729 (2005).
127. J. C. Mitchell, J. R. Harris, J. Malo, J. Bath, A. J. Turberfield, J. Am. Chem. Soc. 126, 16342 (2004).
128. A. Ekani-Nkodo, A. Kumar, D. K. Fygenson, Phys. Rev. Lett. 93, 268301 (2004).
129. N. C. Seeman, Q. Rev. Biophys. 38, 363 (2005).
130. N. C. Seeman, Methods Mol. Biol. 303, 143 (2005).
131. H. Yan, T. H. LaBean, L. Feng, J. H. Reif, Proc. Natl. Acad. Sci. USA 100, 8103 (2003).
132. C. Lin, Y. Liu, S. Rinker, H. Yan, Chem. Phys. Chem. 7, 1641 (2006).
133. H. Wang, Bell System Technical Journal 40, 1 (1961).
134. S. H. Park, C. Pistol, S. J. Ahn, J. H. Reif, A. R. Lebeck, C. Dwyer, T. H. Labean, Angew. Chem. Int. Ed. Engl. 45, 6607 (2006).
135. Y. Liu, Y. Ke, H. Yan, J. Am. Chem. Soc. 127, 17140 (2005).

136. P. W. Rothemund, N. Papadakis, E. Winfree, PLoS Biol 2, e424 (2004).
137. E. Winfree, R. Bekbolatov, DNA Computing 2943, 126 (2004).
138. R. Schulman, E. Winfree, DNA Computing 3384, 319 (2005).
139. D. Soloveichik, E. Winfree, DNA Computing 3892, 305 (2006).
140. P. W. Rothemund, Nature 440, 297 (2006).
141. D. Lubrich, J. Bath, A. J. Turberfield, Nanotechnology 16, 1574 (2005).
142. T. H. LaBean, E. Winfree, J. H. Reif, in DIMACS Series in Discrete Mathematics and Theoretical Computer Science E. Winfree, D. K. Gifford, Eds. (Am. Math. Soc., Providence, RI, 1999) pp. 123.
143. H. Liu, Y. Chen, Y. He, A. E. Ribbe, C. Mao, Angew. Chem. Int. Ed. Engl. 45, 1942 (2006).
144. H. Liu, Y. He, A. E. Ribbe, C. Mao, Biomacromolecules 6, 2943 (2005).
145. Y. He, Y. Tian, Y. Chen, Z. Deng, A. E. Ribbe, C. Mao, Angew. Chem. Int. Ed. Engl. 44, 6694 (2005).
146. R. P. Goodman, Biotechniques 38, 548 (2005).
147. N. C. Seeman, J. Biomol. Struct. Dyn. 8, 573 (1990).
148. C. Vaillant, B. Audit, C. Thermes, A. Arneodo, Eur. Phys. J. E Soft Matter 19, 263 (2006).
149. Y. Ke, Y. Liu, J. Zhang, H. Yan, J. Am. Chem. Soc. 128, 4414 (2006).
150. W. P. C. Stemmer, A. Crameri, K. D. Ha, T. M. Brennan, H. L. Heyneker, Gene 164, 49 (1995).
151. P. J. Paukstelis, J. Nowakowski, J. J. Birktoft, N. C. Seeman, Chem. Biol. 11, 1119 (2004).
152. P. E. Constantinou, T. Wang, J. Kopatsch, L. B. Israel, X. P. Zhang, B. Q. Ding, W. B. Sherman, X. Wang, J. P. Zheng, R. J. Sha, N. C. Seeman, Org. Biomol. Chem. 4, 3414 (2006).
153. D. Reishus, B. Shaw, Y. Brun, N. Chelyapov, L. Adleman, J. Am. Chem. Soc. 127, 17590 (2005).
154. Y. He, Y. Chen, H. Liu, A. E. Ribbe, C. Mao, J. Am. Chem. Soc. 127, 12202 (2005).
155. Y. He, C. Mao, Chem. Comm., 968 (2006).
156. P. E. Constantinou, T. Wang, J. Kopatsch, L. B. Israel, X. Zhang, B. Ding, W. B. Sherman, X. Wang, J. Zheng, R. Sha, N. C. Seeman, Org. Biomol. Chem. 4, 3414 (2006).
157. Y. Chen, C. Mao, J Am Chem Soc 126, 13240 (2004).
158. C. M. Niemeyer, M. Adler, Angew Chem Int Ed Engl 41, 3779 (2002).
159. N. C. Seeman, Trends Biochem Sci 30, 119 (2005).
160. F. C. Simmel, W. U. Dittmer, Small 1, 284 (2005).
161. M. Brucale, G. Zuccheri, B. Samori, Org. Biomol. Chem. 3, 575 (2005).
162. Y. Chen, S. H. Lee, C. Mao, Angew. Chem. Int. Ed. Engl. 43, 5335 (2004).
163. X. Yang, A. V. Vologodskii, B. Liu, B. Kemper, N. C. Seeman, Biopolymers 45, 69 (1998).

164. B. Yurke, A. J. Turberfield, A. P. Mills, Jr., F. C. Simmel, J. L. Neumann, Nature 406, 605 (2000).
165. J. J. Li, W. Tan, Nano Lett. 2, 315 (2002).
166. A. J. Turberfield, J. C. Mitchell, B. Yurke, A. P. Mills, Jr., M. I. Blakey, F. C. Simmel, Phys. Rev. Lett. 90, 118102 (2003). 167. W. B. Sherman, N. C. Seeman, Nano Lett. 4, 1203 (2004).
168. J. S. Shin, N. A. Pierce, J. Am. Chem. Soc. 126, 10834 (2004).
169. Y. Tian, C. Mao, J. Am. Chem. Soc. 126, 11410 (2004).
170. W. Shen, M. F. Bruist, S. D. Goodman, N. C. Seeman, Angew. Chem. Int. Ed. Engl. 43, 4750 (2004).
171. P. Yin, H. Yan, X. G. Daniell, A. J. Turberfield, J. H. Reif, Angew. Chem. Int. Ed. Engl. 43, 4906 (2004).
172. Y. Chen, M. Wang, C. Mao, Angew. Chem. Int. Ed. Engl. 43, 3554 (2004).
173. Y. Chen, C. Mao, J. Am. Chem. Soc. 126, 8626 (2004).
174. G. Seelig, B. Yurke, E. Winfree, J. Am. Chem. Soc. 128, 12211 (2006).
175. G. Seelig, B. Yurke, E. Winfree, DNA Computing 3384, 329 (2005).
176. T. Liedl, M. Olapinski, F. C. Simmel, Angew. Chem. Int. Ed. 45, 5007 (2006).
177. T. Liedl, F. C. Simmel, Nano Lett. 5, 1894 (2005).
178. W. U. Dittmer, F. C. Simmel, Nano Lett. 4, 689 (2004).
179. R. M. Dirks, N. A. Pierce, Proc. Natl. Acad. Sci. USA 101, 15275 (2004).
180. C. Buranachai, S. A. McKinney, T. Ha, Nano Lett. 6, 496 (2006).
181. W. M. Shu, D. S. Liu, M. Watari, C. K. Riener, T. Strunz, M. E. Welland, S. Balasubramanian, R. A. McKendry, J. Am. Chem. Soc. 127, 17054 (2005).
182. L. M. Smith, Nature 440, 283 (2006).

CHAPTER 13

RECENT PROGRESS ON BIO-INSPIRED SURFACE WITH SPECIAL WETTABILITY

Shutao Wang,[a,c] Huan Liu,[b] Lei Jiang[a,b,*]

[a] Key Laboratory of Organic Solids, Institute of Chemistry, Chinese Academy of Sciences, Beijing 100080, P. R. China; [b] National Center for NanoScience and Technology, Beijing 100080, P. R. China; [c] Graduate School of Chinese Academic of Sciences, Beijing 100084, P. R. China; [*]Corresponding author: Tel.& Fax: (+86) 10-82627566, E-mail: jianglei@iccas.ac.cn

Design and creation bio-inspired surface with special wettability has aroused great interest from the various areas of scientific research and industrial applications. Micro- and nanoscale structures have been witnessed to act as the crucial role in realizing unique surface wettability of plants and insects in nature. Inspired by these findings, various methods have been developed to fabricate superhydrophobic surfaces, which has been oriented to practicality and functionality beyond original fundamental simulation. In addition, chemical composition is another main aspect in the control of surface wettability, combined which many stimuli-responsive surfaces with reversible wettability, sometimes even between superhydrophobicity and superhydrophilicity, triggered by temperature, light, electric and pH *etc.*, have also been fabricated both from inorganic/organic compounds via various techniques. Particularly, a few encouraging results on dual-responsive surfaces has also been reviewed in this paper, which are important for practical applications and indicative of a promising future. In this review, recent progresses on bio-inspired surface with special wettability are covered from natural surfaces with special wettability, artificial superhydrophobic surfaces, and reversible stimuli-responsive surfaces.

1. Introduction

In the past decades, many excellent studies have extensively focused on surface wettability, driven by promising industrial applications, such as self-cleaning, antifogging, anticorrosion, anti-snow-sticking, smart responsive devices, lossless liquid transfer, *etc.*, and important theoretical researches as well [1-5]. Surface wettability is one of the fundamental properties of a solid surface, and is governed by surface chemical compositions and surface structures [6-8]. It is from these two aspects that the perfect combination of nanotechnology and chemistry drastically accelerates the understanding and evolution of this field.

The developments of surface wettability closely relates to the new discoveries of biological research on the fascinating surfaces of plants or insects. For example, the water droplets can roll off the lotus leaves easily and bring off dust particles, which originate from the waxy material and micro-and nano- hierarchical structure [9,10]. Inspired by lotus leaves, a lot of artificial lotus-like surfaces have been prepared for realizing the self-cleaning property [11-17]. Cicada's wings are endowed with self-cleaning and transparent properties due to the aligned nanostructures on their surfaces [18], which provide a new clue to prepare transparent glass or other surfaces with the self-cleaning property. Water stride can freely jump and walk on the water surface in river or lake, which owing to surface hydrophobic compounds and micro-and nano- hierarchical structure on stride legs [19]. The large water-repellent force "artificial legs" has been obtained by mimicking stride legs [20,21]. Moreover, desert beetle can collect water by virtue of its back with alternately hydrophilic and superhydrophobic regions [22]. Therefore, researches on the discovery of special wettability surfaces in nature, particularly understanding the structural effect within, directly propel the recent development of bio-inspired surfaces with special wettability.

As known materials in nature are formed after millions of years' evolution and they are much more complex than what we can image. Thus, learning from nature, grasping certain algorithm beneath, and then trying to apply it to the artificial process is right the destination pursued by many scientists. With the comprehensive exploration of responsive materials in recent [23], the stimuli-responsive surfaces with controllable wettability are emerging as a new highlight in the field of surface science and nanotechnology [24-29]. Furthermore, combined with the binary-cooperative effects on the surface [30], just as philosophy about yin and yang that proposed by Chinese ancient philosophers, and inspirations from nature, our group constructs the first surface switch between superhydrophilicity and superhydrophobicity on micro- and nano-hierarchical solid substrates using the thermal responsive polymer [31]. Subsequently, other intelligent surfaces switch between superhydrophilicity and superhydrophobicity have been fabricated, responding to external stimuli, such as light [32], temperature [33], electric field [34], and pH [35]. These switchable wettability surfaces result from the cooperative effect of the appropriate surface geometric structures and stimuli-responsive materials, which would bring a promising future for the construction of new generation intelligent devices.

This review is limited primarily to the bio-inspired surface with special wettability. It includes the following sections: some basic aspects about surface wettability, unique superhydrophobic surfaces in nature, artificial superhydrophobic surfaces, superhydrophilic surfaces, stimuli-responsive surfaces between superhydrophobicity and superhydrophilicity, and conclusion. Likewise, a few issues relating to the manipulation of surface wettability from superhydrophobicity to superhydrophilicity, are also discussed.

2. Some Basic Aspects about Surface Wettability

2.1. Hydrophilicity and Hydrophobicity

Surface wettability is one of paramount properties on solid surfaces and relate closely to fundamental science theory and practical applications. When a water droplet is placed on the solid surface, it always exhibits one of the four cases: hydrophilicity, hydrophobicity, superhydrophilicity, and superhydrophobicity (Figure 1). The wettability on solid surface is commonly determined by CA measurements. The CA is considered as the cooperative result of different tensions at the three-phase line on a flat surface, which is given by Young's equation [36]. Here the water droplet is regarded as a uniform mathematic entity between surface and body. In fact, the water structures and reactivity on both interface and body are very different and change along with the change of contacting surfaces. Learning this point promotes the deep understanding on surface wettability.

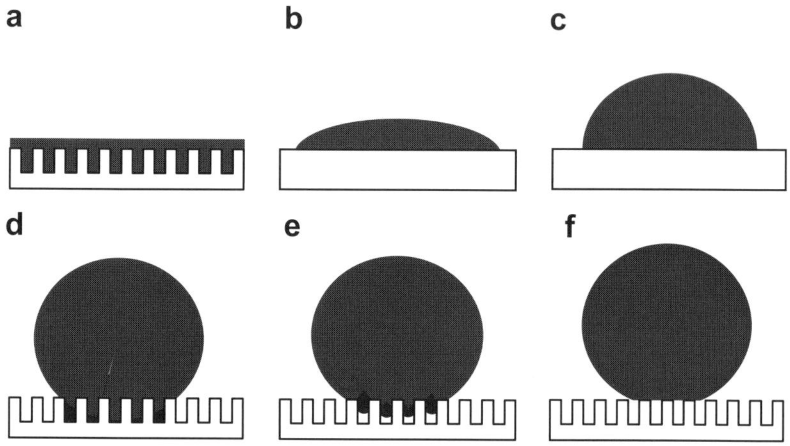

Figure 1. Several models of surface wettability, showing a) superhydrophilicity, b) hydrophilicity, c) hydrophobicity, and d-f) superhydrophobicity (from d to f: Wenzel's model, transition between Wenzel's and Cassie's, and Cassie's model).

By reviewing the recent studies on water structure and reactivity using the surface force apparatus and ancillary techniques, Volger pointed that water structure is a manifestation of hydrophobicity and quantitative definition of hydrophobicity and hydrophilicity [37], although the convergence of experimental observations and theoretical interpretation remains further done. The hydrophobic surface is the surfaces with CA higher than 65° and pure water adhesion tension less than 30 dyn/cm between which there exhibit long-range attractive forces in distance of tens of nanometers. In contrast, the hydrophilic surface is the surfaces with CA less than 65° and pure water adhesion tension less than 30 dyn/cm between which there exhibit repulsive forces. On the surfaces with hydrophobicity and hydrophilicity, water adopts two distinct structures as shown in Figure 2, owing to the difference in self-association between vicinal water and bulk-water. A relatively more-dense water region with a collapsed hydrogen-bonded network forms against hydrophilic surfaces, while a relative less-dense water region with an open hydrogen-bonded network forms against hydrophobic surfaces. Chandler *et al.* also confirmed theoretically that the depletion of hydrogen bonding in water near to hydrophobic surfaces could lead to drying of extended hydrophobic surfaces and long ranged forces at nanometer scale. Looked from another perspective, the distinguished difference of water structures at surfaces also suggests the difference between hydrophobicity and hydrophilicity.

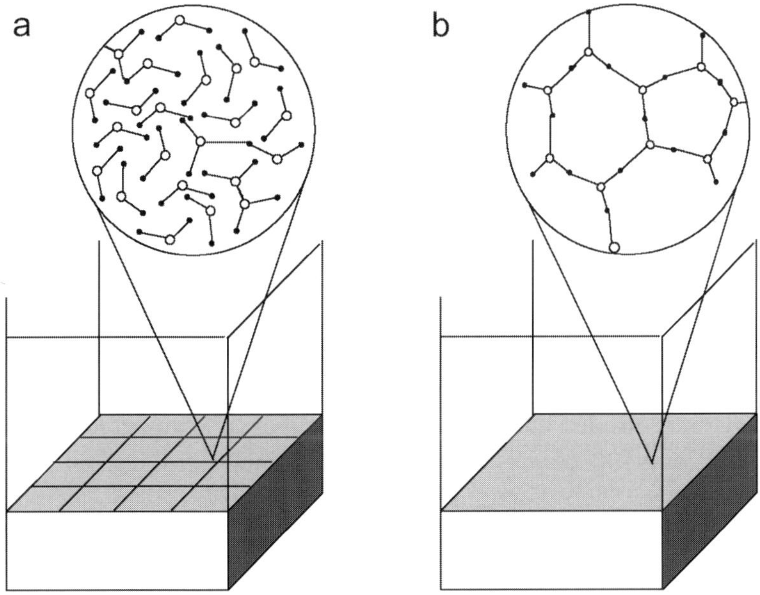

Figure 2. Two dimensional projection of water at or near a) hydrophilic and b) hydrophobic surfaces, illustrating the hypothetical state of self-association. The Lewis sites on the hydrophilic surface (indicated by the lattice site array) competes with water self association resulting to a more dense region near the surface. By contrast, water at the hydrophobic surface bearing no competing Lewis sites forms a region less dense than bulk water.

2.2. Wenzel's Model and Cassie's Model

Different from ideal flat surfaces, the real surfaces are often rough. So surface roughness has to be considered as the other important factor to influence surface wettability, where two main theories are involved: Wenzel's model (Figure 1d) and Cassie's model (Figure 1f). In Wenzel's model [6], it is assumed that the liquid can intrude and contact the concave region completely on solid surfaces, and the equation (Equation 1):

$$\cos \theta_r = r \cos \theta, \quad (1)$$

where θ_r is the apparent CA on rough surfaces, θ is the intrinsic CA on flat surfaces, and r is surface roughness. In Cassie's model [7], it is considered that air can be trapped by the liquid to form a composite surface, thus the apparent CA is modified as Equation 2:

$$\cos \theta_r = r f_1 \cos \theta - f_2, \quad (2)$$

where f_1 and f_2 are the area fractions of the projecting solid and vapor on the surface, respectively, and $f_1 + f_2 = 1$. Viewed from both theories, surface roughness can enhance surface wettability, both hydrophilicity and hydrophobicity, but leading to different states.

In some practical cases, a water droplet on the solid surface maybe adopts certain a transition state between Wenzel's and Cassie's models, which brings out the question about the transitions between both the extremes arises (Figure 1e). Some studies show that a water droplet with apparent CA consistent with Cassie's model will change into a state with Wenzel's model when the droplet bears definite physical pressure [38]. As a result of the transition, part of water fills the grooves of the substrate and surface gets pined with the droplet. A threshold value θ_T exists between the two regimes, given by Equation 3 that is derived by combining Equation 1 and Equation 2:

$$\cos \theta_T = (f_s - 1)/(r - f_s). \quad (3)$$

If the Young CA is lower than θ_T, the trapped vapor pockets are metastable, under which situations the transition possible takes place. Only for $\theta > \theta_T$ or $\cos \theta < -1/r$ the Cassie regime is stable when the vapor pockets is stable too. In other words, the stable Cassie regime relies on rough and hydrophobic enough surface and θ_T value as small as possible.

2.3. Superhydrophilicity and Superhydrophobicity

Surface roughness can enhance the surface wettability, consequently leading to two extreme cases of special wettability: superhydrophilicity and superhydrophobicity. On the superhydrophilic surface, the water droplet can wet and spread out and the static CA inclines to 0° [39]. Some surfaces can show the superhydrophilic property merely depending on the high-surface-energy chemical compositions. Nevertheless, in most conditions, the two- and three-dimensional capillary effect originating from surface structure plays a crucial role in realizing the superhydrophilic surfaces. On the superhydrophobic surface, the water droplet has the static CA larger than 150°. The maximum CA was reported to merely reach about 120° on the flat CF_3-terminated surface with the lowest free energy [40]. So the introduction of surface roughness is necessary to obtain the superhydrophobic surfaces. In practical applications of the superhydrophobic surfaces, the CA hysteresis between advancing and receding angles is also important because it reflects the dynamic behavior of the droplet on the superhydrophobic surfaces. The sliding angle (SA) generally used to describe the CA hysteresis. On the superhydrophobic surfaces governed by the Wenzel's model, the water droplet remains fixed even if the film is slightly tilted to a distinguish angle; however, it rolls off easily on those governed by the Cassie's model [41]. The theoretical simulation reveals that increasing surface roughness can induce the transition from Wenzel's model to Cassie's one [42]. Thus, the CA hysteresis can be lower by increasing surface roughness. Moreover, the short continuous three-phase contact line is favorable for surfaces with a low CA hysteresis. Therefore, surface geometrical structures should be taken into account in designing and constructing the surfaces with special wettability.

3. Unique Superhydrophobic Surfaces in Nature

When the hydrophilic route is used for self-cleaning in daily life, nature brings forth the alternative superhydrophobic route to reach the same aim. The studies have revealed that all these natural superhydrophobicity not only related to their intrinsic property of surface materials, but also may more depend on surface micro- and/or nanostructures. Besides the self-cleaning effect, the natural superhydrophobic surfaces also possess other excellent functions.

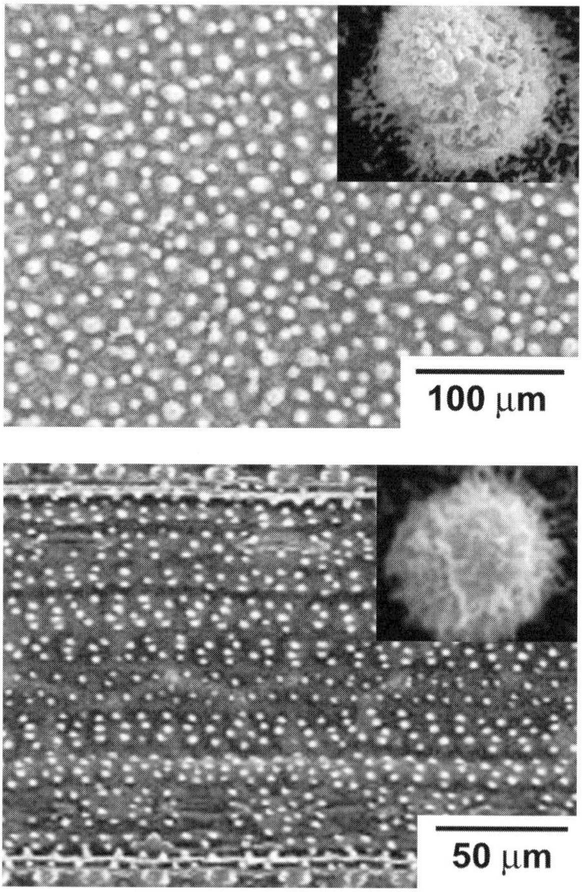

Figure 3. Large scale SEM image of (a) the lotus leaf (*Nelumbo nucifera*) and the magnified image on a single papilla (inset), which show the interesting micro- and nanoscale hierarchical structures; and (b) the rice leaf (*Oryza sativa*) and the magnified image on a single papilla (inset), exhibiting the anisotropic hierarchical structures. Reprinted with permission from [Ref. 10], L. Feng, *et al.* Adv. Mater. 14, 1857 (2002), Copyright @ WILEY-VCH.

The superhydrophobicity of lotus leaves is the most famous case, usually designated as "lotus effect" that water can keep spherical on lotus leaves and roll off easily. Barthlott and Neinhuis first revealed the secret about the superhydrophobicity of lotus leaves that the unique property results from the micrometer sized papillae on the surfaces and the hydrophobic waxy materials as well [9]. However, detailed studies on lotus leaves by scanning electron microscopy (SEM) suggest that there exist micro- and nanoscale hierarchical structures on the surfaces and the micro-papillae (5-9 μm) are covered by fine-branched nanostructures (*ca.* 120 nm) (Figure 3a). Further theoretical simulation and experiment results indicate that the unique superhydrophobicity results from the integration of dual-scale hierarchical structures and waxy materials [10].

Another example is the anisotropic superhydrophobic phenomenon on the rice leaves [10]. The rice leaves show a similar hierarchical structure to that on lotus leaves (Figure 3b). Differently, the hierarchical structure adopts the anisotropic arrangement mode instead of the random mode on lotus leaves. In detail, the papillae are arranged in one-dimensional order parallel to the edge of the leaves. The water droplet can roll off easily along the ordered direction (SA is *ca.* 3-5°), while is harder to move along the perpendicular one (SA is *ca.* 9-15°). The phenomenon indicates that the microscopic anisotropic structure leads directly to the macroscopic anisotropic wettability. In comparison, the homogeneous distribution of papillae on lotus leaves contributes to the isotropic wettability. The experimental mimicry of anisotropic and isotropic hierarchical structures using aligned carbon nanotube (ACNT) also confirms the important influence of hierarchical structures on surface wettability [43].

In addition to plant leaves, many insects also show their special superhydrophobicity. For example, Cicadae possess a pair of transparent and superhydrophobic wings that can be contamination-free and antiwetting [18]. There exist numerous aligned nanocolumns with diameters of ca. 70 nm and a column-to-column space of ca. 90 nm (Figure 4). Surface roughness has reverse effect on surface wettability and optical transparence. Such surface nanostructures on cicada's wings may provide a good example for designing the superhydrophobic and antireflective film.

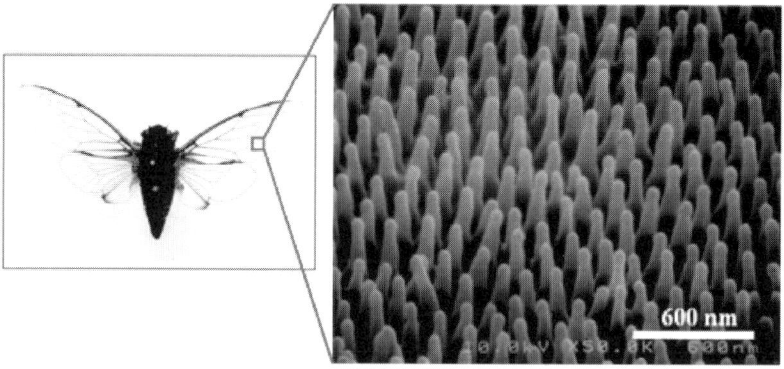

Figure 4. *Cicada orni* with superhydrophobic wings. The SEM image of its surface shows regularly aligned nanocolumns. Reprinted with permission from [Ref. 18], W. Lee, *et al.* Langmuir 20, 7665 (2004), Copyright @ American Chemical Society.

Furthermore, the superhydrophobicity on the surface of iridescent butterfly's wings has also been observed [44]. Recently, our group firstly discovered *directional* adhesion on superhydrophobic wings of butterfly *Morpho aega* [45], as shown in Figure 5. A water droplet rolls off easily along the radial outward (RO) direction of the central axis of body but is pinned tightly against the RO direction. Interestingly, these two distinct states can be tuned by controlling the posture of wings downward or upward, and the direction of airflow across the surface along or against the RO direction, respectively. It is believed that this special property is caused by the direction-dependent arrangement of *flexible* nano-tips on the ridging nano-stripes and micro-scales overlapped on the wings at *one-dimensional* level (Figure 6c) where two distinct contact modes of a droplet with orientation-tunable microstructures occur and thus produce different adhesive forces. This finding may be helpful to design smart fluid-controllable interfaces in novel microfluidic devices and directional easy-cleaning coatings.

Water striders can stand effortlessly and walk freely on water, because of the remarkable superhydrophobic force provided by their nonwetting legs [19]. The maximal supporting of a single leg reaches up to 152 dyn (1 dyn = 10^{-5} N) so that it can support about 15 times weight of the insect. The SEM observation reveals that the leg is covered by numerous oriented microscale needle-shaped setae and many elaborate nano-grooves are on each microseta, forming an interesting hierarchical structure (Figure 6). Such a unique hierarchical structure and the hydrophobic waxy layer contribute to the superhydrophobicity of the water strider's leg.

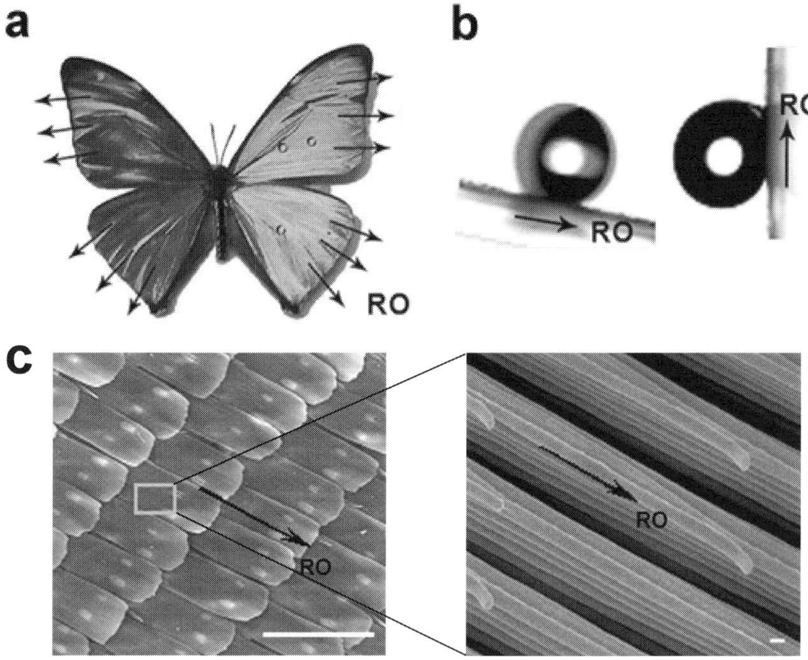

Figure 5. The superhydrophobic butterfly's wings with one-dimensional anisotropic adhesive property. Image of a) an iridescent blue butterfly *M. aega*, where the black arrows denote the radial-outward (RO) directions; b) a droplet that easily rolls off along the RO direction (left), and a droplet that is firmly pinned on the wing that is tilted upward, even at upright (right); and c) SEM images of the periodic arrangement of overlapping micro-scales on the wings and fine lamella-stacking nano-stripes on scales, which exists an interesting hierarchical micro- and nanostructures on the surface of wings. Scale bars: a: 100 μm; b: 100 nm. Reprinted with permission from [Ref. 45], Y. Zheng, *et al.* Soft Matter, Copyright @ The Royal Society of Chemistry.

Moreover, the Stenocara beetle shows the capability of fog-collecting from fog-laden wind on its back. Porker and Lawrence reveal that beetle's back consists of an array of surface bumps decorated on top with hydrophilic non-waxy regions (about 100 μm in diameter) on a superhydrophobic background and the superhydrophobic regions consist of micro-hemispheres in a regular hexagonal array (Figure 7) [22]. Thus, the design of such fog-collecting structure may have promising applications in water-trapping tent and building coatings.

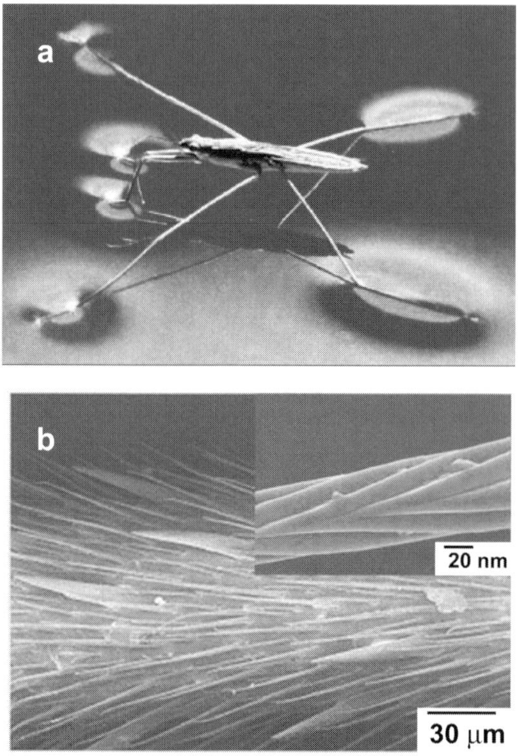

Figure 6. Image of a) a water strider (*Gerris remigis*) with super-hydrophobic legs, standing on the water surface; and b) an SEM image of the leg with oriented spindly microsetae. The inset is a single seta with the nano-groove structure. Reprinted with permission from [Ref. 19], X. Gao, *et al.* Nature 432, 36 (2004), Copyright @ Nature Publishing Group.

In short, these perfect surfaces with superhydrophobicity gestated by nature will always give the inspiration to develop facile approaches, to fabricate special micro- and nanostructures, and to obtain functional and applicable surfaces with special wettability.

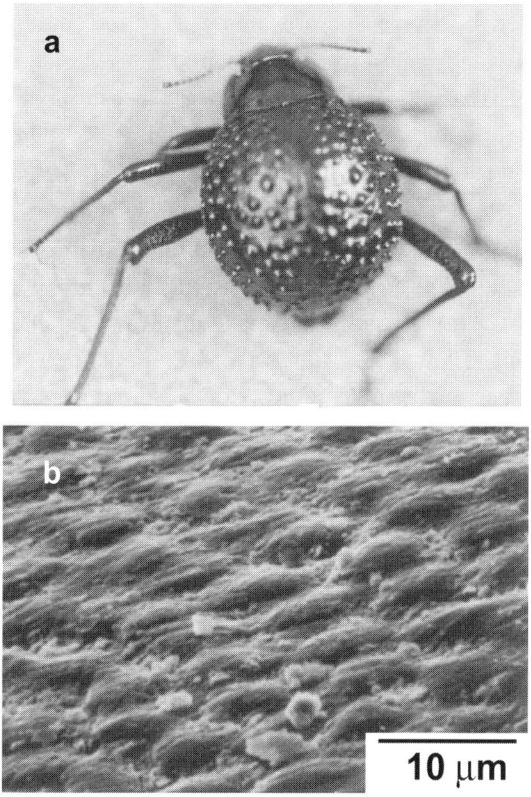

Figure 7. The water-capturing surface of the fused overwings of the desert beetle (*Stenocara*). Image of a) an adult female, showing the dorsal view, where peaks and troughs are evident on the surface of its back; and b) an SEM micrograph of the textured surface of the depressed superhydrophobic areas. Reprinted with permission from [Ref. 22], A. R. Parker *et al.* Nature, 414, 33 (2001), Copyright @ Nature Publishing Group.

4. Artificial Superhydrophobic Surface

As the above-mentioned, the studies on natural surfaces reveal that the realization of surface superhydrophobicity depends on the unique micro- and/or nanostructures, besides the chemical composition. Inspired by these, two main approaches are inferred to construct artificial superhydrophobic surfaces: constructing a micro- and/or nanostructures on the surface of hydrophobic materials, and chemically modifying a micro- and/or nanostructured surface with materials of low surface free energy. Guided by this principle, many excellent methods have been proposed to fabricate superhydrophobic surfaces, such as photolithography-based microfabrication [46], plasma etching/ polymerization of polymer [47,48], chemical vapor deposition (CVD) [17,49] and wet chemical deposition [50-52], template extrusion technique [53,54], glancing-angle-deposition [55], sol-gel method [56] and so on. Currently, the striking development of nanotechnology and material science is favorable to the biomimic construction of micro- and/or nanostructures on the solid substrates. Thus, many new approaches have been proposed to construct the artificial superhydrophobic surfaces, which endow some remarkable features: simple process, environmental stability, and multi-function.

4.1. Towards the Simple Process

Complex procedures and rigorous conditions are not desired in preparation of superhydrophobic surfaces, so one of main tasks in this area is to explore the simple process means always. In recent various elegant methods have been developed, and here we will show some examples to construct superhydrophobic surfaces through the simple process.

Our discussions begin with polymer materials because polymer materials have widely uses in daily life and industrial applications. Recently, a simple template-based rolling press method has been developed [57], by transferring the nanopore to a tubular alumina surface, to prepare large-scale, well-aligned nanopillars polymer surfaces with superhydrophobicity. In comparison with other template techniques [58,59], this method is simple, inexpensive and repeated-usable. However, it is not suitable to the weak hydrophobic polymer owing to the limited height of nanopillars formed on the surfaces.

The introduction of the electrohydrodynamic (EHD) technique successfully overcomes the above-mentioned shortcoming of template [60-62], because it is a versatile technique that has been used to manufacture fibers or particles with the diameter ranging from nanometers to micrometers. A superhydrophobic polystyrene film with a lotus-like structure consisting of porous microspheres and nanofibers has been prepared by the EHD method [60]. The porous microspheres contribute to the superhydrophobicity by increasing surface roughness, while nanofibers interweave to form a 3D network that reinforces the composite film. It is believed that the EHD method is a simple and effective way to prepare large-area superhydrophobic surfaces and is easily extended to a wide variety of materials. For example, Menceloglu *et al.* [63], Alcock *et al.* [64], and Agarwal [65] *et al.* recently have reported the formation of superhydrophobic surface with the similar structures, respectively, by electrospinning of fluorinated polymers under suitable conditions. Rutledge *et al.* have fabricated a superhydrophobic rough poly(caprolactone) surface by the EHD method after further modification of a hydrophobic polymer by CVD [66].

Another simple technique to convert simple polymers into superhydrophobic surfaces is to control polymer crystallization by cooling process or phase separation. It has become widely used to construct micro- and nano-structures, after Tsujii and co-workers prepared the superhydrophobic surfaces firstly by the crystallization of melted alkylketene dimmer [67]. Erbil *et al.* control the crystallization of isotactic polypropylene by choosing suitable solvents and manipulating the temperature, leading to a porous superhydrophobic surface with the CA about 160° (Figure 8a) [68]. Furthermore, Lu *et al.* have fabricated the superhydrophobic low-density polyethylene films by modulating the surface crystallization behavior [69]. Recently, to simplify the crystallization process, Xie *et al.* have reported a one-step route based on phase separation and obtained a porous superhydrophobic coating [70]. By use of phase-separated phenomenon, Han *et al.* prepared a rough superhydrophobic surface through the hydrogen-bonding interaction of supramolecular organosilicane [71].

In addition, a one-step immersing process has recently been proposed to construct a superhydrophobic coating on metal surfaces [72]. In a typical experiment, taking the copper as an example, a copper substrate is immersed in the solution of $CH_3(CH_2)_{12}COOH$ with a appreciate concentration for a period of time, and then the superhydrophobic coating with flower-like microstructures is formed (Figure 8). Here involves the bio-inspired morphology genesis, leading to the one-step formation of new hydrophobic materials $Cu(CH_3(CH_2)_{12}COO)_2$ and surface microstructures. In comparison with other wet chemical routes [50-52], this study may provide a simple technique for the preparation of superhydrophobic coating on metal surfaces.

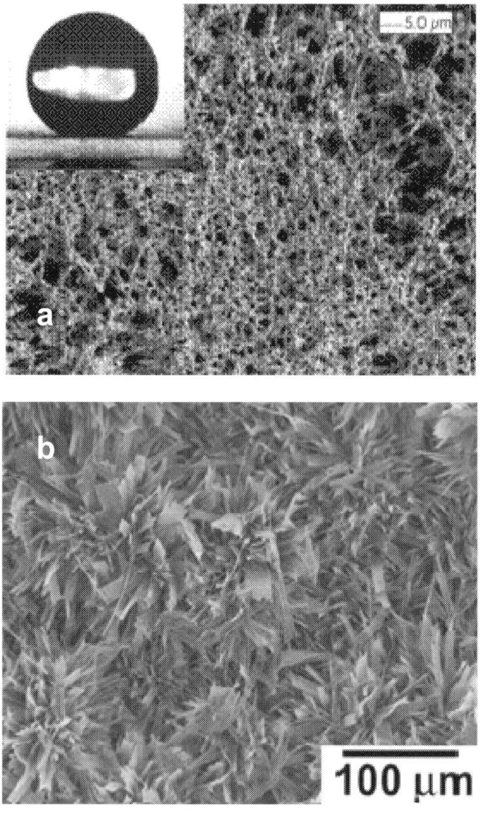

Figure 8. SEM image of a) the superhydrophobic coating of the isotactic polypropylene with CA 160° prepared by a crystallization process, and b) the superhydrophobic $Cu(CH_3(CH_2)_{12}COO)_2$ film with flower-like microstructures fabricated by a one-step immersing process. Reprinted with permission from [Ref. 68], H. Y. Erbil, *et al.* Science 299, 1377 (2003) Copyright @ American Association for the Advancement of Science. Reprinted with permission from [Ref. 72], S. Wang, *et al.* Adv. Mater. 18, 767 (2006) Copyright @ WILEY-VCH.

Therefore, whether the superhydrophobic coating can enter the practical applications depends on the improvement of fabricated methods and techniques. With the continuous invention of advanced technology, it is inevitable that nanotechnology have been the dominant driven force in the construction of superhydrophobic surfaces.

4.2. Towards Environmental Stability

The practical applications of superhydrophobic coatings are always inhibited by the problem of environmental stability when the superhydrophobic coatings meet bad external conditions, such as the acid or basic environmental conditions, the existence of organic solvents, or the temperature change. For example, the superhydrophobic alkylketene dimmer surface is environmentally weak, mainly because of its good solubility in organic solvent and its melting point as low as 67°C [67]. Recently, some studies have attempted to solve this problem by exploring and constructing novel superhydrophobic surfaces.

Recently, our group has given the first example that the film can keep the superhydrophobic property when exposed different pH environments [73]. We obtained the nanostructured graphite-like carbon films by pyrolyzing polycrylonitrile nanofiber films. Owing to the intrinsic chemical and thermal resistances of graphite –like carbon [74], the as-prepared films are superhydrophobic in nearly the whole pH range (Figure 9a). Even if the films were immersed into strong alkaline or acid solutions for 24h, the superhydrophobicity and structure on the films can be kept well. Subsequently, we found that the polyaniline/polystyrene composite film prepared by an EHD method also shows the stable superhydrophobicity over the whole pH range [75]. Liu *et al.* reported the superhydrophobic aluminum or its alloys rough surfaces modified by perfluorononane and vinyl-terminated poly(dimethylsiloxane), which are also suitable different pH conditions [76]. Very recently, Chang *et al.* synthesized a new class of low surface energy polymer materials: polybenzoxazine, and prepared a superhydrophobic composite film of polybenzoxazine and silica, which shows good pH stability [77].

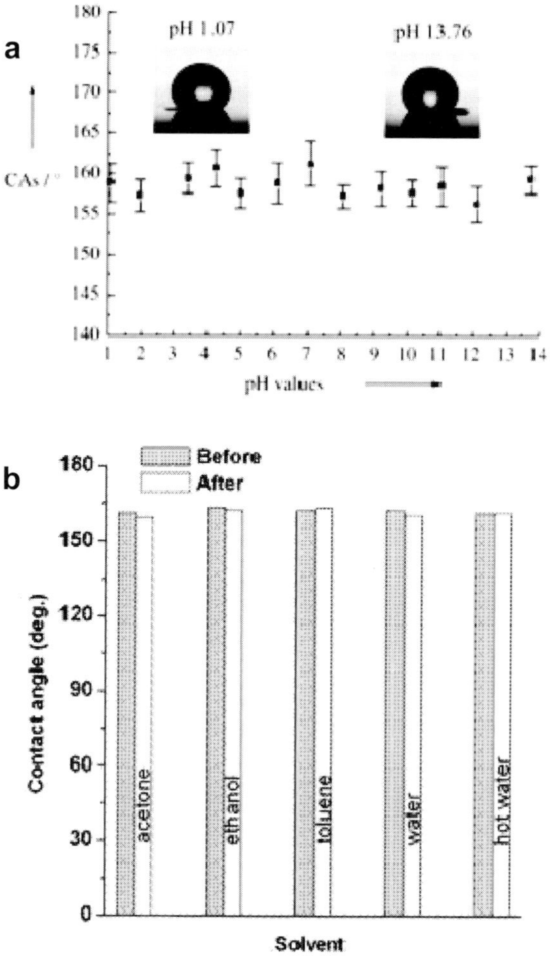

Figure 9. Plot of (a) the function between pH values and CAs on the nanostructured graphite-like carbon films, showing a good stability to different pH environments, and (b) the relationship between solvents and CAs on the microstructured $Cu(CH_3(CH_2)_{12}COO)_2$ films, indicating good solvent stability. Reprinted with permission from [Ref. 73], L. Feng, *et al.* Angew. Chem. Int. Ed. 42, 4217 (2003). Reprinted with permission from [Ref. 72], S. Wang, *et al.* Adv. Mater. 18, 767 (2006). Copyright @ WILEY-VCH.

Recently, Yan and his co-workers successfully fabricated a superhydrophobic poly(alkylpyrrole) films with aligned needle-like microstructures by the electrochemical polymerization technique [78].

This film shows high stability to heating and organic solvents, owing to the thermal stability and poor solubility of conductive polymer consisting of long-range-conjugated aromatic rings. Additionally, the superhydrophobic $Cu(CH_3(CH_2)_{12}COO)_2$ film also shows good environmental stability [72]. After kept in general solvent such as water, acetone, ethanol, and toluene, the surface did not debond and the CA also did not change (Figure 9b). Even after exposed to air for several months, these surfaces also kept a good water-repellent property, with the CAs in the range from 156° to 163°. It is believed that the stable superhydrophobicity originates from the good stability of chemical and geometrical factors of this copper carboxylate. The environmentally stable superhydrophobicity has also been observed on some other superhydrophobic surfaces, such as the porous polypropylene film [68], the rough polyelectrolyte multiplayer surface coated by silica nanoparticles [79], and the fluorinated-polythiophene film [80].

Briefly, the intrinsic chemical and physical properties of surface materials directly affect the surface environmental stability. So the design and synthesis of novel interface materials and composite materials will be the necessary key to build environmentally stable superhydrophobic surfaces.

4.3. Towards Multi-Function

The superhydrophobic surfaces have been well developed; however, the realization of multi-function on the superhydrophobic surface is desired for our daily life and industrial applications, for example, transparent and antireflective windows, colorful coating, antistatic coating, and other unique functions.

Up to now, many studies have been performed on the construction of superhydrophobic surfaces with good optical properties. Form the view of surface roughness, the transparence and hydrophobicity is a pair of competitive properties. The increase of surface roughness can enhance the hydrophobicity but decrease the transparence due to the light-scattering loss. Therefore, only when surface roughness is modulated to an appropriate range, both properties can be realized well [81].

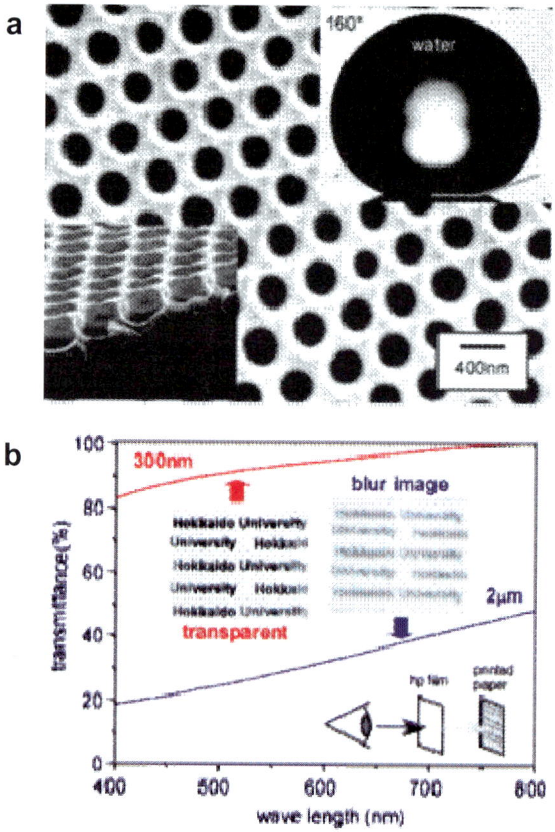

Figure 10. Representation of (a) the SEM image of the transparent and superhydrophobic fluorinated polymer film (top image and cross section) and the water CA on this film, and (b) the transmittance of the 2-μm- and the 300-nm-sized films and photograph of the film with printed papers underneath. Reprinted with permission from [Ref. 86], H. Yabu *et al.* Chem. Mater. 17, 5231 (2005), Copyright @ American Chemistry Society.

Tadanaga *et al.* [82] and Nakajima *et al.* [83] prepared transparent boehmite coating on glass substrates via the sol-gel and high-temperature treatment. Takai *et al.* adopted a microwave plasma-enhanced CVD process to escape from the shortcoming of high-temperature treatment, and obtained a transparent superhydrophobic film at low temperature [84,85]. Very recently, Yabu *et al.* have prepared the superhydrophobic fluorinated polymer films by casting the polymer solution under humid conditions [86]. By controlling the size of pores, a sub-wavelength honeycomb-patterned film is optically transparent and shows superhydrophobicity (Figure 10a). Therefore suitable surface roughness play important role in tuning transparence and hydrophobicity. In addition, Gu *et al.* have successfully combined structure color and superhydrophobicity on the inverse opal films [87]. This work effectively expends the applications of superhydrophobic surfaces in decorative materials, optical devices, and nanomaterials [88-92]. However, only a few cases have been reported on the superhydrophobic surface with antireflective properties. For example, Xu *et al.* fabricated a superhydrophobic and antireflective film using the methyl-modified silica [93,94].

The conductive superhydrophobic surfaces have also aroused much attention because electrical conductivity on the superhydrophobic surfaces may be very important for antistatics, corrosion protection, and conductive textiles [95]. Firstly, the combination of electrical and hydrophobic properties begins with the electrochemical polymerization of pyrrole in the presence of fluorinated complexes with low surface energy [96]. The water CA on the fluorinated polypyrrole fabric was only 110°. Then Shi *et al.* prepared the hydrophobic and conductive polythiophene with the CA about 134° [97]. Through further enhancing the surface roughness, the conductive poly(alkylpyrrole) film with microtube-like structures [78] and the conductive fluorinated polythiophene film with cauliflower-like structures [80] have been prepared through the electrochemical polymerization approach, respectively, both of which show stable good superhydrophobicity. Recently, our group prepared the conductive superhydrophobic ZnO films, which show interesting nanoporous structures (Figure 11a), by the electrochemical deposition (ECD) [98]. The conductivity of the as-deposited films was investigated via conductive atomic force microscopy. As shown Figure 11b, the bright regions suggest that the film is highly conductive. Additionally, metals such as gold, silver, and copper, as good electrical conductors, have been used to construct conductive and superhydrophobic surfaces [13-16]. For example, Zhang and co-workers deposited gold and silver clusters on ITO glasses by combination of ECD and layer-by-layer (LBL) techniques, and then the superhydrophobic surfaces were obtained after the modification of hydrophobic molecule [13,20,99]. However, the potential applications of these metal-based functional surfaces need to be further explored. Besides, to be mentioned, a multifunctional carbon nanofiber film doped by Fe_3O_4 has been prepared by an EDH technique [100], which exhibits conductive, magnetic and superhydrophobic properties. It provides a good example to construct the multifunctional surfaces effectively by use of composite materials.

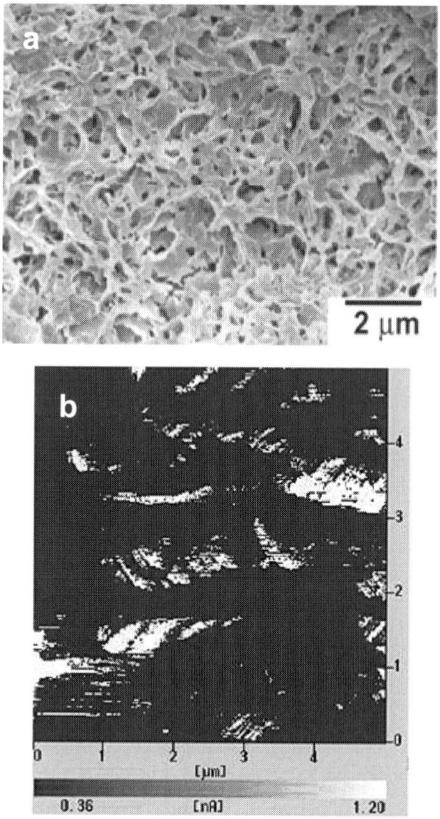

Figure 11. Depiction of (a) the SEM image of superhydrophobic ZnO nanoporous films prepared by ECD, and (b) the simultaneous current scanning image of the as-prepared thin films on ITO glass at a bias voltage of 0.5 V. Reprinted with permission from [Ref. 98], M. Li, *et al.* J. Phys. Chem. B. 107, 9954 (2003) Copyright @ American Chemistry Society.

Moreover, water strider's legs bring scientists new inspiration to construct functional superhydrophobic surfaces [19]. Recently, Zhang *et al.* fabricated the superhydrophobic surfaces on the curved gold threads [20]. By combining ECD and LBL techniques, they deposited dendritic gold clusters on the gold threads (Figure 12a, b), and obtained the superhydrophobic coating after further modification of n-dodecanethiol. The as-prepared thread can float on the water's surface (Figure 12c), which is a superhydrophobic behavior similar to that of water stride's leg. This unique superhydrophobic property originates from the surface special structures and modification of low-surface-energy molecules. Very recently, Shi *et al.* also prepared the copper wire with superhydrophobic submicrofiber coating to mimic water strider's legs [21]. The maximum supporting force of their artificial leg was measured to be about 83 dynes, indicating a strong water-repellent force. These studies can help in understanding the reason why a water strider can walk freely on water, and may help in the design of novel aquatic navigational subminiature robots and other devices.

Recently, an aligned polystyrene nanotube film, which is prepared by the template method, shows superhydrophobicity and a high adhesive property [101]. The maximum adhesive force is about 59.8 μN as assessed by a high-sensitivity microelectromechanical balance system. The film can hold a water droplet even it is tilted upside down. The mechanism that described here is similar to the geckos' foot in nature [102,103]. The difference is that these interactions are between solid and solid. The similar phenomenon was also observed on a methyloctyldimethoxysilane film that self-assembled on silicon surface through the solution deposition [104]. These super-hydrophobic films with high adhesive force to water are expected to be used to transfer or manipulate small liquid droplets as a "mechanical hand" in the future without loss or contamination for micro-volume analysis.

At all, the realization of superhydrophobic surfaces mainly depends on the cooperative effect of low-surface-energy materials and surface geometrical structure. To the creation and synthesis of new low-surface-energy materials is very important, nature prefer the construction of unique micro- and/or nanostructures by simple materials. In coincidence, nanotechnology has currently provided many convenient and elegant means to fabricate micro- and/or nanostructures on the surface. Therefore, we believe that the integration of the inspiration from nature, nanotechnology, and chemistry will bring a promising future for superhydrophobic surfaces.

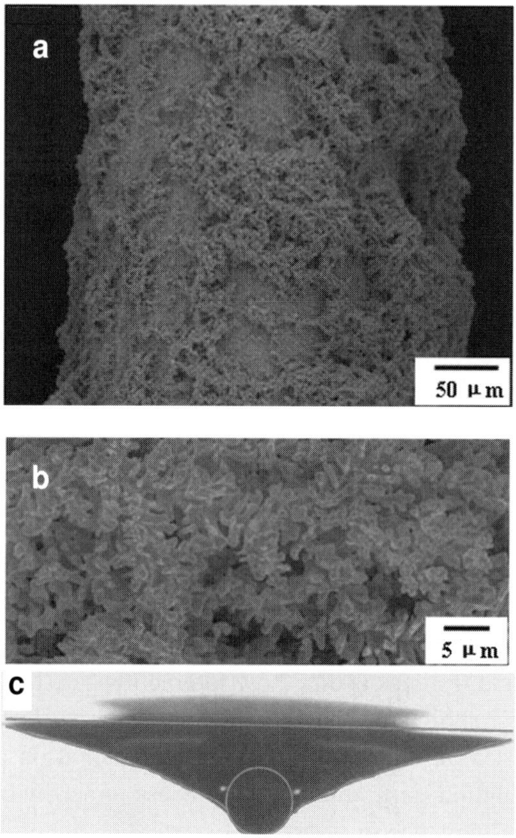

Figure 12. The gold thread coated by gold aggregates to mimic the superhydrophobic water stride's leg. An SEM image of (a) gold thread, and (b) gold aggregates. A photo of (c) the superhydrophobic gold threads floating on the water's surface is also shown. Reprinted with permission from [Ref. 20], F. Shi, *et al.*, Adv. Mater. 17, 1005 (2005), Copyright @ WILEY-VCH.

5. Superhydrophilic Surfaces

Superhydrophilic surfaces [105,106] are those with high surface energy and tend to exhibit the water CA less than 5° after surfaces textured in certain manner, on which water prone to spread completely. Generally, it can be achieved by utilizing capillary effects [107], including two-dimensional [38,108,109] and three-dimensional capillary effects [110].

Researches on superhydrophilicity can date back to 1997, R. Wang [111] etc. have proposed a light-induced superamphiphilic (CAs for both water and oil are all approaching 0°) surface. They fabricated a TiO_2 thin film which gave an initial water CA of 72°±1° (Figure 13a), and the water droplets spread out on the film with the water CA of merely 0°±1° (Figure 13b) after ultraviolet (UV) irradiation,, showing typical superhydrophilicity, which can be ascribed to the photo-generated oxygen vacancies on the surfaces [39]. Such UV-induced superhydrophilic nature of the surface imparts desirable antifogging behavior [112] (Figure 13c-d) by preventing light scattering water droplets from forming on the surface (water instead spreads into a uniform sheet), which dooms to facilitate our daily life in great degree.

Thereafter, many scientists devoted into the research on superhydrophilicity, and they do fabricate many superhydrophilic surfaces successfully both from photochemistry active inorganic (such as TiO_2, ZnO and SnO) and organic (such as polypyrrole [113] and polyaniline [114]) materials. Similar with the theory based on which we fabricated superhydrophobic surfaces, there are two factors (surface topology and surface free energy) influencing the surface wettability when hydrophilicity is refereed. Therefore, both enhancing the surface roughness on the hydrophilic surfaces and increasing the surface free energy through chemical methods can contribute to a superhydrophilic surface. Generally, there involves two approaches in fabrication of superhydrophilic surfaces, one is to use the photosensitive semi-conductor materials that become superhydrophilicity after UV irradiation [115,116], and the other is to use the textures surfaces to promote superwetting behavior [117,119].

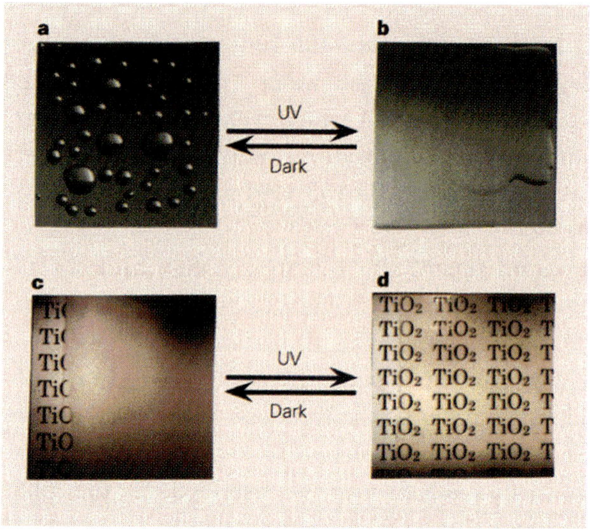

Figure 13. The as-prepared TiO_2 film (a) is hydrophobic with water CA of 72°±1°. After UV irradiation (b), water droplets spread on the film completely and the film shows superhydrophilicity. Exposure of a TiO_2-coated glass to water vapor (c) shows the formation of fog (small water droplets), which hinders the view of the text on paper placed behind the glass. The antifogging effect on UV-irradiated TiO_2-coated glass (d) shows that the text behind can be clearly viewed. Reprinted with permission from [Ref. 111], R. Wang, *et al.*, Nature, 388, 431 (1997), Copyright @ Nature Publishing Group.

ZnO, as a well-known wide-band semi-conductor, is another kind of materials whose surface energy can be tuned by UV-irradiation. It has been confirmed that UV-irradiation can change the wettability of ZnO film from hydrophobicity to superhydrophilicity [119], and the mechanism is similar to that of TiO_2. Besides ZnO, almost all the semi-conductor materials which can show superhydrophilicity after UV irradiation follow the similar mechanism where photo-generated oxygen vacancies results in the increasing of the surface free energy efficiently. As mentioned above, textured surfaces can also contribute a superhydrophilic surface, building from which it has been demonstrate that both lithographically textures surfaces [120] and microporous surfaces [121] can be rendered superhydrophilicity.

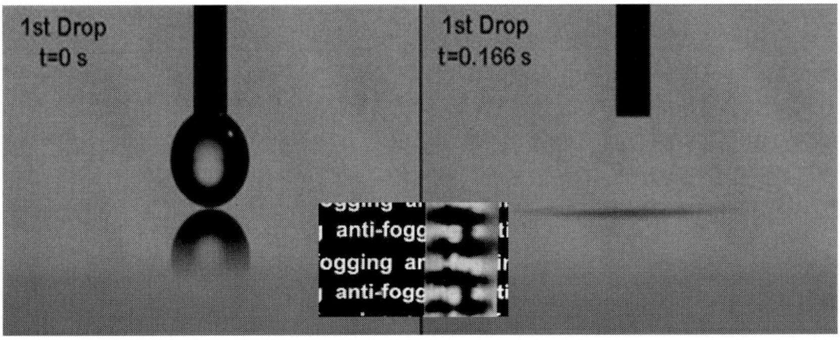

Figure 14. Still images from video contact angle measurements for the water droplet with the volume of 0.5μl. The inset is the comparison of the fogging behavior of a bare glass slide (right-hand slide) with a slide partially coated (left-hand side) with a superhydrophilic polyelectrolyte multilayer film. Reprinted with permission from [Ref. 122], F. C. Cebeci, *et al.*, Langmuir 22, 2856 (2006), Copyright @ American Chemistry Society.

For example, based on the nano-porous structures, a stable superhydrophilic surface was prepared from layer-by-layer assembled SiO_2 nanoparticles and a polycation under proper conditions [122]. Water droplets are spread immediately upon it contacting with the film (Figure 14) for its interconnected pore morphology which is readily invaded by water. Moreover, the film also exhibits well antifogging characteristics (inset in Figure 14). Based on this work and inspired by the natural phenomenon that the wing surface of the Namib Desert beetle can harvest water, they recently [123] develop a novel patterned surface with dissimilar wetting properties, that is with hydrophilic or superhydrophilic patterns on an otherwise superhydrophobic surfaces [124], created by selectively delivering polyeletrolytes to the surface in a mixed water/2-propanol solvent. The as-prepared film show the similar water harvest properties with those of Namib Desert beetle, which offered us more opportunity in designing and fabricating of novel devices such as planar microcanals [125].

Not limited to inorganic compounds, some organic compounds or polymers can also be used to construct superhydrophilic surfaces, and some progresses on this area have been achieved. W. Yang's group has made many excellent achievements in this area. They develop a facile method to modify the polymer (such as poly(propylene) (PP) and poly(ethylene terephthalate) (PET)) films into highly hydrophilic through UV-initiated surface photografting of an N-vinyl pyrrolidone/N,N'-methylenebisacrylamide (NVP/MBA) inverse microemulsion [126]. After modification on PP film, the CA can decreased from 102° (on smooth blank PP film) to 5° (inset in Figure 15a), and the water droplet can spread completely on the surfaces (Figure 15b left). The situation on the PET film is similar, and the modified film shows good antifogging effect, which is obvious from Figure 3b where the picture behind the transparent PET film can be clearly distinguished after the PET was modified.

Figure 15. Representation of (a) an AFM topographic image of the PP film grafted with 5wt% NVP/MBA inverse microemulsion, where the CA on the film is 5° from the inset image; and (b) photographs showing the change of surface wettability and transparency before and after modification. Reprinted with permission from [Ref. 126], Y. Wang, *et al.*, Macromol. Rapid Commun. 26, 1788 (2005), Copyright @ WILEY-VCH.

Based on this work, they also [114] developed a surface of PP film with Polyaniline (PANI) nanowires and sub-micro/nanostructured dendrites via routine oxidative polymerization of aniline under different conditions, and the as-prepared film is superhydrophilic (Figure 16) even with different poly-(acrylic acid) (PAA) grafting percentage. On the film, the immobilized PANI enhances the hydrophilicity, and simultaneously the as-formed hierarchical micro- and nano- structures effectively enhanced the hydrophilicity, therefore resulting in a superhydrophilic surface. Moreover, the rapid spreading of water is very important in various applications such as biological examination.

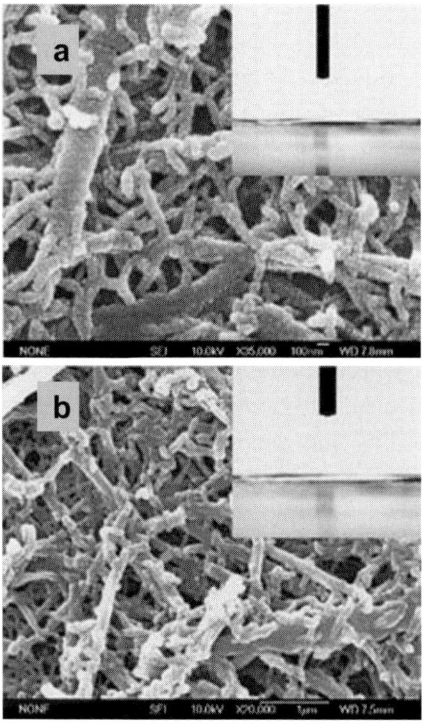

Figure 16. SEM images of PANI sub-micro/nanostructured dendrites immobilized on PP films' surfaces (the inset shows the shape of water droplet). The PAA grafting percentage is (a) 1.1wt% (CA is 3°), and (b) 1.5wt% (CA is 0°). Reprinted with permission from [Ref. 114], W. Zhong, *et al.*, Macromol. Rapid Commun. 27, 563 (2006), Copyright @ WILEY-VCH.

To realize the rapid spreading, both the textured surface and the high surface energy are needed. Very recently, our group fabricated a hierarchical structured TiO_2 nonwoven mat where micro-fibers with nano-channels on each fiber are abundant, and the water droplet can spread completely within 15ms, showing untra-fast wetting behavior. It is of profound importance in many areas, especially in areas related to micro-cave reaction, for example, as papers in ink-jet printing and rapid biological examination. In addition, such special wetting behavior may inspire us in designing many functional commodities, such as diapers and napkins.

In conclusion, the superhydrophilic surfaces can ensure the complete wetting of water on certain surface, which has many promising applications on areas such as chemical sensor, antifogging, antireflection and even bacteria-resistant. Optimized the design and fabrication of superhydrophilic surfaces must open a new perspective for the functional-oriented new nano-materials.

6. Surfaces with Tunable Wettability from Superhydrophobic to Superhydrophilic

As two extreme cases of wettability, superhydrophilicity and superhydrophobicity are capable of bring forward many new properties, such as self-cleaning, anti-fogging etc, from which we can benefit much both in our daily life and practical industry/agriculture applications. Therefore manipulation the wettability from superhydrophilicity to superhydrophobicity is of paramount importance, especially for the micro-fluid based researches. Up to day, many research groups have realized the manipulation of the surface wettability between these two opposite extreme states by various techniques based on self-assembly and photo-catalysis. Changing the topology nature or the free energy of surfaces, or changing both, can help to control the wettability.

Recently, a hierarchical structured copper film [16] was successfully fabricated by two-step electrochemical deposition method, and after modified by n-alkanoic acids, $CH_3(CH_2)nCOOH$ (n=1, 2, 3, …, 16) their wettability can be tuned from superhydrophilicity to superhydrophobicity with the increasing of the carbon number of the n-alkanoic acids. In this case, the surface roughness provided by the hierarchical structures amplified the transition range of water CA offered by the self-assembled monolayer.

Following the similar mechanism but from different technique, a gradient surface [127] with a wetting property that can change from superhydrophobicity (156.4°) to superhydrophilicity (less than 10°) (Figure 17) was fabricated continuously by utilizing the controlled adsorption of a thiol self-assembled monolayer on a rough gold surface (the top-morphology can be seen form Figure 17a and 17b), where the micro-/ nano- hierarchical gold clusters amplify the wetting property offered by the controlled self-assembled monolayer with a density gradient along the surface. Here, the gradient can be sound guaranteed by gradual increased immersion time along the altitude direction from the bottom to the upper of the gold substrate when it immersing into the solutions vertically. This kind of material would provide a larger oriented driving force for many important biological and physical processes and might have potential applications in water droplet movement, oriented axonal specification of neurons, protein adhesion, and so on.

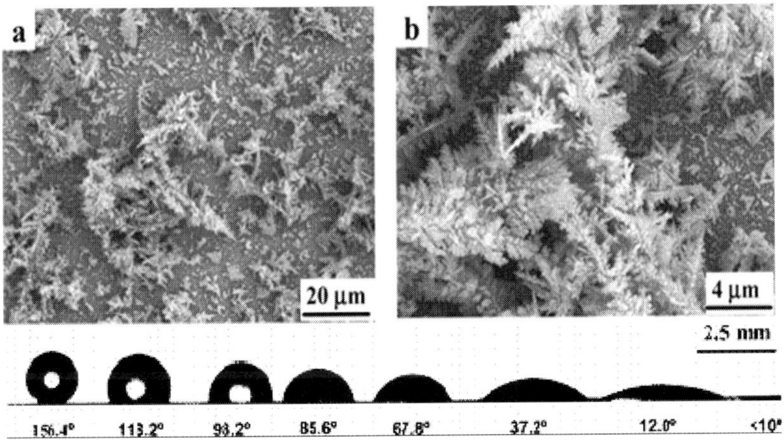

Figure 17. SEM image of (a) deposited gold structures, and (b) magnified gold clusters. The bottom is the photograph of the CAs along the gradient surface. The volume of the water droplets was kept at ~3 μL. Reprinted with permission from [Ref. 128], X. Yu, *et al.*, Langmuir 22, 4483 (2006), Copyright @ American Chemistry Society.

From the facts mentioned above, it is evident that self-assembled monolayer is an effective method to manipulate the surface wettability for their ability in tuning the surface free energy, which has been confirmed by many researches [13] and the notable molecular applied are generally fluorinated compounds, silanes and thiols [40,128,129].

In addition, control the other experimental factors can also contribute surfaces with different wettability, such as the fabrication temperature and the pH value. Recently, a facile method was developed to control the wettability of the colloidal crystals by just modifying the assembly conditions, where the colloidal crystals were assembled into latex spheres with hydrophobic core and hydrophilic shell [130]. The wettability of the film can be controlled from superhydrophilicity to superhydrophobicity by raising assembly temperature based on the phase reversion of the hydrophobic or hydrophilic polymer segments as a result of lowering the interfacial free energy. More importantly, the wettability transition temperature of the film can be finely controlled between 90°C and 20°C by modifying the ratio of soft/hard segments of the polymer (Figure 18a).

On the other hand, the wettability of the colloidal crystals can be modified by assembly pH by covering the latex surface by surfactant of sodium dodecylbenzenesulfonate (SDBS) as well [92]. The superhydrophobic or superhydrophilic film can be obtained when modifying the assembly pH from 6 to 12, which leads to the presence or absence of hydrogen bonding between $SO_3^-Na^+$ of SDBS and hydrophilic COOH around the latex surface (Figure 18b).

However, the wettability of the surfaces mentioned above manipulated from superhydrophobicity to superhydrophilicity are not on the same surface, not less the reversibility, which will limit their applications. If the superhydrophobicity and superhydrophilicity can be realized on the same surface, especially can be realized reversibly, and the switching effects are therefore obtained which be used in many fields such as selective adsorption of special molecular and lossless transport of the liquid.

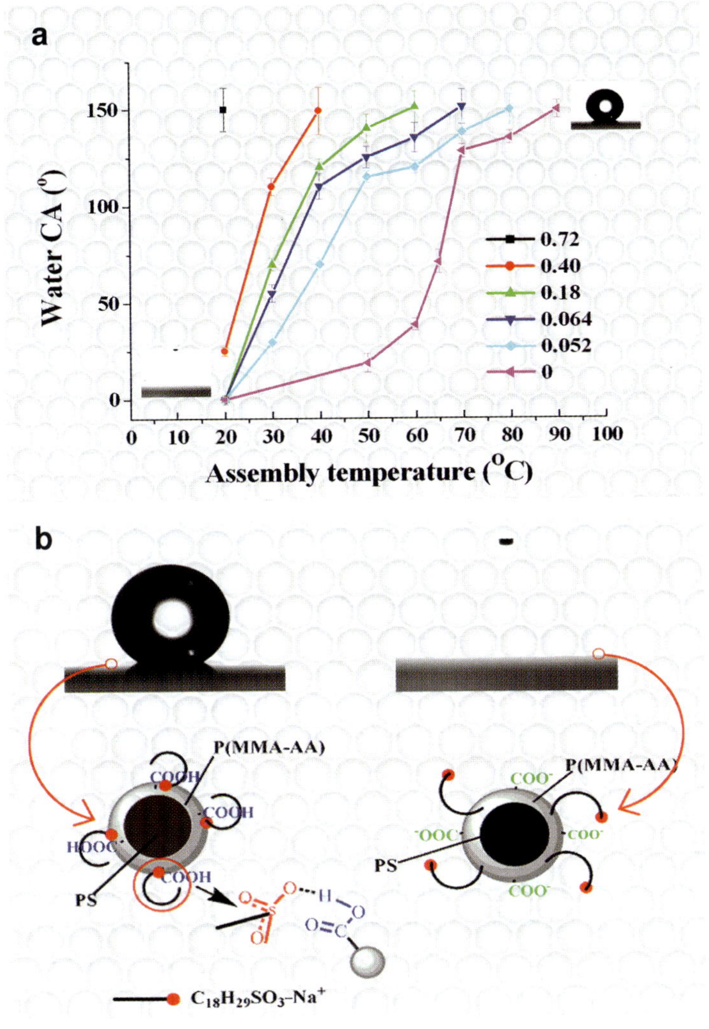

Figure 18. Images of (a) the relationship of assembly temperature with water CA of the as-prepared films assembled from latex spheres with varying ratio of nBA/St, and (b) the water droplet shape on the films assembled from suspension with pH of 6.0 (left) and 12 (right) respectively; the bottom is the according illustrative structure of the latex sphere in the films with pH 6.0 and 12, respectively. The conformation of hydrogen bonding is noted by arrow. Reprinted with permission from [Ref. 130], J. Wang, *et al.*, Macromol. Rapid Commun. 27, 188 (2006), Copyright @ Wiley-VCH. Reprinted with permission from [Ref. 92], J. Wang, *et al.*, Chem. Mater. 18, 4984 (2006), Copyright @ American Chemistry Society.

7. Responsive Surfaces between Superhydrophobicity and Superhydrophilicity

7.1. Single Stimuli-Responsive Surfaces

Recently, researches on the smart switching between superhydrophilicity and superhydrophobicity have attracted intensively interest worldwide, and many scientists devoted into this area. Within several years, much responsive wettability [9] stimulated by single outer-stimulus, such as temperature [31], light [131,132], electric field [133], pH [35], solvent [134] and magnetism, have been successfully prepared from both organic and inorganic materials. Herein, we will review the recent progresses with reversible wettability between superhydrophilicity and superhydrophobicity both on the inorganic and organic film with different geometric nature.

7.1.1. Photo-Responsive Surfaces

As a kind of photo-responsive materials, inorganic semi-conductor oxides (including SnO_2, TiO_2 and ZnO) are well-known capable of reversible switching the surface chemical environment between the states that is more favorable for hydroxyl adsorption (oxygen vacancies) [135] and that is more favorable for oxygen adsorption (hydroxyl groups) [136] by the alteration of UV irradiation and dark storage or heating approach, which make it possible to create smart surfaces with controllable wettability. Inspired by this, combining with the amplify effects endowed by proper surface geometric morphology, making surfaces with wettability switched between superhydrophobicity and superhydrophobicity reversibly is feasible, and in fact we do obtain a series films like this described as follows.

Based on our former work that nanostructured surfaces [10], especially those with hierarchical micro- and nanostructures [60], are capable to endow the superhydrophobicity, we have developed many superhydrophobic films from semi-conductor oxides. Through the alteration of UV irradiation and storage in dark for certain time, their wettability can be reversible switched between superhydrophobicity and superhydrophilicity, exhibiting typical photo-switch effect. Although the similar behavior has been observed by Sun where the as-prepared smooth ZnO film show the water CA of merely 109° (far from superhydrophobic) and 10° after exposed to UV irradiation, the reversible switching between two extreme states on the oxide are first fulfilled by our group on the ZnO nanorod films [137]. Thereafter, some other semi-conductor oxide films with different surface morphologies are fabricated, on which the wettability are reversible between tow extreme states (Figure 19d) stimulated by UV light, such as hierarchical micro- and nanostructured ZnO film from CVD technique (Figure 19a) [131] and ZnO nanorod films from various methods [138-140], flower-like TiO_2 nanostructured film from hydrothermal technique (Figure 19b) [141] and photoelectrochemical etching method [142], and aligned SnO_2 nanorod film prepared by crystal-seed-based hydrothermal method (Figure 19c) [143]. It is a successful example that making good use of the photocatalysis property of the semi-conductor oxide in controlled wettability area, which must give us many inspirations in multi-functional materials design and fabrication. Moreover, besides inorganic compounds, some organic compounds can also show the similar behavior in photo-responsive wettability.

Compared with the inorganic compounds, organic materials have many advantages (such as easy manipulation, quick response and further chemical modification), and it can be widely used not only in photo-switchable-based researches and message storage. The favored photo-responsive organic materials are those materials that have a reversible photo-induced transformation between two physical and/or chemical states, ranging from surface energy, absorption spectra, refractive index, dielectric constant, oxidation/reduction potential, to geometrical structure [26,144].

Azobenzene and its derivatives, as a classic of photoresponsive organic compounds, can undergo a reversible transition from trans- and cis-isomers when irradiated with UV/visible light [145,146], resulting the changes in both geometry and dipole moment, and as well the surface wettability. The trans has a low surface free energy for its small dipole moment, and on the contrary the cis possess a high surface free energy for its large dipole moment. Thus, the isomerization of azobenzene from trans to cis can result in the wettability change from more hydrophobic to more hydrophilic.

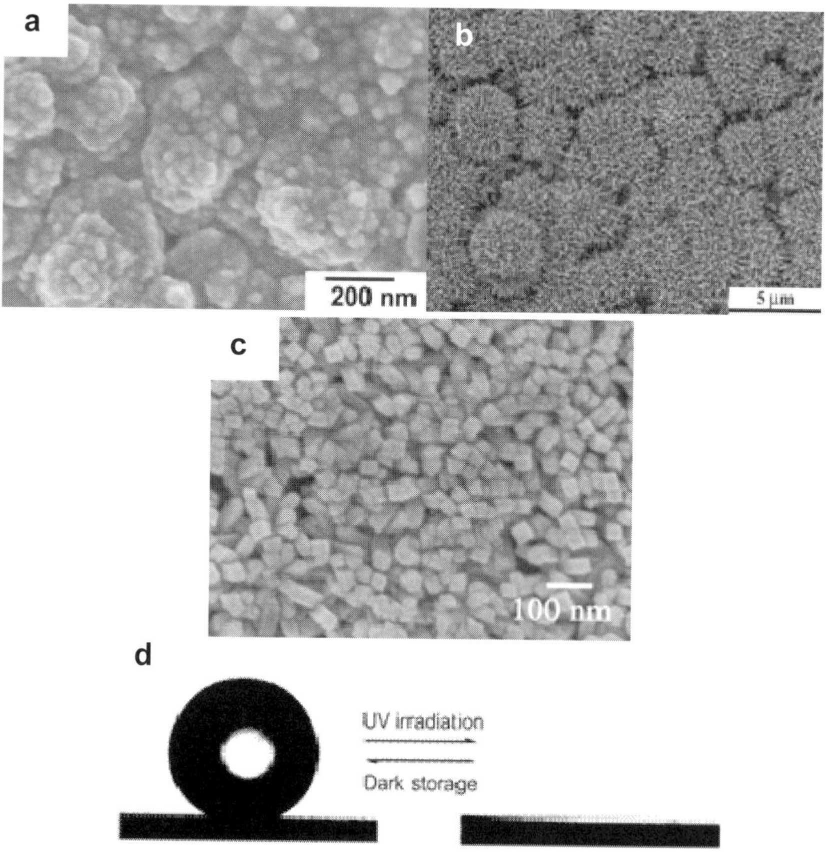

Figure 19. SEM image of (a) ZnO film prepared from chemical vapor deposition method, (b) TiO_2 prepared from wet-chemical method and (c) SnO_2 film. The reversible wettability switched by the UV illumination and dark storage between superhydrophobicity and superhydrophilicity is shown in (d). Reprinted with permission from [Ref. 131], H. Liu, *et al.*, Langmuir, 20, 5659 (2004) Copyright @ American Chemistry Society. Reprinted with permission from [Ref. 141], X. Feng, *et al.*, Angew. Chem. Int. Ed. 44, 5115 (2005), Copyright @ WILEY-VCH. Reprinted with permission from [Ref. 143], W. Zhu, *et al.*, Chem. Commun. 26, 2753 (2006), Copyright @ The Royal Society of Chemistry.

In this respect, many techniques are developed using the azobenzene and their derivatives to control the surface wettability, and some of them can realized the wettability switched between two extreme states. Self-assembling an azobenzene monolayer on a flat substrate can merely bring a small CA change less than 10° via UV irradiation [147,148], and if on a smooth polymer film the CA change by UV irradiation can up to 11° [149]. However, the results above are all based on the smooth surface where the amplification effect of the surface roughness to the wettability was not been well addressed. Therefore, using the aligned silicon column substrate [132] or inverse opal [150] as the rough substrates, we recently prepared the azobenzene polyelectrolyte monolayer through a simple electrostatic self-assembly technique. The reversible CA change can up to 66° on the proper textured surface after UV and visible light irradiation. Although great efforts have been done in order to obtaining an organic film with the reversible wettability between superhydrophobicity and superhydrophilicity, little progress is achieved and it remain a great challenge for the scientists majored in materials and chemistry.

Only very recently, Kilwon Cho and his coworker [151] prepared an organic photo-switchable nanoporous multilayer film with wettability that can be reversibly switched between superhydrophobicity and superhydrophilicity with UV/visible irradiation. As shown in Figure 20a, first a hierarchical structured porous organic-inorganic hybrid multilayer film was fabricate on silicon substrate through layer-by-layer technique, and then modification by a kind of azobenzene derivatives, 7-[(trifluoromethoxyphenylazo)phenoxy]pentanoic acid (CF_3AZO), which has the fluorine atoms on its chain tails. The CA can reversible changed between 152±3° and less than 5° when the deposition cycle are proper chosen (Figure 20b), realizing the wettability switched between superhydrophobicity and superhydrophilicity reversibly on organic surface. This is a stirring result in organic compounds areas which perfectly considering all the influence factors, both viewed from enhancing the surface roughness and tailor the surface free energy to the maximum degree by grafting some functional groups, but still follow the traditional Wenzel and Cassie equations, which provide us versatile possibilities to tailor the surface wettability, not only limited in the photo-responsive area.

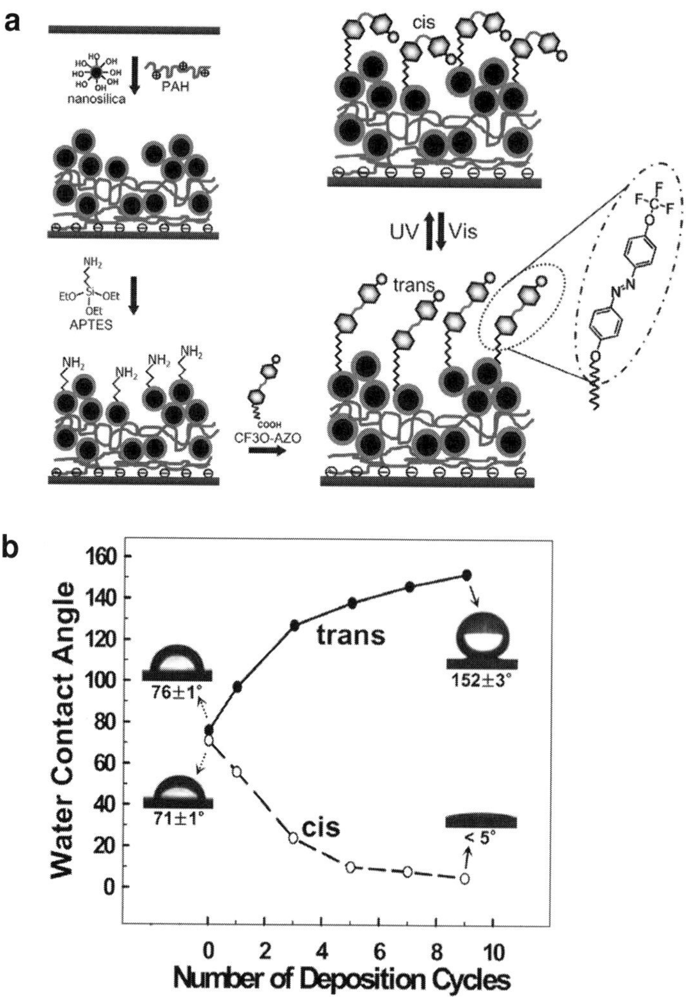

Figure 20. Schematic of (a) the reversible photoisomerization of a roughness-enhanced photoswitchable surface, and (b) the relationship between the number of deposition cycles and the water CA on smooth substrate (dotted arrows) and on (PAH/SiO_2) multilayer (solid arrows) after UV/visible irradiation. Reprinted with permission from [Ref. 151], H. S. Lim, *et al.*, J. Am. Chem. Soc., ja0655901, Copyright @ American Chemistry Society.

In addition, spiropyrans [152] is another organic compound that undergo reversible transition between a closed, nonpolar form, and an open, polar form, when irradiated with visible (450 - 550 nm) / UV (366 nm) light. As know the nonpolar form is hydrophobic, whereas the polar form is hydrophilic. Therefore it can also be applied to tailor the surface wettability, for example the reversible light-induced CA change can up to 23° [144] on fractally rough Si nanowires surfaces. Particularly, on such surfaces the UV-induced advancing CA is lower than the receding CA under visible irradiation, which allows the movement of water droplets under the influence of a UV-visible light gradient. In conclusion, besides these two typical photo-responsive organic compounds, it is reasonable to believe that there are many other photo-responsive organic compounds that can be used to tailor the surface wettability, such as diarylethene derivatives.

Very recently, a kind of photochromic diarylethene molecule was successfully applied to switch the surface wettability reversible between superhydrophobicity (with water CA of 163°) and general hydrophobicity (with water CA of 120°) via the alteration of UV and visible light irradiation [153]. However, different from the mechanism followed by azobenzene and spiropyrans where the surface chemical state is sensitive to the light, it is the photo-induced reversible change in surface morphology that provides the wettability change, which is obtained by the photo-induced reversible formation of fine fibril structures on coated microcrystalline surfaces caused by the photo-isomerization.

7.1.2. pH-Responsive Surfaces

Surface wettability that is responsive to certain conditions of water, such as pH value, electrolyte *etc.*, is of profound importance in many practical areas. For example, surfaces with superhydrophobicity to the liquid in all pH value [73] can effectively be used in anti-corrosion area. Moreover, surface with particular responsive to the liquid with certain pH value can also be used in controlled micro-fluidic switches, controllable separation systems *etc.*

In this aspect and based on work from Whitesides' group [154,155] that the pH-responsive behavior can be observed on a smooth surface for the deprotonation of the surface carboxylic acid group, Zhang *et al.* [34] fabricated a pH-responsive surface combining the technique of electrodeposition for the micro-/nano-composite structured rough gold surface and self-assembly for the surface modification by mixed monolayer containing $HS(CH_2)_9CH_3$ and $HS(CH_2)_{10}COOH$. The surface can indicate the pH property of the liquid dropped on it because it is superhydrophobic to acidic droplets with pH of 1 and is superhydrophilic for basic water droplets with pH of 13. Furthermore, they also modified a rough gold substrate with 1-(11-mercaptoundecanamido) benzoic acid, and a similar pH-responsive property was also observed [156]. The pH-responsive wettability on such surfaces are results from the reversible protonation and deprotonation of the surface carboxylic acid groups.

7.1.3. Thermal-Responsive Surfaces

Poly(N-isopropylacrylamide) (PNIPAAm) [157,158] is a well known temperature responsive polymer which can be used to chemical modified various substrates via surface-initiated atom transfer radical polymerization [159,160] for some thermal-responsive properties. In this respect, our group makes an innovative contribution [31] in thermal-responsive wettability which can switch reversibly between superhydrophobicity and superhydrophilicity triggered by temperature change. After being grafted of PNIPAAm on the silicon column alignment with proper column spacing, the surface shows high water CA of about 149.3° when the temperature was increased to 40 ^{o}C and show superhydrophilicity when temperature was decreased to 25 ^{o}C, and the wettability can be manipulated reversibly between these two extreme states simply by the heating and cooling the substrate. Thus thermal responsive water spreading is considered to result from the competition between intermolecular and intramolecular at temperature of about 32 - 33°C [161] originated from the uniform layer of the polymers tightly attached on the rough silicon surfaces.

Thereafter, Qiang Fu and his co-workers [33] further develop this technique where they modified a brush of PNIPAAm on the nanoporous aluminum oxide surface. Similarly, they also observed the surface wetting behavior can be reversible switched between superhydrophobicity and superhydrophilicity when the pore diameter was proper chosen (seen from Figure 21, the proper pore diameter is 200 nm) simple by changing the temperature. However, to be noticed, they also observe the topography change in nano-scope, which might associate with the expansion and contraction of the grafted polymer be atomic force microscopy (AFM). Therefore, the mechanism of such thermal-responsive surface follows not only the surface free energy change, but also that of topologic nature.

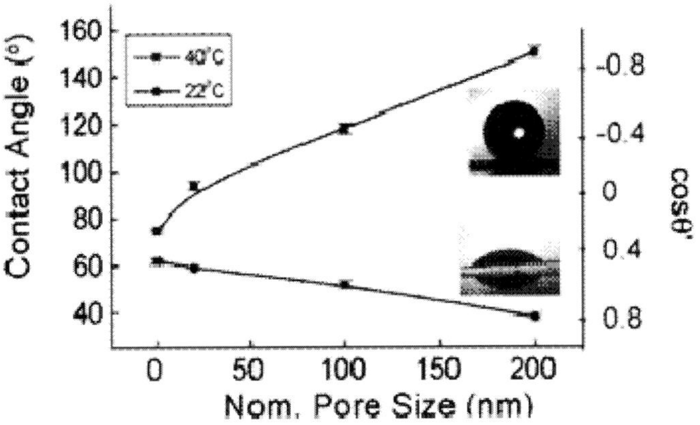

Figure 21. Water CA data (sessile drop) measured at 22°C and 40°C for PNIPAAm grafted to aluminum oxide surfaces with varying nominal pore size. Reprinted with permission from [Ref. 33], Q. Fu, *et al.*, J. Am. Chem. Soc. 126, 8904 (2004), Copyright @ American Chemistry Society.

In addition to polymer films, many hard materials also show thermal responsive properties. For example, hard foamy film [121], prepared from condensing organo-triethoxysilane in a mixture of an organic solvent and water, can show the reversible wettability between superhydrophobicity and superhydrophilicity, but show the changing tendency of CA from superhydrophobicity to superhydrophilicity as the temperature was elevated which is right the opposite to that mentioned above. It might originate from the reversible cross-linking of the silica backbone.

7.1.4. Electric-Field Responsive Surfaces

Switching the surface wettability via the electrical potential is of great importance and has received special attentions especially in recent years. Self-assembly monolayer of various compounds which can undergo the conformational transition triggered by electric potential is efficient in construction surfaces with electric-field responsive wettability. In this aspect, Lahann *et al.* [24] realized the switching of wettability but within a narrow scale (20° to 30° in water CA) on a gold surface that was modified by a low-density carboxylate-terminated monolayer. Self-assembly monolayer of ionizable alkanethiolate [162] on gold surface can also induce surface wettability change for the reversible conformational reorientation under negative and positive potentials.

Reversible switchable wettability between superhydrophilicity and superhydrophobicity stimulated by electric field was firstly realized by Yan *et al.* [34] on a porous conductive polypyrrole (PPy) film made from simple electrochemical process. It can be well attributed to the reversible switching between doped (oxidized) state and dedoped (neutral) state of PPy (upper in Figure 22b), and the former and latter state are generally hydrophobic and hydrophilic, respectively. If the porous structure (Figure 22a) is introduced, the wettability are amplified and switched between two extreme states (lower in Figure 22b).

Electro-wetting [163-166], based on electrostatic effect, has been proved to be another method to control the surface wettability. It is a dynamic wetting and the external electricity is applied throughout the whole process. Charges were built up both at the liquid side and at the solid electrode by applying a voltage between the liquid and a conducting layer, resulting in decrease in interfacial free energy. Water behavior (including wetting and nonwetting) on the surface is therefore different via the variation of external electricity, leading an electric-responsive surface. There is a long way for electric-field responsive surfaces to realize the reversible switching between two extreme states, and some liquid-crystal molecular are promising capable to play an important role in this area.

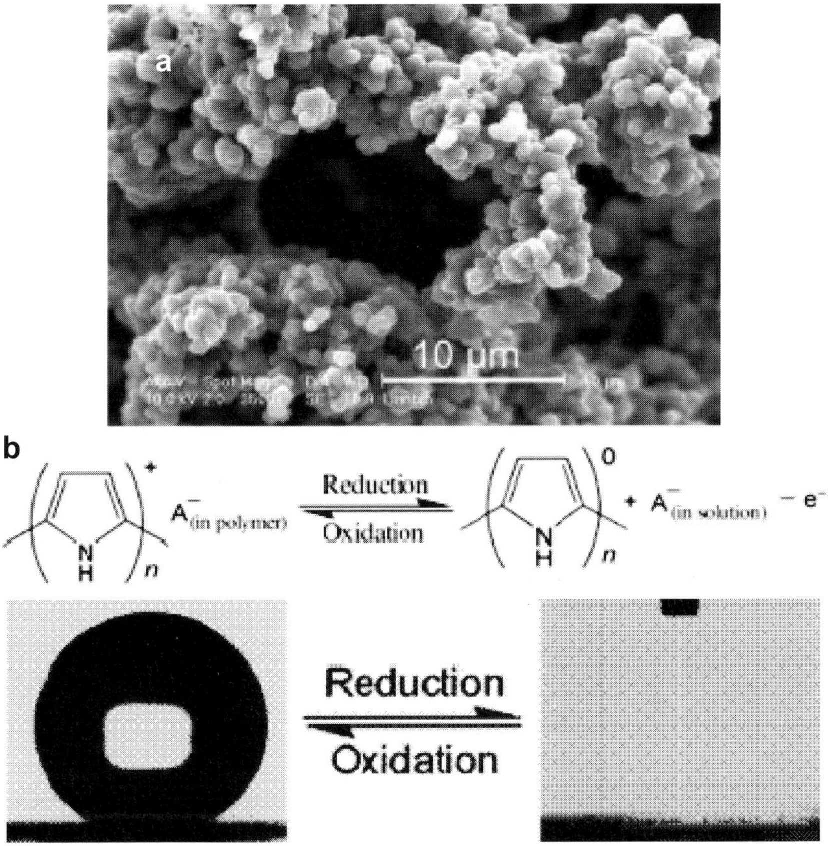

Figure 22. Representation of (a) the SEM image of porous PPy film; and (b) the switching process (upper schematic) between the doped state and dedoped state of PPy, and the profile (lower schematic) of a water droplet on a highly porous PPy film. Reprinted with permission from [Ref. 34], L. Xu, *et al.*, Angew. Chem. Int. Ed. 44, 6009 (2005) Copyright @ WILEY-VCH.

7.1.5. Mechanical Force Responsive Surfaces

The switch-effects in surface wettability mentioned above are ensured by the change in surface chemical states or in surface morphology, or in both of them, triggered by outer stimulus, and they all undergo certain chemical process. Till very recently, scientist suggested that the reversible wettability between superhydrophobicity and superhydrophilicity can be fulfilled purely by a physical approach. As we known, the so-called pure physical approach can only alter the surface morphology, i.e. surface roughness, which is right one of the effective factor to influence the surface wettability.

Han's group [117] recently proposed a facile method to realize the reversible switching between superhydrophobic and superhydrophilic wettability on a polyamide film with a triangular net-like structure by biaxially extending and unloading the elastic film. After extending the elastic polyamide films biaxially to an extension ratio larger than 120%, the film show superhydrophilicity (Figure 23b), while when the stress was unloaded, the surface microstructures recover to its original state and the wettability therefore returned to super-hydrophobicity (Figure 23a). In this case, the good elasticity of the polyamide film is very crucial, which make it possible for the morphology to change after being loaded external mechanical force. In addition, the wettability of elastic PTFE can also be increased from hydrophobic (118°) to super-hydrophobic (165°) when PTFE was extended perpendicularly to the fibrous crystals direction with an extension ratio changes from 5 to 190% [167].

External mechanical force can do effectively alter the surface morphology of the solid surfaces, especially those elastic polymer surfaces, and therefore alter their surface wettability. For example, scraping-induced rough surface can enhance the inherent wettability of polymer films. If the morphology can be reversible switched between two states through the exerting and unloading of external mechanical force, the reversible wettability is therefore foreseeable.

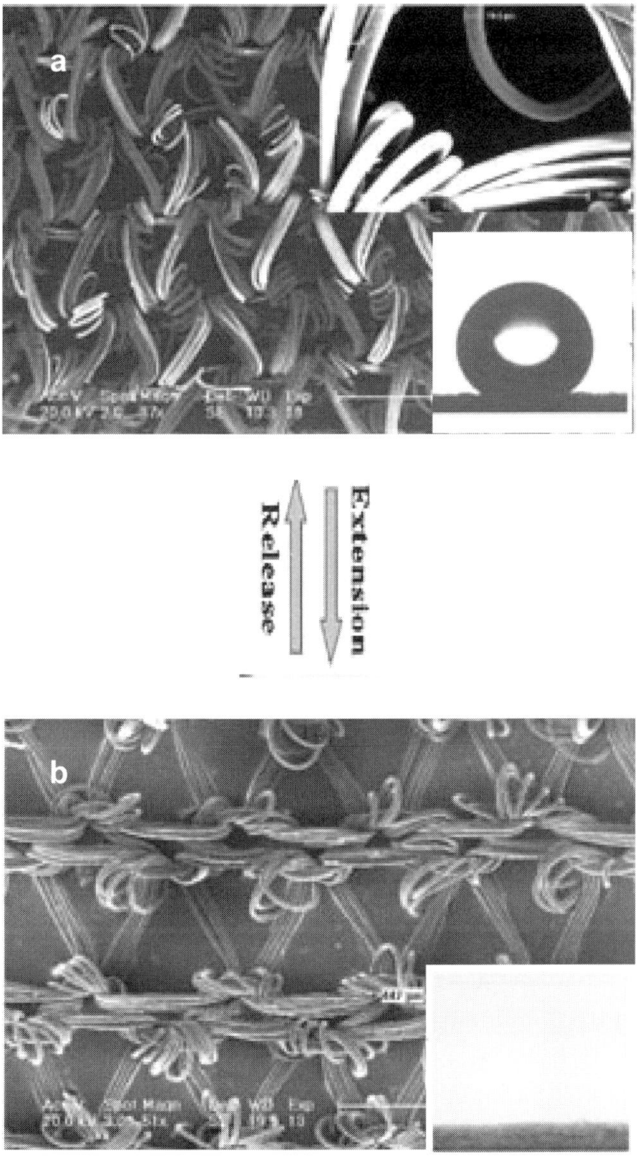

Figure 23. Typical SEM images and the wettability of the triangular net-like structure of the elastic polyamide film (a) before or unloading and (b) after biaxial extension. Reprinted with permission from [Ref. 117], J. L. Zhang; *et al.*, Macromol. Rapid Commun. 26, 477 (2005), Copyright @ WILEY-VCH.

7.2. Multi Stimuli-Responsive Surfaces

Reversibly controlling the surface wettability has aroused great interest and been realized by modifying the surface with stimuli-responsive materials. Under external stimuli such as temperature, pH and light, surface wettability can reversibly change between hydrophilic and hydrophobic on smooth surface, and super-hydrophilicity and super-hydrophobicity on rough surface. Although many stimuli-responsive surfaces have been fabricated and exhibit reversible wettability, they are responsive to only one kind of external stimuli. In some applications, such as surfactants, intelligent micro-fluidic switching, controllable drug release, and thermally responsive filters, multiple-responsive materials, which are independent on responsiveness to several factors, are quite indispensable. For instance, a certain drug needs to be delivered to some part of a body, the temperature and pH of which differ from those of other parts, so the carrier of the drug must be exactly responsive to both temperature and pH.

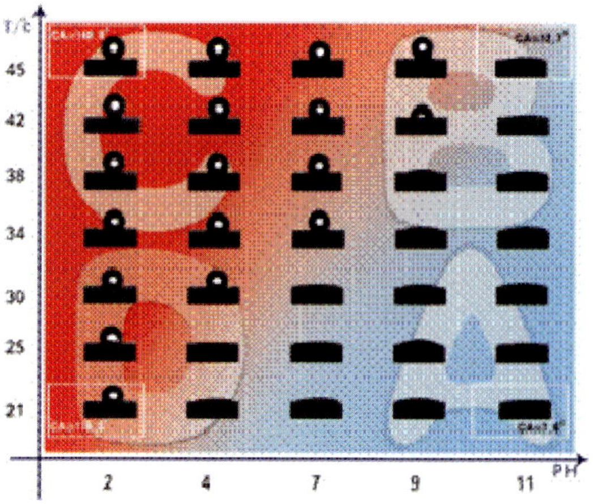

Figure 24. The relationship between CAs and pH and/or temperature on the dual responsive P-(NIPAAm-co-AAc) film. Reprinted with permission from [Ref. 168], F. Xia, *et al.*, Adv. Mater. 18, 432 (2006), Copyright @ WILEY-VCH.

The dual-stimuli-responsive surface is obtained by fabricating a poly(*N*-isopropyl acrylamide-*co*-acrylic acid) [P(NIPAAm-*co*-AAc)] copolymer thin film on rough substrates [168], which shows tunable wettability between superhydrophilicity and superhydrophobicity, and responsivity to both temperature and pH. The dual responsivity of the P(NIPAAm-*co*-AAc) films is due to the effective addition of the thermal-responsive component, NIPAAm, and the pH-sensitive one, acrylic acid (AAc). The surface wettability conversion relates closely to the H-bond reversible change between the two components (NIPAAm and AAc) and the water as suggested in Figure 24. In region A, the loosely coiled conformation of P-(NIPAAm-co-AAc) chains and intermolecular hydrogen bonding with water molecules leads to high surface free energy and a small water CA. In region C, the AAc chain will become non-ionized and the compact and collapsed conformation [169] of P-(NIPAAm-co-AAc) chains induced by intramolecular hydrogen bonding leads to low surface free energy and large water CAs. The region B and D are the transition phases. Thus, the circle of four regions corresponds to the change of surface wettability.

Another example is the photo-stimulated dual-responsive tungsten oxide nanofilm prepared by a simple electrochemical deposition process that exhibits wettability conversion between superhydrophobicity and superhydrophilicity, and photochromism (Figure 25) [32]. The as-prepared tungsten oxide film exhibits a pebble-beach-like morphology (Figure 25a), on which is superhydrophobic with a CA about 151°. Upon UV irradiation, the CA decreases to less than 5° (the insert in Figure 25b). Long-period dark storage makes these irradiated films recover originally superhydrophobic. Interestingly, after UV irradiation, the as-prepared film became yellowish green from the initial brown, and when this film was placed in the dark it recovered to brown. It revealed that the switching process involved the reversible changes in tungsten values, oxygen vacancies, oxygen gas, and adsorbed water molecules [170-172].

Briefly, the key to the manipulation of surface wettability is to manipulate the surface chemical or topological nature, or both. When stimuli-responsive surfaces whose wettability can be switched between two different states are involved, it is enough that either of the surface chemical and topological nature is stimuli-responsive. The previous work suggests us that surfaces made from stimuli-responsive inorganic compounds or grafted by stimuli-responsive organic compounds is an efficient approach, while still with some limitation. For example, the switched wettability triggered by magnetism still remain a challenge now, although it has been testified that droplet with certain magnetism can be moved through exerting and releasing of magnetism field.

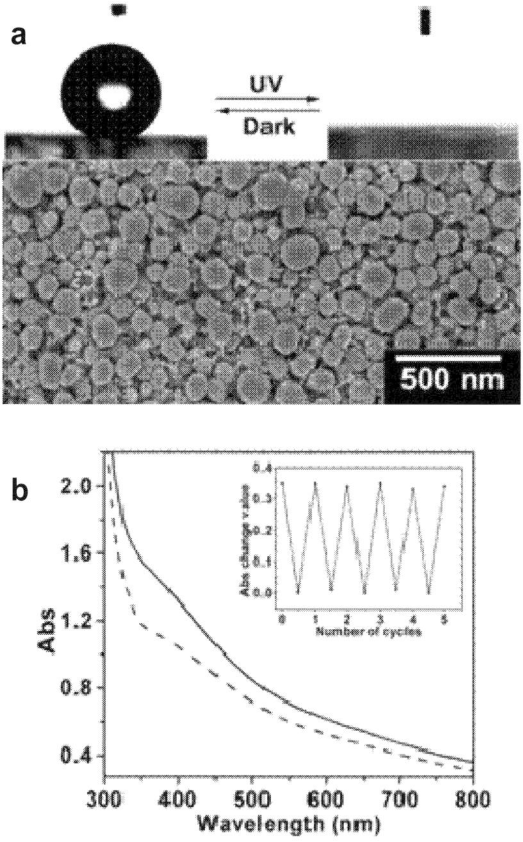

Figure 25. Schematic of (a) the top-view SEM image of the dual-responsive tungsten oxide nanostructured films, and the photo-responsive reversible surface switch between superhydrophobicity and superhydrophilicity; and (b) the absorption spectra of the as-prepared film before (solid line) and after (dash line) irradiation with 365 ± 10 nm UV light. The insert exhibits a good reversibility. Reprinted with permission from [Ref. 32], S. Wang, *et al.*, Angew. Chem. Int. Ed. 45 1264 (2006), Copyright @ WILEY-VCH.

8. Conclusions and Outlook

In this review, we have summarized the recent progress in the field of bio-inspired surfaces with special wettability, from natural to artificial surfaces, from special wettability to switchable wettability, and from fundamental theories to practical applications as well, and some of them have show good potentials in market-oriented industrial fabrication. However, surface wettability is a complex problem involving many subtle changes occurred on surfaces, and vast related work remains on the theoretical analysis for new phenomena, therefore lots of experimental explorations are needed for practical applications. For superhydrophobic surfaces, several main aspects need further implements: to explore simple and facile methods, to synthesize low-cost materials, to develop large-scale preparation, to obtain good environmental stability, and to realize important multi-function. The solution of these problems will pave the way for superhydrophobic surfaces to practical applications. As for responsive switchable surfaces between superhydrophilicity and superhydrophobicity, it remains at the early stage, and bio-inspired smart surfaces will become one of recent focuses in surface science and nanotechnology. Especially, the complex real conditions require the intelligent surfaces with special wettability that can response to dual- or multi-stimuli from the environment, such as temperature, solution, light, electric, and so on. Therefore, the synthesis of new responsive materials, the construction of heterogeneous surface, the combination of multi-scales, the bistable states and the cooperative effect of multi-weak-interaction will become the main integrants to construct novel responsive surfaces with special wettability. Moreover, it must be noted that the liquid or gas also occupies the equal important place to the solid at the surface wettability. Although most of current attention is paid on the solid surface, we believe that the story of the liquid or gas will be more attractive than that of the solid. Learning from nature is our constant principle for there are the numerous mysterious properties in nature, which formed after millions of evolution, will give us inspiration to develop novel interface materials.

References

1. Nakajima, K. Hashimoto and T. Watanabe, Monatsh. Chem. 132, 31 (2001).
2. R. Blossey, Nat. Mater. 2, 301 (2003).
3. Y. Liu, L. Mu, B. H. Liu and J. L. Kong, Chem. Eur. J. 11, 2622 (2005).
4. I. P. Parkin and R. G. Palgrave, J. Mater. Chem. 15, 1689 (2005).
5. T. L. Sun, L. Feng, X. F. Gao and L. Jiang, Acc. Chem. Res. 38, 644 (2005).
6. R. N. Wenzel and Ind. Eng. Chem. 28, 988 (1936).
7. A. B. D. Cassie and S. Baxter, Trans. Faraday Soc. 40, 546 (1944).
8. D. Öner and T. J. McCarthy, Langmuir 16, 7777 (2000).
9. W. Barthlott and C. Neinhuis, Planta. 202, 1 (1997).
10. L. Feng, S. H. Li, Y. S. Li, H. J. Li, L. J. Zhang, J. Zhai, Y. L. Song, B. Q. Liu, L. Jiang and D. B. Zhu, Adv. Mater. 14, 1857 (2002).
11. S. Shibuichi, T. Onda, N. Satoh and K. Tsujii, Langmuir 12, 2125 (1996).
12. J. T. Han, X. R. Xu and K. W. Cho, Langmuir 21, 6662 (2005).
13. X. Zhang, F. Shi, X. Yu, H. Liu, Y. Fu, Z. Q. Wang, L. Jiang and X. Y. Li, J. Am. Chem. Soc. 126, 3064 (2004).
14. N. J. Shirtcliffe, G. McHale, M. I. Newton and C. C. Perry, Langmuir 19, 5626 (2003).
15. N. J. Shirtcliffe, G. McHale, M. I. Newton, G. Chabrol and C. C. Perry, Adv. Mater. 16, 1929 (2004).
16. S. T. Wang, L. Feng, H. Liu, T. L. Sun, X. Zhang, L. Jiang and D. B. Zhu, ChemPhysChem. 8, 1475 (2005).
17. K. K. S. Lau, J. Bico, K. B. K. Teo, M. Chhowalla, G. A. J. Amaratung, W. I. Milne, G. H. McKinley and K. K. Gleason, Nano Lett. 3, 1701 (2003).
18. W. Lee, M. K. Jin, W. C. Yoo and J. K. Lee, Langmuir 20, 7665 (2004).
19. X. F. Gao and L. Jiang, Nature 432, 36 (2004).
20. F. shi, Z. Wang and X. Zhang, Adv. Mater. 17, 1005 (2005).
21. X. Wu and G. Shi, J. Phys. Chem. 110, 11247 (2006).
22. A. R. Parker and C. R. Lawrence, Nature, 414, 33 (2001).
23. T. P. Russell, Science, 297, 964 (2002).
24. J. Lahann, S. Mitragotri, T. Nran, H. Kaido, J. Sundaram, I. S. Choi, S. Hoffer, G. A. Somorjai and R. Langer, Science 299, 371 (2003).
25. S. H. Anastasiadis, H. Retsos, S. Pispas, N. Hadjichristidis and S. Neophytides, Macromolecules 36, 1994 (2003).
26. S. Abbott, J. Ralston, G. Reynolds and R. Hayes, Langmuir 15, 8923 (1999).
27. D. M. Jones, J. R. Smith, W. T. Huck and C. Alexander, Adv. Mater, 14, 1130 (2002).
28. K. Ichimura, S. K. Oh and M. Nakagawa, Science 288, 1624 (2000).
29. J. Berná, D. Leigh, M. Lubomska, S. Mendoza, E. Pérez, P. Rudolf, G. Teobaldi and F. Zerbetto, Nat. Mater. 4, 704 (2005).

30. L. Jiang, R. Wang, B. Yang, T. J. Li, D. A. Tryk, A. Fujishima, K. Hashimoto and D. B. Zhu, Pure Appl. Chem., Vol. 72, Nos. 1–2, pp. 73–81 (2000).
31. T. Sun, G. Wang, L. Feng, B. Liu, Y. Ma, L. Jiang and D. Zhu, Angew. Chem. Int. Ed. 43, 357 (2004).
32. S. Wang, X. Feng, J. Yao and L. Jiang, Angew. Chem. Int. Ed. 45 1264 (2006).
33. Q. Fu, G. V. Rama Rao, S. B. Basame, D. J. Keller, K. Artyushkova, J. E. Fulghum and G. P. López, J. Am. Chem. Soc. 126, 8904 (2004).
34. L. Xu, W. Chen, A. Mulchandani and Y. Yan, Angew. Chem. Int. Ed. 44, 6009 (2005).
35. X. Yu, Z. Wang, Y. Jiang, F. Shi and X. Zhang, Adv. Mater. 17, 1289 (2005)
36. A. W. Adamson, A. P. Gast, Physical chemistry of surfaces, sixth ed. John Wiley & Son, Inc. New York (1997), pp.353.
37. E. A. Vogler, Adv. Coll. Inter. Sci. 74, 69 (1998).
38. A. Lafuma and D. Quéré, Nat. Mater. 2, 457 (2003).
39. R. Wang, K. Hashimoto, A. Fujishima, M. Chikuni, E. Kojima, A. Kitamura, M. Shimohigoshi and T. Watanabe, Adv. Mater. 10, 135 (1998).
40. Nishino, T.; Meguro, M.; Nakamae, K.; Matsushita, M. and Ueda, Y. Langmuir 15, 4321 (1999).
41. M. Miwa, A. Nakajima, A. Fujishima, K. Hashimoto and T. Watanabe, Langmuir 16, 5754 (2000).
42. R. E. Johnson and R. H. Dettre, Adv. Chem. Ser.43, 112 (1963).
43. Sun, T.; Wang, G.; Liu, H.; Feng, L.; Jiang, L. and Zhu, D. J. Am. Chem. Soc. 125, 14996 (2004).
44. T. Wagner, C. Neinhuis and W. Barthlott, Acta Zool 77, 213 (1996).
45. Y. Zheng, X. Gao and L. Jiang, Soft Matter, in press.
46. B. He, N. A. Patankar and J. Lee, Langmuir 19, 4999 (2003).
47. W. Chen, A. Y. Fadeev, M. C. Hsieh, D. Öner, J. Youngblood and T. J. McCarthy, Langmuir 15, 3395 (1999).
48. I. Woodward, W. C. E. Schofield, V. Roucoules and J. P. S. Badyal, Langmuir 19, 3432 (2003).
49. S. H. Li, H. J. Li, X. B. Wang, Y. L. Song, Y. Q. Liu, L. Jiang and D. B. Zhu, J. Phys. Chem. B. 106, 9274 (2002).
50. X. D. Wu, L. J. Zheng and D. Wu, Langmuir 21, 2665 (2005).
51. B. T. Qian and Z. Q. Shen, Langmuir 21, 9007 (2005).
52. E. Hosono, S. Fujihara, I. Honma and H. Zhou, J. Am. Chem. Soc. 127, 13458 (2005).
53. L. Feng, S. H. Li, H. J. Li, J. Zhai, Y. L. Song, L. Jiang and D. B. Zhu, Angew. Chem. Int. Ed. 41, 1221 (2002).
54. L. Feng, Y. L. Song, J. Zhai, B. Q. Liu, J. Xu, L. Jiang and D. B. Zhu, Angew. Chem. Int. Ed. 43, 800 (2003).
55. S. Tsoi, E. Fok, J. C. Sit and J. G. C. Veinot, Langmuir 20, 10771 (2004).

56. Shirtcliffe, N. J.; McHale, G.; Newton, M. I. and Perry, C. C. Langmuir 19, 5626 (2003).
57. C. W. Guo, L. Feng, J. Zhai, G. J. Wang, Y. L. Song, L. Jiang and D. B. Zhu, Chem. Phys. Chem. 5, 750 (2004).
58. M. E. Abdelsalam, P. N. Bartlett, T. Kelf and J. Baumberg, Langmuir 2005, 21, 1753.
59. G. Zhang, D. Y. Wang, Z. Z. Gu and H. Möhwald, Langmuir 21, 9143 (2005).
60. L. Jiang, Y. Zhao and J. Zhai, Angew. Chem. Int. Ed. 43, 4338 (2004).
61. Z. M. Huang, Y. Z. Zhang, M. Kotaki, S. Ramakrishna and Compos. Sci. Technol. 63, 2223 (2003).
62. J. M. Deitzel, J. Kleinmeyer, D. Harris and N. C. B. Tan, Polymer 42, 261 (2001).
63. K. Acatay, E. Simsek, C. Ow-Yang and Y. Z. Menceloglu, Angew. Chem. Int. Ed. 43, 5210 (2004).
64. A. Singh, L. Steely and H. R. Allcock, Langmuir 21, 11604 (2005).
65. S. Agarwal, S. Horst and M. Bognitzki, Macromol. Mater. Eng. 291, 592 (2006).
66. M. Ma, Y. Mao, M. Gupta, K. K. Gleason and G. C. Rutledge, Macromolecules 38, 9742 (2005).
67. S. Shibuichi, T. Onda and N. Satoh, K. Tsujii, J. Phys. Chem. 100, 19512 (1996).
68. H. Y. Erbil, A. L. Demirel, Y. Avci and O. Mert, Science 299, 1377 (2003).
69. X. Y. Lu, C. C. Zhang, Y. C. Han and Macromol. Rapid. Commun. 25, 1606 (2004).
70. Q. D. Xie, G. Q. Fan, N. Zhao, X. L. Guo, J. Xu, J. Y. Dong, L. Y. Zhang, Y. J. Zhang and C. C. Han, Adv. Mater. 16, 1830 (2004).
71. J. T. Han, D. H. Lee, C. Y. Ryu and K. Cho, J. Am. Chem. Soc. 126, 4796 (2004).
72. S. Wang, L. Feng and L. Jiang, Adv. Mater. 18, 767 (2006).
73. L. Feng, Z. L. Yang, J. Zhai, Y. L. Song, B. Q. Liu, Y. M. Ma, Z. H. Yang, L. Jiang and D. B. Zhu, Angew. Chem. Int. Ed. 42, 4217 (2003).
74. K. Esumi, M. Ishigami, A. Nakajima, K. Sawada and H. Honda, Carbon 34, 279 (1996).
75. Y. Zhu, J. Zhang, Y. Zheng, Z. Huang, L. Feng and L. Jiang Adv. Funct. Mater. 16, 568 (2006).
76. Z. Guo, F. Zhou, J. Hao and W. Liu, J. Am. Chem. Soc. 127, 15670 (2005).
77. C. Wang, T. Wang, P. Tung, S. Kuo, C. Lin, Y. Sheen and F. Chang, Langmuir 22, 8289 (2006).
78. H. Yan, K. Kurogi, H. Mayama and K. Tsujii, Angew. Chem. Int. Ed. 44, 3453 (2005).
79. L. Zhai, F. C. Cebeci, R. E. Cohen and M. F. Rubner, Nano. Lett. 4, 1349 (2004).
80. M. Nicolas, F. Guittard and S. Géribaldi, Angew. Chem. Int. Ed. 45, 2251 (2006).
81. A. Nakajima, J. Am. Ceram. Soc. 112, 533 (2004).
82. K. Tadanaga, N. Katata and T. Minami, J. Am. Ceram. Soc. 80, 1040 (1997).
83. A. Nakajima, A. Fujishima, K. Hashimoto and T. Watanabe, Adv. Mater. 11, 1365 (1999).

84. A. Hozumi and O. Takai, Thin Solid Films 334, 54 (1998).
85. Y. Y. Wu, H. Sugimura, Y. Inoue and O. Takai, Chem. Vap. Deposition 8, 47 (2002).
86. H. Yabu and M. Shimomura, Chem. Mater. 17, 5231 (2005).
87. Z. Z. Gu, H. Uetsuka, K. Takahashi, R. Nakajima, H. Onishi, A. Fujishima and O. Sato, Angew. Chem. Int. Ed. 42, 894 (2003).
88. J. Y. Shiu, C. W. Kuo, P. Chen and C. Y. Mou, Chem. Mater. 16, 561 (2004).
89. M. E. Abdelsalam, P. N. Bartlett, T. Kelf and J. Baumberg, Langmuir 21, 1753 (2005).
90. G. Zhang, D. Y. Wang, Z. Z. Gu and H. Möhwald, Langmuir 21, 9143 (2005).
91. J. L. Zhang, L. J. Xue and Y. C. Han, Langmuir 21, 5 (2005).
92. J. Wang, J. P. Hu, Y. Wen, Y. Song and L. Jiang, Chem. Mater. 18, 4984 (2006).
93. Y. Xu, W. H. Fan, Z. Li, D. Wu and Y. Sun, Appl. Optics 42, 108 (2003).
94. Y. Xu, D. Wu, Y. H. Sun, Z. X. Huang, X. D. Jiang, X. F. Wei, Z. H. Wei, B. Z. Dong and Z. H. Wu, Appl. Optics. 44, 527 (2005).
95. G. Kousik, S. Pitchumani and N. G. Renganathan, Prog. Org. Coat. 43, 286 (2001).
96. D. Mecerreyes, V. Alvaro, I. Cantero, M. Bengoetxea, P. A. Calvo, H. Grande, J. Rodriguez and J. A. Pomposo, Adv. Mater. 14, 749 (2002).
97. Z. Zhang, L. Gu and G. Shi, J. Mater. Chem. 13, 2858 (2003).
98. M. Li, J. Zhai, H. Liu, Y. L. Song, L. Jiang and D. B. Zhu, J. Phys. Chem. B. 107, 9954 (2003).
99. N. Zhao, F. Shi, Z. Q. Wang and X. Zhang, Langmuir 21, 4713 (2005).
100. Y. Zhu, J. Zhang, Y. Zheng, Z. Huang, L. Feng, and Lei Jiang, Adv. Funct. Mater. 16, 568 (2006).
101. M. H. Jin, X. J. Feng, L. Feng, T. L. Sun, J. Zhai, T. J. Li and L. Jiang, Adv. Mater. 17, 1977 (2005).
102. K. Autumn, Y. A. Liang, S. T. Hsieh, W. Zesch, W. P. Chan, T. W. Kenny, R. Fearing and R. J. Full, Nature 405, 681 (2000).
103. A. K. Geim, S. V. Dubonos, I. V. Grigorieva, K. S. Novoselov, A. A. Zhukov and S. Y. Shapoval, Nat. Mater. 2, 461 (2003).
104. X. Y. Song, J. Zhai, Y. L. Wang and L. Jiang, J. Phys. Chem. B., 109, 4048 (2005).
105. H. Irie, W. Washizuka, N. Yoshino and K. Hashimoto, Chem. Comm. 11, 1298 (2003).
106. Machida M., Norimoto K., Watanabe T., Hashimoto K. and Fujishima A., J. Mater. Sci. 34, 2569 (1999).
107. G. McHale, N. J. Shirtcliffe, S. Aqil, C. C. Perry and M. I. Newton, Phys. Rev. Lett. 93, 036102 (2004).
108. J. Bico, C. Marzolin, D. Quere and Europhys. Lett. 47, 743 (1999).
109. J. Bico, U. Thiele and D. Quere, Colloids Surf. A 206, 41 (2002).
110. J. Bico, C. Tordeux and D. Quéré, Europhys. Lett. 55, 214 (2001).

111. R. Wang, K. Hashimoto, A. Fujishima, M. Chikuni, E. Kojima, A. Kitamura, M. Shimohigoshi and T. Watanabe, Nature, 388, 431 (1997).
112. T. Ogawa, N. Murata and S. Yamazaki, J. Sol-Gel Sci. Technol. 27, 237 (2003).
113. W. Zhong, S. Liu, X. Chen, Y. Wang and W. Yang, Macromolecules, 39, 3224 (2006).
114. W. Zhong, X. Chen, S. Liu, Y. Wang and W. Yang, Macromol. Rapid Commun. 27, 563 (2006).
115. A. Hattori, T. Kawahara, T. Uemoto, F. Suzuki, H. Tada and S. J. Ito, Colloid Interface Sci. 232, 410 (2000).
116. X. T. Zhang, O. Sato, M. Taguchi, Y. Einaga, T. Murakami and A. Fujishima, Chem. Mater. 17, 696 (2005).
117. J. L. Zhang; X. Y. Lu, W. H. Huang and Y. C. Han, Macromol. Rapid Commun. 26, 477 (2005).
118. K. C. Song, J. K. Park, H. U. Kang and J. Kim, Sol-Gel Sci. Technol. 27, 53 (2003).
119. R. D. Sun, A. Nakajima, A. Fujishima, T. Watanabe and K. Hashimoto, J. Phys. Chem. B, 105, 1984 (2001).
120. G. McHale, N. J. Shirtcliffe, S. Aqil, C. C. Perry and M. I. Newton, Phys. Rev. Lett., 93, 036102/1 (2004).
121. N. J. Shirtcliffe, G. Mchale, M. I. Newton, C. C. Perry and P. Roach, Chem. Commun. 25, 3135 (2005).
122. F. C. Cebeci, Z. Wu, L. Zhai, R. E. Cohen and M. F. Rubner, Langmuir 22, 2856 (2006).
123. L. Zhai, M. C. Berg, F. V. Vebeci, Y. Kim, J. M. Milwid, M. F. Rubner and R. E. Cohen, Nano Lett. 6, 1213 (2006).
124. K. Tadanaga, J. Morinaga, A. Matsuda and T. Minami, Chem. Mater. 12, 590 (2000).
125. H. Gau, S. Herminghaus, P.Lenz and R. Lipowsky, Science 283, 46 (1999).
126. Y. Wang, J. Deng, W. Zhong, L. Kong and W. Yang, Macromol. Rapid Commun. 26, 1788 (2005).
127. X. Yu, Z. Q. Wang, Y. G. Jiang, and X. Zhang, Langmuir 22, 4483 (2006).
128. H. Tada and H. Nagayama, Langmuir 11, 136 (1995).
129. D. Anton, Adv. Mater. 10, 1197 (1998).
130. J. Wang, Y. Wen, X. Feng, Y. Song and L. Jiang, Macromol. Rapid Commun. 27, 188 (2006).
131. H. Liu, L. Feng, J. Zhai, L. Jiang and D. Zhu, Langmuir, 20, 5659 (2004).
132. W. Jiang, G. Wang, Y. He, X. Wang, Y. An, Y. Song and L. Jiang, Chem. Commun. 28, 3550 (2005).
133. T. N. Krupenkin, J. A. Taylor, T. M. Schneider and S. Yang, Langmuir 20, 3824. (2004).
134. A. Sidorenko, S. Minko, K. Schenk-Meuser, H. Duschner and M. Stamm, Langmuir 15, 8349(2000).

135. M. Miyauchi, A. Nakajima, T. Watanabe and K. Hashimato, Chem. Mater. 14, 2812 (2002).
136. L. Wang, D. Baer, M. Engelhard and A. Shultz, Surf. Sci. 344, 237 (1995).
137. X. Feng, L. Feng, M. Jin, J. Zhai, L. Jiang and D. Zhu, J. Am. Chem. Soc. 126, 62 (2004).
138. L. Huang, S. P. Lau, H. Y. Yang, E. S. P. Leong, S. F. Yu and S. Prawer, J. Phys. Chem. B. 109 7746 (2005).
139. X. Q. Meng, D. X. Zhao, J. Y. Zhang, D. Z. Shen, Y. M. Lu, L. Dong, Z. Y. Xiao, Y. C. Liu and X. W. Fan, Chem. Phys. Lett. 413, 450 (2005).
140. Y. Li, W. P. Cai, G. T. Duan, B. Q. Cao and F. Q. Sun, F. Lu, J. Colloid Interface Sci. 287, 634 (2005).
141. X. Feng, J. Zhai and L. Jiang, Angew. Chem. Int. Ed. 44, 5115 (2005).
142. H. Irie, T. S. Ping, T. Shibata and K. Hashimoto, Electrochem. Solid-State Lett. 8, D23 (2005).
143. W. Zhu, X. Feng, L. Feng and L. Jiang, Chem. Commun. 26, 2753 (2006).
144. R. Rosario, D. Gust, A. A. Garcia, M. Hayes, J. L. Taraci, J. W. Dailey and S. T. Picraux, J. Phys. Chem. B. 108, 12640 (2004).
145. G. S. Kumar and D. C. Neckers, Chem. Rev. 89, 1915 (1989).
146. A. Natansohn and P. Rochon, Chem. Rev. 102, 4139 (2002).
147. L. M. Siewierski, W. J. Brittain, S. Pdtrash and M. D. Foster, Langmuir 12, 5838 (1996).
148. N. Delorme, J. Bardeau, A. Bulou and F. Poncin-Epaillard, Langmuir 21, 12278 (2005).
149. C. Feng, Y. Zhang, J. Jin, Y. Song, L. Xie, G. Qu, L. Jiang and D. Zhu, Langmuir 17, 4593 (2001).
150. H. Ge, G. Wang, Y. He, X. Wang, Y. Song, L. Jiang and D. Zhu, Chem. Phys. Chem. 7, 575 (2006).
151. H. S. Lim, J. T. Han, D. Kwak, M. Jin and Kilwon Cho, J. Am. Chem. Soc. ja0655901.
152. R. Rosario, D. Gust, M. Hayes, F. Jahnke, J. Springer and A. A. Garcia. Langmuir 18, 8062 (2002).
153. K. Uchida, N. Izumi, S. Sukata, Y. Kojima, S. Nakamura and M. Irie, Angew. Chem. Int. Ed. 45, 6470 (2006).
154. S. R. Holmes-Farley, C. D. Bain and G. M. Whitesides, Langmuir, 4, 921 (1988).
155. C. D. Bain and G. M. Whitesides, Langmuir 5, 1370 (1989).
156. Y. G. Jiang, Z. Q. Wang, X. Yu, F. Shi, H. P. Xu and X. Zhang, Langmuir, 21, 1986 (2005).
157. H. G. Schild, Prog. Polym. Sci. 17, 163 (1992).
158. Z. Hu, Y. Chen, C. Wang, Y. Zheng and Y. Li, Nature, 393, 149 (1998).
159. X. Huang and M. J. Wirth, Macromolecules, 32, 1694 (1999).
160. D. M. Jones and W. T. S. Huck, Adv. Mater. 13, 1256 (2001).
161. S. Lin, K. Chen and R. Liang, Polymer, 40, 2619 (1999).

162. Y. Liu, L. Mu, B. H. Liu, S. Zhang, P. Y. Yang and J. L. Kong, Chem. Commun. 10, 1194 (2004).
163. W. J. J. Welters and L. G. J. Fokkink, Langmuir 14, 1535 (1998).
164. H. J. J. Verheijen and M. W. J. Prins, Langmuir 15, 6616 (1999).
165. T. B. Jones, Langmuir 18, 4437 (2002).
166. R. A. Hayes and B. J. Feenstra, Nature, 425, 383 (2003).
167. J. L. Zhang, J. Li and Y. C. Han, Macromol. Rapid. Commun. 24, 1105 (2005).
168. F. Xia, L. Feng, S. Wang, T. Sun, W. Song, W. Jiang and L. Jiang, Adv. Mater. 18, 432 (2006).
169. M.K. Yoo, Y.K. Sung, C.S. Cho and Y.M. Lee, Polymer 38, 2759 (1997).
170. C. Bechinger, G. Oefinger, S. Herminghaus and P. Leiderer, J. Appl. Phys. 74, 4527 (1993).
171. J. Zhang, D. Benson, C.Tracy, S. K. Deb, A. W. Czanderna and C. Bechinger, J. Electrochem. Soc. 144, 2022 (1997).
172. T. He and J. N. Yao, Res. Chem. Intermed. 30, 459 (2004).